Servicing Digital HDTV Systems

by Robert Goodman

Servicing Digital HDTV Systems

by Robert Goodman

PROMPT® Publications is an imprint of Sams Technical Publishing, 5436 W. 78th St., Indianapolis, IN 46268.

International Standard Book Number: 0-7906-1223-2
Library of Congress Catalog Card Number: 2001091654

Acquisitions Editor: Deborah Abshier
Senior Editor: Kim Heusel
Editor: Jim Cauley
Typesetting and Interior Design: Debbie Berman
Proofreaders: Jan Zunkel, Cricket Franklin
Cover Design: Christy Pierce
Graphics Conversion: Christy Pierce
Illustrations: Courtesy the author

PRINTED IN THE UNITED STATES OF AMERICA

9 8 7 6 5 4 3 2 1

Contents

9 Audio Systems Digital Signal Processing and Troubleshooting Techniques 203

10 Digital HDTV Reception Problems and Antenna Requirements 223

11 HDTV Power Supplies 243

Acknowledgments

An electronics service book of this magnitude is not possible without first obtaining technical information, diagrams, and photos from many sources. For this book, these sources include high-definition television electronics consumer manufacturers, service and sales dealers, technical engineering personnel, test instrument companies, service technicians from around the country, and, of course, my own years of actual on-the-bench work with conventional TV service and now HDTV troubleshooting and repairs.

To the following individuals and electronics manufacturers, I express my gratitude for providing information that has made this book possible:

Sencore, Inc. — Al Bowen, president; Don Multerer, technical training; Cindy Bowen, graphics; Jeff Murray, director of sales; Stan Warner, technical writer

Zenith Electronics Corp. — John L. Taylor, vice president of communications

Microtune, Inc. — Rick Blumberg, Kathleen Padula, Alice M. Perkins

DIRECTV — Ms. Barbara Chen

Thomson multimedia — Mike Begala, Denis Boutwell

Toshiba of America — Tod Hill, John Snyder, Scott Schock

Winegard Company (Antennas) — Burlington, Iowa

Introduction

Information in this technical service book will help the electronics service technician meet the new challenges of digitally controlled and digital signal transmissions that make up the high-definition television system of today and into the future.

HDTV is composed of a steady stream of bits and bytes. The HDTV signals of video and audio are digitized separately, then arranged into streams, put into packets, processed for transmission, and then transmitted via a TV station over the air or sent over a wide-band cable system to a viewer's home HDTV receiver.

This book introduces an overall view of how the digital HDTV system operates. There will be various explanations of HDTV circuit operations and functions. Some chapters will contain ways to use specialized test instruments to troubleshoot HDTV receiver problems, and one chapter is devoted to special equipment that is useful in testing, and making adjustments, alignments, and other repairs of these state-of-the-art HDTV receivers and various connected components.

HDTV gives the viewer a sharper picture by broadcasting signals in a digital format that eliminates ghosts, picture noise, and other types of terrestrial interference. A chapter in this book will also look at and explain the various HDTV scanning techniques. The three common scanning systems are:

- 480p Standard Definition
- 720p Progressive Scanning
- 1080i Interlaced Scanning

TV viewers have been very impressed with the sharp, stunning picture and sound quality produced by the digital HDTV system. Consumers also will appreciate the many additional features that will become available over the digital TV super highway.

The TV broadcast stations are "on track" converting to HDTV picture transmission and upgraded wide-band cable systems are being installed at a "fast clip" to deliver digital, HDTV programs. Also, more high-definition programs are coming on-line, too. And as more high-definition sets are produced and sold, it appears the manufacturers are lowering the price tag. Pricing of HDTV sets will be the big factor driving consumers' purchase of a new digital receiver.

The digital "small dish" viewers are now an impressive number that are wanting HDTV sets. DIRECTV and EchoStar now offer high-definition reception channels. RCA has the model DTC-100 — the first DIRECTV IRD (Integrated Receiver Decoder) equiped for high definition reception from satellite and terrestrial HDTV broadcasts. It also tunes conventional satellite over-the-air TV station reception and cable TV channels. Thus, the high-definition TV evolution is on a roll.

CHAPTER 1

Digital HDTV System Operation

DIGITAL TELEVISION OVERVIEW

In this chapter, you will find out what Digital Television (DTV) is and how it works. Basically, this will be an introduction to the Advanced Television System Committee (ATSC) format. Although, the complete conversion to an all-digital HDTV system will take more than ten years to implement, the FCC has already assigned a new digital transmitter frequency to all U.S. television broadcast stations. The new ATSC format allows terrestrial transmission of digitally coded program material that will have a higher video resolution, as well as CD-quality audio. This digital format also has a wide-screen format with a 16x9 aspect ratio, as opposed to the 4x3 ratio found in today's standard analog format. The highest resolution program material, or picture content, is referred to as High-Definition Television or HDTV.

The ATSC format also provides the capability of broadcasting multiple, lower resolution programs simultaneously, should the program material not be broadcast in high definition. These multiple programs are transmitted on the same RF carrier channel used for one HDTV program. This is referred to as multicasting. Compression technology allows the simultaneous broadcast of several digital channels. The present analog system does not have this capability. The standard definition signal will be noise-free, similar to the picture quality viewed on a digital satellite system, and quite an improvement over the present NTSC broadcast standard.

Using digital technology, broadcasters can supplement DTV programs with additional data. Utilizing the unused, or "opportunistic" bandwidth, broadcasters can deliver computer information or data directly to a computer or TV receiver. In addition to creating new services, digital broadcasting allows the broadcaster to provide multiple channels of digital programming in different resolutions, while providing data, information, and/or interactive services.

HDTV PICTURE QUALITY

HDTV produces much better picture quality than conventional TV. Lines of resolution increase from the 525 interlaced; to 720 lines; and up to 1080 lines. Additionally, the ratio between picture width and height increases to 16:9, whereas conventional television's aspect ratio is 4:3. This improvement in picture quality provides many new options. The HDTV resolution and format specifications are shown in **Figure 1-1**.

Digital television refers to any TV system that operates on a digital signal format. DTV is classified under two categories: HDTV and SDTV.

Standard Definition TV (SDTV) refers to DTV systems that operate off the 525 line interlaced or progressive standard. This will not produce as high of quality video as HDTV does.

Vert Lines	*Horz Pixels*	*Aspect Ratio*	*Pix Rate*
1080	**1920**	**16:9**	**60I, 30P, 24P**
720	**1280**	**16:9**	**60P, 30P, 24P**
480	**704**	**16:9, 4:3**	**60P, 60I, 30P, 24P**
480	**640**	**4:3**	**60P, 60I, 30P, 24P**

FIGURE 1-1

In addition to the higher quality picture that HDTV delivers, it also has an advanced sound system. This audio system is supported by Dolby Digital audio compression, which also includes surround sound provisions.

Set-Top Converter Box

The set-top converter box is used to receive many different signals including high-definition digital, standard digital, satellite digital, analog cable, and the standard UHF/VHF signals. For several years TV stations will continue to transmit more analog than digital signals, since analog TV receivers are still the most popular.

The set-top converter decodes an 8-level vestigial sideband (VSB) digital signal that is transmitted by the TV station. VSB is the digital broadcasting system now being used in the U.S.

The set-top converter can decode the digital signal for a standard TV receiver; however, the picture quality will only improve slightly. Without a high-resolution screen to detect the digital enhancements, the VSB to analog conversion is the only truly useful function performed by the box.

In the future, built-in digital systems will be included within new HDTV sets. With so many standard television sets in use, it will take a while for built-in digital TV receivers to become the majority.

Most sets now marketed use terms such as HDTV-ready, digital-ready, or HD-compatible. These terms do not indicate that the TV set can produce a digital signal, only that they have a jack available in which to plug a set-top decoder. Most of those sets do, however, have enhanced screen resolutions.

Digital Video Formats

There are several video formats available; however, the most common are those with 720 or 1080 lines of resolution. The majority of formats employ either interlaced or progressive resolution and vary the number of frames per second. HDTV and set-top box manufacturers will supply units that will read these formats. In addition, they will supply equipment to decode the complex audio signals.

Digital Television Signal

Terrestrial HDTV transmission is accomplished on an 8-level vestigial sideband, or 8-VSB. It is derived from a 4-level AM VSB and then trellis coded into a scrambled B-level signal (cable will use an accelerated data rate of 16-VSB). A small pilot carrier is then added and placed in such a way that it will not interfere with other analog signals. A flow chart illustrating these data stream events is shown in **Figure 1-2.**

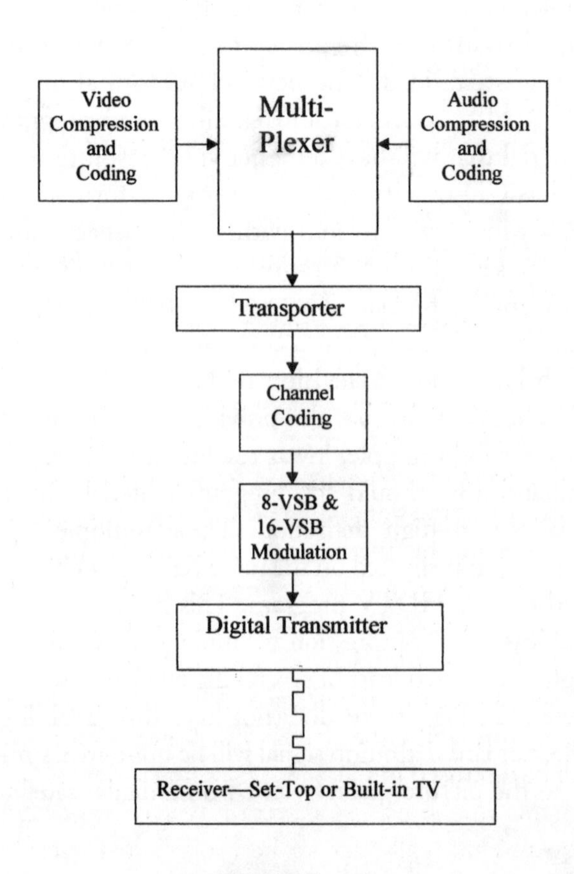

FIGURE 1-2

Satellite systems are already transmitting digital HDTV signals. DIRECTV has two HDTV channels now and plan more in the near future. Digital satellite systems will have a head start over conventional TV stations in transmitting high-definition television.

Notes On Compatibility

Electronics service technicians and TV viewers are concerned as to whether new digital TV systems will be compatible with standard VCRs, camcorders, DVD players, and other entertainment products. In most cases, it appears that they will be. Just about all equipment manufacturers are equipping their products with composite video and analog inputs on their digital HDTV receivers.

INTRODUCTION TO THE DTV DELIVERY SYSTEMS

Some of the first DTV programs will appear on terrestrial broadcast stations. DTV is also available via the Direct Broadcast Satellite (DBS) dish system. You will be seeing more HDTV programs on cable as more companies convert to a wide-band digital system.

Digital broadcasting provides many new challenges and opportunities for the professional electronics technician. Typically, an outdoor antenna will be required to receive HDTV station programs. (Refer to chapter 10 for HDTV antenna requirements). It's very important to remember that the reception characteristics of a digital signal and an analog signal are quite different. A DTV receiver does not behave like a standard NTSC analog television receiver. When receiving an analog NTSC broadcast, as the signal strength decreases, the amount of noise in the picture increases. Eventually, the picture is either full of snow or is blanked out. In contrast, a digital broadcast is completely noise free until the signal level is too low for the receiver to decode. Once the digital signal threshold is reached, the picture will either freeze, blank out or both. As will be discussed in Chapter 10, pointing the HDTV antenna in the precise direction is very important. The antenna will have to be positioned to receive the best average sum of all digital signals within the viewing area. In some instances, the HDTV viewer may need more than one antenna due to the varied locations of the transmitter towers. A signal strength indicator will be built into many receivers to help position the antenna.

The digital signal is transmitted using a standard 6 MHz bandwidth. This is the same bandwidth that NTSC analog TV broadcasters use. The DTV signals are also broadcast in the same spectrum or range of frequencies that NTSC is broadcast in; primarily UHF. In most applications, the same antenna can be used for both HDTV and NTSC reception. Some new antenna designs are currently being developed. They will blend in with their surroundings and be less noticeable than traditional rooftop antennas.

The new satellites that broadcast DTV signals do not use the same satellites currently used for DIRECTV and USSB. However, the DTV satellite broadcasts will be similar enough to allow the same dish to receive both regular DBS programming and DTV programming. A new dish and receiver will be needed for these dual-purpose systems. In addition to reception (antenna), considerations must be given to signal distribution. Again, it is necessary to remember that the signal does not become noisy as the signal strength weakens. The signal levels and picture quality that are tolerable to the viewer with the present analog system may have too much noise to operate within the digital systems. The installation of a low-loss, high-quality signal distribution system may be required.

HDTV Formats and Modes

There are more than a dozen possible standards for transmission of digital video. The number of lines per picture, the number of pixels per line, the aspect ratio, the frame rate, and the scan type are used to define these standards. The resolution chart in **Figure 1-3** provides a listing of the DTV standards provided by the ATSC format. Some of the higher resolution formats have not yet been implemented and not all of the formats are considered high definition. However, the use of digital technology means that all formats will result in a vast improvement in video and audio quality.

Digital Transmission Formats				
	Horz Pixels	Vert Lines	Aspect Ratio	Scan/ Frame
1.	1920	1080	16:9	30i
2.	1920	1080	16:9	30p
3.	1920	1080	16:9	24p
4.	1280	720	16:9	60p
5.	1280	720	16:9	30p
6.	1280	720	16:9	24p
7.	704	480	16:9	60p
8.	704	480	16:9	30i
9.	704	480	16:9	30p
10.	704	480	16:9	24p
11.	704	480	4:3	60p
12.	704	480	4:3	30i
13.	704	480	4:3	30p
14.	704	480	4:3	24p
15.	640	480	4:3	60p
16.	640	480	4:3	30i
17.	640	480	4:3	30p
18.	640	480	4:3	24p

FIGURE 1-3

Generally, the 1080p, 1080i and 720p formats are considered high-definition formats. However, limitations in current receiver technology prevent these formats from being available in current consumer televisions. No doubt, in the future years, and as technology improves, these higher resolutions (above 1080p) may become available. Although the format of the broadcast material might change, the receiver will use the same digital processing to convert the various formats.

Will NTSC Broadcasts Disappear?

The transition to digital TV is expected to take up to ten years, possibly longer. At some point in the future, no analog broadcast stations will be on the air. Once that point is reached, all of the analog channels will be reallocated to the spectrum of other user services.

During the transition period, set-top converter boxes will be available that will decode the digital signal and allow it to be displayed on a standard analog NTSC receiver. However, with this configuration, there will be a decrease in picture resolution. NTSC set owners will be able to use their current NTSC analog TV sets until the NTSC transmitter has to shutdown. At that point in time, in order to watch TV programs, the viewer will have to purchase a digital TV or a converter box that converts the digital signal to analog.

Standard-Definition and High-Definition Basics

Picture resolution can be specified in pixels or "lines". Resolution is the maximum number of transitions possible on the screen in a horizontal and vertical direction. The maximum resolution that a cathode ray tube (CRT), or picture tube, can display is determined at the time the CRT is produced. The greater amount of horizontal and vertical pixels, the greater the resolution capability. The resolution of a computer monitor is most often specified by the number of pixels it can display. This is listed in both horizontal and vertical direction, e.g. 1920(h) x 1080(v). Pixels are also used to rate the resolution of the new digital ATSC and HDTV formats.

In broadcast television, the resolution of the studio camera that captures the video is what determines the highest resolution possible. The picture resolution produced by the camera is given in pixels, similar to a CRT. This is the current resolution limitation as the transition to high-definition digital TV takes place. In NTSC, the ability of an analog signal voltage to quickly transition from low to high is comparable to a pixel changing from black to white.

The number of lines transmitted in the current NTSC analog format is 525. This is considered standard definition (SD) transmission. A standard definition transmission of a 525-line, NTSC signal can be transmitted in the analog or digital (ATSC) mode. A high-definition transmission can be transmitted only in the digital television mode.

- SDTV or SD — Standard definition has 525 lines of resolution transmitted, but only 480 of these lines are viewable. SD can be sent as an NTSC analog or a digital (DTV) transmission.
- HDTV or HD — A high-definition transmission contains 720 or more horizontal lines and can only be transmitted in a digital format.
- DTV — A digital TV transmission refers to the digital encoding of a picture signal that may contain either a high- or low-resolution picture. The digital picture is not viewable on an analog NTSC TV set without a set-top box.

Digital Transmission Formats

The new ATSC format provides eighteen digital transmission formats. The first six offer high-definition signals with a 16x9 aspect ratio. The remaining 12 formats are standard-definition signals with progressive (P) or interlaced (i) scan. Although they do not provide maximum resolution, they do offer a significant improvement over the conventional analog NTSC signal. The quality of an SD signal is slightly better than that of a DVD player. Standard-definition transmissions leave enough space for an additional digital video signal to be transmitted in the same frequency spectrum (channel). This allows a broadcast station to transmit more than one program at a time on the same 6 MHz digital channel.

HDTV Encoder/Transmitter Block Diagram

The National Television Systems Committee created the analog TV standard known as NTSC. The new digital standard, HDTV, was developed by the Advanced Television System Committee (ATSC). The primary objective of ATSC was to develop a digital transmission format that would fit within a 6 MHz bandwidth. Another major goal in developing the ATSC format was to allow expansion and versatility in the transmission of additional content such as electronic program guide (EPG) information and digital data such as text copy. Using this new digital transmission technique, a broadcast station can transmit multiple digital programs simultaneously within a 6 MHz bandwidth. However, in some situations, and in order to broadcast multiple digital programs within the 6 MHz bandwidth, the maximum picture resolution may have to be compromised.

To better understand how a high-resolution digital picture is transmitted in a 6 MHz bandwidth, a review of the digital encoder block diagram in **Figure 1-4** would be useful. The HDTV transmitter block diagram consists of two parts. The Packet Generation section multiplexes compressed video and audio, along with additional services data, into a single digital bit stream. The vestigial side band (VSB) transmission section then scrambles this digital data to allow for error correction during decoding and reconstruction of the signal. The VSB transmission section also adds the data sync and transmits the data via the RF power amplifiers and antenna.

Transport Packet Generation

When video and audio analog signals are converted to a digital format, a data bit stream is generated. Without compression, this digital data bit stream would not fit in the allowed 6 MHz channel bandwidth. The digital video compression is accomplished using the latest Motion Picture Experts Group (MPEG) encoding techniques. The digital audio compression is accomplished using the new Dolby Digital encoding system.

The additional services data can contain information pertaining to the program being viewed. It can also contain unrelated data such as Internet or computer information. Data relevant to the TV program is structured according to a Program and System Information Protocol, or PSIP. It can also contain additional content such as data sync, real time, station ID, alternate channel number information and much more.

As discussed previously, the video and audio output from the encoders is first multiplexed with the additional services data into a single data bit stream by the data multiplexer. The data is then

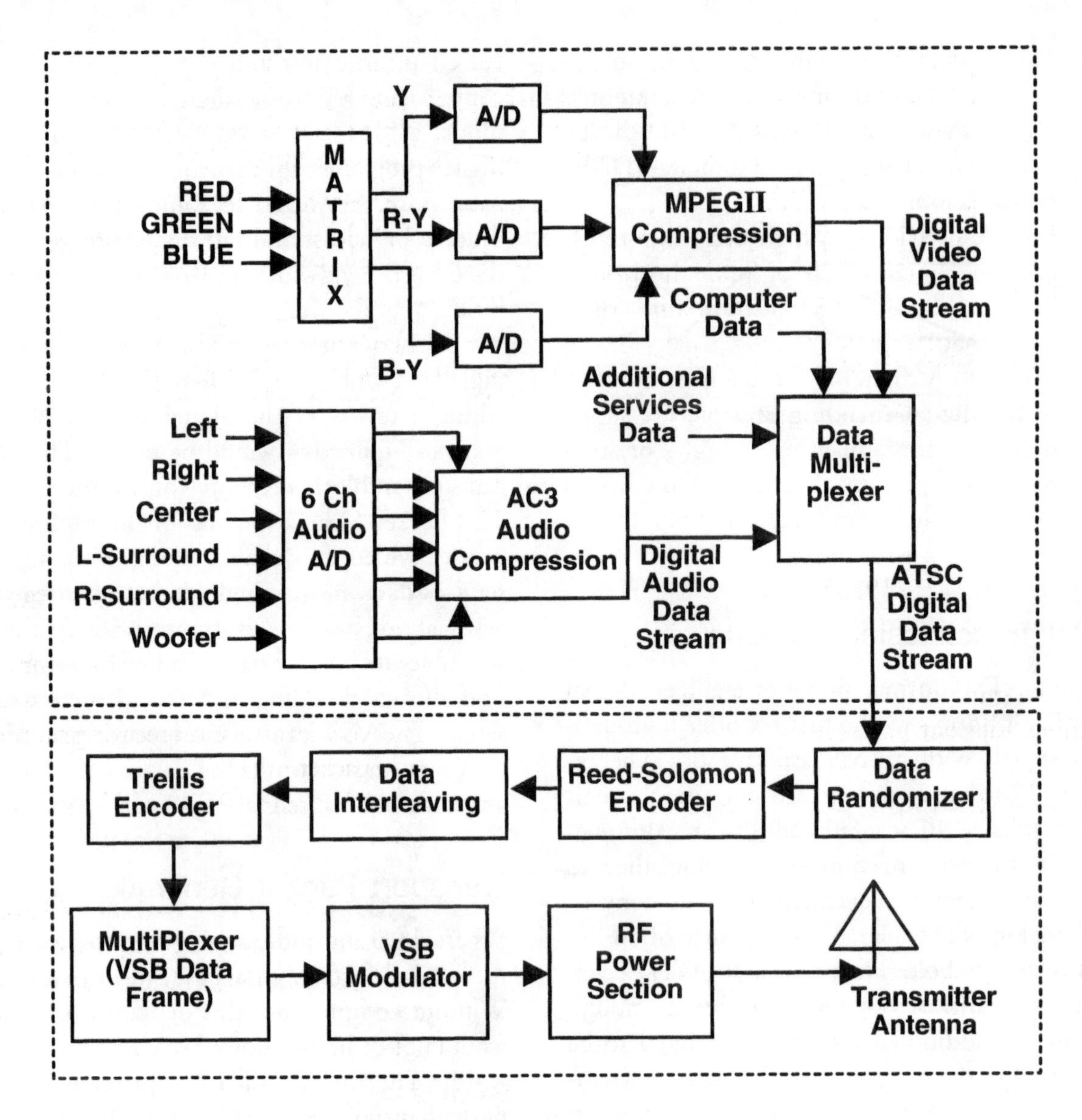

FIGURE 1-4

output in sections called transport packets. They are then applied to the VSB transmission part of the block diagram.

VSB TRANSMISSION OPERATIONS

- Data scrambling
- Data grouping
- Adding data sync

Data Scrambling

The data randomizer, the Reed-Solomon Encoder, and the data interleaver sections all scramble the data, but do so in different ways. *Scrambling* is a technique used to recover lost, damaged, or missing data during decoding at the receiver end. Scrambling scatters the data in a predetermined way so that it is no longer in the normal 1, 2, 3, 4, 5, 6 sequence. Refer to the data scrambling illustration in **Figure 1-5**. For example, the scrambled data may have a packet sequence such as: 6, 12, 18, 24, 30, 32, 34, 1, etc. If data packets 1, 7, 13, 19, 25, and 31 were lost or damaged during transmission, it would not be as significant or noticeable than if sequential (7, 8, 9, 10, 11, 12, etc.) data packets were corrupted. The data can be reconstructed. Here's how: when the decoder returns the packets to the linear order,

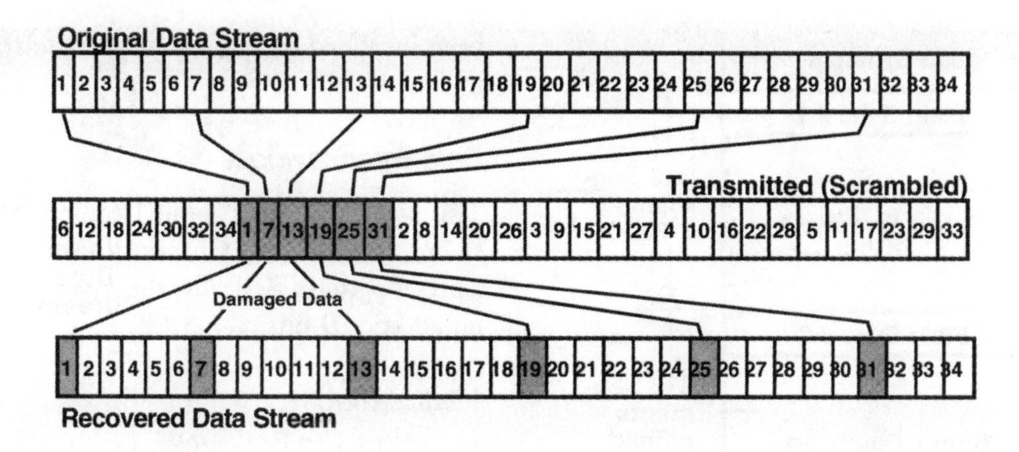

FIGURE 1-5

data packets 1, 7, 13, 19, 25, and 31 are separated by a number of unaffected, or good packets. The larger loss is actually divided among smaller losses that can be interpolated, or recalculated, from the remaining adjacent packets.

Grouping the Data

The trellis encoder works with the VSB modulator to reduce transmission bandwidth. The encoder takes a stream of bits and groups them into 3-bit words called symbols. The purpose is to ultimately reduce the bandwidth by reducing the frequency.

Data Sync

Segment sync and field sync are added to the transport packets in the multiplexer. The segment sync marks the beginning of each transport packet. The field sync identifies a single group of transport packets. The multiplexer's data output is referred to as a VSB data frame.

After MPEG and AC-3 compression, the audio and video data are multiplexed with additional services data into a 188-byte transport packet. The transport packet is output at a maximum data bit rate of 19.4 Megabits/second (Mbits/sec). The packet consists of a link header and a payload. Refer to ATSC signal configuration illustration in **Figure 1-6**.

Link Header

Each packet is preceded by a link header that contains the following information:

- SYNC — Identifies the beginning of the packet and a sample of a 27 MHz clock signal.
- CONTINUITY COUNTER or ERROR CONTROL — The identifying number for each packet.
- TRANSPORT SCRAMBLING CONTROL — Identifies if this packet is scrambled.
- PACKET ID (PID) — Links related and marks duplicated, or imported, packet. Marks the location of the unscramble packet, if any.

Payload

The payload consists of 184 Bytes of multiplexed audio, video and additional services data.

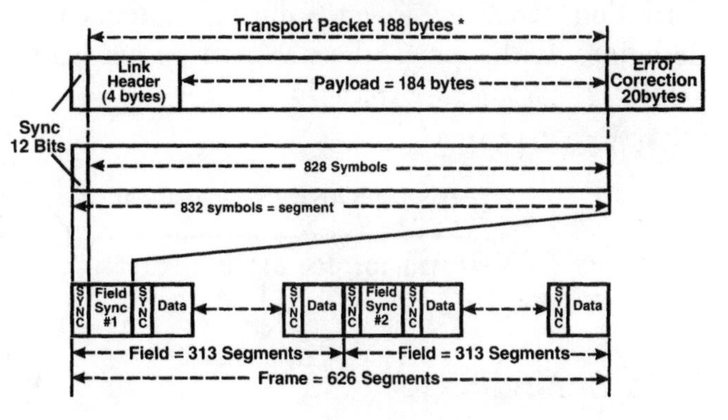

FIGURE 1-6

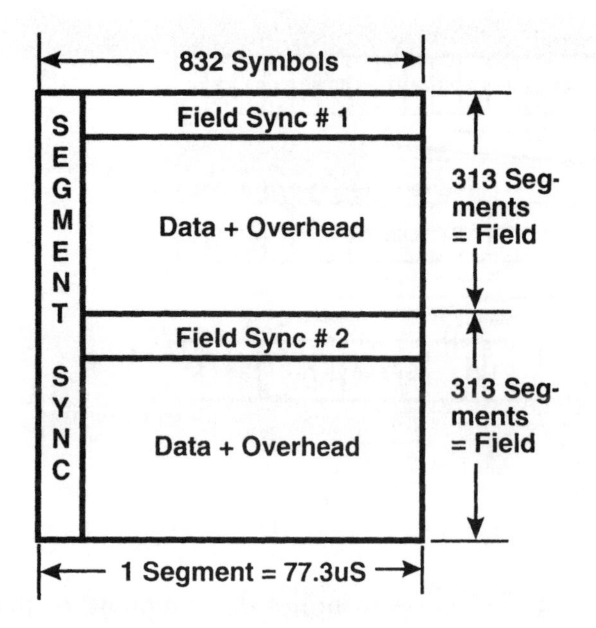

FIGURE 1-7

VSB Transmission

The VSB transmission is processed in the following stages: data randomizer, Reed-Solomon encoder, data interleaving, multiplexing and Trellis encoding.

Data Randomizer

The construction of the VSB data frame begins at the data randomizer. This block (see Figure 1-4), scrambles only the payload data in a set pattern so it can be descrambled at the decoder.

Reed-Solomon Encoder

The Reed-Solomon encoder scrambles the entire transport packet except for the early data sync in the link header. It adds 20 parity bytes for error correction to ensure accuracy during subsequent decoding. At this point, the transport packet has been increased to 208 bytes. Refer to the drawing in **Figure 1-7**.

Data Interleaving

This block scrambles about 1/6th of a field of data (more than a single transport packet) according to a preset pattern. Basically, the data interleaver adds no additional bits.

Multiplexing

The multiplexer combines the symbols with segment sync to form what is referred to as a segment. Each segment contains four symbols of segment sync, plus 828 symbols of payload data. Thus, each segment contains the equivalent of a scrambled transport packet plus sync. A field of digital data is made up of approximately 312 segments. The multiplexer adds additional field sync to the beginning of the 312 segments, which then brings the field up to 313 segments. Two fields equal a frame or 626 segments which is equivalent to a complete vestigial sideband (VSB) data frame. This frame is then routed to a VSB modular for transmission. Each segment contains data sync, data and overhead information.

Trellis Encoding

The following is a simplified explanation of the trellis encoding process. The trellis encoder adds additional bytes to the data stream and then converts the data bit stream to 3-bit words. These 3-bit words represent one of eight different levels and these 3-bit words are referred to as symbols. Refer to the drawing in **Figure 1-8**.

The trellis encoder works with the VSB modulator in a way that reduces the bandwidth by approximately 1/3. The encoder then converts the

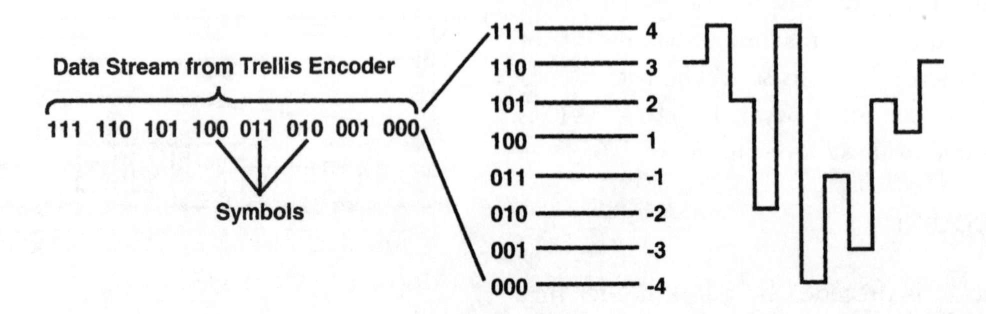

FIGURE 1-8

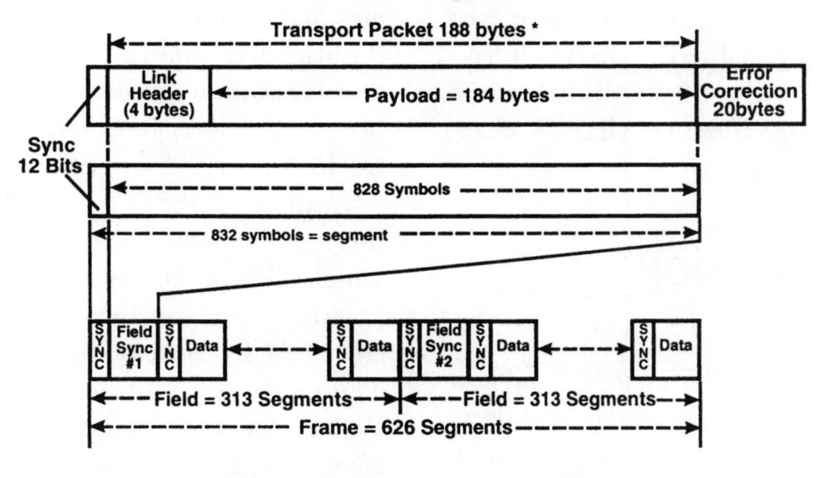

FIGURE 1-9

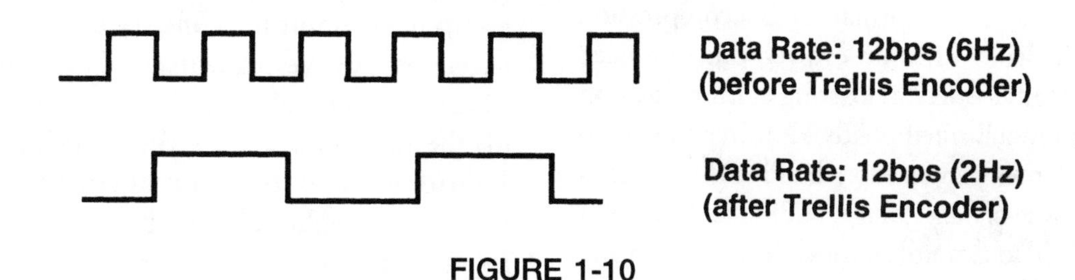

FIGURE 1-10

incoming stream into word groups. The 3-bit words are assigned in a mapping table that is designed to reduce transitions, or bandwidth.

The VHS modulator converts each 3-bit symbol to a corresponding DC level for transmission. Three bit symbols can designate up to 2 to the power of 3 (2^3) or 8 analog levels. For example, at the modular input of the Trellis scheme diagram, the data is 110, 111, 101, 010, 111, 000, etc. That corresponds to a DC level of +4, +3, +2, +1, etc., as shown at the modulator output in Figure 1-8. By replacing the 3-bit word with a single DC voltage, the digital information sent to the AM transmitter is reduced by one third.

Bandwidth Reduction Using Trellis Encoding

To better understand how a trellis encoder reduces bandwidth, here is a worst case example that uses a high-frequency digital signal. The signal is a waveform that is alternating between digital 0 and 1. Twelve of these alternating bits are shown, in **Figure 1-9**, sampled over a one second period for a data rate equal to 12 bits/sec. By examining this waveform, you can see that two bits are the equivalent of one cycle. Therefore the data rate of 12 bits/sec is equivalent to 6 cycles/sec or 6 Hertz (Hz).

The trellis encoder generates the 3-bit symbols and the VSB modulator translates each symbol into a DC level for transmission. The worst-case transmission occurs when the DC level is alternating between low and high. However, since a single DC level represents the three bits, after modulation, the input data rate of 12 bits/sec results in a transmission rate of only 2 cycles/sec vs. 6 cycles/sec. The drawing in **Figure 1-10** illustrates these data rates.

The data rate of 32.28 million bits per second (Mbps) is reduced by 1/3 to 10.76 Mbps using the trellis encoding method. Since two bits are the equivalent of one cycle, the data rate is halved to arrive at the equivalent frequency. Therefore the audio, video and additional services data will have a maximum frequency of to 10.76 Mbps/2 = 5.38 MHz. This is within the current 6 MHz TV channel bandwidth.

MPEG COMPRESSION INTRODUCTION

In order to understand the ATSC standard and how it works, you need a basic knowledge of MPEG video compression and why it is needed. The acronym MPEG refers to the Motion Picture Experts Group, named after the committee or group that developed and standardized the process. MPEG, and MPEG2, are very broad standards that encompasses many types of video compression including those used in the ATSC format. IT is a compression method for video, which has evolved due to a need to transmit digital video on existing communication channels within a limited bandwidth. In addition to the limits of the transmission bandwidth, there is the obvious issue of data storage. MPEG and MPEG2 compressed video can utilize much less space on any given storage medium.

Communication Media	Typical Bandwidth
ISDN Line	144 Kbps (kilobits per second)
T1 Line	1.5 Mbps
T3 Line	45 Mbps

With the large capacity of relatively inexpensive digital storage media available today, including high-density hard drives, DVD'S, etc., it would seem that full motion video storage would be of no concern. However, digital video requires a considerable amount of disk space. Uncompressed, full bandwidth digital video requires an unbelievable 252 Mbps of bandwidth. This equates to approximately 31 Mbytes (million bytes) per second of video. This does not include the audio portion that must be stored along with the video.

Component Video

Let's now look at the basic digital video content. In its most basic form, video is made up of four components. Luminance (Y) provides the brightness and contrast level. Basically this translates to the black and white component. The chrominance, or color portion of video, consists of three components called R-Y, B-Y, and G-Y signals. Luckily, we do not need to save the G-Y signal as it can be re-created from the R-Y and B-Y signals. This leaves three components for video: Y, R-Y, and B-Y. A sample of each component can be represented by an 8-bit digital word that translates to 256 different levels for each of the three video elements. Thus, a single pixel is made up of 3 components, Y, R-Y, B-Y, as is illustrated in **Figure 1-11**. Mathematically, this is represented as 2^8 (Y) ∞ 2^8 (R-Y) ∞ 2^8 (B-Y) = 2^{24}, equal to 24-bit color or 16 million colors.

The resolution, or the amount of detail in the picture, is also important when considering bandwidth and storage requirements. Common computer resolution consists of 800 horizontal pixels by 600 vertical pixels. Each one of these pixels must contain the Y, R-Y, and B-Y portion of the signal. Refer to the drawing shown in **Figure 1-12**. If these numbers are multiplied, you calculate the following: one component (Y) is equal to 800 ∞ 600, or 480,000 pixels per frame. Each of the three components is an 8-bit word. Therefore, 3 ∞ 480,000 ∞ 8 = 11.52 Mbits are in one frame. So, thirty frames per second is

FIGURE 1-11

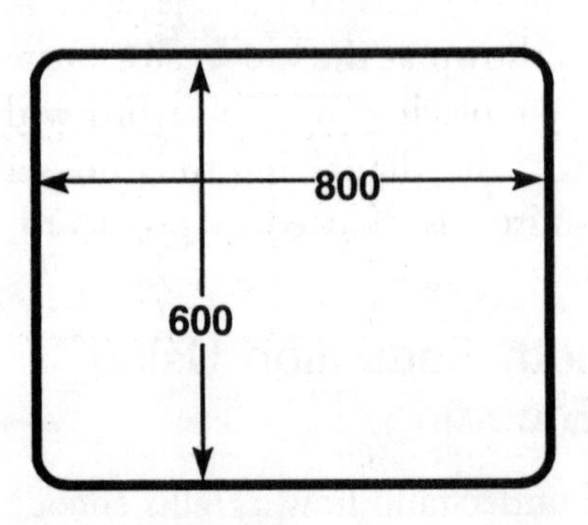

FIGURE 1-12

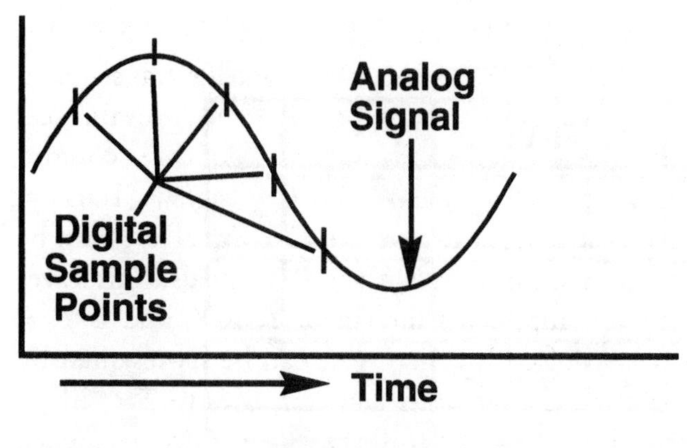

FIGURE 1-13

equivalent to 30 ∞ 11.52 Mbits = 345.6 Mbits/sec (per second of video).

To calculate the amount of storage needed on a hard drive or CD ROM, simply divide the bandwidth by 8 bits per byte to get 43.2 Mbytes for each second of video. As you can see, given the amount of data — and without some sort of compression — the transmission and storage of digital video would be very difficult. This is why MPEG video provides an advantage.

MPEG Compression Process

The following basic steps are performed during the video compression process:

1 . Digital sampling
2 . Discrete cosine transform (DCT)
3 . Predictive and motion encoding
4 . Hoffman encoding

Digital Sampling

The number of times per second a signal is sampled is called its sampling frequency. In audio, a common sampling frequency is 44,000 samples per second. This is approximately two times the highest audio frequency, or 20,000 Hz. In contrast, video is sampled at four times the highest frequency or 13.5 megahertz (MHz).

Each term, i.e. 4, in the sampling structure 4:4:4, represents the sampling ratio of Y to R-Y to B-Y, respectively. The same number of samples are taken of Y as they are of R-Y and B-Y. Note the digital sampling drawing shown in **Figure 1-13**.

The eye is less sensitive to color changes than it is to luminance. Significant reduction of data can be accomplished during the sampling process if we sample the color components (R-Y and B-Y) half as much as the luminance component (Y). The result is a 1/3 reduction or a bandwidth of 166 Mbits/sec. Video sampled using this technique is sampled using a 4:2:2 sampling structure. Y is sampled normally with R-Y and B-Y sampled at half the normal rate.

Other methods include a 4:1:1 sampling ration, which further reduces color sampling to 1/4 of the Y component, compressing by 1/2 or reducing bandwidth requirements to 125 Mbits/sec.

Another method in use is a modified 4:2:2 as shown in **Figure 1-14**. This sampling structure is referred to as 4:2:0. The 4:2:0 samples R-Y half as much as Y and skips B-Y on the last line. This sampling is illustrated in **Figure 1-15**. However, on the next horizontal line, B-Y is sampled half as much as Y, and R-Y is skipped. This sequence is repeated, effectively reducing the color components

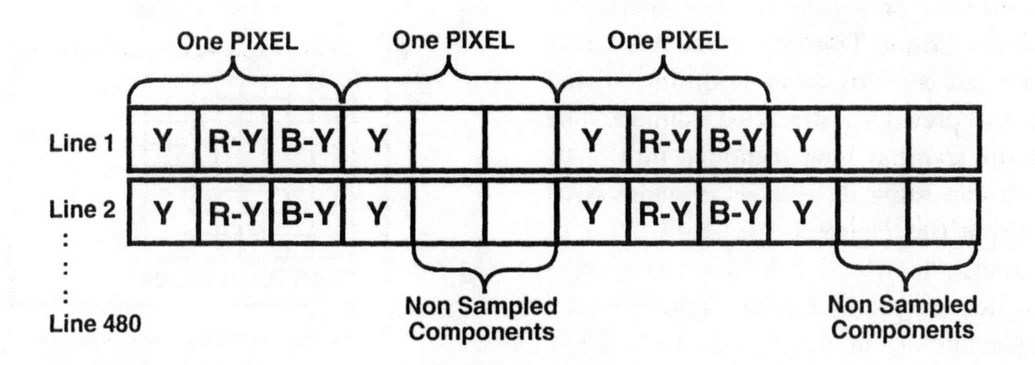

FIGURE 1-14

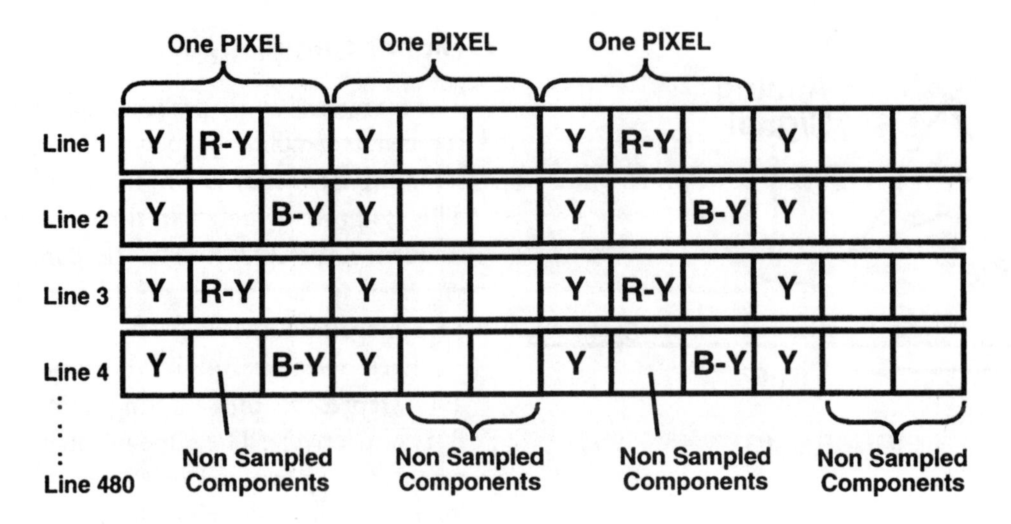

FIGURE 1-15

by another 50 percent, achieving 1/2 compression or a 125 Mbit/sec bandwidth Figure 1-15). Through interpolation, the 4:2:0 is reconstructed into 4:2:2 without the extra bandwidth requirement.

Discrete Cosine Transform

After the selective sampling process is complete, the next step in the MPEG process is to remove very fine or high-frequency picture detail, imperceptible to the human eye. It is undetectable because it is typically masked by other picture content. In a video frame, the very fine picture details consist of high-frequency information that is basically rapidly changing luminance and color content. In contrast, the low-frequency information contains slow changing luminance and color picture details.

The high-frequency information consumes most of the data stream and is the area of focus in the next stage, compressing content. The process of eliminating the imperceptible information is called Discrete Cosine Transform, or DCT. This sampling process converts information into digital data using the previously described method. The digital picture frame is then sectioned into 5400 blocks, each consisting of 8 pixels in width by 8 pixels in height (see **Figure 1-16**).

DCT transforms the 8 x 8 group of pixel + values into frequency components. Although pixel values vary randomly in the 8 x 8 block, DCT repositions low-frequency components on the upper left corner and high-frequency components on the lower right of the block. Through an additional numerical conversion process called *quantization*, frequency component values are assigned. High-frequency components are identified by the zero values (lower right) and low-frequency components by the larger values (upper left). Data compression is accomplished through elimination of the high-frequency components designated by zeros. Refer to the 8 x 8 pixel blocks in **Figure 1-17**.

Special Redundancy

Within a frame, there are many redundant pixels. An example would be those found in the image of a blue sky. This type of redundancy within the

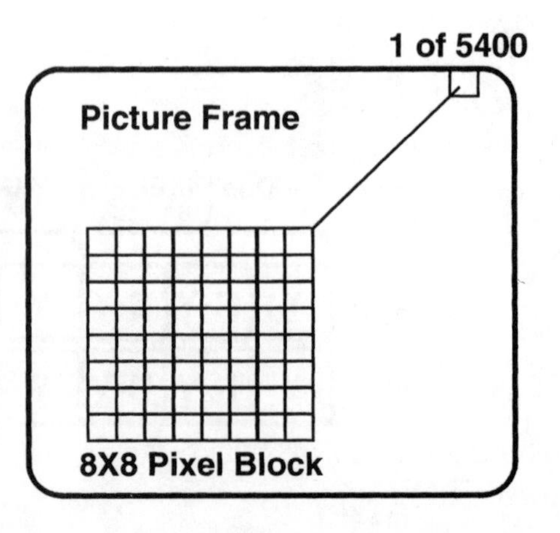

FIGURE 1-16

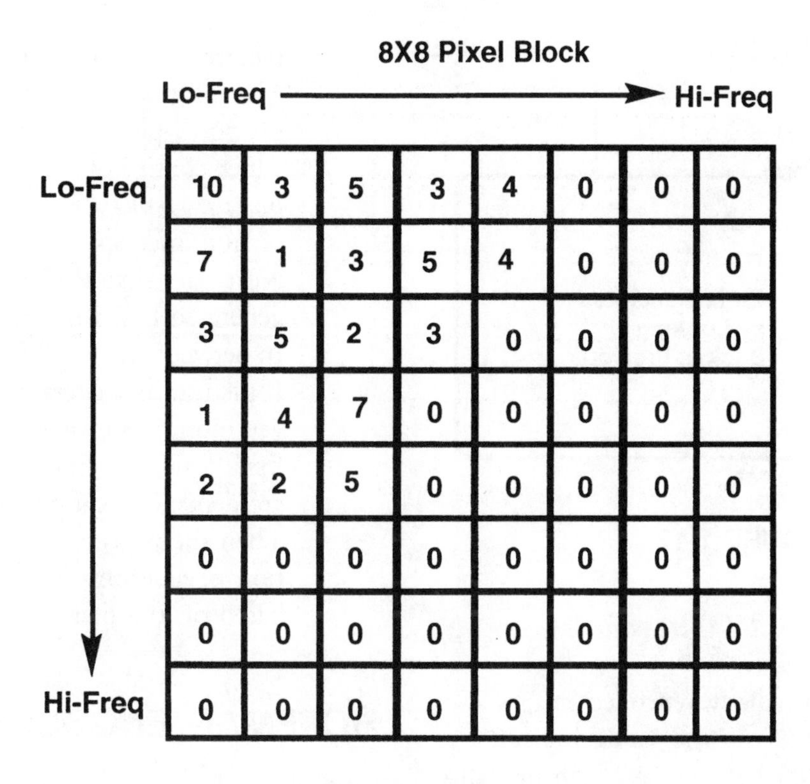

FIGURE 1-17

horizontal and vertical plane of a frame is called *special redundancy*. One pixel could be stored with repetition information for the remaining pixels both in the vertical as well as the horizontal direction. This would eliminate the need to store every pixel in the frame. Frames having these type of characteristics are referred to as *intraframes*.

Temporal Redundancy

Within video scenes, many parts of a frame are redundant. An example would be a coach sitting on the sideline. With the exception of arm movement, the other portions of the frame remain unchanged over time. This type of redundancy over time is called *temporal redundancy*. The first frame could be stored as the reference, or non-changing portion, while remaining frames carry the arm movement information. This would eliminate the need to store several full frames. Frames using temporal redundancy, which predict information based on preceding frames, are called *interframes* or motion predicted images.

PREDICTIVE AND MOTION ENCODING

Predictive and motion encoding is the next step in the MPEG compression process. It takes advantage of both spatial and temporal redundancy to begin the compression process.

I-Pictures

To begin the process of using spatial and temporal redundancy techniques in compression, you need a reference or a starting frame that is not dependent on previous frames. This beginning or starting frame would make use of spatial redundancy within itself and is termed an I-picture. I-pictures (intraframes) have zero dependency on previous frames. However, they do provide information to preceding frames.

The other picture types used by MPEG are called P-pictures (predictive) and B-pictures (bidirectional). I-pictures carry the most amount of scene data content. These frames are three times the size of a P-picture and five to six times the size of a B-picture.

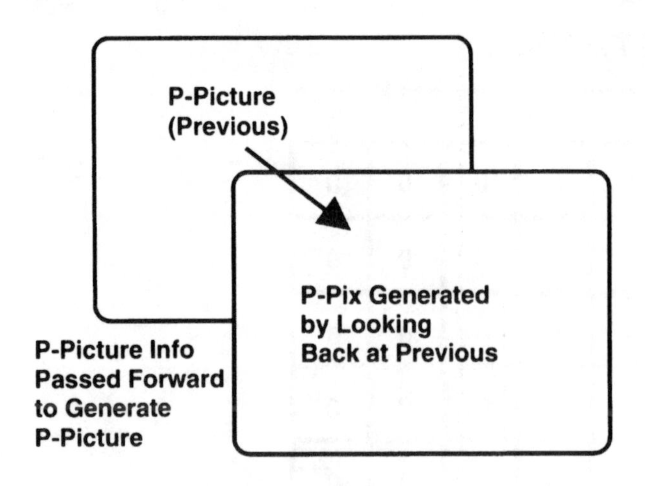

FIGURE 1-18

P-Pictures

The P-pictures are predictive encoded images also known as interframes. As the name indicates, a P-picture is a predicted Image based on previous I-pictures or P-pictures. The P-picture is dependent on past Images in order to exist. This technique is illustrated in **Figure 1-18.**

B-Pictures

The B-pictures have bidirectional dependency (previous and preceding) and are called bidirectional predicted images. The B-picture is also a predicted image but it is based on preceding I- and P-pictures. B-pictures are typically made up of motion information and consist of the least amount of data. The illustration in **Figure 1-19** should give you a better understanding of this technique.

GROUP OF PICTURES (GOP)

I-, P-, and B-Picture Generation Process

To more clearly understand the relationship between the I-, P-, and B-pictures we need to understand how they are generated. Refer to the GOP drawing shown in **Figure 1-20.**

1. The start of an entirely new scene would require an I-Picture or a reference for other pictures to follow until the next I-picture. The information between an I-picture and the next I-picture reference is called a GOP, which consists of one I- and many P- and B-pictures. I-pictures typically reoccur at 15-picture intervals.
2. Next, the first P-pictures in the GOP is generated based on the I-picture.
3. In between the I and first P-picture, several B-pictures are generated as necessary to convey motion information from the I to the first P-picture. For this reason B-pictures are dependent on past and future pictures.
4. Then the process repeats with the generation of a second P-picture, which is now based on the first P-picture.

HOFFMAN ENCODING

One of the last steps in the MPEG process is a statistical method that compresses the data even further. This is called Hoffman encoding. Basically, this process interprets a string of MPEG data and replaces it with information that allows regeneration. A good example is a string of eight binary "1's" (11111111) replaced by (1x8) which indicates that the 1 is repeated eight times. In this way the digital video data stream can be compressed even further.

MPEG vs. MPEG2

Many of the MPEG processes previously talked about are common to both MPEG and MPEG2. This explains the backward compatibility between the two. The main difference between MPEG and MPEG2 is that MPEG2 uses a 720 x 480 resolution, while MPEG1 uses a resolution of 350 x 240. This difference alone accounts for a 75 percent reduction in data when using MPEG rather than MPEG2 and accounts for the major differences in video resolution. MPEG2 compresses data by a factor of 40 to 1 on average, while MPEG compresses data by an average factor of 140 to 1. Therefore, 250 Mbits/s rates are reduced to 6.25 Mbits/s, on average, using MPEG. MPEG2 also uses variable rate compression (720 x 480 = 350,000 pixels)

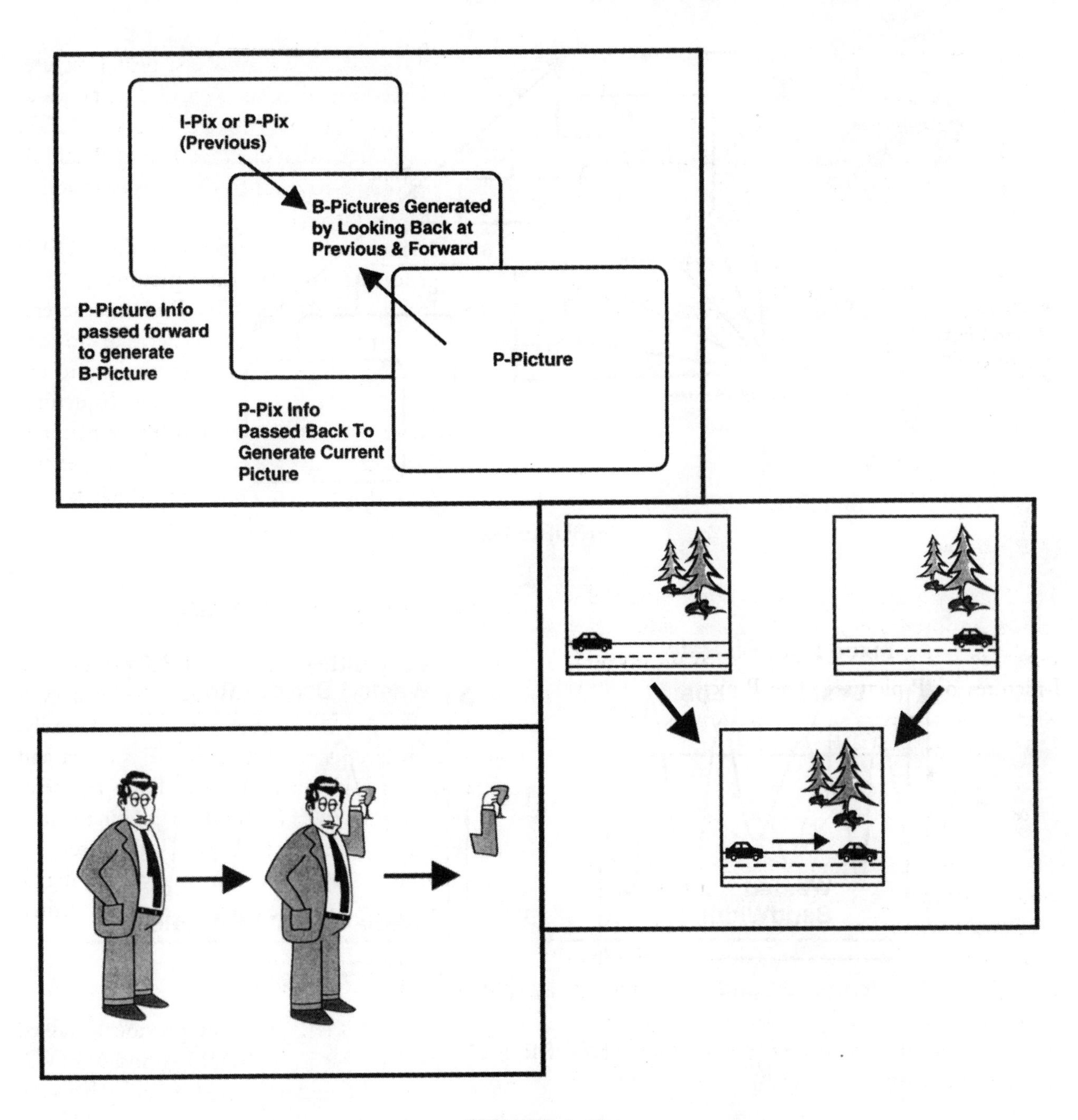

FIGURE 1-19

while MPEGI uses a fixed rate (350 x 240 = 84,000 pixels). So, MPEG represents 84,000/350,000, or 25 percent of MPEG2.

Fixed vs. Variable Data Bit Rate

A variable or changing transmission data bit rate provides a tremendous advantage when digitizing video. If the data bit rate were fixed, it could not accommodate itself to the changing needs of different video scenes. Consider the fast-paced action of a football game with a player running for a touchdown as the camera pans the crowded field. Scenes that are full of motion equate to an extremely demanding video scene, one that requires the bit rate to be very high. Now picture the same football player sitting on the bench and moving very little. There is little movement or change in

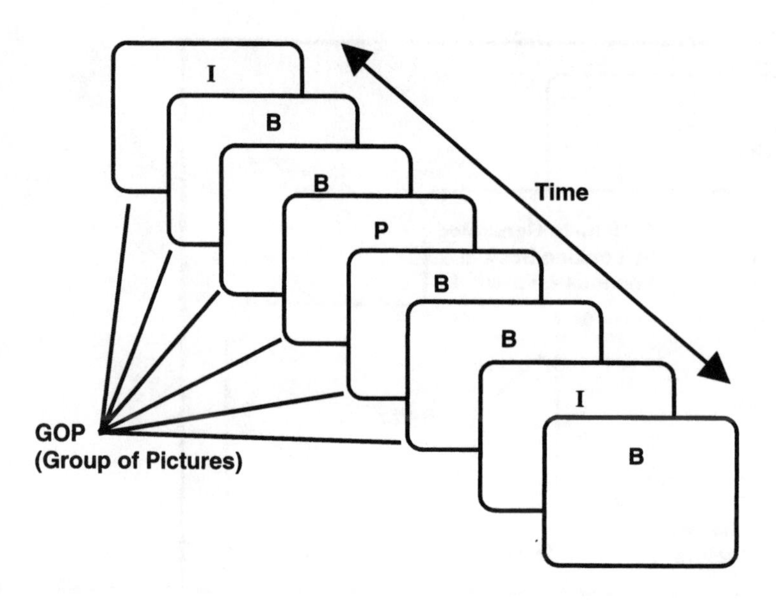

FIGURE 1-20

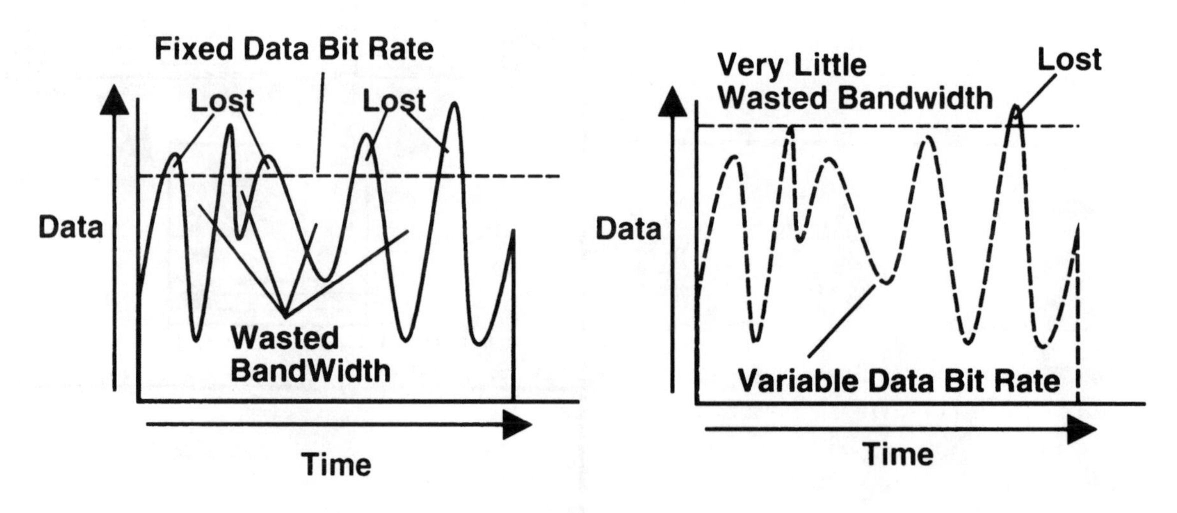

FIGURE 1-21

the scene so the data bit rate needed to transmit that image can be quite low. Refer to the fixed and variable data bit rate drawing in **Figure 1-21**.

If the data bit rate were fixed, the system might be inadequate when recording the football scene. In addition, it would waste bandwidth filming the bench scene. A variable data bit rate accommodates both scenes by varying the bit rate as the scene changes. With a fixed data bit rate, the 6 MHz bandwidth provided for HDTV would have to be much higher to maintain the same high quality picture. Transmitting data at a varying rate is one of the key technologies that makes ATSC and HDTV possible.

24p HDTV Mode (24 Frames-Per-Second)

In video, what appears to be a continuously moving image is actually a series of discrete still pictures, called frames. Every frame of a video consists of two interlaced fields, each of which contains half the frame's scanning lines. A standard NTSC or ATSC video picture runs at roughly 30 frames per second.

STATIONS BROADCASTING DIGITAL SIGNALS

Station	City	Date On	Owner	Format	Time of Day
WCBS (CBS)	New York	Nov. 98	CBS	1080i	24 hours
WNYW (Fox)	New York	Apr. 99	Fox	480p	24 hrs. May 1
KTLA (WB)	Los Angeles	Oct. 98	Tribune	1080i **	12 hours
KCBS (CBS)	Los Angeles	Nov. 98	CBS	1080i	24 hours
KNBC (NBC)	Los Angeles	Oct. 98	NBC	1080i	24 hours
KABC (ABC)	Los Angeles	Nov. 98	ABC	720p	24 hours
KCOP (UPN)	Los Angeles	Dec. 98	Chris Craft	480p	24 hours
KTTV (Fox)	Los Angeles	May 99	Fox	480p	24 hours
WFLD (Fox)	Chicago	May 99	Fox	720p	24 hours
WPVI (ABC)	Philadelphia	Nov. 98	ABC	720p	24 hours
KYW (CBS)	Philadelphia	Nov. 98	CBS	1080i	24 hours
WTXF (FOX)	Philadelphia	Oct. 98	Fox	720p	24 hours
WCAU (NBC)	Philadelphia	Nov. 98	NBC	1080i	24 hours
KGO (ABC)	San Francisco	Nov. 98	ABC	NA	NA
KPIX (CBS)	San Francisco	Nov. 98	CBS	1080i	24 hours
KTVU (Fox)	San Francisco	Nov. 98	Cox	720p	24 hours
KRON (NBC)	San Francisco	Nov. 98	Chronicle	NA	NA
KBHK (UPN)	San Francisco	Jun. 99	Chris Craft	480p	24 hours
WCVB (ABC)	Boston	Oct. 98	Hearst-Argyle	720p	24 hours
WMUR (ABC)	Boston	Nov. 98	Imes	480i	
WHSH (HSN)	Boston	Nov. 98	USA Broadcst.	480p	8 hours
WHDH (NBC)	Boston	Apr. 99	Sunbeam Tel.	480i	24 hours
WFXT (Fox)	Boston	Apr. 99	Fox	720p	24 hours
WBZ (CBS)	Boston	May 99	CBS	1080i	24 hours
WFAA (ABC)	Dallas	Nov. 98	Belo	1080i	24 hours
KDFW (Fox)	Dallas	Nov. 98	Fox	720p	24 hours
KXAS (NBC)	Dallas	Nov. 98	NBC	1080i	24 hours
KTVT (CBS)	Fort Worth	May 99	Gaylord	1080i	24 hours
WJLA (ABC)	Washington	Nov. 98	Allbritton	480p	24 hours
WTTG (Fox)	Washington	July 99	NA	NA	NA
WUSA (CBS)	Washington	Nov. 98	Gannett	NA	NA
WRC (NBC)	Washington	Nov. 98	NBC	1080i	24 hours
WETA (PBS)	Washington	Nov. 98	Public	1080i	24 hours
WXYZ (ABC)	Detroit	Oct. 28	Scripps Howard	720p	17 hours
WWJ (CBS)	Detroit	July 99	CBS	1080ip	24 hours
			upconverting; analog will pass thru later		
WJBK (Fox)	Detroit	May 99	Fox	720p	24 hours
WDIV (NBC)	Detroit	Apr. 99	Post Newsweek	1080i	24 hours
WSB (ABC)	Atlanta	Nov. 98	Cox	720p	24 hours
WXIA (NBC)	Atlanta	Nov. 98	Gannett	1080i	24 hours
WGNX (CBS)	Atlanta	NA	Tribune/Meredith	1080i	24 hours
WAGA (Fox)	Atlanta	Apr. 99	Fox	720p	24 hours
KHOU (CBS)	Houston	Nov. 98	Belo	1080i	24 hours
KOMO (ABC)	Seattle	Oct. 98	Fisher	720p	24 hours
KCTS (PBS)	Seattle	Sept. 98	Public	1080i	24 hours
KING (NBC)	Seattle	Oct. 98	Belo	1080i	24 hours
WKYC (NBC)	Cleveland	Jun. 99	Gannett	1080i	24 hours
WPLG (ABC)	Miami	May 99	Post Newsweek		
KATU (ABC)	Portland	Oct. 98	Fisher	720p	24 hours
KOPB (PBS)	Portland	Nov. 98	Public	1080i	24 hours
KMGH (ABC)	Denver	NA	McGraw Hill		
KCRA (NBC)	Sacramento	May 99	Hearst Argyle	1080i	24 hours
WTHR (NBC)	Indianapolis	Sept. 98	Dispatch	1080i	24 hours
WISH (CBS)	Indianapolis	Dec. 98	LIN	1080i/480i	24 hours
WBTV (CBS)	Charlotte	Nov. 98	Jefferson-Pilot	1080i	24 hours
WLWT (NBC)	Cincinnati	Nov. 98	Gannett	1080i	24 hours
WCPO (ABC)	Cincinnati	Nov. 98	Scripps Howard	720p	10-6 (M-F)
WKRC (CBS)	Cincinnati	Nov. 98	Jacor	1080i	Varies
WRAL (CBS)	Raleigh	May 98	Capitol	1080i	24 hours
WBNS (CBS)	Columbus	Nov. 98	Dispatch	NA	NA
WKOW (ABC)	Madison	Nov. 98	Shockley	720p	18 hours
KITV (ABC)	Honolulu	Jan. 98	Hearst-Argyle	720p	24 hours
WITF (PBS)	Harrisburg	Nov. 98	Public	1080i	24 hours
KCPT (PBS)	Kansas City	Nov. 98	Public	1080i	24 hours
WMVT (PBS)	Milwaukee	Oct. 98	Public		Varies
WMPN (PBS)	Jackson	Nov. 98	Public	1080i	24 hours
WTNH (ABC)	New Haven	Dec. 98	LIN	720p	24 hours
WTAE (ABC)	Pittsburgh	Jan. 99	Hearst Argyle	720p	24 hours
WNOU (NBC)	South Bend	Dec. 98	Michiana Telecasting Corp.	1080i	Varies
KXLY	Spokane, WA	NA	NA	NA	NA
WMFD (Ind.)	Mansfield, OH	Feb. 99	Independent	480i	24 hours

FIGURE 1-22

In contrast, movie films you see on television, cable or video cassettes have all had their frame rates converted by a special machine called a *telecine.*

The telecine converts the 24 film frames into video fields. However, video requires 30 frames or 60 fields, while film requires 24. The telecine performs this process by converting 12 film frames to 24 fields (2 fields/film frame) and another 12 film frames to 36 fields (3 fields/film frame). It is kept seamless by converting one film frame to two fields and the next one to three. This cycle is repeated (2, 3, 2, 3) until the 60 fields have been completed. This telecine process is referred to as *112-3 Pull Down.* To eliminate the process used to convert movies to a video format, a provision was made during the development of the new ATSC format. It allowed the broadcasting of 24-frame film directly with no conversion, and for the film to be viewed in the original format in which it was recorded. Therefore, you will see the 24p format listed in the ATSC format charts.

DIGITAL TELEVISION QUESTIONS AND ANSWERS

Q: What is digital delevision? And what's the current status of high-definition television (HDTV)? Are HDTV and DTV the same thing?

A: The FCC, its Advisory Committee on Advanced Television Service, and the Advanced Television Systems Committee, a consortium of companies, research labs, and standards organizations, have defined 18 different transmission formats within the scope of what it broadly calls the *Digital Television Standard.* DTV is the umbrella term for all 18 formats.

Six of these formats are considered high definition because they constitute a significant improvement over the resolution quality of current TV, referred to as NTSC format and established nearly 50 years ago. Most TV viewers will see a great improvement in image quality even with the other twelve formats because of digital transmission. The TV viewers will also benefit from DTV formats such as wide-screen theater-like displays, enhanced audio quality, and new data services.

Q: Besides better resolution, audio performance, and data services are there any more reasons to have a HDTV set?

A: One of the basic improvements with HDTV is the way it is transmitted. Digital transmission can deliver a near perfect signal — free of ghosts, interference, and picture noise.

Q: Will you be able to view the new HDTV broadcasts on a current conventional TV set?

A: Yes. You can watch HDTV broadcast programs by using a special HDTV decoder device. These set-top-boxes, which connect as easily as VCRs, will receive digital transmissions and convert all 18 formats to standard TV, allowing you to see and hear digital programming.

Q: Will you be able to watch high-definition TV using this set-top box and a standard TV set?

A: Not high definition, but a big improvement that will provide better images and sound using A/V equipment that you may already own. The decoder box will output HDTV broadcasts with Dolby Digital audio providing more precise localization of sounds and a more convincing, realistic ambiance. You may already have a multi-channel, multispeaker audio system allowing you to take advantage of Digital TV's enhanced sound quality.

Many HDTV decoders will also provide three high quality connections for monitors. Component video outputs will allow you to connect the box to most home theater LCD projectors and direct-view sets with component inputs to provide optimum image quality. Many large-screen TVs can be connected via S-video, which maintains high image quality by separating the luminance and chrominance signals. You can even connect this box to a standard VGA computer monitor, which provides a more crisp and detailed picture than the conventional TV.

Q: How are these HDTV signals received?

A: In most locations you should be able to receive HDTV with any standard UHF antenna. The exact style of antenna that is required for optimal reception may vary depending on your geographic location and distance from the TV tower. Consult your electronics distributor for advice in selecting the optimal antenna for your location.

Q: What's the schedule for various cities to start HDTV broadcasts?

A: The chart in **Figure 1-22** lists cities and dates that stations will start broadcasting HDTV. This is the schedule as of January 1, 2000.

Q: When will HDTV direct view sets, HDTV decoders and projection TV sets become available?

A: By late 1999 and the first half of 2000 a good selection of set-top boxes, projection sets, and some direct-view sets that produce HDTV pictures became available. Some manufacturers have prepared the market in advance by selling HDTV-ready projection TV receivers with multiple high-quality video inputs and direct-view large-screen TVs with component video inputs.

Q: What about digital signals from cable systems or satellite TV? Aren't some cable and satellite systems already transmitting digital signals at this time? Will digital HDTV sets display signals from these systems?

A: That is correct. Some cable and satellite systems already use digital technology to transmit their TV programming. These systems require the customer to use a converter box for that service. Many of these standards are not compatible with each other or the ATSC standard for digital television. However, some HDTV-ready sets will allow the viewing of output signals from the various services when using a proprietary digital converter. Also, some HDTV sets and decoder boxes have a port in which to plug a device to interface with future digital cable and satellite set-top boxes.

As DTV becomes increasingly available, set makers will offer a wide selection of DTV and HDTV receivers, decoders, and additional components. These innovations will enable the TV viewer to enjoy significantly better video and audio than received from today's NTSC-TV system.

ADVANCED DIGITAL ENTERTAINMENT

Digital television (DTV) is an exciting new digital broadcast standard that will provide vastly improved picture and sound quality compared to the current analog NTSC technology. The DTV standard allows broadcasters the choice of transmitting single programs in HDTV or multiple standard-definition television (SDTV) programs. DTV will eventually replace analog television and will enable broadcasters to send a choice of more varied information over the airwaves, cable and satellite, that includes the following:

- More detailed picture information
- Clearer images without ghosts or snow
- Dolby digital CD-quality sound
- Wide-screen pictures
- On-screen data (text and graphics)

Brilliant Quality

1080i interlaced HDTV, the highest level of DTV, will deliver up to 1,080 scan lines, more than double the scan lines of analog TV signals with six times the resolution. HDTV will also make it possible for TV screens to have the wide 16:9 aspect ratio of movie theater screens — about a third wider than the 4:3 aspect ratio of NTSC format — allowing you to view movies as originally intended. Add Dolby Digital surround sound, and HDTV will deliver an entirely new experience when viewing movies, sports events and other TV programs.

Cutting-Edge Features

HDTV (either 1080i interlaced or 720 progressive scan) and SDTV (480 interlaced) will offer the exciting experience of clearer, more detailed digital video and audio than today's NTSC signal. The broadcasters can choose the type and number of signals they transmit within their allotted bandwidth (6 MHz) and transmission rate (19.3 Mbps). In the future, as new DTV products are sold with advanced features, the broadcasters will be rolling out new services. A few of the possible benefits are as follows:

- Up to four SDTV programs broadcast from one TV station simultaneously, where you currently only receive one. These pictures will be clearer than NTSC, free of interference like snow and ghosts, but will not have as much resolution as an HDTV picture.
- On-screen data, such as educational materials, or team statistics during a game.
- Pay-per-view movies and premium channels, as on present satellite TV and cable TV systems. Access to web sites related to the program you are watching.
- Home shopping using your remote control to make your choices.

Portions of this chapter information courtesy of Thomson multimedia.

CHAPTER 2

HDTV Test Equipment Requirements

INTRODUCTION

This chapter contains information on test equipment required for troubleshooting, repairing, and aligning high-definition color television receivers. An overview and features of this HDTV equipment will be covered.

The beginning of this chapter features the Thomson multimedia's Chipper® Check test unit that is used to troubleshoot the GE/RCA MM101 HDTV chassis. This model CCF002 chipper check interface box, when connected to your PC or laptop, will provide you with a quick, convenient way to perform adjustments, troubleshooting and circuit diagnostics.

CHIPPER CHECK OVERVIEW FOR GE/RCA MM101 HDTV CHASSIS

The evolution from early analog television to modern, digitally controlled television, has produced a number of new challenges for the service technician. These new microcomputer-controlled televisions can exhibit a variety of symptoms that are not identifiable using the same thought process as that used for analog TVs. In addition, alignment is no longer performed by adjusting a potentiometer until the desired DC voltage is obtained. Instead, a digital value, stored in memory, is converted to the DC voltage required by the alignment circuit. At first, this type of system can be confusing since it is not always obvious which adjustment is being performed simply by looking at the display on the set's screen. Some adjustments are not incorporated until the receiver is turned off and back on, which makes it very difficult to know when the adjustment is correct. Chipper Check was developed to address these differences and to provide the technician with a convenient method of performing adjustments and diagnosing problems. The Chipper Check system is composed of two major components. The first component, and the most visible, is the hardware interface. The interface is responsible for physically connecting the TV and personal computer together. The second part is the instructional software that runs on the personal computer. This software provides the instructions for performing adjustments and diagnostics. The Chipper Check block diagram and connections are shown in **Figure 2-1**.

Chipper Check Hardware

The hardware consists of small adapter boards that attach directly to the chassis via a communications port, an interconnect cable, and the Chipper Check interface box. These items are shown in **Figure 2-2**. A standard parallel printer cable is used to connect the interface box to the parallel port on the personal computer; however, this cable is not part of the Chipper hardware package. The adapter board allows the flexibility of connecting the interface box to a number of different chassis families.

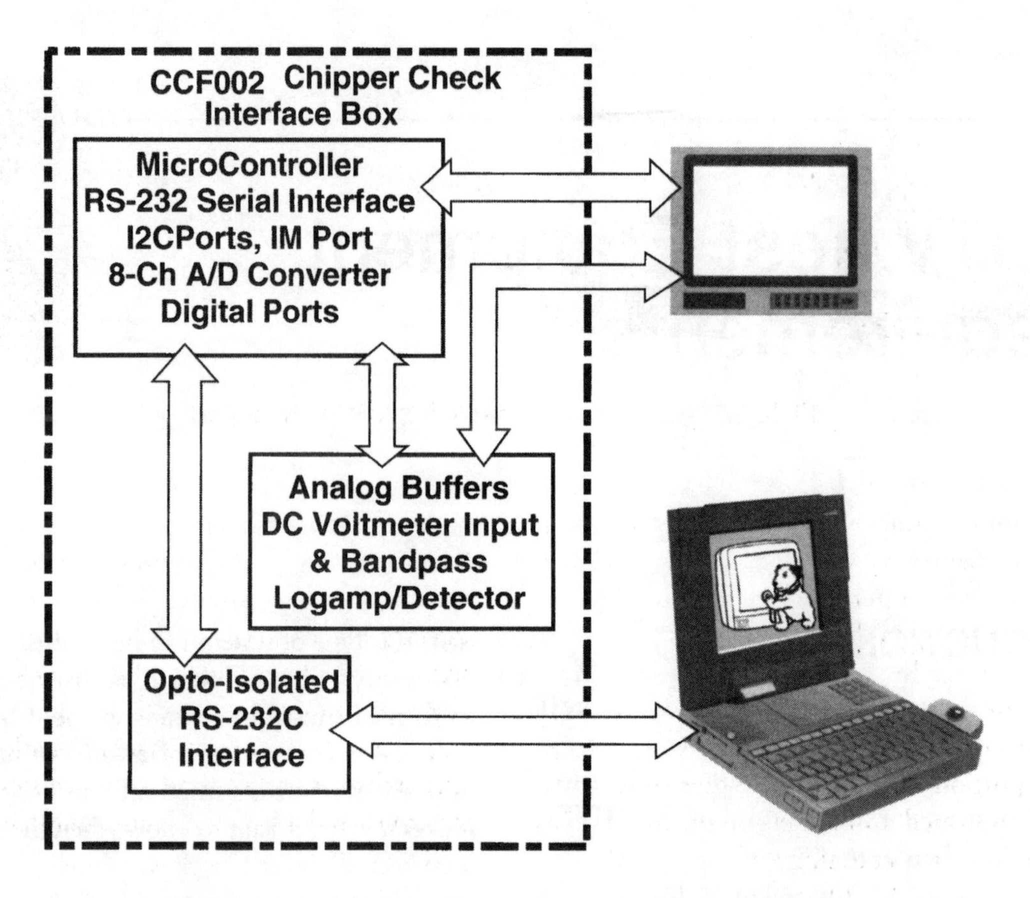

FIGURE 2-1 Chipper Check Simplified Block Diagram

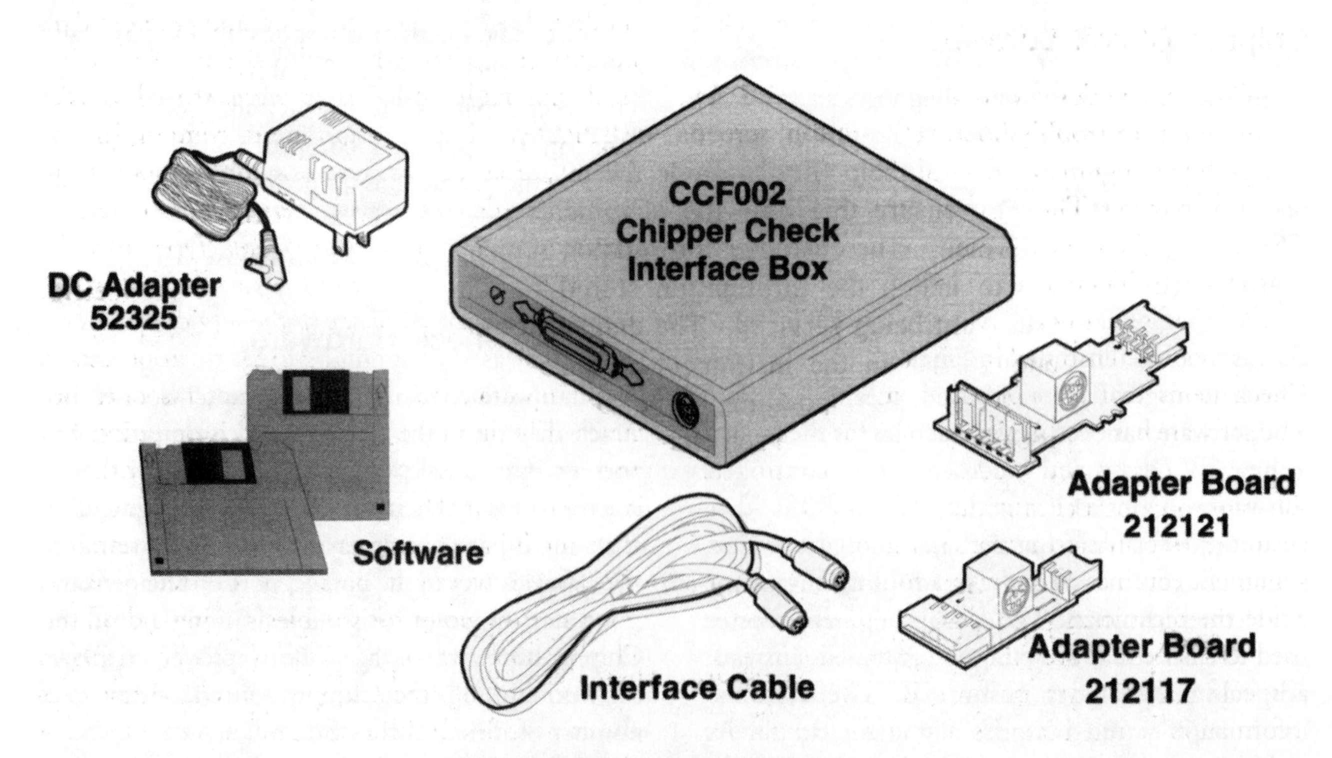

FIGURE 2-2 Chipper Check Items

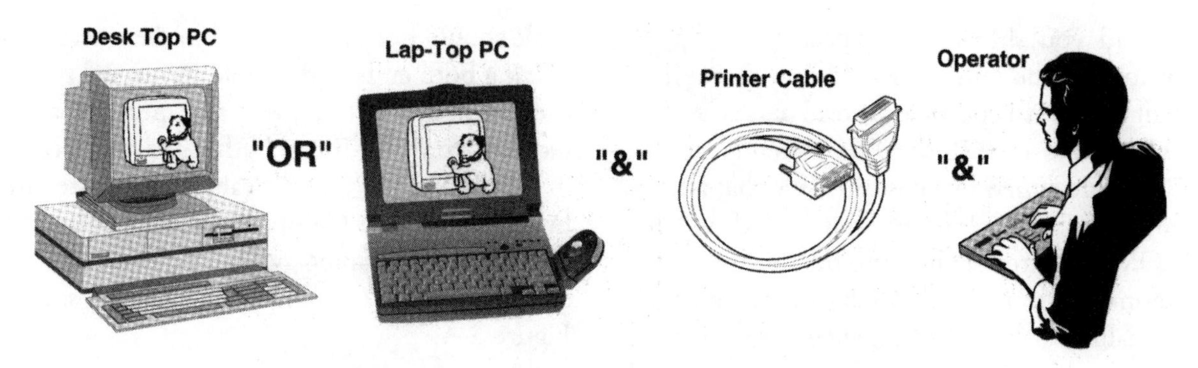

FIGURE 2-3 Chipper Check Items Not Supplied

The communications format may vary from one chassis model to the next. So, rather than developing a new interface box for each chassis, changing the adapter board will allow the same interface box to be used. The Chipper Check interface box provides electrical isolation between the PC or laptop and the HDTV receiver under test. The portion of the interface that connects to the PC is powered from the computer's parallel port. The interface contains a power supply that provides power to the TV receiver. The interface then converts the signal from the parallel port of the computer into the correct protocol for the HDTV's receiver microprocessor.

Chipper Check Software

The Chipper Check software allows the technician's PC or laptop to troubleshoot, communicate with, and perform alignments on a digitally controlled television receiver. The software also provides many different routines that are unique to each major chassis family. There is also a set of standard drivers for the interface hardware used for all of the different chassis models. Illustrated in **Figure 2-3** are Chipper Check items that are not supplied with the unit. The software has been designed to be easily updated as new TV chassis and models are introduced. The software contains a chassis auto-detection function, customer information screens, diagnostic routines, alignment routines along with on-line help files to guide the technician. The chassis auto-detection is used to ensure that only the alignments required for a specific chassis are performed. The customer information screen provides a way for the service technician to match the information stored by the Chipper Check software to the specific instrument being serviced. The diagnostic routines are used to read any error codes stored in the instrument and to identify which integrated circuit in the chassis is not responding. The alignment routines provide all alignment procedures needed for each chassis. The help files provide instructional information including user information, alignment procedures, test point locations, and a list of various troubleshooting tips.

Using the Chipper Check for Troubleshooting

The Chipper Check has two basic modes of operation, *Dead Set* and *Normal.* The Dead Set mode is helpful when the television receiver does not power on. The Chipper Check can be used to read the fault codes that were stored in the EEPROM. These fault codes remain in the EEPROM as long as there is standby power to it. In this mode, the chassis auto-detection feature does not function, so the chassis type must be manually selected from a list. When the fault codes are recovered, they indicate to the service technician which IC was not responding to the microprocessor. It is important to remember that this does not necessarily mean the IC indicated is defective, but merely that it is not communicating with the microprocessor. The cause could be something other than the IC itself, such as a component external to the IC. However, it does give the technician a good starting point for troubleshooting a dead set. Furthermore, when the standby power supply is not functioning, the Chipper Check cannot read the error codes. If this is the case, it would indicate that the problem is likely in the standby power supply and you should not expect the Chipper Check to find the failure. In addition to reading the fault codes, it is possible to read and store the contents of the EEPROM. The contents of the EEPROM are stored in a customer file that allows the original settings to be reinstalled in the EEPROM after troubleshooting the dead set. The picture-in-picture EEPROM data cannot be read or reinitialized in the dead set mode.

Chipper Check Setup Connections

To use the Chipper Check in the normal mode, it is necessary to place the television microprocessor in the *slave* mode to prevent communication problems with the interface hardware. The adapter board and cable should be connected to the Chipper Check port on the television chassis before the television set is turned on, but the other end of the cable should not be connected to the interface box. Having both ends of the cable connected may load the communications lines and prevent the TV receiver from turning on. After the TV receiver is in the slave mode, and the cable is connected to the interface box, the Chipper Check software can be started. The adapter board location and setup connections for the Chipper Check are shown in **Figure 2-4**.

Chipper Check Operation

The first thing the program does is to check communications between the TV receiver and the computer. Part of this process is to detect which chassis is connected. If the wrong chassis is detected, it is possible to manually select the correct chassis type. Use caution when selecting the chassis model because the alignment requirements can differ.

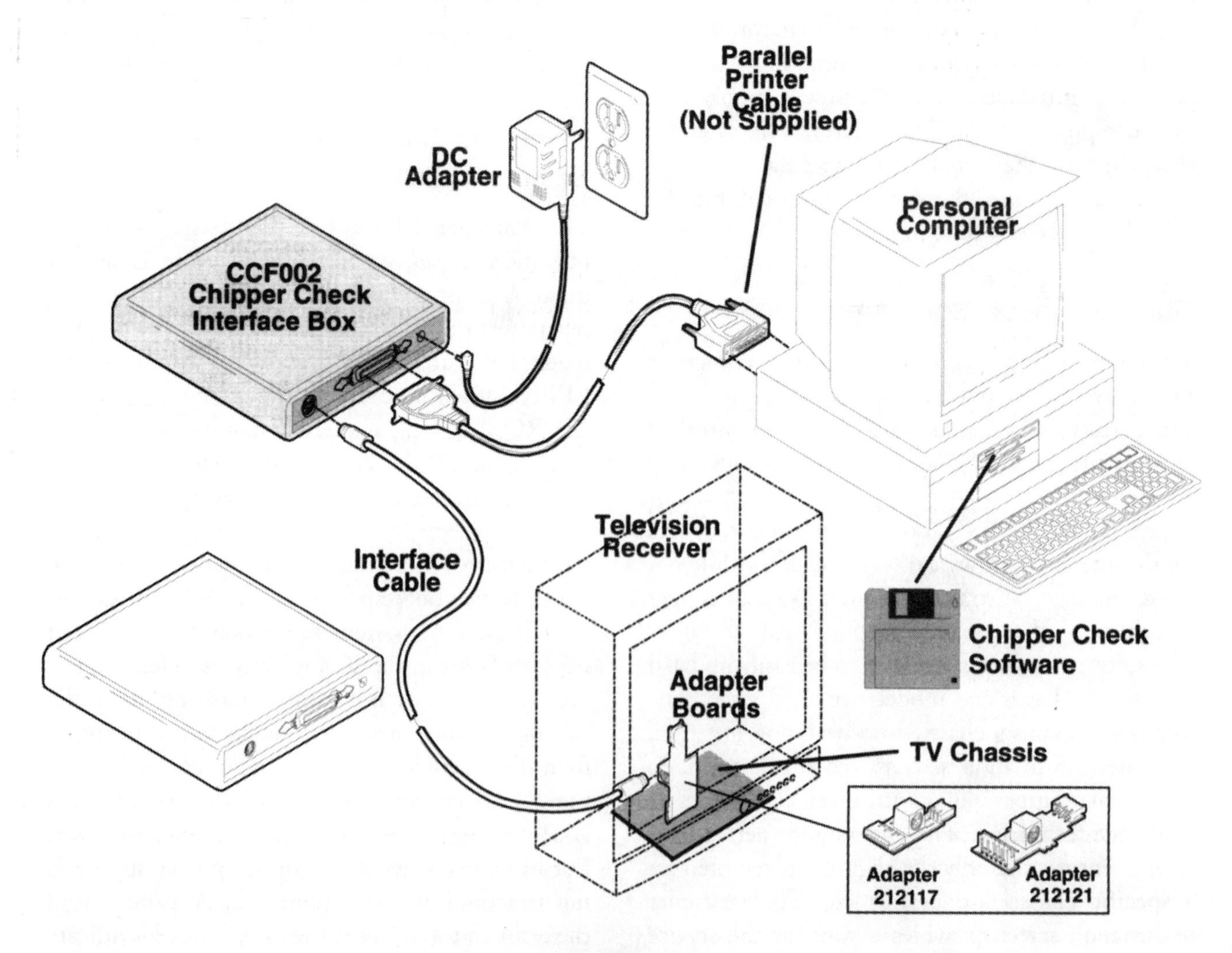

FIGURE 2-4

Selecting the wrong chassis can cause the computer to lock up or store incorrect information in the TV receiver's EEPROM.

After communication is established, a customer information screen appears. This allows customer information such as name and model/serial number data to be stored. It is important to note that when the information on this screen is saved, the contents of the EEPROM are not yet associated with the file. This screen is placed here as a convenience to the service technician. The customer information, or job ticket number, is stored at the beginning of the process when the information is still readily available.

Diagnostic Function

The screen after customer information allows three choices: diagnostics, alignment, or major part replacement.. The diagnostic portion gives the option of reading fault codes, checking the EEPROM, or reinitializing the EEPROM. The fault code again tells you which major component or IC is not responding. This simply indicates the area of the trouble. When the EEPROM is checked, data is *written to* and *read from* every location in the EEPROM. If the computer can read and write to every location, the EEPROM is functioning correctly and should not be replaced. However, this does not mean that the data stored in the EEPROM is correct. For this reason, the option of re-initializing the EEPROM is provided. Initializing the EEPROM will write the *factory values* to certain locations in the EEPROM. These items need to be set to certain values to ensure proper initial operation. None of the alignment data is modified nor is customer information, such as scan lists and channel labels changed. During initialization, the customer controls are set to the factory preset values. Included are the convergence settings are included on projection receivers.

Alignment Functions

The service alignments are grouped into circuit sections or by the effect each has on the picture. The steps for each different alignment must be performed in the proper order. That is, once a particular alignment is attempted, the steps should be performed in the order indicated. However, the order in which each alignment is performed does not matter. The highlighted text in the alignment procedure shows test point locations on the chassis and lists other helpful tips for performing the alignment. When activated, the help button on the alignment screen will prompt instructional information to be displayed if the alignment cannot be properly adjusted.

Part Replacement Function

The last option is *replacement part.* When this option is selected, the technician must enter which major part has been replaced. The Chipper Check software then walks the technician through all alignments that should be performed or checked after that part has been changed. For example, if a component was changed in the PIP (picture-in-picture) tuner, all PIP tuner alignments should be checked. However, it is not necessary to perform the PIP color and tint alignments so these are not displayed on the screen.

Error Codes

There are three specific EEPROM locations that hold error codes generated by the TV's software routine of the chassis. Upon entering the security address, a digital readout of the error code is displayed on the front panel in positions one, two, and three. The Chipper Check also reads these three locations and produces a text translation for each code.

The error code location, possible error code, and the text explanation are as follows:

Error Code Location	Possible Error Code	Text Explanation
1	7	Scan Loss
2	3	+12 V Run Supply
3	192	2nd Tuner PLL

In all cases, error location 1 is the first code logged by the system control microprocessor since the last time all three error code locations were cleared. Location 3 is the most important location since it contains the most recent code logged by system control. In theory, there may have been hundreds of codes in each of the three locations that, at some point, were reset to zero. Then, upon attempting to restart the set, a current error code list would be generated, provides more useful and current data to the technician troubleshooting the problem.

In the above example, errors in locations one and two could have occured any time between the original manufacture date of the receiver and the most recent startup attempt. Generally, without knowing when the first two errors were logged, only error code three — a second tuner PLL (phase locked loop) IC failure in the case of our explanatory table — could be considered current and useful. After the successful completion of any repair, all three error code locations should be set to zero.

A second important point regarding the error code list is the manner is which they are generated and how they are checked. When Chipper Check or system control sends information to an IC device, it is in the form of complete words. Most IC devices acknowledge the receipt of data by sending it back to the original microprocessor address, but only after adding a binary value of one (0001) to the data. This is known as a *parity* check. For example, in the RCA CTC179 chassis, EEPROM data sent to the T2 register 09, would look like this in digital: 1011, 1010, 0000, 1001. The first byte contains the T2 address, 1011, 1010 (BAh). The next is the address of the register inside the T2, 0000, 1001 (09h). This would be followed by the actual data byte, in our example 1010, 0111. When the T-chip acknowledges it has received new data for the register, it changes the address DA, incrementing it by one to become BB. The T-chip register address and data remain the same. System control then compares the addresses in the outgoing data with that of the returning data. If the two addresses match, normal operation is continued. If they do not, an error code is logged. The chassis may or may not go into the *batten* routine depending upon which error is logged. In addition, unidentified errors could be generated on the service menu. Depending on the software, the error code could recorded as either the outgoing address or the parity address. In the latter case, the T-chip address is BA, and is stored as an error code in decimal notation, 186. The returning address is BB, or decimal 187. Either code may show up in the error information. Regardless, it should be interpreted as a communication error with the T-chip. The system control and T-chip codes are shown in **Figure 2-5**.

NOTES ON ERROR CODES

The error codes do not necessarily mean a specific IC failure. They simply indicate data was sent to a specific address and not written into the register correctly, according to the parity check routine. This could mean something as simple as a temporary read/write error, as severe as a catastrophic device failure, or something in between. Keep in mind that many other factors can cause a read/write error. The error codes are simply meant to aid in failure diagnosis by steering the technician in the right direction. The codes cannot troubleshoot a chassis. Begin by checking power supplies to the device. Then check the IIC clock and data lines. Finally, check all signal lines. As the troubleshooting technician, you must still prove device failure. Neither, the Chipper Check nor the error codes are capable of identifying specific hardware failures.

INITIALIZATION AND CUSTOMER FILES

When checking an EEPROM for possible failure or corrupt data the technician has two options prior to performing the EEPROM test. Chipper Check allows the technician to either initialize the EEPROM or store/retrieve a customer file. There are important differences between the two files.

First, there are several locations for data in the EPROM that are not used in field service. This is data loaded at the time the set was manufactured and does not require adjustment, alignment, or testing by the technician. However, as with any EEPROM, the data in those locations can become corrupt for any number of reasons while the EEPROM is otherwise fully functional. The initialization file, different for each chassis version, only resets initialization locations to factory settings for that specific chassis. The file affects no other data.

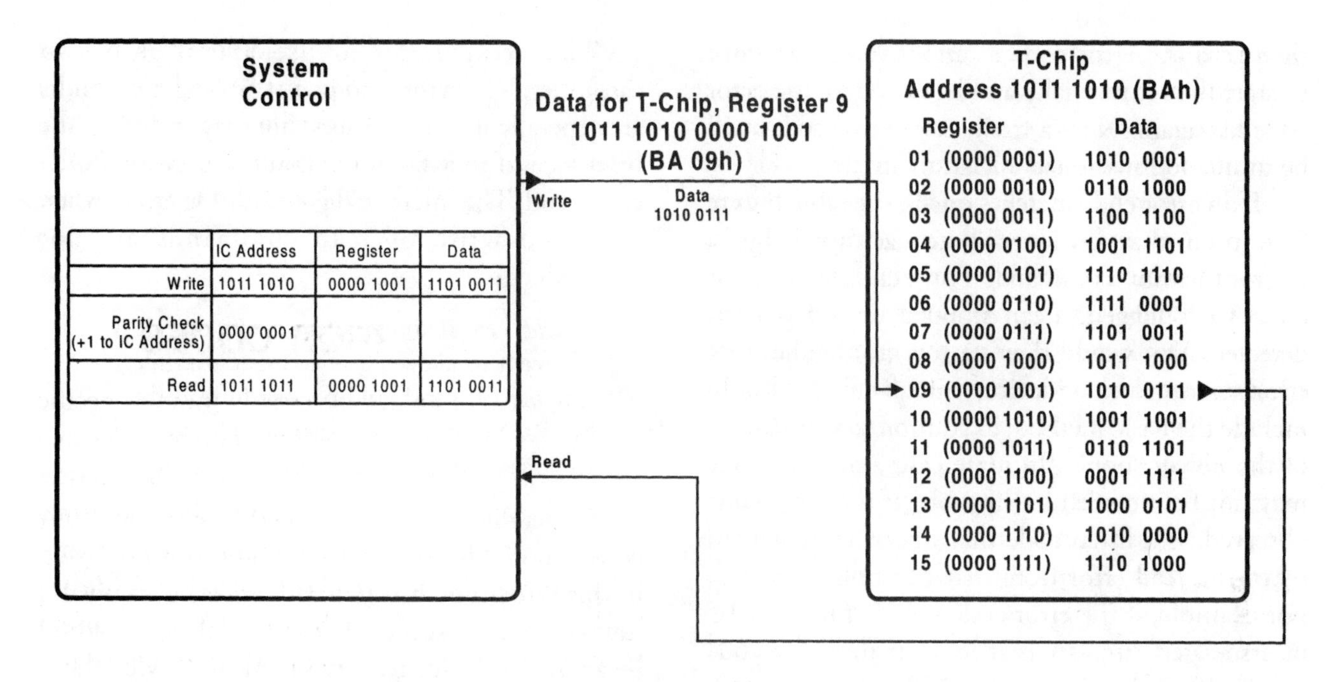

FIGURE 2-5

A customer file saves the data in every location of the EEPROM to the PC's drive for future recovery. This includes data in the same locations as that in the initialization file. Unfortunately, Chipper Check can not determine whether the data is corrupt or not. The customer file is useful if an EEPROM must be replaced or for use in troubleshooting faulty EEPROM data. It should be saved to a unique file name. If an EEPROM is replaced, the customer file can be loaded back into the EEPROM. All alignments, customer settings, and the initialization string can be recovered. Since the EEPROM was replaced, some of its data may be corrupt. The technician should check the set very carefully prior to returning it to the customer. The proper order to restore the EEPROM is to download the customer file first, then reinitialize the receiver.

PLEASE NOTE: The Chipper Check does not replace a troubleshooter and his or her technical expertise.It is used as a tool, just like a video analyzer, scope or multimeter by the service technician to troubleshoot and performing tests and alignments of HDTV chassis.

Error Codes

Under normal conditions, failure of an IC device will prevent the set from turning on and an error code will be logged in the first three parameter locations of the service menu. (These error codes and their meaning will be discussed later in this chapter) It is important to note that if a failed bus device is the cause for service, the normal microprocessor routine that acknowledges the IC device is disabled in the service mode.

As with previous RCA chassis, the MM101 logs error codes in the EEPROM during certain failure scenarios. In general, there will be an error code for each IC bus device, plus a few additional devices. When one of these devices fail, an error code will be stored in the EEPROM. There is room for three failure codes. They are referred to as the first error, the second error and the last error.

Note that the first two error locations do not update with continuous errors, only the third does. Because the first two do not, there is no way to know exactly when the error occurred. However, the third location will always have the latest code. It is a good practice to write the three codes down, then zero the error code locations. Next, attempt to start the set normally and let it update the error code list again. Now a fresh set of error codes will be available for the technician to use.

If an error code matches one in the table Figure 2-6, then that device did not acknowledge a microprocessor command. For example, if error code 13 is logged, a fan stopped error has been detected. You would then narrow troubleshooting efforts around that device. Initial efforts should include the basic method, in addtion to verification of the power supply. In some cases, an error code may not be included in the table. If the error code observed, is a listed code that's been incremented by one, a read error from that device is indicated. For example, if the error code is 136, the TVB IC or associated circuitry is at fault. If the error code is 137, the read register of the TVB did not respond. Failure of the TVB or TVB circuitry might still be indicated. In these cases, IC failure or interruption of supply power to the IC is more likely than faulty associated circuitry.

Other error codes may indicate a failure condition in different areas of the chassis, such as the cooling fans or power supplies. It is important to understand how these error codes are detected by the system control circuitry so they can be interpreted correctly and used as clues in successfully troubleshooting the receiver.

Many error codes will also place the set in power fatal mode. The codes are listed in **Figure 2-6**. This means the microprocessor will always perform a three strikes, you're out routine. After the initial failure, system control will attempt to restart the set twice. If the second attempt (third shutdown) to restart is unsuccessful, the microprocessor will shut the set down and override any operator attempt to restart. At this point, the set can only be restarted in the service mode, which ignores any error code shutdown. Thus, "yes" in the power fatal column indicates that the fault condition will always cause a three srikes shutdown routine to run. "Maybe" indicates it will run when power fatal mode is enabled. This would occur under normal operation, but not during service mode. "Never" indicates the fault will not cause a shutdown routine to run.

The error codes are separated into two categories, IC devices and power faults. Power faults are generally non-IC devices, but may include some IC devices if they fail the initial power reset (POR) command. The micro will generally log errors when requested write or read commands are not acknowledged.

> *TECH TIP:* If the problem is one that does not seem to allow a service mode override, Chipper Check should be able to read the EEPROM error code locations. However, the +5-volt standby supply must be operational.

Generally, it can be stated that if an error code is noted in the service mode and is not listed in this error code list, it could either be an invalid number due to some software glitch, or it could be an active device read error. All IC devices have a corresponding write and read error code, plus confirmation of the event. For example, error 160 refers to the main EEPROM and indicates a write command whose execution could not be verified by the microprocessor. Error 161 indicates the micro requested information or a status condition from the EEPROM, but was unable to read it. Also note that some error codes seem to indicate two different device errors. Upon closer inspection, the two devices share an error code, but are on separate bus lines. At this time, there is no way to determine which, if not both, component is faulty. Both should be regarded as defective until one can be eliminated. A list of the error codes for the RCA/GE MM101 chassis is shown in **Figure 2-6.**

> *TECH TIP:* Under certain circumstances, the +15-Volt standby supply can drop out. Then the set will shut down intermittently or even go through a "batten down the hatches" routine. It is possible that no error code will be logged. If the symptom is intermittent shutdown and no error codes are shown, monitor the +15-Volt standby supply, associated circuitry, and any devices powered by this supply.

CHIPPER CHECK NOTES

Most alignments can be handled using the Chipper Check. All service menu alignments and

Error Code	Power Fatal	Error	Bus	Device	Condition
0	Yes	Power Fail	--	+15V Standby	Check +15Vs Supply
1	Yes	Initial Power Fail	--	+12V Run Dropout	Check +12V Run
3	Yes	+12V Run	--	+12V Run Monitor	Check +12V Run Regulation
9	Yes	Video IC	RUN 1	U22300	IC or Supply Failure
7	Yes	Scan Loss	--	Deflection	Loss of Scan
10	Maybe	F2PIP POR	RUN 1	U18100	IC or Supply Failure
11	Maybe	Stereo Decoder	RUN 2	U31701	IC or Supply Failure
12	Maybe	AVR Latched	--		AVR Active
13	Yes	Fan Stopped	--	U13101	Fan or Fan Detect Circuit Failure
16	Yes	IIC Run1 or Run2 Bus Error	--	U13101	IIC clock or data failure
18	Yes	IIC Standby Bus	--	U13101	IIC clock or data failure
20	Yes	Software Stack Overflow	--	U13101	Various Conditions. Reset system.
6	Maybe	S-Video Switch	RUN 1	U16500	IC or Supply Failure
44	Maybe	F2PIP IC	RUN 1	U18100	IC or Supply Failure
52/53	Maybe	USB Hub IC	USB	U13201	IC, Supply or USB downstream device failure
64/65	Maybe	PIP IF DAC	RUN 2	U27902	IC or Supply Failure
66/67	Maybe	Main Tuner DAC	RUN 1	U32602	IC or Supply Failure
68	Maybe	Deflection DAC	RUN 2	U24800	IC Failure
72/73	Maybe	SYNC Proc	RUN 2	U38300	IC or Supply Failure
128/129	Maybe	Stereo Decoder	RUN 1	U31701	IC or Supply Error
134	Maybe	Composite AV Switch	RUN 1	U16501	IC or Supply Failure
136	Maybe	NTSC Decoder	RUN 1	U22300	IC or Supply Failure
136	Maybe	TVB	RUN 2	U11800	IC or Supply Failure
140	Maybe	Deflection Processor	RUN 2	U14350	IC or Supply Failure
160/161	Maybe	I/O EEPROM	STDBY	U13102	IC or Supply Failure
164/167	Maybe	Main Tuner EEPROM	RUN 1	U32601	IC or Supply Failure
164/167	Maybe	Deflection EEPROM	RUN 2	U14354	IC or Supply Failure
168/169	Maybe	2nd Tuner EEPROM	RUN 2	U27903	IC or Supply Failure
186/187	Maybe	OSD IC	RUN 1	U13202	IC or Supply Failure
192	Maybe	PIP Tuner PLL	RUN 2	U17401	IC or Supply Failure
194	Maybe	PIP IF DAC	RUN 2	U27902	IC or Supply Failure
196	Maybe	Main Tuner PLL	RUN 1	U25501	IC or Supply Failure
198	Maybe	Main IF DAC	RUN 1	U32602	IC or Supply Failure

FIGURE 2-6

adjustments are also included and can be performed using the Chipper Check. Except for color, temperature, and screen geometry adjustments, Chipper Check is the preferred method for all service diagnosis and adjustments.

HDTV TEST EQUIPMENT REQUIREMENTS

In the following section test equipment for HDTV testing, alignment, and troubleshooting is covered. Some of the test instruments can also be used for conventional NTSC analog color TV receiver service as well as HDTV receivers.

Sencore SC301 MHz Multifunction Oscilloscope

This microprocessor-controlled scope is designed for a multitude of applications for the electronics service technician. For ease of operation the AutoSetup function allows for automatic setup of signal measurement settings. Other functions include on-screen alphanumeric readout and cursor functions for voltage, time, and frequency measurement. Ten different user, defined instrument settings can be saved and recalled without restrictions. The built-in RS-232 serial interface allows for remote controlled operation via PC. The SC301 scope is shown in **Figure 2-7.**

The SC301 scope includes two vertical input channels and a second time base with the ability to magnify extremely small portions of the input signal 1000 times. The second time base has its own triggering controls, including level and slope selection, to allow a stable and precise display of asynchronous or jittery signal segments.

Because the accuracy of the display is so important when viewing pulse or square wave signals, the SC301 has a built-in switchable calibrator, which checks the instrument's transient response characteristics from probe tip to the scopes CRT screen. The essential high-frequency compensation of wide-band probes can be performed with this calibrator, which features a rise time of less than 4 ns.

The Sencore SC301 offers the right combination of triggering control, frequency response, and time base versatility to facilitate measurements in a wide range of applications for troubleshooting HDTV receivers.

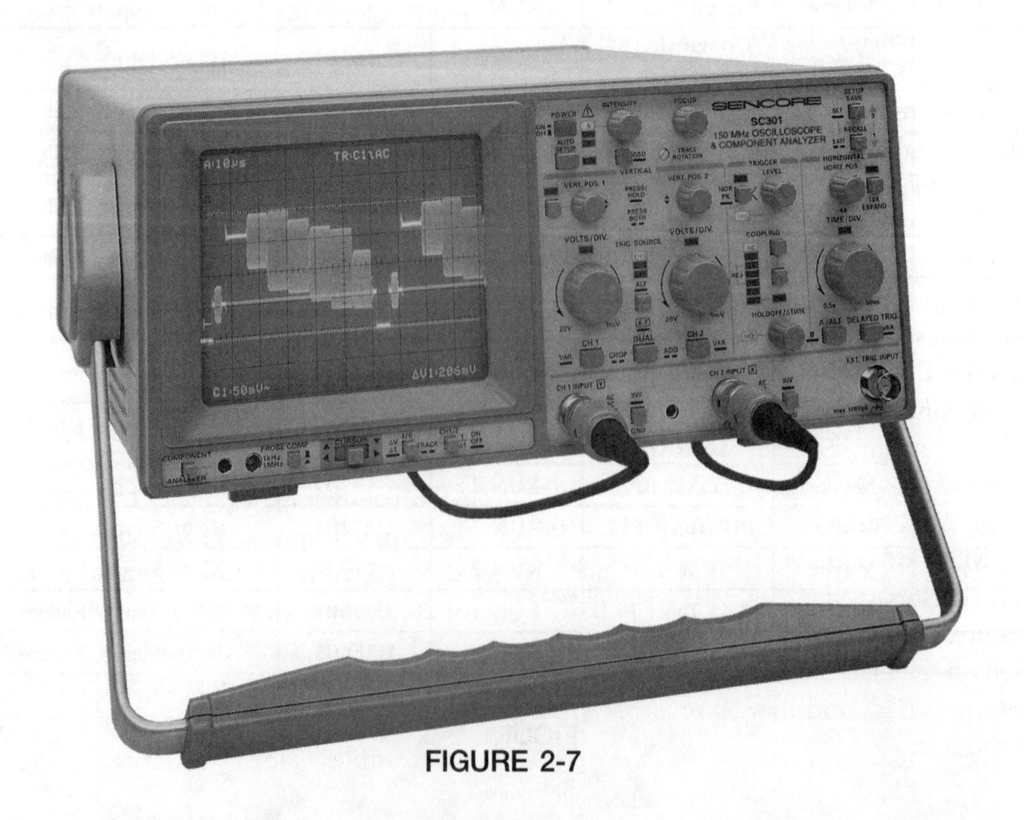

FIGURE 2-7

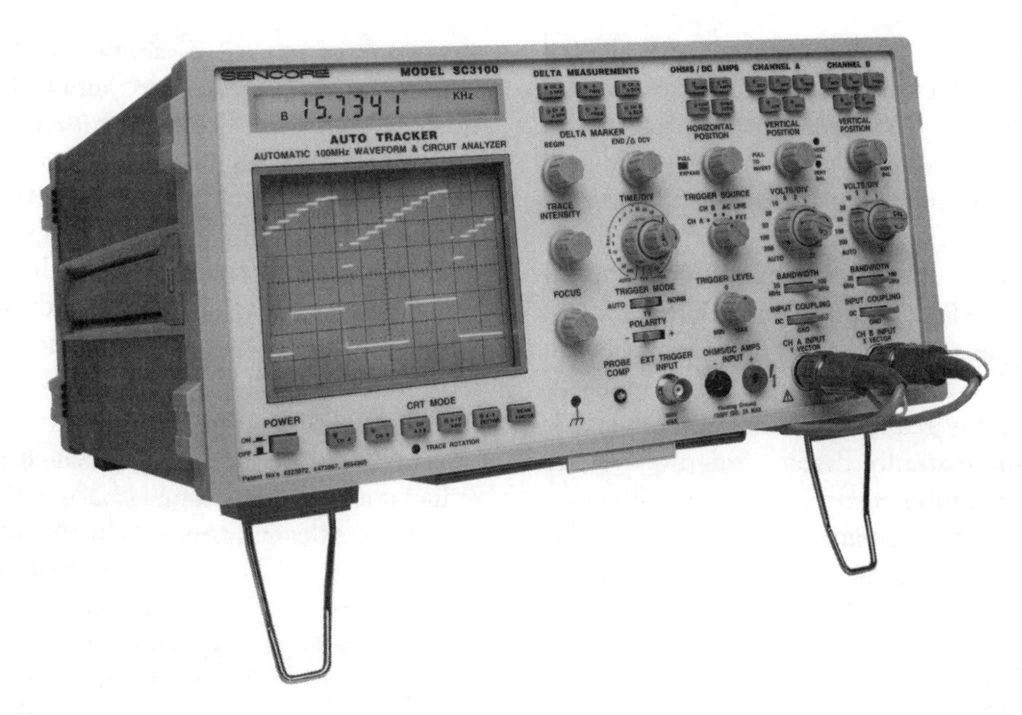

FIGURE 2-8

Sencore SC3100 Auto-Tracker Scope

The Sencore SC3100 waveform and circuit analyzer offers a whole new concept in electronic troubleshooting instruments. It provides all of the measurements needed to completely analyze a waveform or circuit faster, and more accurately than other scopes. The Sencore SC3100 Auto-Tracker is shown in **Figure 2-8**.

Of course, there are other digital oscilloscopes that provide a digital readout of the signal displayed on the scope's CRT. But these units fall short of the waveform analyzing capabilities of the SC3100 in two very important areas: ease of use and accuracy. Other digital scopes only do the graticule counting for you. You must still properly lock in the waveform and obtain a CRT display. Then, you need to find the menu that contains the function you want and select the measurement channel before you take a reading. Both the input coupling and horizontal and vertical verniers must be set correctly to make the reading. Many other digital scopes even require you to adjust cursors to the area of the signal you want to measure. On top of all this, the measurements are still based on the inaccurate analog CRT circuits and user interpretation, and may have errors of 15 percent or more.

Unlike conventional analog scopes and other digital scopes, the SC3100 allows you to measure all waveform parameters at the push of a button, without time consuming setups, graticule counting or cursor setting. For most measurements the waveform doesn't even need to be displayed on the CRT. All of the measurements are made by digital circuits that are independent of the CRT circuitry, for virtually error-free measurements.

The SC3100 also includes ohms, current, and continuity tests to analyze and track down problems in the entire circuit. It is a complete waveform and circuit analyzer for troubleshooting and alignment of HDTV receivers.

SC3100 Features

The SC3100 scope employs a full-feature, high-performance, dual-trace oscilloscope. It has a 100-MHz bandwidth, useable to 120 MHz, to accurately view digital signals. The input range extends from a 2 volt sensitivity with a direct input, to a maximum of 2000 volts when using the high-voltage probes.

The sync circuit's controls are streamlined to just four, yet they provide solid triggering on all signals. Sync separators provide a solid trigger lock on complex video waveforms. Special circuits remove halfline shifting to prevent vertical sync ghosting when viewing a line of interlaced video.

Conventional scope measurement capabilities are greatly improved upon when two types of digital techniques to analyze waveforms. The Auto-Tracking feature allows the measurement of frequency, DC volts, or AC voltage of the applied signal independent of any CRT control setting. For AC voltage readings, you can select between peak-to-peak, average RMS, or dBm (1 milliwatt into 600 ohms reference). The Auto-Tracking function automatically detects and displays any changes in the input signal.

The delta measurement feature allows the analysis of time, frequency, and amplitude of any portion of the waveform. A delta DC voltage test automatically determines the absolute DC level of any waveform point referenced to ground to confirm proper logic levels or analyze trip points.

All of the Auto-Tracking and delta measurements are made through the same probe that applies the signal to the CRT display. The measurements are unaffected by the vertical and horizontal vernier and position controls. Because all of the waveform measurements are made using digital circuits that are separate from the analog CRT circuits, the readings are much more accurate and faster than those made with conventional oscilloscopes .

You can measure DC current up to 2 amps with minimal voltage burden. The ohmmeter has full autorange for resistance measurements up to 100 megohms. A special continuity test provides a nearly instantaneous audible continuity indication for quick circuit tracing.

These tests are performed via an input separate from the oscilloscope's input to prevent measurement errors and to keep high currents from the sensitive scope input. All of these functions are fuse protected against overload.

Sencore LC103 ReZolver Tester

In today's HDTV receivers, high-performance circuits are required. Component parameters such as leakage, dielectric adsorption, and ESR (effective series resistance) are good indicators of a capacitor's ability to perform properly incircuit. Inductors too, must have correct specifications and high quality. Unless these parameters are thoroughly analyzed, troubleshooting can become a guessing game.

The LC103 takes the guesswork out of capacitor and inductor testing. It provides in-circuit capacitor value and ESR testing along with automatic, out-of-circuit analysis of capacitor value, leakage, ESR, and dielectric absorption. Inductors are measured automatically in-circuit for value and quality analysis.

The LC103 (**Figure 2-9**) is a complete, automatic, microprocessor-controlled capacitor and inductor analyzer. Its features make it ideally suited for individual component analysis during service or maintenance and HDTV receiver troubleshooting.

LC103 Analyzer Features

The Sencore LC103 is a dynamic, portable, automatic, analyzer that measures capacitors and inductors in or out of circuit. It is designed to quickly identify defective capacitors and inductors by providing quick in-circuit tests and complete analysis, out-of-circuit. The test results are shown on an easy-to-read fluorescent display in easily understood terms. All capacitor and inductor test results may also be displayed as good/bad compared to standards adopted by the Electronic Industries Association (EIA).

In addition to testing capacitor values up to 20 Farads, the LC103 also checks for leakage at their rated working voltage, up to 1000 Volts. The ESR check measures effective series resistance.

Automatic test lead zeroing nulls test lead capacitance, resistance, and inductance for accurate test results. The LC103 is protected from external voltages applied to the test leads by a fuse located on the front panel. Stop-testing circuitry locks all test buttons when voltage is sensed on the test leads. And battery operation makes the LC103 completely portable for on-site troubleshooting for all types of electronic servicing.

The LC103 Front Panel

The front panel of the LC103 is shown in **Figure 2-10** and the descriptions of the controls are as follows:

FIGURE 2-9

1. Power off-on button.
2. Component type button to select the specific component under test.
3. Test lead fuse.
4. Test lead jack.
5. Test lead zero buttons.
6. Pull chart for operating instructions.
7. Component parameters buttons.
8. Warning LED to indicate leakage voltage.
9. Stop testing LED; flashes with an audible alarm when test lead fuse is blown.
10. In-circuit test buttons.
11. Out-of-circuit test buttons.
12. Component test results display panel.
13. Component setup display; allows the usesr to enter component type, value, tolerance, and voltage rating.

Sencore VP300 Video Pro

The VP300 delivers the HDTV, NTSC, and computer display video signals you can use for accurate alignment of front and rear projectors, monitors, direct view displays, and video walls for all operating modes.

VP300 Video Pro Features:

- ATSC HDTV and Standard Definition formats (1080i, 720P)
- 4x3 and 16x9 aspect ratios
- Component Video (Y, Pb, Pr) and RGB Video Output
- Using the VP300, shown in **Figure 2-11**, the HDTV technician has a signal source that can be used to align a home theater video display.
- Every home theater installation requires signals for driving the display's HDTV, NTSC, and computer modes.

The VP300 Video Pro Multimedia Generator provides the following:

- Composite and S-Video NTSC/PAL outputs
- Monitor output (SVGA, XGA)

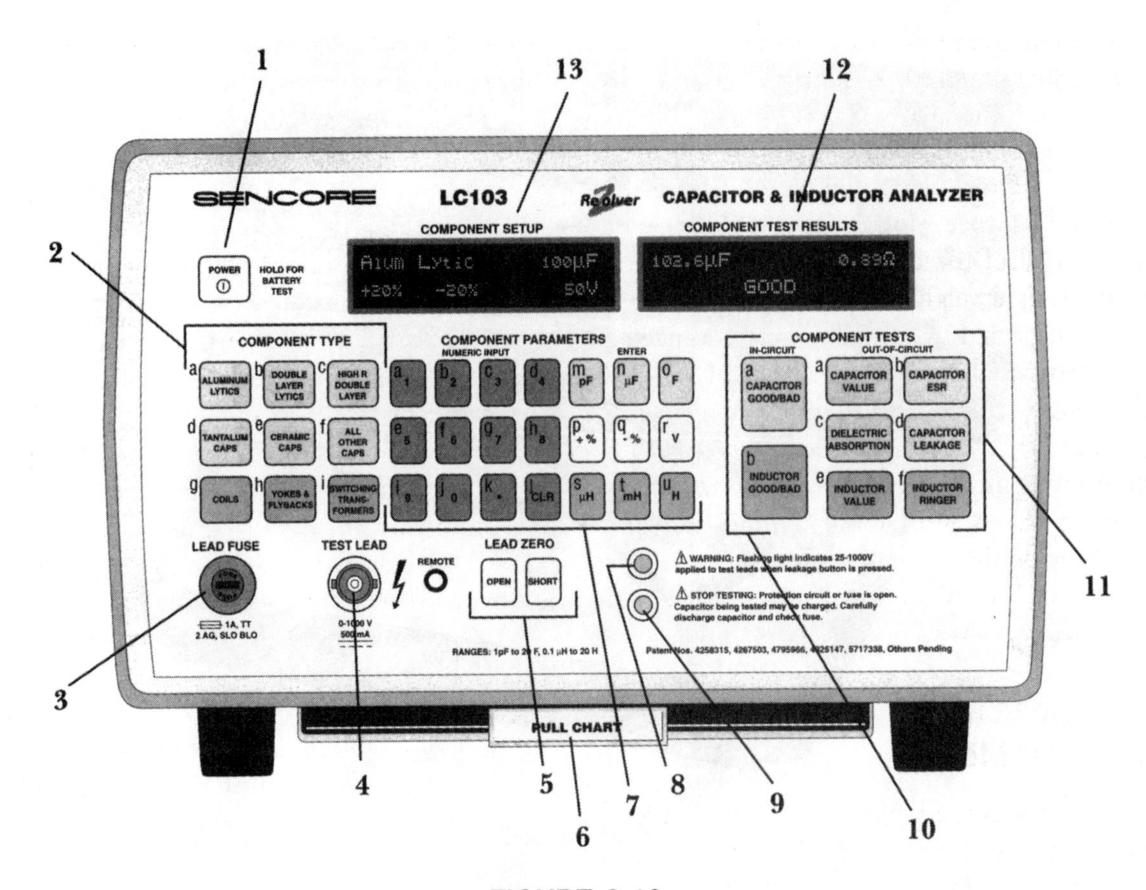

FIGURE 2-10

FIGURE 2-11

- Setup and alignment test Patterns
- Handheld, portable, battery operated (8 hours on one charge)

The Video Pro provides the patterns for black level, white balance, focus, and convergence alignment in all of the display's operating modes. It is portable, lightweight, and battery operated, so it can be transported to any job where you need a reliable video signal source.

In summary, the VP300 can generate signals for servicing HDTV displays, standard televisions and computer monitors. Provide component, composite, S-Video and computer monitor video outputs. Provide the particular signal for the type of display that needs service. Supply setup and alignment test patterns.

Sencore SL754D "Channelizer" Digital Signal Level Meter

The SL754D is a rugged, hand-held, weatherproof 863-MHz (including sub-band) signal-level meter that measures both analog and digital signal levels. The SL754D, shown in **Figure 2-12**, also features single button data collection of all system parameters.

The SL754D has the following features:

- All channel/frequency tuning from 5 to 863 MHz (including sub-band and UHF) to meet system testing requirements.
- Auto attenuation over the entire -35 dBmV measuring range.
- One touch auto inspection and data collection for quick and reliable system testing.
- Backlit digital display can be viewed in low-light conditions.
- The unit will run for over six hours on a single charge and can be fully recharged in less than three hours.
- Capable of measuring dBmV or uV/M to facilitate CLI testing in your system.

Sencore CP288 Color Pro Analyzer

This ColorPro Analyzer is ideal for testing and adjusting computer monitors, rear screen projection TV sets, front screen projection sets, HDTV display screens, and video walls.

The CP288 color analyzer features are as follows:

- Luminance level alignment and color tracking on video displays.
- Easy-to-read, graphical displays (CIE and RGB) illustrate which colors need to be adjusted to achieve the optimum picture.
- Portable operation.
- Laboratory specifications are National Institue of Standards and Technology (NIST) traceable and allow precise measurement capabilities and assures that the display is aligned to industry standards.
- Display calibration report printout provides documentation of its performance for the display owner and a calibration reference.
- Sensor pod attaches to the CRT displays.
- Color pod holder accessory and 25 foot extension cable for video wall and front projector adjustments.

Sencore CP290 Video Luminance Level and Video Display Tracker

CP290, shown in **Figure 2-14**, lets you quickly and accurately adjust luminance level and color

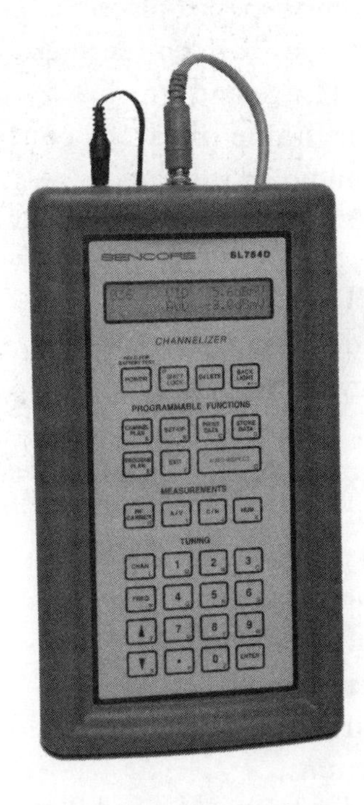

FIGURE 2-12

tracking of any display to the highest performance standard.

CP290 ColorPro features the following:

- Luminance level alignment and color tracking on video displays.
- Easy-to-read xyY, RGB, color temperature, and the most common CIE chromaticity references.
- Handheld, battery operation lets you align displays on location.
- Laboratory specifications (NIST traceable) deliver precise measurement capabilities and give the assurance that the display is aligned to industry standards.
- Wide luminance measurement range lets you make critical alignments at low and high brightness extremes.
- Sensor pod attaches to the display (CRT) freeing your hands to make adjustments.
- 30-foot extension cable lets you work from the back side of video walls.

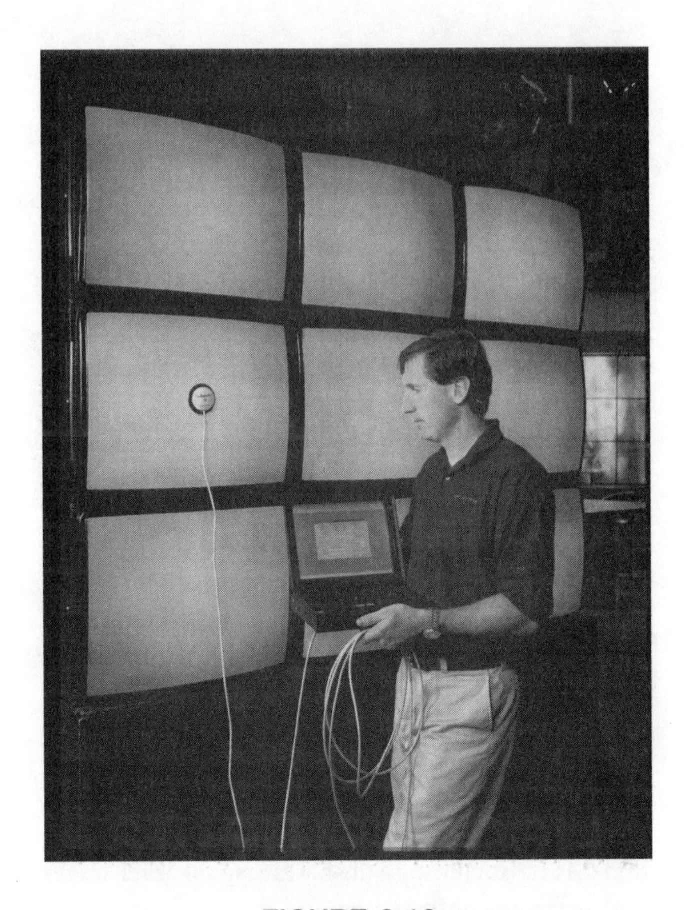

FIGURE 2-13

Sencore SP295 Sound Pro Audio Analyzer

The Sencore SP295 is a professional-quality audio analyzer that lets you quickly equalize room acoustics, optimize speaker placement, and calibrate system settings for concert-quality audio installations. The Sound Pro has features custom designed for home theater, consumer, and commercial audio installations.

The SP295 Sound Pro has the following features:

- Real-Time Analyzer
- Sound Pressure Level Meter
- Energy-Time Graph
- Noise Criteria Test
- Integrated Audio Signal Generator
- Exclusive Speaker Polarity Test
- Speaker Distortion Meter
- Dynamic Cable Checker
- Signal Level Meter
- Digital Sample Scope
- Report Templates for RTA and ETG Spectrums
- Handheld, Portable, and Battery Operation

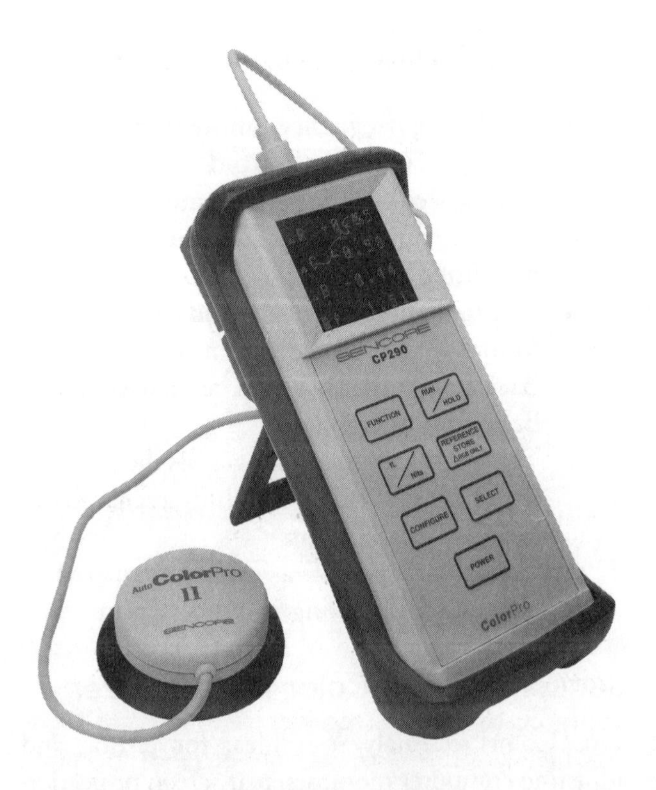

FIGURE 2-14

Real-Time Analyzer

The Sound Pros Real-Time Analyzer (RTA) uses sophisticated Fast-Fourier Transform (FFT) analysis to graphically display the energy content across the audio spectrum. This allows you to fine-tune the audio system, equalize the room acoustics, and correctly place speakers. The SP295 SoundPro is shown in **Figure** 2-15.

Sound Pressure Level Meter

The SP295's autoranging microphone and SPL meter measure levels from a whisper-quiet 35 dB to an extreme high volume level 125 dB SPL. Use the supplied microphone attached to the SoundPro for quick measurements and portability, or extend the microphone for precise remote acoustic analysis.

Energy Time Graph

The SoundPros unique Energy-Time Graph (ETG) function excites the room with a short-duration audio pulse and analyzes the reflected wavefronts. It then graphically displays the pulse decay time and calculates the RT60 reverberation time, helping the technician identify room resonances and locate the source of reflections. The delay finding feature identifies delay times in multiple-speaker layouts.

Integrated Audio Signal Generator

The SP295's signal generator provides all the test signals that are required for complete system component testing and audio analysis, including white and pink noise, pulse, sine, and square waves. The output levels are fully adjustable as indicated on the digital readout. Output connector options include RCA phono, 1-41, mono, it, stereo, and XLR.

Sencore VG91 Universal Generator

The Sencore VG91 Universal Video Generator is a multichannel, multipattern, all-purpose conventional video and MTS audio generator. It provides all the off-air TV and cable channels to verify customer TV receiver problems, re-create trouble symptoms, and isolate receiver reception problems. Each test channel is variable in signal

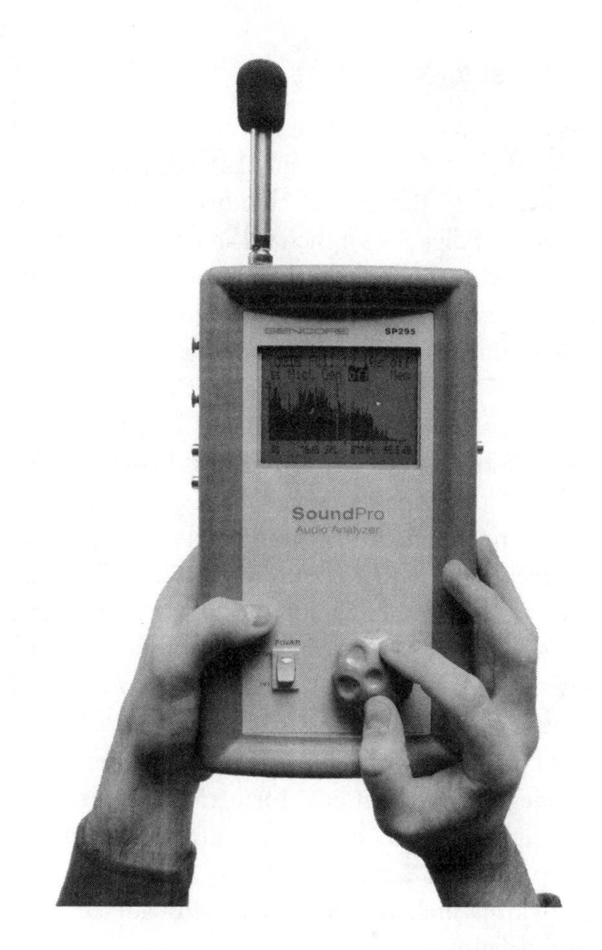

FIGURE 2-15

level and can be modulated with video patterns and mono or MTS audio. Video IF signals, standard S-Video (YC), composite video, and baseband audio signals are provided to isolate problems for any video component. All of the test signals that are required to thoroughly and accurately test, align, and troubleshoot NTSC TV-Video systems are provided in one complete, easy-to-use instrument.

The VG91 Universal Video Generator, shown in Figure 2-16, serves as the signal source for the Sencore VC93 All-Format VCR Analyzer and the TV92 "TV Video Analyzer. Its signals meet FCC and NTSC specifications to provide accurate tests for troubleshooting. As video systems change, the VG91 can be updated with other accessories or companion analyzers to meet the servicing needs of new video systems such as HDTV.

VG91 Features

The VG91 produces all of the allocated VHF and UHF off-air TV channels and 125 standard, HRC, or ICC cable designated channels. The RF TV channels are digitally synthesized and match all off-air and cable channel designated frequencies to test cable-ready tuners. An RF attenuator provides accurate signal levels to duplicate weak or strong signals.

A 45.75-MHz IF generator with the accuracy of a crystal provides a modulated IF carrier as a tuner substitute to isolate defects in video-IF circuits. Exclusive trap signals dynamically check adjacent channel rejection and provide quick, accurate trap alignment. A modulated 4.5 MHz audio IF signal is provided for audio troubleshooting. An AFT test provides a quick performance check of the receiver's AFT circuitry to simplify troubleshooting and alignment.

The VG91's NTSC video pattern generator provides standard patterns for performance testing and alignment. An exclusive, Luma/Chroma Bar Sweep pattern analyzes circuits with extended bandwidths such as comb filters and wideband "I" color circuits.

Sencore TVA92 Video Analyzer

The TVA92 Video Analyzer is designed to be used with the VG91 Video Generator. The VG91 serves as the heart of a multifaceted video analyzer system and also provides the signals needed by additional analyzers. Operating with the VG91 and/or addtional components, the TVA92 isolates faulty stages in the sync, video, chroma, deflection, and audio sections of TV receivers and monitors. It generates fast signals and performs analysis that cuts troubleshooting time and increases troubleshooting efficiency.

TVA92 Features

The TVA92 analyzer consists of three sections:

1. The audio, video, and chroma drive signals
2. The signal monitor and DVM
3. The horizontal output tests

The drive signals provide a known, good signal that can be injected into any audio, video, or color stage after the video or audio detectors. The drive signals are low impedance and overide the existing signal(s). This eliminates the need to disconnect components or disable the signal already present in the circuit. The drive level is adjustable and can be monitored with an internal digital meter so you can match the substitute signal to the level required by the circuit.

A separate output drive provides the 3.58-MHz color oscillator and vertical and horizonatal blanking (sandcastle signal) drive signals so they can be used simultaneously with the video drives. This allows you to quickly isolate luminance, color, or blanking problems.

The vertical yoke drive signal puts the yoke in normal operating mode. This provides fast, conclusive visual results on the raster of the vertical yoke's operation.

Dynamic horizontal output tests are provided to analyze the horizontal output and deflection stages. These tests locate abnormal loading or timing conditions before power is applied to the chassis. The horizontal output dynamic tests analyze the chassis horizontal output transistor, B+ voltage, flyback pulse PPV and duration, and drive signal for a quick diagnoses of startup/shutdown problems, intermittents, or incorrect parameteric values. A *ringer* test can help to find shorted turns among flybacks, yokes, and switching transformers.

The horizontal output device sub and drive emulate the horizontal output device, allowing the horizontal output and flyback circuits to operate under typical operating conditions. The output current is monitored and controlled for safe operation. This unique TVA92 feature allows you to substitute for any horizontal output device and isolate problems before you damage expensive replacement parts.

An autoranged digital meter allows you to monitor all of the drive signals to prevent over driving. The meter also allows you to measure external DC and peak-to-peak voltages up to 2 kilovolts (kV).

Information, drawings, and photos in this chapter courtesy of Thomson multimedia (RCA/GE) and Sencore, Inc. test instruments.
*Auto-Tracking and "AUTO TRACKER" are trademarks of SENCORE, INC.

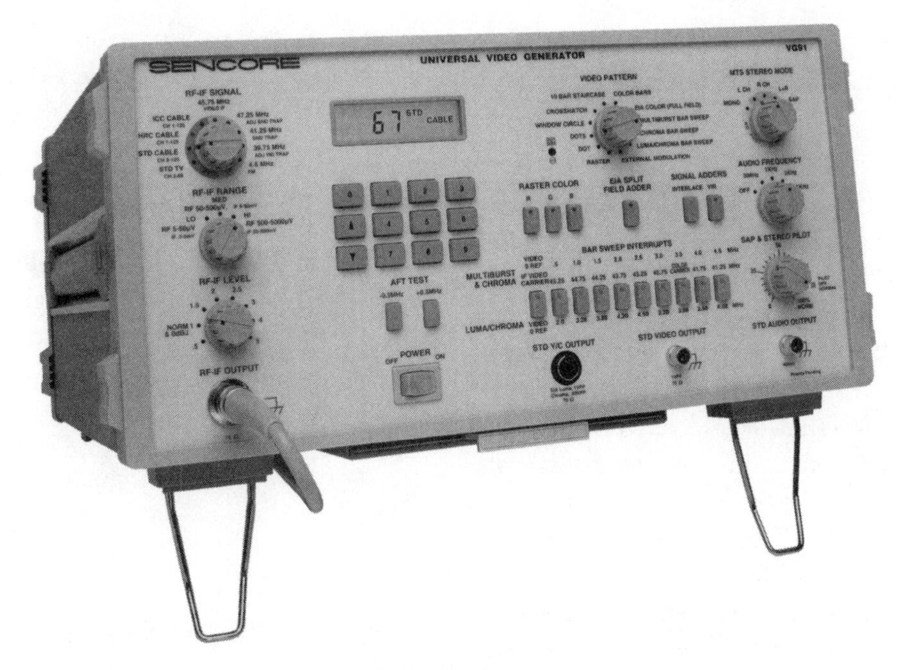

FIGURE 2-16

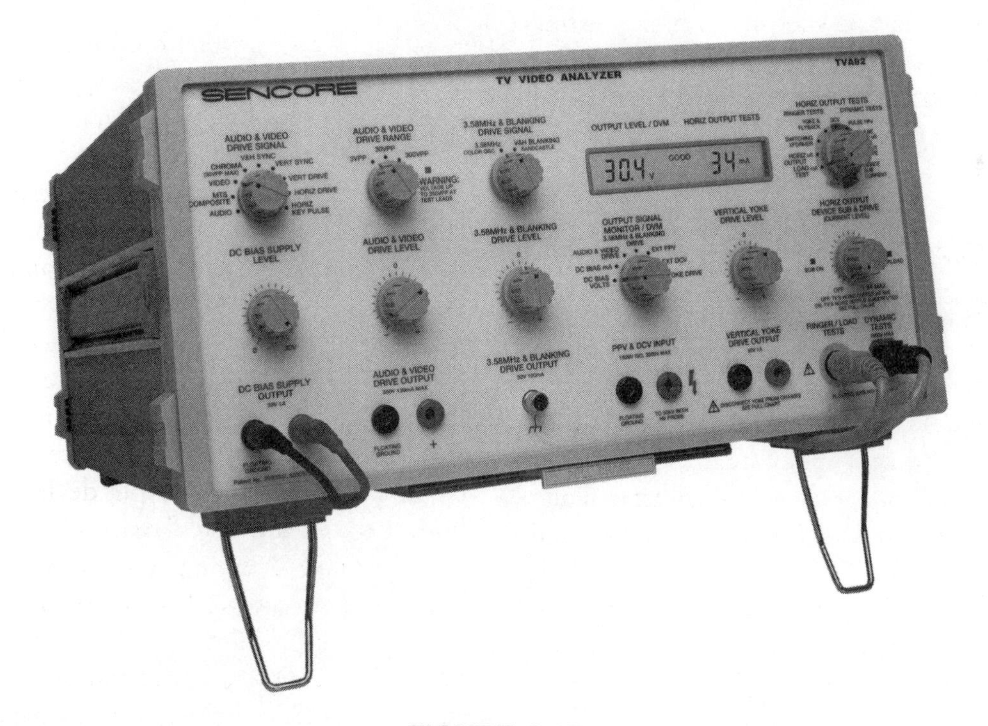

FIGURE 2-17

CHAPTER 3

HDTV Video and Digital Circuit Operation

INTRODUCTION

Some of the topics in this chapter are:

- Video signal compression
- Video Preprocessing
- Sampling Technique
- Understanding Digital Video
- Basic Microprocessor operation
- Memory Devices
- Microprocessor Troubleshooting Tips
- Microprocessor Reset (Boot-Up)
- Zenith HDTV Receiver Block Diagram

This chapter begins with video signal compression and digital signal processing that is utilized for HDTV systems. Digital circuit theory covers sampling, aliasing, quantization, and composite digital video.

The chapter continues with digital circuit operation that can be used to understand and troubleshoot HDTV receivers. These operations include digital to analog (D/A) converters, digital-ramp converters, serial data, parallel data, memory RAM/ROM, and microprocessor operation. Also, a microprocessor troubleshooting guide is discussed.

VIDEO SIGNAL COMPRESSION

The technique for compression of the HDTV video bit stream is obtained by mathematical algorithms that are beyond the scope of what can be presented in this book. Let's keep it simple and think of compression in the realm of motion picture production.

The motion picture format consists of a series of still 35mm frames that are slides projected on a screen, but put into a fast sequence. These frames are displayed at a rate of 30 per second when used for television transmission. Because the frames change very rapidly, your eye and brain interpret them as a smooth, continuous motion.

Let's now look at how these picture frames can be compressed. If you watch a movie long enough you may notice that many scenes are the same and have very little motion change.

The Temporal Compression Mode

When the data for a scene that has very little motion is compressed, it can then be transmitted using very few frames. If the camera stays in a fixed position, then subsequent frames need only contain as much information as necessary to update the few changes that have occurred in these series of frames. This type of compression is referred to as *temporal* compression, which means that the compression is based only on that portion of a frame that does not change over a period of time.

The Spatial Compression Mode

Of course, when you have scenes (frames) with a lot of changing action, temporal compression cannot be utilized. With *spatial* compression, a scene may have the same sky and ocean view that are consistent. And, it may not be necessary to transmit the color and brightness for every small bit of information contained in these frames. The technique used by the MPEG standard of video compression is to save the digital information for the areas of the frames that are changing while reusing information for ones that do not change. This is a bit of an oversimplified explanation of the digital video compression operation, but it's enough information for our purposes.

More Detail Compression Explanation

Generally, the MPEG2 standard is organized into a system of profiles and levels so that it will operate for terrestrial, but also for cable, satellite and computers with proper design and software. The digital television standard portion is based on the MPEG main profile. In order to achieve data compression as defined by the MPEG main profile, there are three types of frames for prediction as was mentioned earlier: I-frames, P-frames, and B-frames.

You may recall, a frame contains lines of spatial information from a video signal. For progressive video, these lines contain samples starting from one instant of time, at frame top, and continuing through successive lines to the bottom of the frame. For interlaced video, a frame consists of two fields, a top field and a bottom field. One of these fields will appear one field later than the other. The MPEG main profile also includes luminance (brightness) and chrominance (color) samples within the frame.

Video Compression Overview

The analog video source signal is fed to the video compression system. The output of the system is a compressed digital signal that contains information that can be decoded to produce a sequence of images that are very near to the original video signal. The process devised by the ATSC eliminates any differences between the original and reconstructed signals for the majority of video images.

A block diagram of video coding showing signal flow is illustrated in **Figure 3-1**. The analog signals that are input to the system are digitized and then sent to the encoder for compression. The compressed data is then transmitted over a TV or cable channel. Once the signal has been received on a TV set, it is decompressed in the decoder. It is very possible that the transmission medium may have corrupted the signal. If that's the case, the

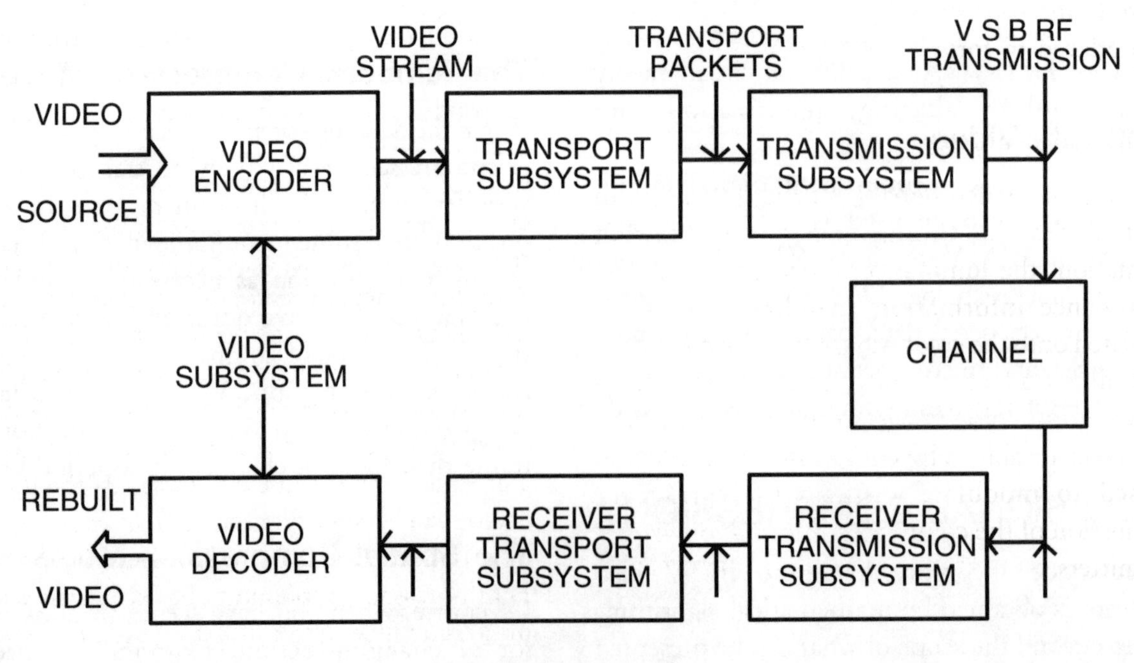

FIGURE 3-1

decoder uses the additional information that was encoded at the transmitting end to reconstruct the image for a corrected display.

VIDEO PREPROCESSING

The analog video signals that are presented to the preprocessor are the red (R), green (G), and blue (B) signals. In video preprocessing, these analog signals are converted to digital samples in a form that can be used in the compression mode.

Video Compression Formats

There are four formats allowed in the digital television standard. HDTV receivers are expected to be able to accept any of these formats, convert them, and display the video content in the receiver's display format.

In HDTV terminology, vertical lines refer to the number of active lines in the picture. Pixels refers to the make up an active line. Aspect ratio describes the ratio of the picture width to picture height.

In the designation for picture rate, the numeric value describes the number of frames or fields per second. Note that each number is followed by the letter I or P. The letter I indicates interlaced scanning, while the letter P indicates progressive scanning. In the case of interlaced scanning, each picture frame consists of two picture fields, which are interlaced to form a complete picture.

Composite Video

The composite video signal is a complex waveform that includes the timing and synchronization information, the luminance information, and the chrominance information that has been phase modulated onto the color subcarrier. The composite signal allows all of the information necessary to produce a full color image to be transferred on a single coaxial cable. The composite signal can also be used to modulate a single RF carrier for transmission of the color video information via TV transmitters.

Digital Video Theory

The process of digitizing analog video waveforms is not a new process. The process usually involves some type of pre-filtering of the analog signal followed by the digitization process. Once in the digital domain, the signal may be transferred, edited, modified, or copied without any degradation or losses. This offers many possibilities for control and processing that were once quite difficult. The key is that if the parameters of the A/D process are known, the process can be reversed and the original analog waveform can be reconstructed.

One of the characteristics of an analog waveform is that it is continuous. The first step in the analog to digital process is to break the continuous waveform into discrete time intervals. This process is referred to as sampling. The next step is to take these individual samples and assign a number to each of them. This operation is called quantization. These conversion steps are shown in the **Figure 3-2** drawings.

SAMPLING TECHNIQUE

The continuous analog waveform is sampled in the domain at discrete, uniform time intervals. This allows the analog signal to be viewed as a sequence of discrete voltage levels. These voltage levels can then be stored, modified, and recovered as needed. To accurately recover the analog waveform, the clock used to sample the signal must be fast enough to produce steps that can reasonably track the original waveform. As the frequency of the waveforms increase, the frequency of the sample clock must also increase to ensure an accurate reproduction of the original signal. The Nyquist Sampling Theorem states that the sample rate must be at least twice that of the highest frequency to be sampled. The accuracy of the sample clock is also important, as variations in the sample rate directly effect the linearity of the reconstructed waveform. The same is true of the clock used in the digital to analog conversion process, as frequency variations of this clock can also produce linearity problems. For this reason, caution must be used in the selection of a reference clock for the digital to analog conversion.

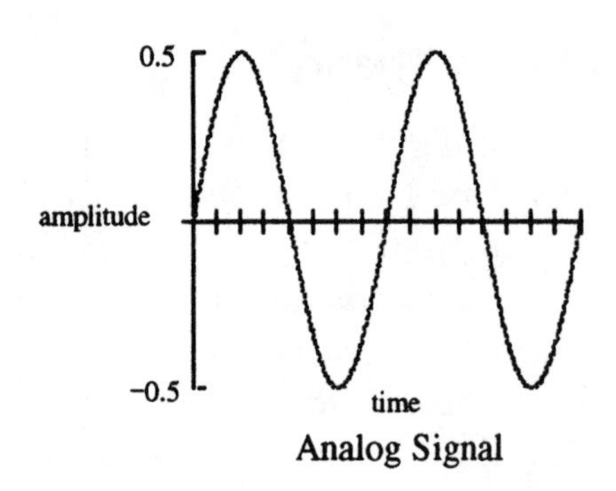

Analog Signal

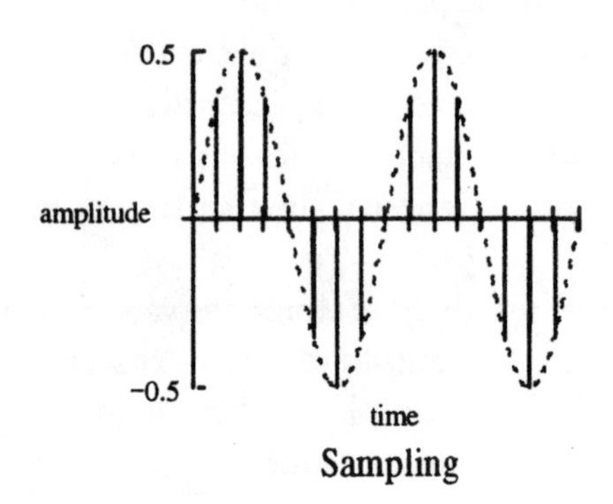

Sampling

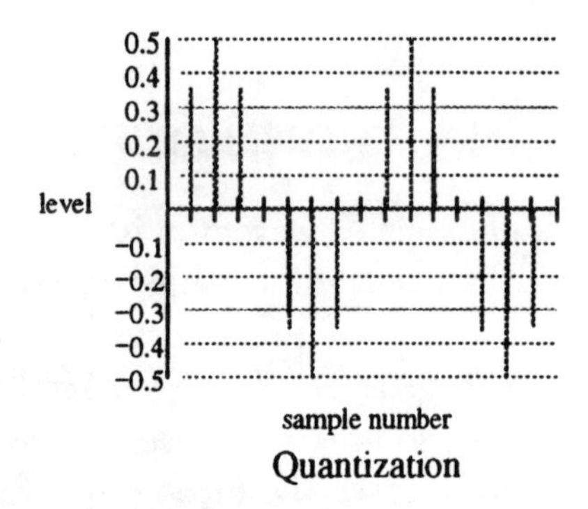

Quantization

FIGURE 3-2

Remember, the Nyquist Sampling Theorem states that the frequency of the sample clock must be at least two times the highest frequency in the signal to be sampled. In reality, the sample clock frequency must be slightly higher than twice the incoming frequency for an accurate reproduction to be made. If this requirement is not met, a phenomenon known as aliasing occurs.

The illustration in **Figure 3-3** shows 2 ms of a 5 KHz sine wave. The sample rate is 4 KHz, which is inadequate for this signal. This causes a signal to alias into the band at what appears to be a 1 KHz rate.

To ensure that no frequencies above the Nyquist threshold reach the A/D, a low-pass filter is usually placed in the signal path prior to the A/D circuit. This filter keeps the frequency content of the waveform below the critical frequency, and is known as an anti-aliasing filter.

Quantization Techniques

The second step in the digitization process is to assign a number corresponding to the voltage level of each sample. The analog signal, shown in **Figure 3-4**, has an amplitude of 1 volt peak-to-peak (V p-p) and is offset +0.500 V (to eliminate negative excursions). After the signal is sampled, the discrete voltage level corresponding to each sample point of the first cycle would be: 0.5 V, 0.854 V, 1.0 V, 0.854 V, 0.5 V, 0.146 V, 0.0 V, 0.146 V, and 0.5 V.

In an 8-bit system, there are 255 (2 to the power of 8) discrete values that can be assigned to the range of the input signal. With 0 equal to 0.0 V and 255 equal to 1.0 V, the quantized values of the first cycle of the sampled signal would be: 128, 218, 255, 218, 37, 0, 37, 128. These quantized values may now be stored, manipulated, or transferred as digital information with the many advantages found in the digital domain.

UNDERSTANDING DIGITAL VIDEO

Many of the original tests done with digital video systems used a composite analog video source. However, to improve the systems performance and quality, the component system was quickly adopted. Thus, the first digital video standards were component. The composite system regained popularity when some of the tape equipment manufacture's introduced a composite digital video recording system called D2.

Digital video uses linear quantization, which means that each voltage step is the same over the entire input signal. Digital video uses both 8-bit and 10-bit schemes, which allocates 256 discrete levels in 8-bit mode and 1024 discrete levels in 10-bit mode. Both 8-bit and 10-bit systems are used in today's systems, and the transmission standards

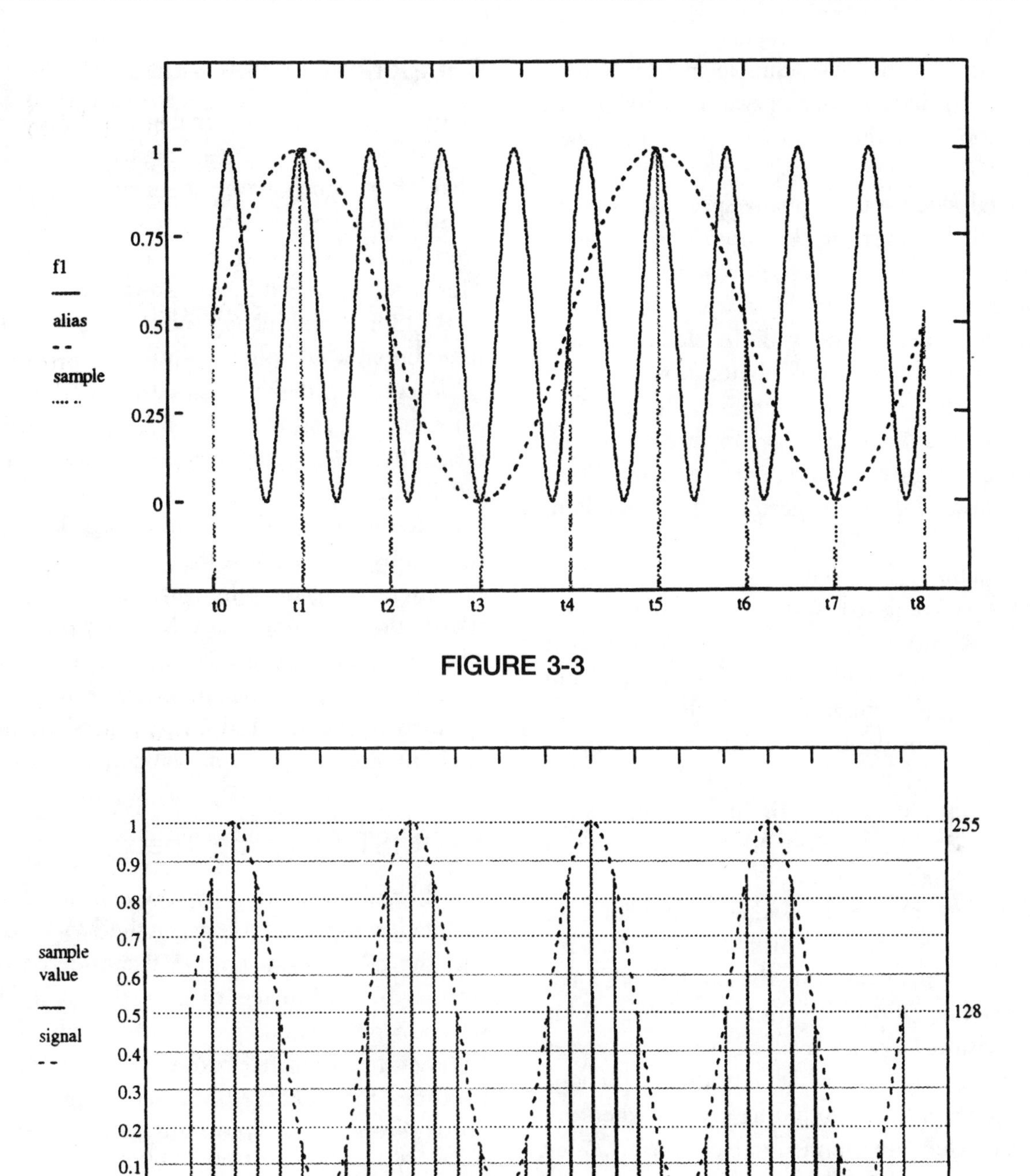

FIGURE 3-3

FIGURE 3-4

allow both to be transferred on the same path. This is accomplished by treating them both as 10-bit systems. The difference is that two additional Least Significant Bits (LSBs) are added to an 8-bit signal to emulate a 10-bit system. The two additional bits are tied to a low-level (0.0 V) value.

The human eye is capable of resolving luminance differences of about 1 percent. Since a television system has an available contrast ratio of about 100 to 1, it would appear that an 8-bit system would be adequate to express all of the necessary luminance levels to produce a good picture. However, low-noise sources such as electronically generated video may produce observable banding in an 8-bit system. Special effects devices, which perform mathematical manipulations on the video data, may also cause problems.

Digital video was originally specified as an 8-bit system, however, at the prompting of some drafting committee members, provisions were made to allow 10-bit data to be transferred on the component interface as well. In the composite digital system, some of the quantization levels are used to digitize the synchronization signals. This reduces the number of levels, or picture resolution, that can be assigned to the actual video data. Thus, the composite document specifies 10-bit precision, with provisions for 8-bit systems.

Compatibility between 8-bit systems and 10-bit systems is accomplished by assigning the same binary weights used in the 8-bit system to the upper eight bits of the 10-bit system. The two additional bits in the 10-bit system do not change the magnitude relation, but provide more resolution, or a smaller step size of the signal. Another way to look at this is that the two additional LSBs in a 10-bit system provide the fractional information, or the data to the right of the radix point.

Care must be exercised when conversions are made between 8-bit and 10-bit systems, as problems may arise. Simply truncating the two LSBs of a 10-bit signal may introduce distortion in an 8-bit system. Distortion can be avoided by using a process called *dynamic rounding* which introduces a pseudo-random term into the rounding process when the fractional values are rounded to eight bits. If an 8-bit signal is to be used in a 10-bit system, the two LSBs added to the 8-bit signal must be set to their low or 0 condition.

Both the component and the composite digital systems are still in use, however, the component digital systems are becoming more prevalent. In the 8-bit vs. 10-bit race, the tendency is toward 10-bit systems, as the cost of digital processing devices continues to drop. However, much equipment in the field is still 8-bit, and new 8-bit equipment is still being produced.

Component Digital Video

A form of the component analog Y, R-Y, B-Y, signals are used for the component digital system. The R-Y and B-Y signals have ranges of ± 0.701 and ± 0.886, respectively. Thus, the color difference signals must first be scaled to produce the proper range. This is accomplished by multiplying the R-Y signal by 0.713 and multiplying the B-Y signal by 0.564. In addition, a DC offset is introduced so the zero-color level is at digital 512 (10-bit) or 128 (8-bit). After the R-Y and B-Y signals have been scaled and offset, they are typically referred to as the Cr and Cb signals. A fully saturated, 100 percent white color bar waveform and the resulting scaled and offset waveforms are shown in the **Figure 3-5**.

After scaling and offsetting, the color difference signals also have nominal values between 0 and 1. The negative excursions of the R-Y and B-Y signals have been offset so that no negative numbers are required in the digital domain. The signal is now ready for the analog to digital conversion.

NYQUIST Sampling Theory

The Nyquist Sampling Theory and the desire to maintain compatibility between the 525-line system and the 625-line system were considered by the standards committee during the sample rate selection process. Sample rates between 12 MHz. (2 times the luminance bandwidth) and 14.3 MHz (4 times the NTSC subcarrier frequency) were considered. One frequency that met these requirements is 13.5 MHz, as 2.25 MHz. (a sub-multiple of 13.5 MHz.) is a common factor to both the 525/59.94 and the 625/50 line systems. Therefore, 13.5 MHz. is used to sample the luminance information. The color difference signals do not require the same bandwidth as the luminance signal, so 6.75 MHz (one-half of 13.5 MHz) was selected as the sample rate. This sampling structure is known as 4:2:2 (the "4" being held over from the four times subcarrier sample rate system considerations and each "2" from the color-difference sample rates at one half the luminance sample rate).

Samples that produce both luminance and color-difference values are said to be co-sited samples,

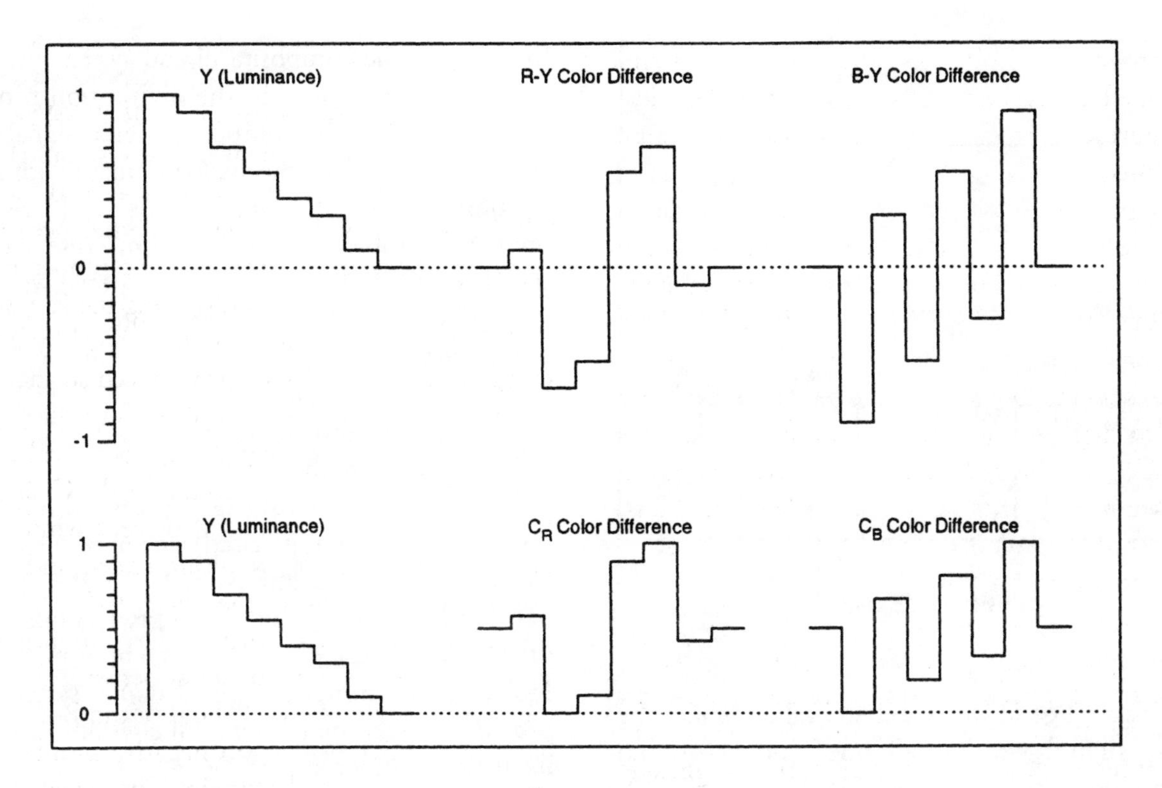

FIGURE 3-5

	Sample 0	Sample 1	Sample 2	Sample 3	Sample 4						
Line n	Y C_RC_B	Y	Y C_RC_B	Y	Y C_RC_B	Y	Y C_RC_B	Y	Y C_RC_B	Y	Y C_RC_B
Line n+1	Y C_RC_B	Y	Y C_RC_B	Y	Y C_RC_B	Y	Y C_RC_B	Y	Y C_RC_B	Y	Y C_RC_B

FIGURE 3-6

which occur every other sample. All even-numbered samples are co-sited and all odd-numbered samples are luminance-only samples. It is important to note that the sampling of the component analog signal is line-locked. This provides an orthogonal sampling grid in which the samples of one line fall directly over samples on other lines. Another important feature is that a co-sited sample point is coincident with the 50 percent amplitude point of the falling edge of analog sync. These two features are illustrated in **Figures 3-6** and 3-7.

Quantization

The luminance is quantized so that the range from black to white extends from 64 (16 for 8-bit) to 940 (235 for 8-bit). This produces a range of 877 (220) quantization levels to represent the 1-Y (luminance) information with black at 64 (16) and 100 percent white at 940 (235). Values from 4 (1) to 63 (15) and 941 (236) to 1019 (254) are not recommended for use. These nonrecommended ranges were established so occasional signal excursions would not create errors in the system and should be avoided in regular use.

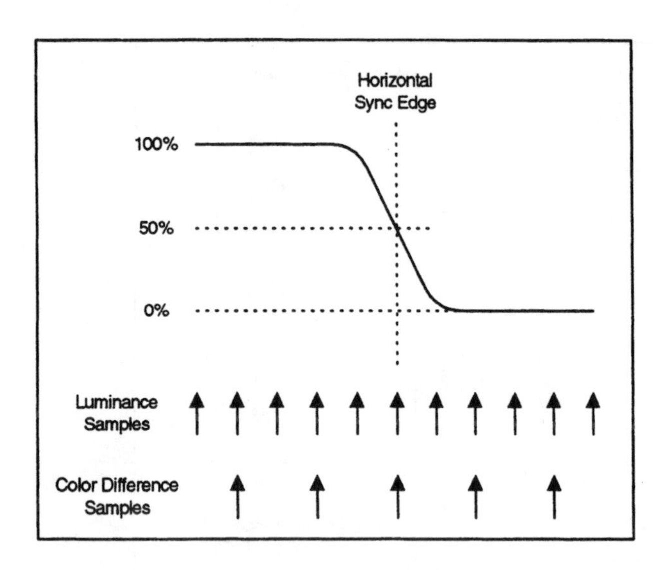

FIGURE 3-7

Reserved codes represent the values where the binary digital word is all ones and zeros. To accommodate the existence of both 8-bit and 10-bit systems, data levels 0 to 3 (8-bit level 0) and 1020 to 1023 (8-bit level 225) are reserved to indicate timing references. These reserved codes may not be used at any other time in the active video signal, as false synchronization may occur.

Composite Digital Video

After the component digital video standards were established, another set of standards was developed to interface digitized composite video. Being a digital format, the composite interface yields some of the same benefits as the component system, such as near loss-less dubbing and noise immunity at what was initially a much lower cost. The composite format is still popular among some broadcasters and smaller post-production houses, as much of their work is in analog NTSC.

The composite digital signal is produced by sampling the 3-14 baseband composite signal at a frequency of four times the chrominance subcarrier. In the component digital video system, only the active part of the analog waveform is digitized. Therefore, a special timing reference signal (TRS) must be inserted into the parallel data stream to maintain proper synchronization. The TRS includes special words that are used to keep the scramble serial data stream properly framed at the deserializer.

In contrast, the composite digital system includes the analog sync pulses in the digitization process. This means that the composite digital signal may be converted back to the analog domain with all of the timing information intact. This does, however, present a problem when the composite parallel signal is serialized, as there are no special words in the data stream to maintain the word framing. For this reason, a set of sync words is known as the timing reference signal identification (TRS-ID), and is inserted into the data stream just after the falling edge of the horizontal sync.

The composite digital system does have limitations, as it is a digitized version of the analog signal. Thus, it is still subject to the narrow bandwidth color, the dot crawl, and other drawbacks associated with the composite system. However, the composite signal digital system still provides a much more powerful environment than its analog counterpart.

NTSC Signal Notes

The composite analog NTSC signal, which is encoded from YIQ data, is sampled at a rate of four times the color subcarrier frequency of 3.58 MHz. This gives a nominal sample frequency of 14.31818 MHz. As covered in the composite digital section, the entire analog waveform, including the synchronization signal, is sampled. Thus, a reference signal, the TRS-ID, must be added to the parallel data before it is serialized to maintain proper word sync in the deserializer. The location of the TRS-ID is shown in the **Figure 3-8**.

Vertical Timing Notes

The drawing in **Figure 3-9** illustrates the states of the F, and V, and the H bits throughout the 625-line frame. This is the information that is transferred in the XYZ word of the timing reference signal.

NTSC Sampling

In the NTSC system, there are 910 samples every horizontal line, and they are numbered 0 to 999. There are 768 samples (0 – 767) during the digital active line, and 142 samples (768 – 909) during the digital blanking interval. The sample points are along

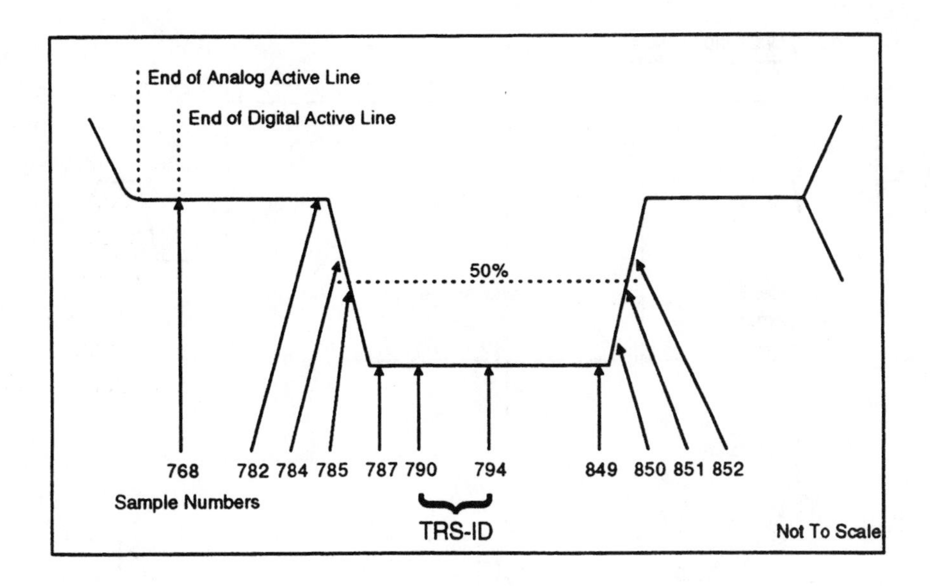

FIGURE 3-8

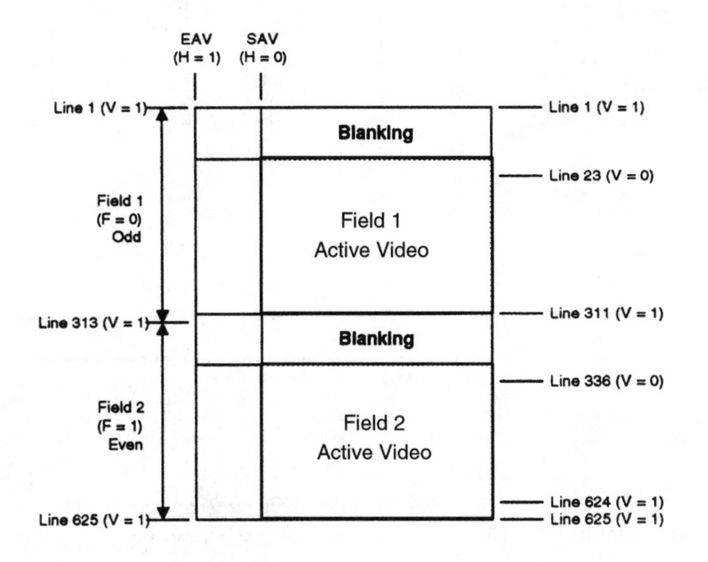

Line Number	F	V	H (EAV)	H (SAV)
1-22	0	1	1	0
23-310	0	0	1	0
311-312	0	1	1	0
313-335	1	1	1	0
336-623	1	0	1	0
624-625	1	1	1	0

FIGURE 3-9

the ±I and ±Q axis. To maintain the proper SC/H (subcarrier to horizontal) phase, sample number 784 occurs at 33 degrees of the subcarrier phase, or 25.6 nS before the 50 percent point of the falling edge of the horizontal sync. **Figure 3-10** shows the relations of the samples to the analog line.

Quantization

The NTSC system uses IRE levels to describe the amplitude of the video signal, with 100 IRE equal to 100 percent white, or 714.3 millivolts. The blanking level is set at 0 IRE, or 0 millivolts, and the sync level is at -40 IRE, or -285.7 millivolts. The NTSC composite digital video system uses linear quantization, so there is a direct relation between the IRE value and the quantized value. Some of the main quantization levels for the composite NTSC system are shown in **Figure 3-11**.

PAL Signal

The composite analog PAL signal, which is encoded from YUV data, is sampled at a rate of four times the color subcarrier frequency of 4.43 MHz. This gives a nominal sample frequency of 17.73447

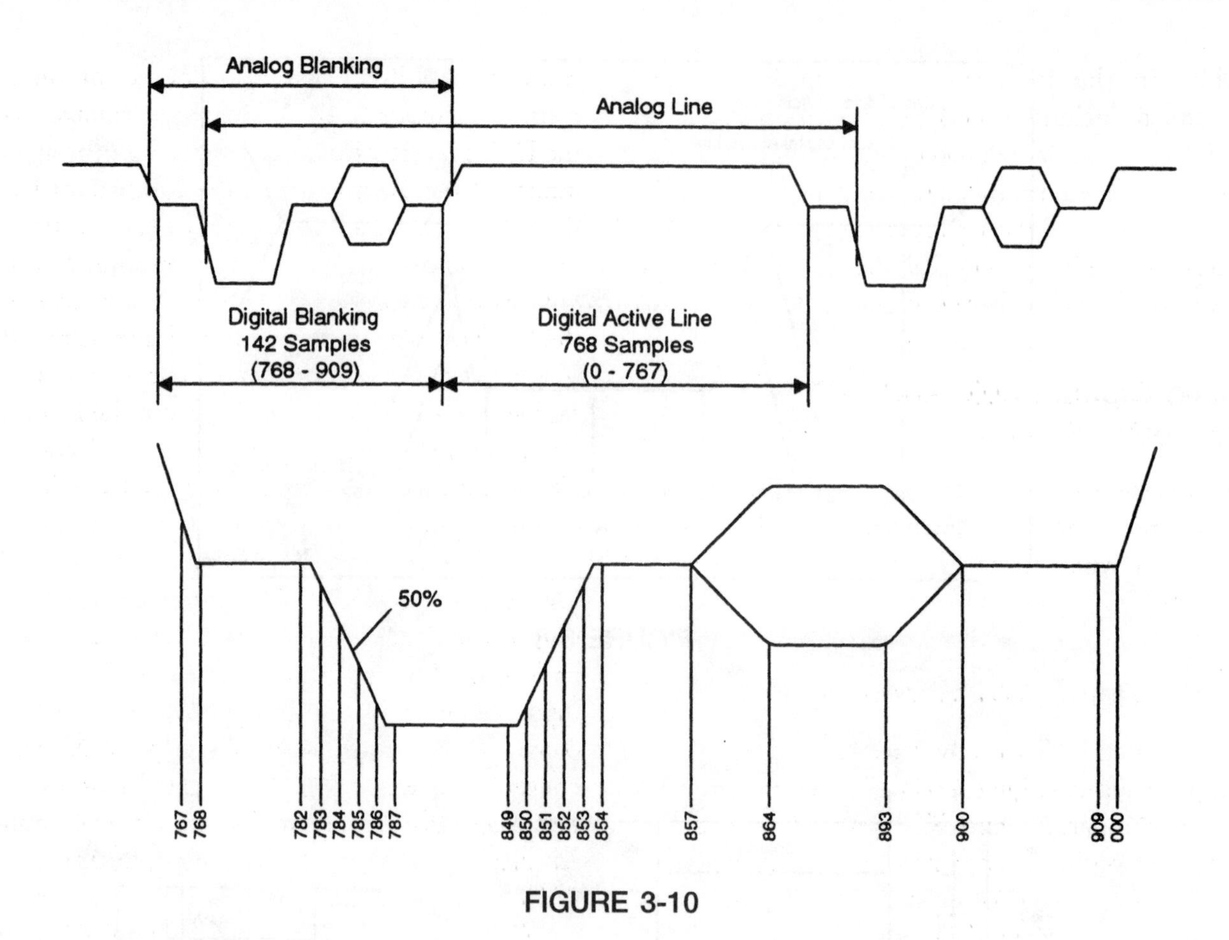

FIGURE 3-10

Waveform Value	Voltage Level	IRE Units	8-Bit Decimal Value	8-Bit Hex Value	8-Bit Binary Value	10-Bit Decimal Value	10-Bit Hex Value	10-Bit Binary Value
Reserved	998.7 mV	139.8	255	FF	1111 1111	1023	3FF	11 1111 1111
Reserved	994.9 mV	139.3				1020	3FC	11 1111 1100
100% Chroma	937.7 mV	131.3	244	F4	1111 0100	975	3CF	11 1100 1111
Peak White	714.3 mV	100	200	C8	1100 1000	800	320	11 0010 0000
Blanking	0.0 mV	0	60	3C	0011 1100	240	0F0	00 1111 0000
Sync. Tip	-285.7 mV	-40	4	04	0000 0100	16	010	00 0001 0000
Reserved	-302.3 mV	-42.3				3	003	00 0000 0011
Reserved	-306.1 mV	-42.9	0	00	0000 0000	0	000	00 0000 0000

FIGURE 3-11

MHz. In the PAL system the entire analog waveform, including the synchronization signal, is sampled. Thus, a reference signal, the TRS-ID, must be added to the parallel data before it is serialized to maintain proper word sync in the deserializer. The location of the TRS-ID on the waveform is shown in the **Figure 3-12**.

Basic Digital Logic and A/D Converter Review

In this section we review digital logic and microprocessor circuit operation, which you encounter when troubleshooting HDTV receiver circuits.

Digital-Ramp A/D Converter

The drawing in **Figure 3-13** shows a basic digital-ramp analog-to-digital (A/D) converter. It has a voltage comparator, an AND gate, a BCD counter and D/A converter. Let's now see how it operates. The analog voltage is applied to point A in the diagram. The output from the D/A converter is fed to point B of the comparator, where it is compared with the analog input. If the analog input voltage is greater than that at point B of the comparator, the clock is allowed to increase the count of the BCD counter. The count on the counter increases until the feedback voltage from the D/A converter becomes greater than the analog input voltage. At this point, the comparator stops the counter from advancing to a higher count.

Assume that there is a logic 1 at point X at the output of the comparator. Then, let's assume that the BCD counter is set at 0000 and that there is 0.7 V applied to the analog input. The 1 at point X enables the AND gate, and the first pulse from the clock appears at the CLK input of the BCD counter. The counter advances its count to 0001, which is output at points A, B, C, and D as 0001, and fed into the D/A converter. According to the truth table, **Figure 3-14**, a binary 0001 input gives a 0.2 V output from the D/A converter. This 0.2 V is fed back to the B input of the comparator. Since the A input is higher (0.7 V as opposed to 0.2 V), the comparator maintains a logic 1 at point X, which continues to enable the AND gate. This lets the next clock pulse through to the counter. The counter advances its count by 1, bringing it to 0010. The 0010 is fed to the D/A converter.

According to the truth table, the 0010 input gives an output of 0.4 V. The 0.4 V is fed back to the comparator, where it is compared again with the input voltage. The A input is still greater, and the comparator maintains a logic 1. The counter

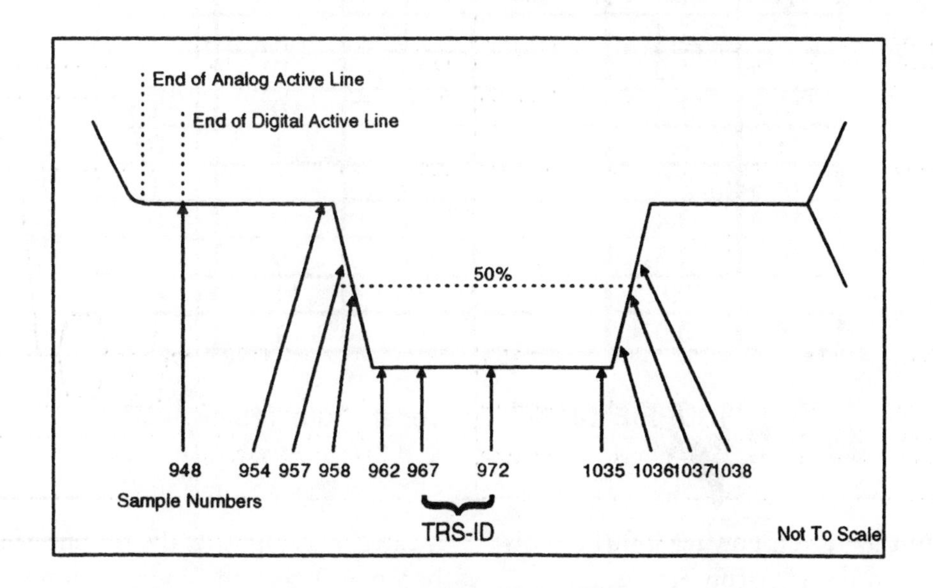

FIGURE 3-12

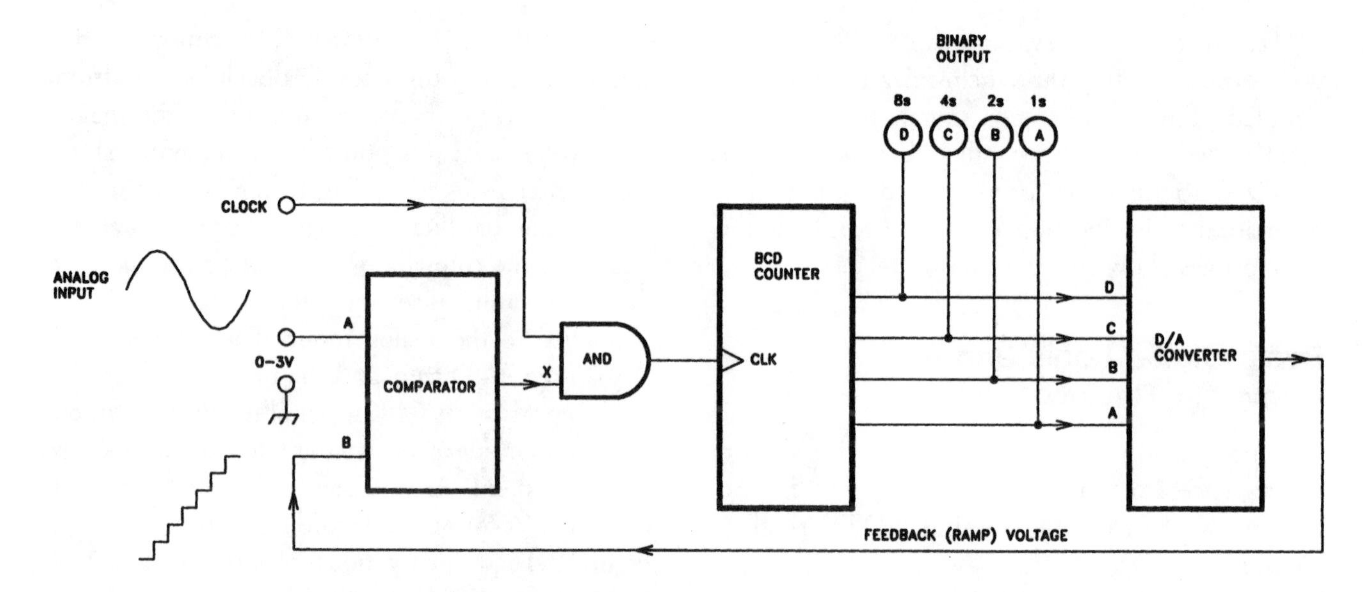

FIGURE 3-13

	Analog Input	Binary	Output		
		8s	4s	2s	1s
	Volts	D	C	B	A
1	0	0	0	0	0
2	0.2	0	0	0	1
3	0.4	0	0	1	0
4	0.6	0	0	1	1
5	0.8	0	1	0	0
6	1.0	0	1	0	1
7	1.2	0	1	1	0
8	1.4	0	1	1	1
9	1.6	1	0	0	0
10	1.8	1	0	0	1
11	2.0	1	0	1	0
12	2.2	1	0	1	1
13	2.4	1	1	0	0
14	2.6	1	1	0	1
15	2.8	1	1	1	0
16	3.0	1	1	1	1

FIGURE 3-14 A/D Converter Truth Table.

increases its count by 1, so it now reads 0011. This 0011 is fed back to the D/A converter.

This operation continues until the output of the counter is 0100 causing 0.8 V to be fed back to input B of the comparator. Now, input B is greater than input A, the output of the AND gate becomes 0, and the counter stops its operation at 0100. Binary 0100 must then equal 0.7 V.

If the voltage fed back from the D/A converter is plotted, a gradual increase of the voltage is

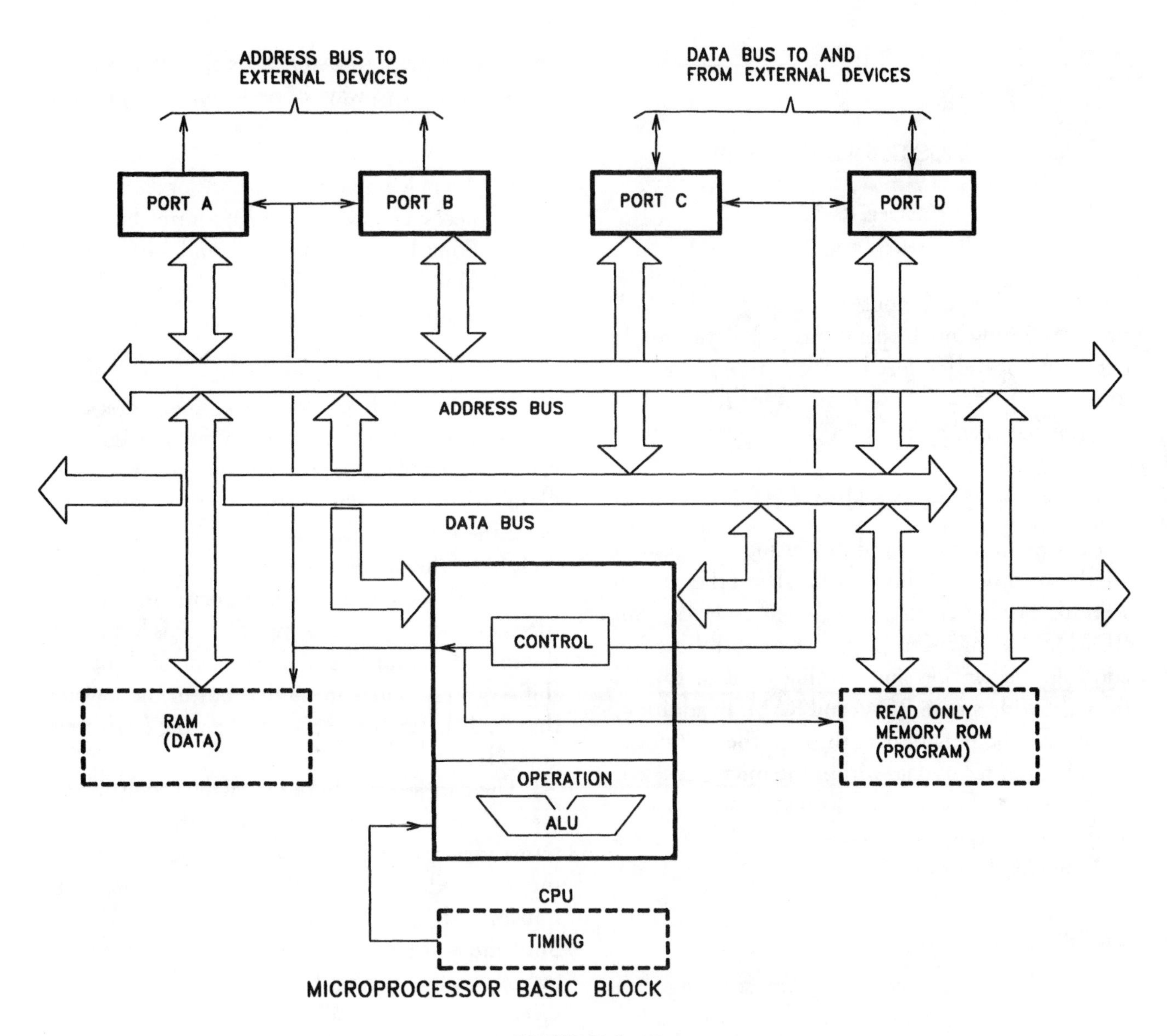

FIGURE 3-15

presented; thus, the name "digital-ramp A/D Converter."

The accuracy of the converter depends on the clock rate and the number of bits used. Note that 0.3, 0.5, 0.7, etc. are not plotted. Higher clock speeds and larger bit words will increase the sample and produce a closer representation of the original input analog signal. However, the binary output will then have more bits and the truth table will be larger.

BASIC MICROPROCESSOR OPERATION

The microprocessors used in HDTV receivers have a lot in common with those used in computers. Because of the many variations of functions within HDTV sets and what they have to control, the microprocessors used in these sets are usually custom designed. However, specially designed chips do not take away from the basic structure of the microprocessor.

A basic block diagram of a microprocessor is illustrated in **Figure 3-15**. The basic blocks for the microprocessor are the CPU, ROM, RAM, data

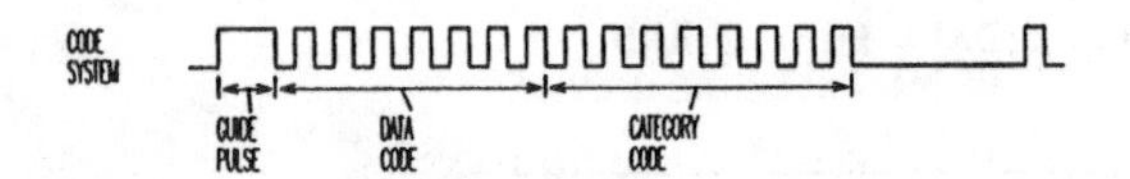

SIRCS Serial Data Frame

FIGURE 3-16

bus, address bus and input/output ports. The clock provides timing for all operations and plays a major part in the operating speed of the central processing unit (CPU). Now follow along as we review each section that makes up the microprocessor system.

Central Processing Unit (CPU)

Located within the central processing unit (CPU) are the arithmetic logic unit (ALU) and the control section. The ALU processes data it receives from ROM, RAM or input ports. The control section controls the writing and reading of data to and from the addresses in ROM and RAM. In addition, it controls the flow of data through the I/O ports.

The transfer of data within the microprocessor and to external devices are through the data bus. There are two types of data busses: serial data busses and parallel data buses.

Serial Data Bus

The serial data bus operates by transmitting data bits along a single wire in series with each other and placing them in a shift register in the receiving IC. There are two ways of performing this transfer of data:

1. A single line is used to transfer data and the necessary codes to synchronize the data.
2. Four lines are used to send information. One line each for Data, Clock, Latch, and Chip Enable pulses.

For the single line technique, serial data is synchronized with clock pulses which clock the data into shift registers in the receiving IC. This can only take place when the Chip Enable (CE), or Chip Select (CS) lines activate the receiving IC. Note the waveform in **Figure 3-16**. At the end of each byte, the latch pulse causes the shift registers to shift the data in parallel form, out to the succeeding stage of processing. The CE is low during the time that four bits of data are clocked into the shift registers. They are shifted out to the registers that feed the microprocessor when the latch pulse goes HIGH.

The data stream, in the serial data transmission method, is transmitted in packets or frames. Each frame will have a start or guide pulse followed by the data code and any other codes that are to be transmitted. This is followed by a "no signal" space until the end of the frame. In this system a guide or start pulse, lasting 12 ms, is sent followed by seven data bits and eight category code bits. With

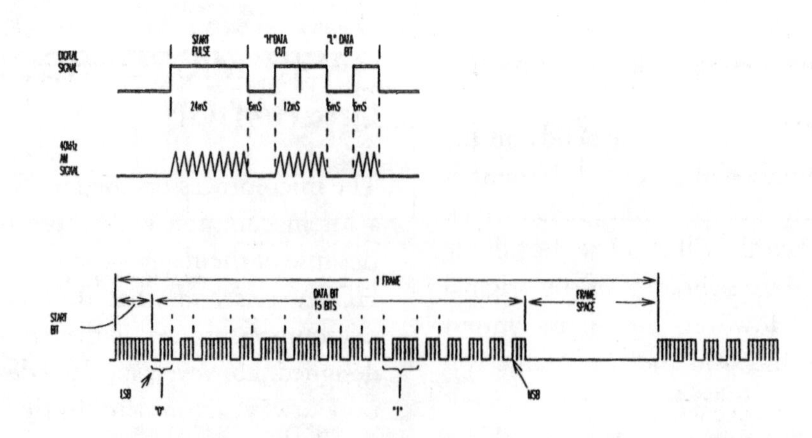

SIRCS Waveform (Serial Data)

FIGURE 3-17

this method, the received data is also placed in a shift register before it is sent to the next processing stage.

Some TV remote control units use the SIRCS waveforms (serial data) shown in **Figure 3-17**. The SIRCS digital code is made up of a start pulse 2.4 ms wide, followed by a series of 15 pulses representing ones and zeros. The whole frame is repeated about 22 times per second, having a frame period of 45 ms.

A signal "0" is represented by 0.6 ms — no signal present — and 0.6 ms of a 40 KHz carrier signal present. A data "1" is represented by 0.6 ms — no signal — and 12 ms of the 40KHz carrier. After the 15 bits, comprised of 7 data bits for the function followed by 8 bits that are used to denote a certain device to be controlled, there is no signal until the end of the frame. As you will note in the **Figure 3-18**, each serial data system will have its own coding system.

Parallel Data Bus

In a parallel data bus or address bus, multiple wires are used, each one carrying one bit of the data word that is transmitted. Thus, a 4-bit data bus will be transmitting 4-bit data words at timed intervals.

Looking at a single data line on a scope, you can see a data stream similar to those shown in **Figure 3-19**. However, the logic level shown at any time is only one bit of a transmitted 4-bit word. For example, at time T1, the 4-bit word is 0010. At time T5, it is 1101. Therefore, when a parallel bus is used, complete data words (bytes) are transferred to an IC at one time for processing.

Referring to the diagram in **Figure 3-19**, should the CPU want to write data 1111 to address 1101 in the RAM, it will pull the read/write line low and place 1101 on the 4-bit address bus. This address will be located in the RAM, and the CPU will send the data 1111 along the 4-bit data bus to the address and all bits will then be sent simultaneously.

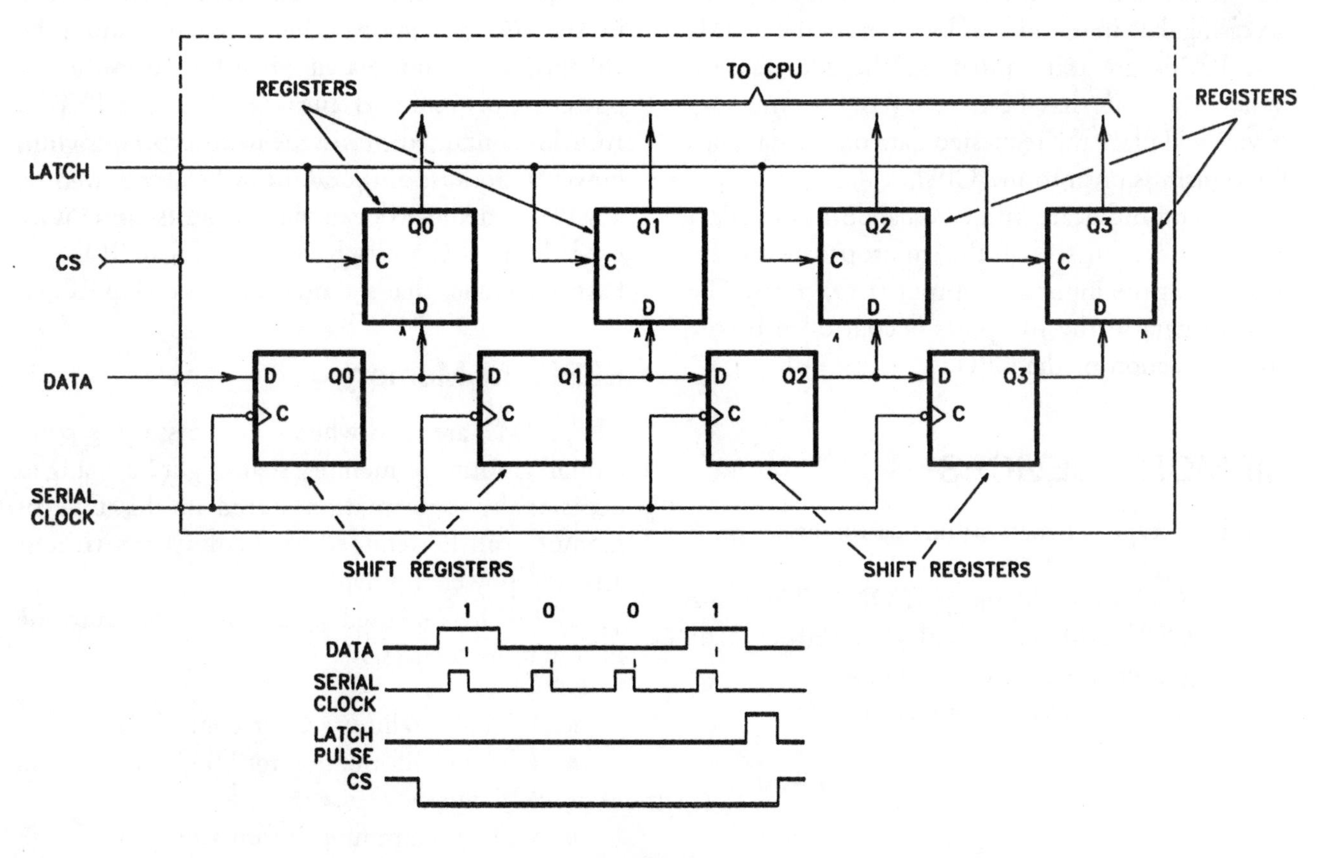

FIGURE 3-18

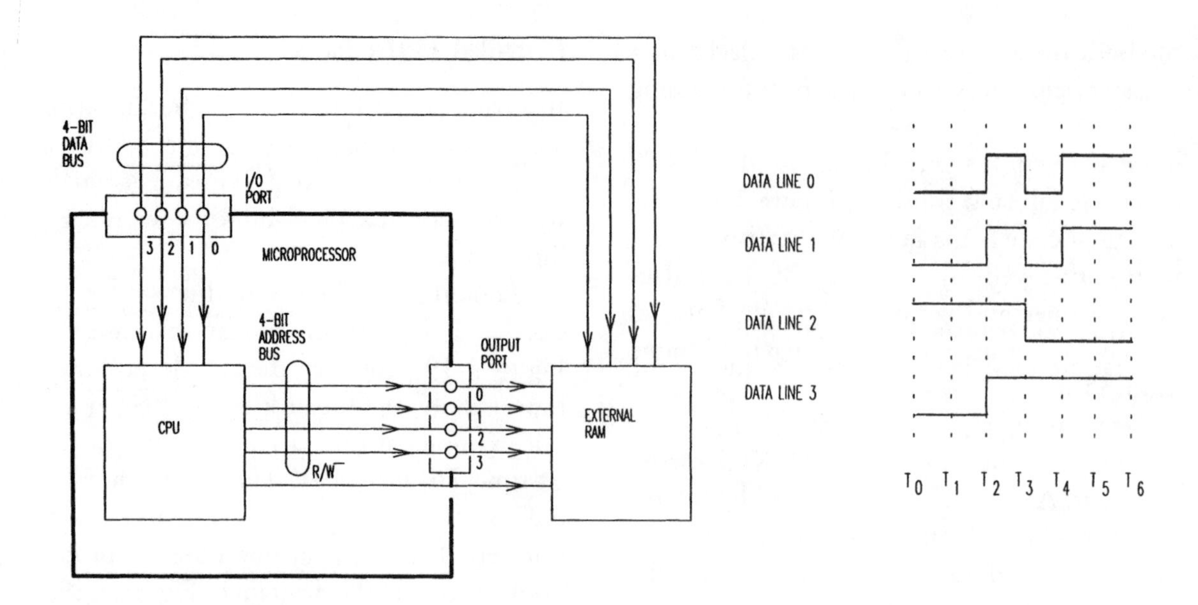

PARALLEL DATA COMMUNICATION

FIGURE 3-19

Address Bus

In addition to the data busses, there is the address bus, which can be serial or parallel. The codes for accessing data in the address locations in the ROMs and RAMs are transmitted on the address bus. When the code has been received, the memory device will place the requested data on the data bus for transmission into the CPU.

To communicate with external devices the data and address busses of the microprocessor are accessed at the input and output (I/O) ports. The flow of data in the I/O ports is controlled by the control section of the CPU.

MEMORY DEVICES

The basic types of memory are as follows:

1. Read Only Memory (ROM). These are EEPROM electrically erasable programmable read only memory.
2. Random Access Memory (RAM).

ROM Memory

The program instructions for running the microprocessor functions are written permanently in the ROM section. This program cannot be changed and is not erased when the device has its power supply removed. In some TV sets and VCRs ROM is contained within the microprocessor chip; however, because of the additional features in some devices, a dedicated external ROM is sometimes used. Some PC units have an external ROM or have a sequence that's written for a Boot-Up ROM.

EEPROM Memory

EEPROMs, are used when it is necessary to store a small amount in memory that might be changed occasionally. Information stored in this type of memory can be retained for many years without a steady power supply.

EEPROMs are used to store user programmable information such as:

- Digital satellite receiver control data.
- Channel information for HDTV tuners and PIP tuners.
- CD programming information.
- VCR programming information.

- User information on various consumer products.

Memory Control

In addition to the data and address lines to a ROM, the microprocessor controls its read and write operations via the CE or CS line.

Referring to **Figure 3-20**, you will see the circuits that are used in the boot-up sequence for a typical microprocessor. These boot-up sequences are stored in the ROM. After the HDTV set is turned on, and reset takes place, the main CPU activates the boot-up ROM via the CE and OE lines and then accesses the addresses with the required data via the address bus. The data is then sent to the main CPU on the data bus. The main CPU then sends the data to the processing circuits that the HDTV receiver needs to operate properly.

RAM Memory

A RAM is a temporary memory device in which data can be stored and retrieved as long as it has a constant power supply. There are two basic types of RAM:

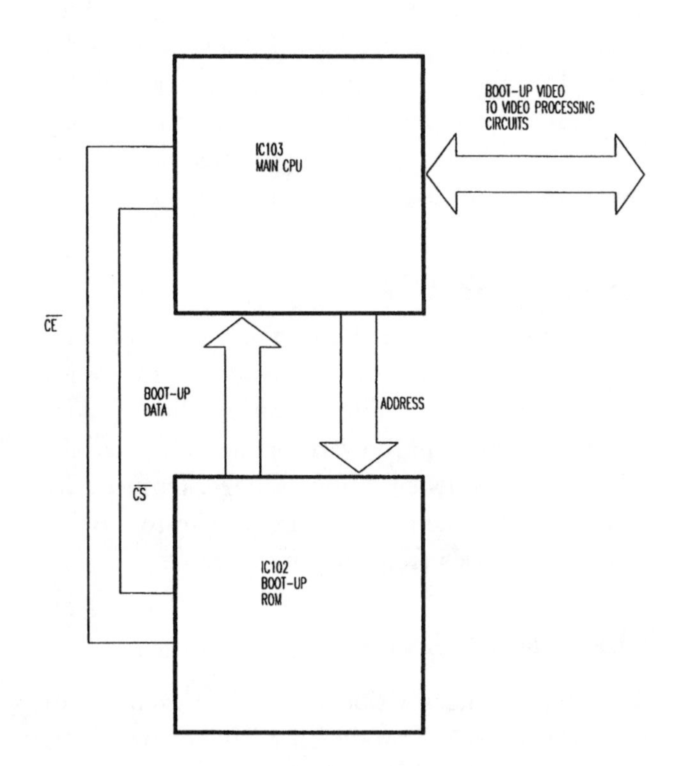

FIGURE 3-20

1. Dynamic RAM (DRAM)
2. Static RAM (SRAM)

DRAMs have the advantage of increasing memory space by the use of one transistor memory cell; however, the data in them must be refreshed (rewritten) at regular intervals or it will be lost. An example of a DRAM is a RAM used in the main CPU of a computer device.

SRAMs use four or six transistor memory cells, which are constructed as flip-flops so they do not hold as much data as a DRAM. However, SRAMs do not require refreshing and have a much lower power consumption.

Memory Control

The microprocessor controls the read and write operations of the RAM via the following control signals:

- Chip Select
- Chip Enable
- Output Enable
- Write Enable
- Column Address Strobe
- Row Address Strobe

The system illustrated in **Figure 3-21** shows the DRAM and main CPU relationship. The main CPU uses the WE and OE lines to tell the DRAM whether it wants to write to it or read data from it. The addresses that the main CPU wants to access are conveyed on the RAS and CAS lines.

MICROPROCESSOR TROUBLE SHOOTING TIPS

Troubleshooting procedures for a microprocessor controlled system will vary with the device that the system controls. However, each microprocessor must have power and ground connections, clock, and reset to operate. Once these are present there should be activity on the data and address busses. Proper power, clock operation and reset should be the first items you check when troubleshooting a microprocessor controlled unit. Refer to the block diagram circuit shown in **Figure 3-22**.

Power Supply Checks

Supply voltages to the microprocessor are applied to pins labeled Vcc and Vdd. Vcc indicates the supply voltage to bipolar transistors collector, and Vdd is to the drain of field effect transistors. In addition to providing power for the internal circuits of the microprocessor, the power supply sets the logic level of the system. In the case where the supply voltage is 5 V, the microprocessor considers voltages above 2 V as a logic high, and those below 0.8 V as logic low. Should the supply decrease for any reason, or there is ripple voltage, or noise glitches, logic levels could change and the system will not be functioning correctly.

Troubleshooting Digital Circuit Power Supplies

Because digital circuit operation is very critical, always check the DC supply voltages with a scope so that both the DC and AC levels are observed.

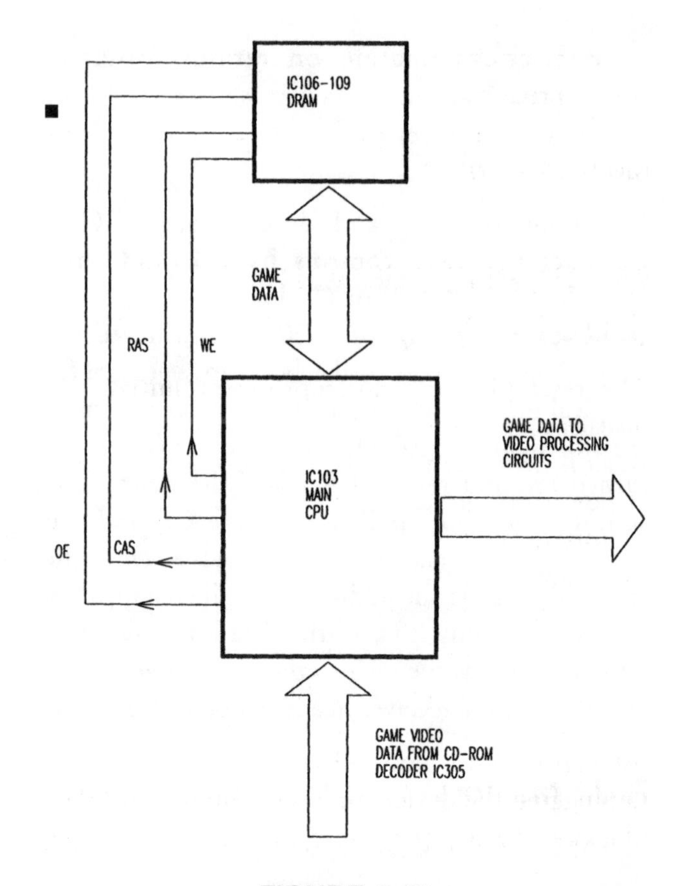

FIGURE 3-21

CLOCK OPERATION

The microprocessor clock is a high-frequency oscillator that synchronizes the internal operations of the controller. The frequency of the clock determines the speed at which the microprocessor executes instructions. Each bit of data is transferred to the leading or trailing edge of the clock. If the clock stops or frequency changes, the microprocessor will not operate correctly, if at all.

Usually microprocessors have an internal clock circuit with external connections for a resonant circuit. In most cases, a crystal is used as the resonant circuit, or an LC circuit may also be used.

Troubleshooting the Clock

If the clock is defective, the microprocessor will not operate, or it may operate intermittently. There are three parameters of the clock that should be checked: frequency, amplitude and waveshape.

Clock Amplitude

The amplitude of the clock signal is usually the same as that of the microprocessor supply voltage. The amplitude of the clock shown in **Figure 3-22** is 3.6 V peak-to-peak (Vpp). Be sure to consider circuit loading by the scope probe. A clock problem is indicated when the clock amplitude is significantly lower than the supply voltage, after taking any effects of circuit loading into account.

Clock Frequency

The frequency of the clock should be stable and accurate at all times. A fluctuation in frequency may result in unreliable operation. Some older computer systems have a variable capacitor to adjust for correct frequency. However, age, temperature and humidity changes may affect them to a point where the clock's frequency may change.

Clock Wave Shape

A crystal produces a sine wave when excited. An LC network also produces a sine wave. It is appropriate then, to see a waveform reperesentative of a sine wave on the clock pins.

If the clock is not functioning properly, replace the crystal. In most cases this should solve the problem since microprocessors are usually very reliable.

MICROPROCESSOR RESET (BOOT-UP)

The reset pin on a microprocessor initiates an instruction to return to the beginning of a program. Reset is performed at power up,and in some cases, at power on. Some computers have an externally mounted reset button to boot-up the device manually.

If reset does not occur at turn on, the HDTV receiver may not work, or the start-up program may start in the middle of the sequence. You can sometimes solve many problems with microprocessor controlled products by just unplugging the device for a few minutes and then plugging the power back in to restart the system. Also, any memory locations and I/O buffers that had data loaded into them may still contain data and may cause the receiver to behave strangely.

The block diagram in **Figure 3-23** shows the power-up reset circuit. These power-up reset circuits have one thing in common; they insert a delay on the reset line when power is applied. The delay time is determined by an RC time constant and is necessary to release reset so that the initial program can start. In most cases, reset is an active low.

Microprocessor Troubleshooting Tips

When troubleshooting microprocessors, you can generally use the same techniques as for other digital systems. When analyzing or troubleshooting, it is very helpful to become familiar with circuit operation.

Troubleshooting microprocessors does have some unique problems. Because the control information is within ROM or software, the signal flow can be difficult to trace. And of course, everything occurs so rapidly it's tough to see in real time. In most cases, a microprocessor system, unlike most logic circuits, cannot be stopped and manipulated. Measurements must be taken while the microprocessor is running. The equipment of choice would the oscilloscope,

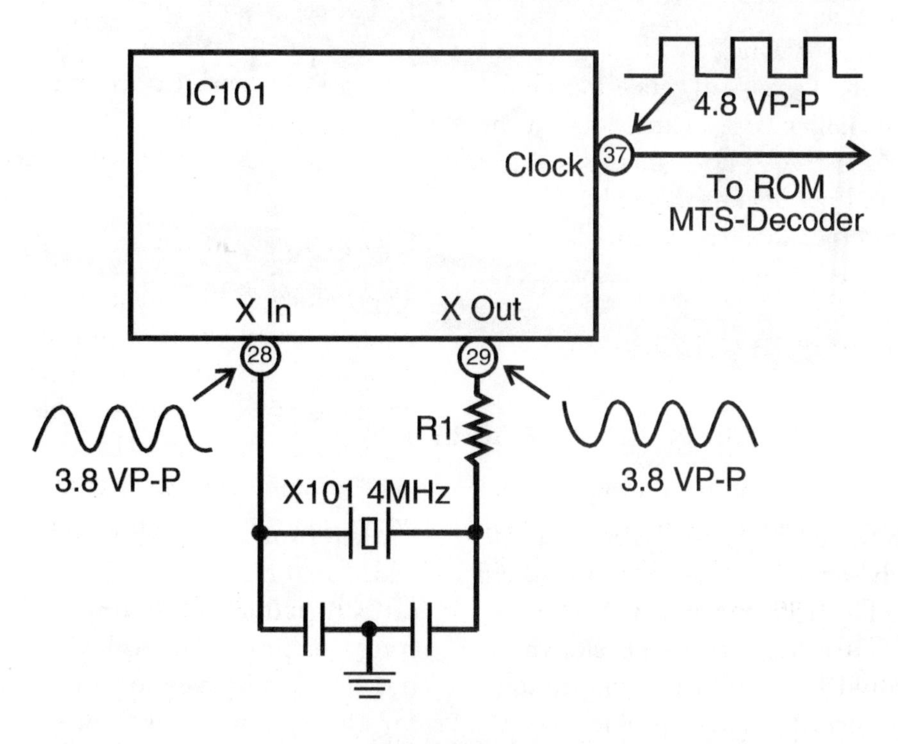

FIGURE 3-22

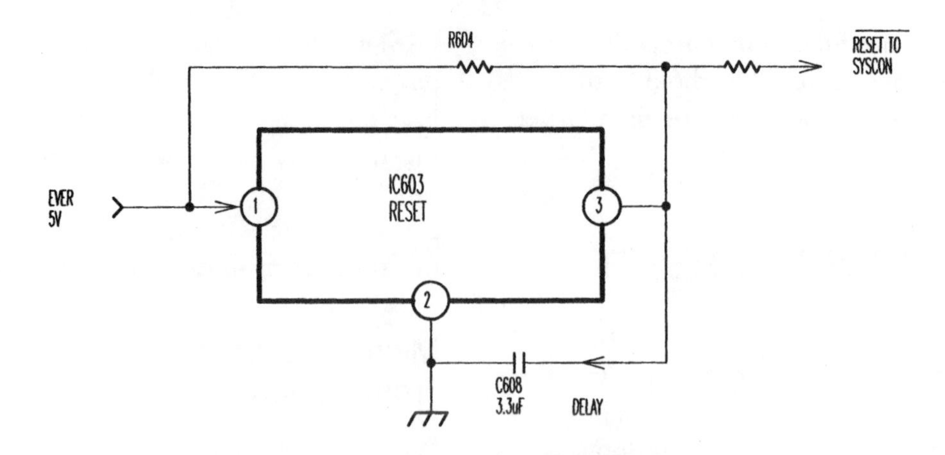

Power Up Reset Circuit.

FIGURE 3-23

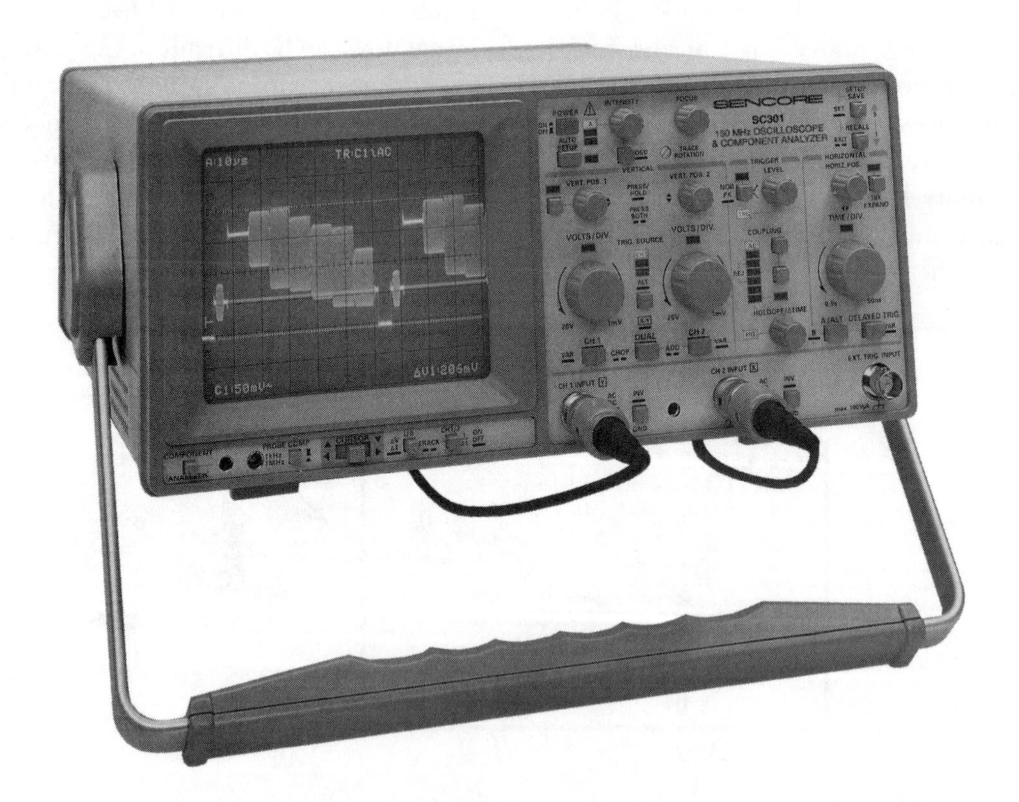

FIGURE 3-24

signature analyzer or logic analyzer as these instruments rely on circuit activity for their measurements. The 150 MHz dual-trace wide band model SC301 Sencore scope shown in **Figure 3-24** would be very helpful in troubleshooting microprocessor system problems.

Microprocessor bus structures can pose additional difficulties. Data on these buses are often unstable or meaningless because of three-state outputs, multiplexing, and switching transients. These conditions cause no problems for the system itself, since it is synchronous

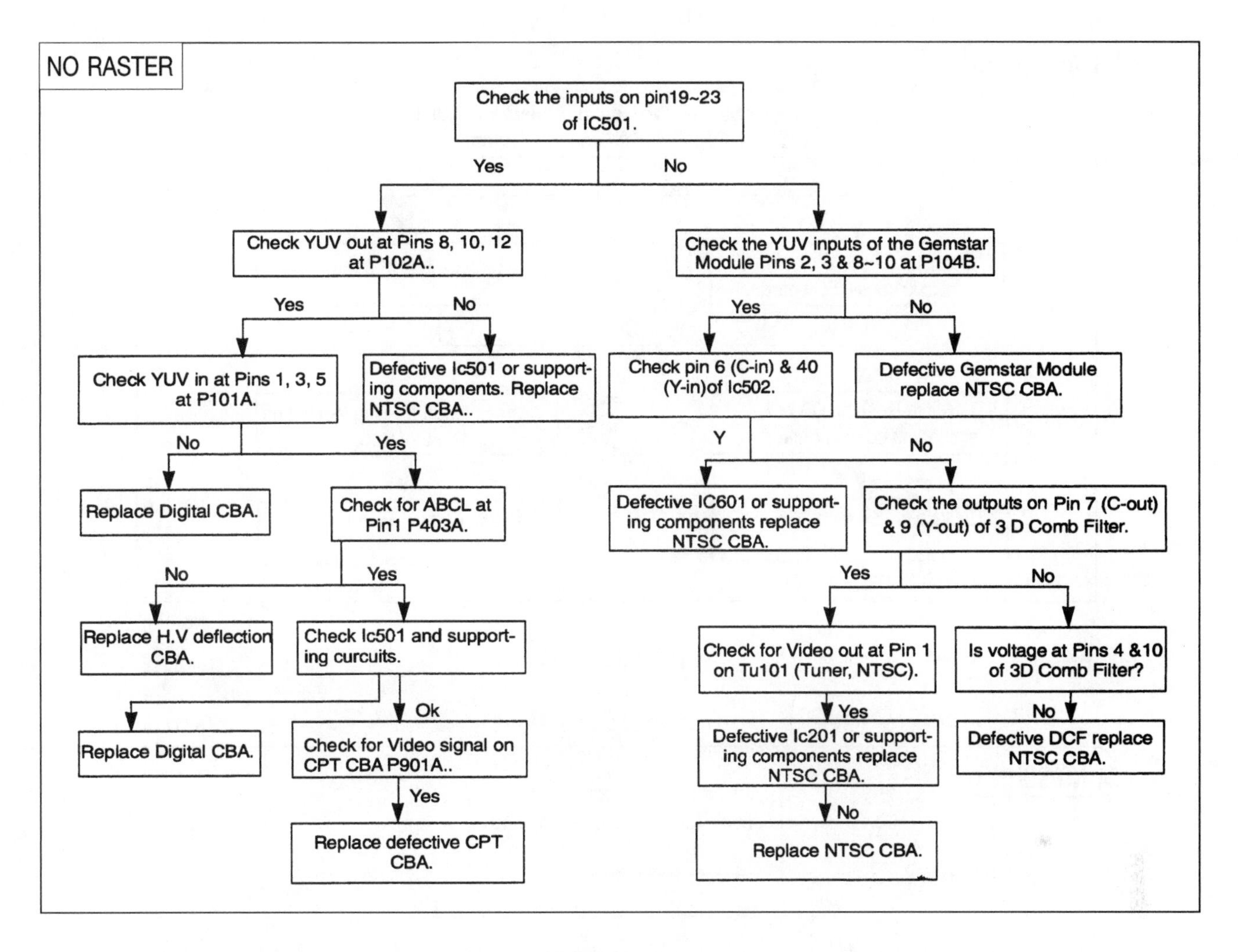

FIGURE 3-25

and knows when the bus lines contain stable signals. The signature analyzer and the logic analyzer also sense when these lines are valid, because of the clock signals provided to them. The oscilloscope does not have this capability. It provides little quantitative information, but is useful for examining qualitative factors, such as general activity, logic levels, waveform timing and bus conflicts.

Complex devices are often connected to the microprocessor buses. It is difficult to test these devices using simple stimulus response testing. The correct operation of these devices can be verified by swapping them with a known good chip, or by observing that the function they perform for the system is being performed correctly.

Another tool needed for digital troubleshooting is good service information with waveforms and voltages provided. Also, a good DMM (digital multimeter) and an oscilloscope with built-in frequency counter are great to have.

If your scope does not have a good frequency counter it is very desirable to have one. The counter is useful for checking the clock frequency of the microprocessor.

Zenith Microprocessor Troubleshooting Guides

Figure 3-25 is the troubleshooting guide for the Zenith M37280 main microprocessor for a no raster condition. **Figure 3-26** is a troubleshooting guide for the microprocessor in a no power-up condition. For a no sound symptom in the Zenith M37280 HDTV receiver, use the troubleshooting guide shown in **Figure 3-27**.

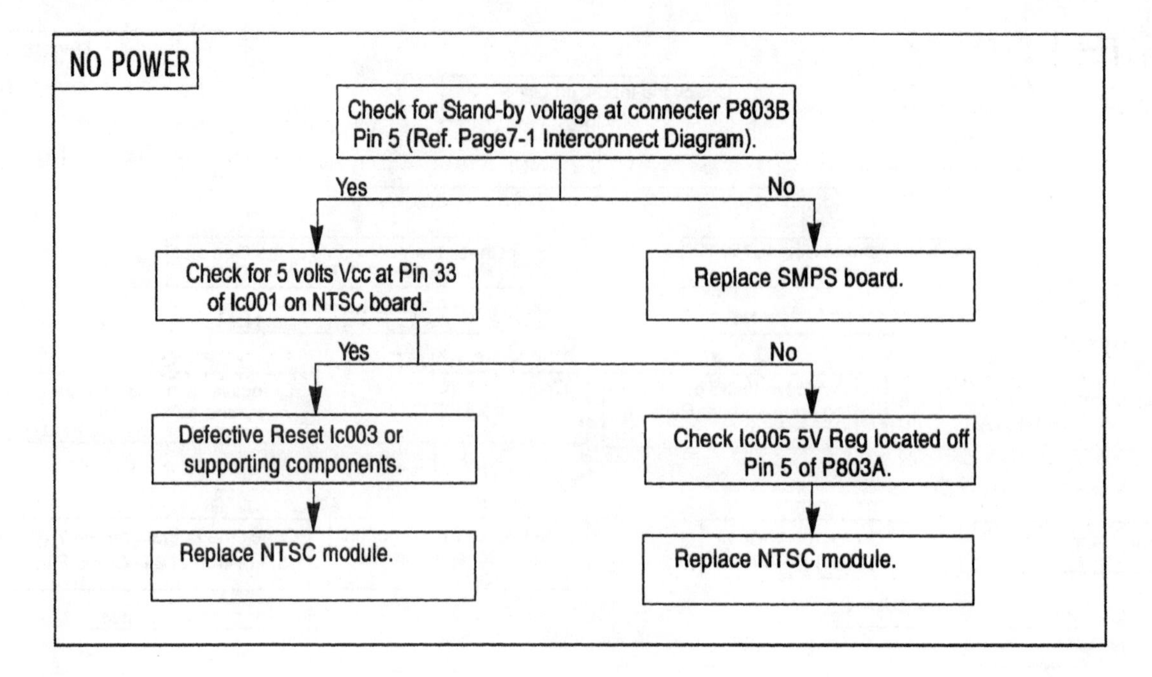

FIGURE 3-26

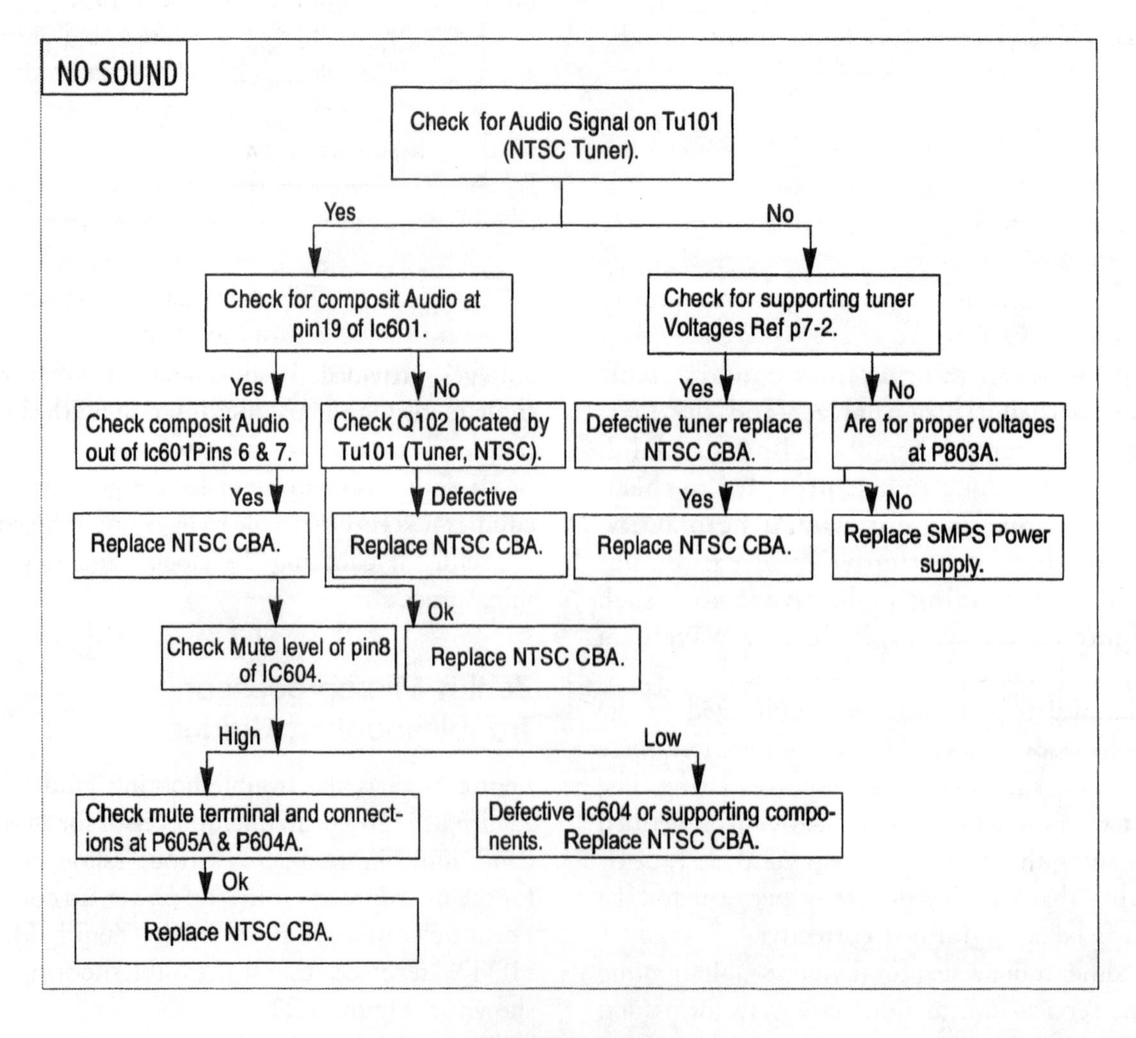

FIGURE 3-27

ZENITH HDTV RECEIVER BLOCK DIAGRAM

The block diagram for a Zenith HDTV set is shown in **Figure 3-28**. The HDTV receiver features are as follows:

- Tuner for high-definition digital television signals
- Input for computer SVGA or any RGB format that the external monitor can handle
- RS 232 service input jack
- Software upgrade port
- R, G, B, and H/V BNC video outputs for 33.75 KHz horizontal scan rate monitor
- Left and right variable stereo outputs
- Dolby Digital encoded audio outputs

HDTV RECEIVER NOTES

The HDTV receivers have a digital tuner, thus the only channels that may be received are digital channels. Because the way in which digital channels are numbered does not directly correspond to the frequency at which the channel is located, a channel search must be performed at initial set turn-on, or the system cannot properly tune any given channel. The first time the unit is turned on, the viewer will be prompted to run the channel search with a display message stating that the system must initialize the channel lineup before normal operation begins. Press the right arrow key to begin setup. The HDTV receiver assumes that over-the-air broadcast is the operating environment and will not search cable frequencies or allow cable signals to be used.

MAJOR AND MINOR CHANNEL NUMBERS

Unlike tuning in an analog system, digital system tuning is completely dependent on the channel search in order to perform channel navigation. Channel search indicates to the system which frequencies contain digital signals and by what channel number they are to be indicated to the set viewer. Each digital channel is known by a major and a minor channel number, indicated as X-Y. The major channel number may be 2-99, and the minor channel number may be 1-199.

A minor number of 0 is used to indicate an analog channel, but no such channels may be received by the digital tuner. The number of minor channels which may be active at any one time is variable, and may run from a single, full-resolution HDTV program to several lower resolution SDTV programs. As each of these programs is from the same digital stream, they may be identified as different subchannels (minor channels) of the same network structure (major channel). The major channel number will likely be the same as the

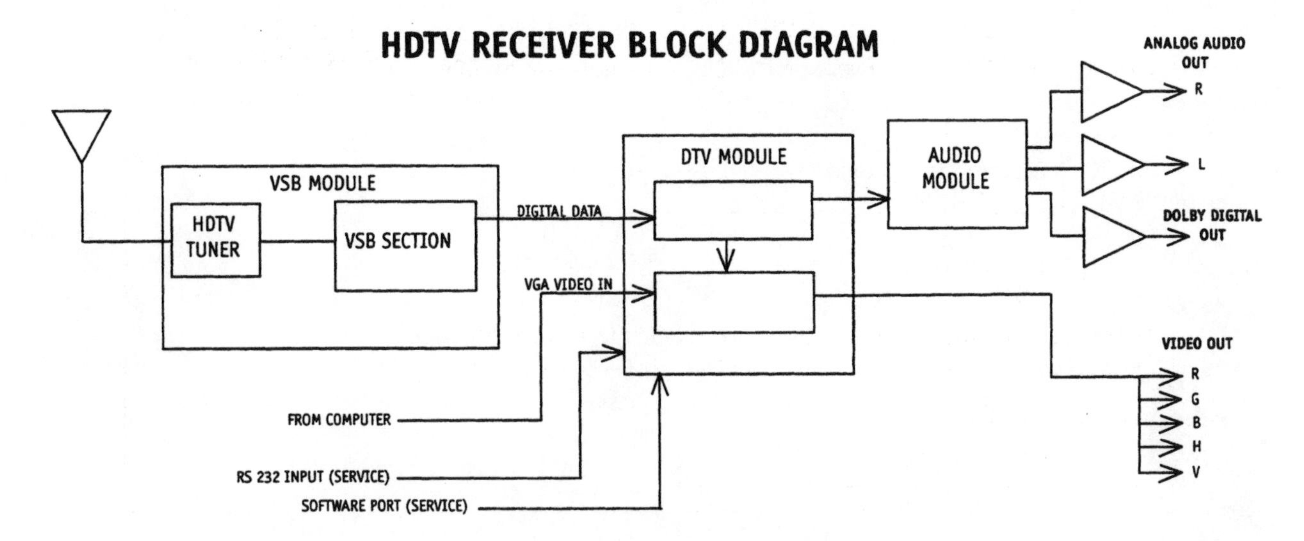

FIGURE 3-28

digital broadcaster's existing analog channel number, as many have incorporated this number into their station's identity. For instance, a broadcaster whose analog channel is 5 may be allocated channel 17's frequency for digital broadcasting, but the programs found there will be identified as 5-1, 5-2, etc.

ANALOG SIGNAL RECEPTION NOTES

Because the HDTV receiver does not contain an analog tuner, all commands for direct channel entry which request a minor channel zero (reserved for analog programs) will tune the lowest numbered digital program at the indicated major channel. This includes the case where a user does not specify a minor channel number by pressing ENTER or allowing a time-out after entering some digits without using the "-" key at all.

> *NOTE:* More information on electronic tuners for HDTV receivers is found in Chapter Eight (8).

Some information and drawings in this chapter are courtesy of Sencore Instruments, Inc. and Zenith Electronics Corp.

CHAPTER 4

HDTV Receiver Picture Viewing Devices

INTRODUCTION

In this chapter we will review various HDTV picture-viewing devices used to present HDTV color video information. These range from the conventional picture tube (CRT), large screen projection screens, the "flat-panel," "hang-on-the-wall" LCD screen, and the flat-panel plasma displays.

You will find out about projection HDTV set construction and operation, plus the service and convergence adjustments required. Then, we'll take a look at these sets' optical light path, the Fresnel lens, and liquid-cooled projection tubes.

We'll then review the flat-screen LCD. This device passes polarized light, via liquid crystals, that controls light intensity by way of polarized filters to produce a color picture. LCDs are generally found on monitors used with personal computers.

Also, we'll discuss details of Pioneer Electronics' basic flat-screen plasma technology that is now being manufactured for HDTV displays. Finally, we'll study the basic structure and operation of these "wall-mounted" hanging screens.

The main advantage of the plasma flat screen is its focus and illumination qualities. Because each cell is activated separately, the focus stays very sharp. And the dark or "hot spots" have been eliminated due to the plasma's even illumination. Additionally, the plasma screen is not susceptible to magnetic distortion.

OVERVIEW OF HDTV PROJECTION AND FLAT-HANGING WALL SCREENS

Analog TV viewing is being affected by the advent of HDTV programs on big-screen receivers. An example is Sony's second-generation Digital Reality Creation (DRC) technology for near high-definition pictures from standard definition sources. It's a new technique for a high-performance display for viewing HDTV programs.

These big-screen sets are the first to incorporate the latest generation of DRC technology, a technique that converts analog signals to near high-definition quality, resulting in clean, sharp images. DRC technology bit-maps the original NTSC and standard definition (480i) signals in real-time, doubling the vertical resolution and horizontal pixel count, creating an image with 960i X 1440 pixels.

When this is paired with a digital television (DTV) set-top decoder box compatible with HD component connections, the Hi-Scan sets will have the ability to display 480i/480p plus high-definition signals at the 1080i resolution.

This technique includes such enhancements as an Extended Definition CRT and optimized positioning of the velocity modulation coil, both of which will contribute to improved picture quality. Other picture enhancements include an advanced flash focus, a full digital auto-convergence feature that automatically aligns the picture tube in just five seconds. A newly designed Lenticular screen, along with other screen improvements, allows for clear viewing from any angle, while providing higher contrast and less surface glare.

Flat-Screen Technologies

Flat-Screen technologies have been developed to decrease the size and to increase the aesthetics of these display, without sacrificing picture quality. The liquid crystal display (LCD) is the predominant flat-panel display (FPD). The LCD works with a light source, polarizing filters and rod-shaped molecules that flow like a liquid and bend light. As the energized crystals move, they modulate light. Since the movement of the liquid takes time, the modulation is imperfect.

Another significant limitation of the LCD panel is its dynamic contrast range, which is about 0.05 to 5 percent. Thus, you cannot make it completely black, and it's not really bright.

If you have ever used a laptop computer, you are familiar with the limitations of the viewing angle. Viewed straight on, the image is great, but viewed from an angle, the thickness of the liquid crystals is greater, causing the image to appear distorted.

With LCDs, response time is deficient, but acceptable, because many computer applications use static or slow images and repetitive information on the screen. HDTV video displays, on the other hand, require extremely fast image responses.

Plasma Flat-Screen Concepts

The plasma screen displays are complex and, at this time, expensive. The technique to create the light is similar to that of the fluorescent light tube. A gas cell is excited into a plasma state. The plasma then emits an ultraviolet light that excites a phosphor coating on the display's back.

You can picture a display of subpixels (red, green, blue) as tiny fluorescent tubes. These are lined up side by side inside a diagonal piece of glass. Later in this chapter we will have an in-depth look at the Pioneer plasma flat-screen HDTV system.

THE FILM SSD SANDWICH

The solid-state display (SSD) has no liquid, no gas, no vacuum, but only solid films. Basically, it's a sandwich with an electrical conductor on the top and bottom. Between these layers are an insulating layer and a semiconductor layer. When a voltage is applied across the two conductors, the semiconductor layer emits light.

The company who developed this SSD screen is Westaim Advanced Display Technologies (WADT). WADT began developing this proprietary SSD technology in 1991 when they started expanding an existing thick-film, hybrid-circuit project into research for displays. This project has developed into WADT's patented SSD technology: a combination of conventional thin-film phosphor with reliable thick-film screen printing. Simply stated, it's a printed circuit board that emits light. The large market for these SSD units will be for reliable, rugged, lightweight displays for airline seatbacks, oscilloscopes, medical instruments, navigation systems, segmented-numeric displays, and alphanumeric type panels. Additionally, the SSD panels could find a market in the home TV receiver/monitor systems.

PROJECTION TV SYSTEMS

For a projection TV system, a number of basic circuits and devices are not found in a direct-view TV receiver. In this chapter we will cover basic components of a television projection system that you will not find in a conventional direct view receiver:

1. An image tube to display the television picture.
2. The optics to project and magnify the image on the display tube.
3. A screen upon which the magnified image is focused and viewed.

The drawing in **Figure 4-1** illustrates a projection TV system in its most basic form. A small-screen, direct-view television receiver is optically coupled to a viewing screen. The only change re-

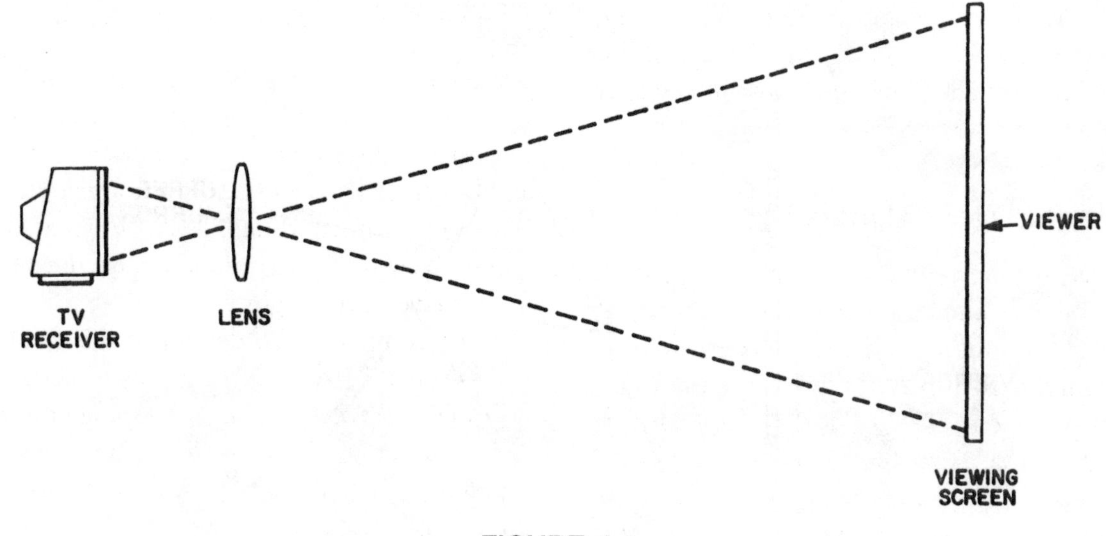

FIGURE 4-1

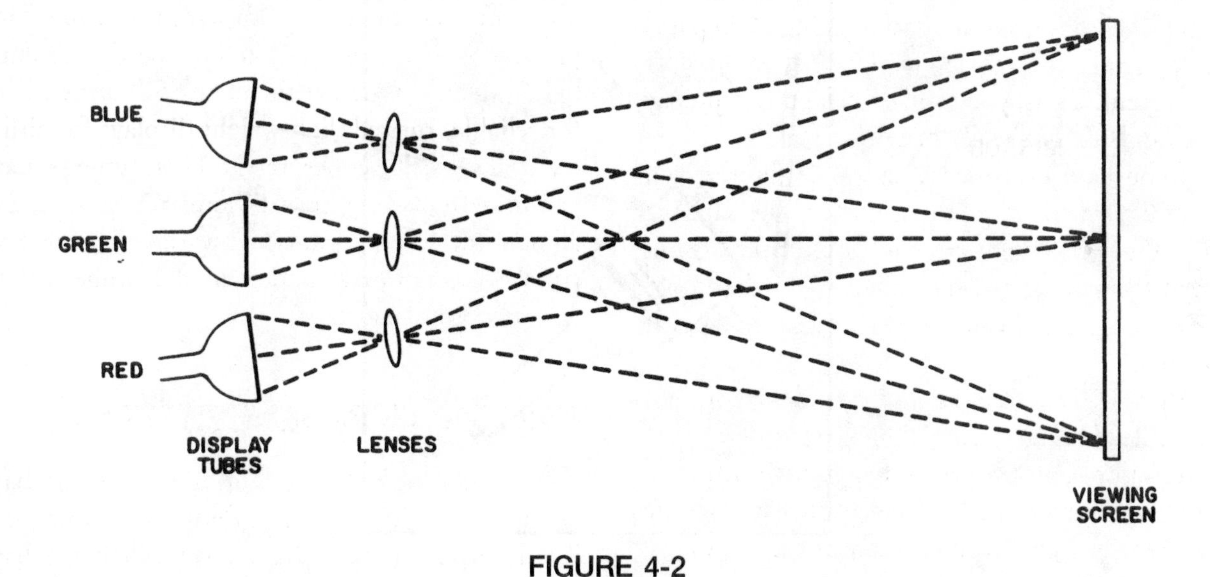

FIGURE 4-2

quired in the normal receiver's electronics is a reversal of the vertical sweep. This sweep reversal is required due to the image inversion by the optics of the lens system.

Illustrated in **Figure 4-2** is a three tube in-line projection system. For simplicity, no optical folds or mirror bends are shown in the optical path. And, all tubes and their respective optical axes are perpendicular to the screen's vertical axis. In this way, optical distortions of the vertical plane, such as vertical nonlinearity and keystone, are not generated.

Basic Projection TV Information

Most projection TV receivers use a projection system that is referred to as a three tube, in-line refractive system. This in-line projection system consists of three direct-view display tubes — one red, one green, and one blue tube.

The three display tubes are mounted in-line on the horizontal axis. The red, green, and blue images on the three display tubes are then optically projected with lenses and mirrors to a viewing screen. This is illustrated in the **Figure 4-3**

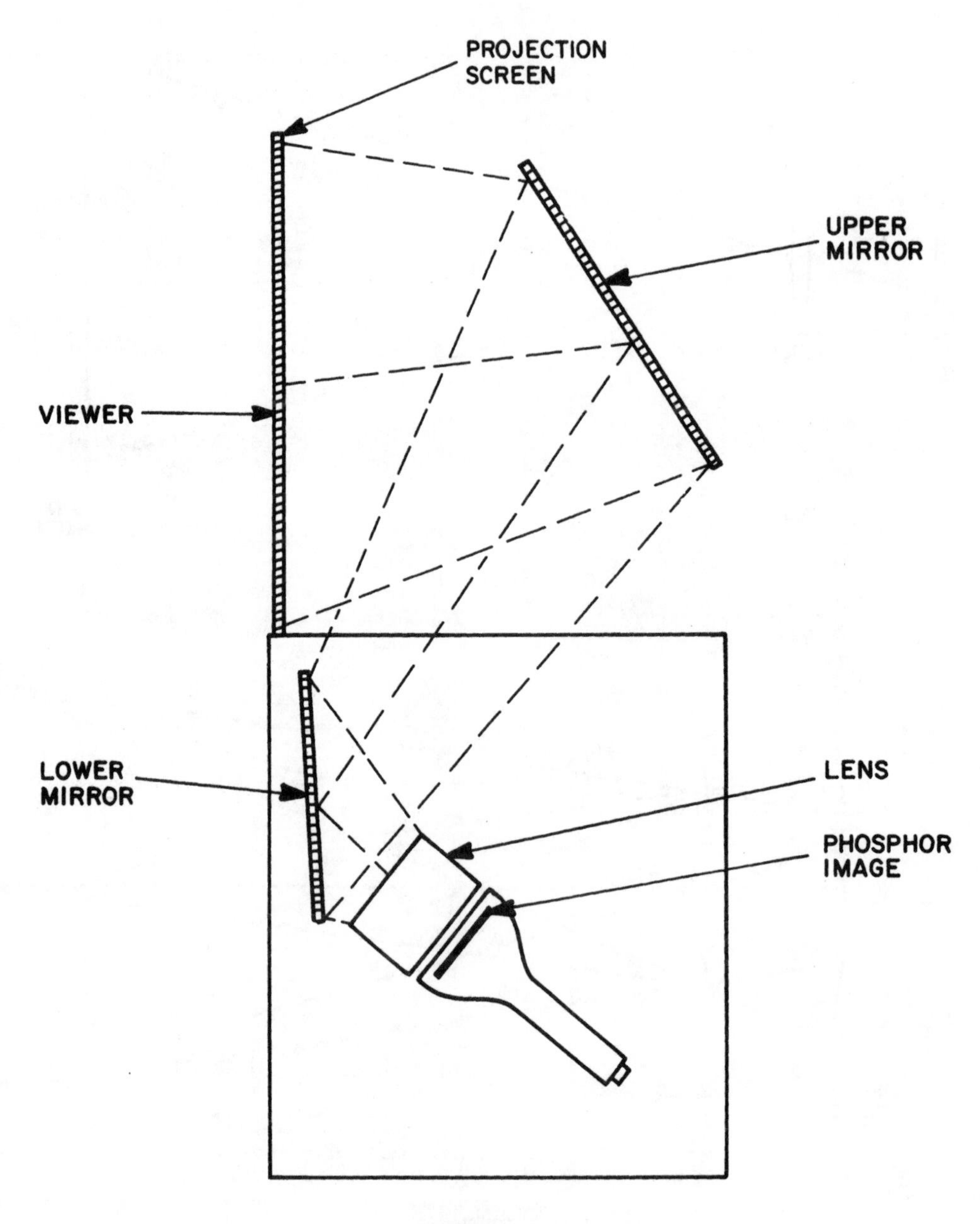

FIGURE 4-3

drawing. At the viewing screen, all three projected images are properly registered and converged to produce the correct color rendition.

The Optical Light Path

The projection system is a combined electronic, optical, and mechanical system. The three individual electronically formed images are combined optically on the projection viewing screen. The original images are optically magnified, approximately 10 times, and aimed through two mirrors in a folded light path to the viewing screen.

The basic elements in the light path consist of a projection screen, an upper or second mirror, a lower or first mirror, projection lenses, and the red, green, and blue CRTs that form the three individual images.

Optical Lens Details

For their optics, many projection sets use the U.S. Precision Lens (USPL) compact delta 7 lenses. This lens, designed by USPL, incorporates a lightpath fold, or bend, within the lens assembly. Refer to drawing in **Figure 4-4** for this USPL CRT and

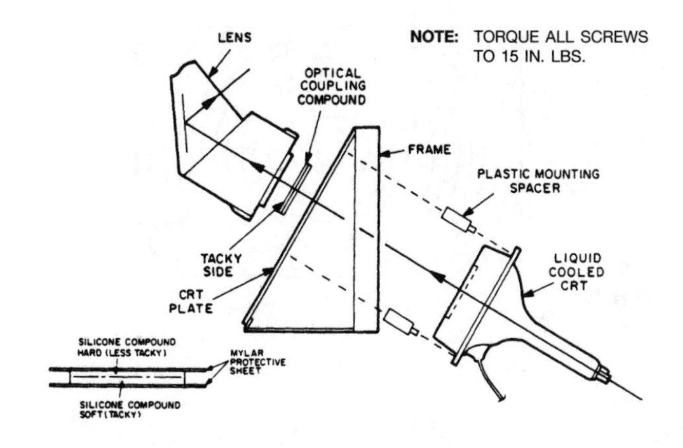

FIGURE 4-4

optical compound assembly. This is accomplished with a front surface mirror that has a lightpath bend angle of 72 degrees. Because of this light path bend, the outward appearance of the lens resembles, somewhat, that of the upper section of a periscope. The lens elements and the mirror are mounted in a plastic housing. Optical focusing is accomplished by rotating a focus handle with wing nut locking provisions. Rotation of the focus handle changes the longitudinal position of the lens element.

Liquid-Cooled Projection Tubes

These TV projection sets use three CRTs (R, G, and B) in a horizontal in-line configuration. The details of this component arrangement is shown in **Figure 4-5**. There are two

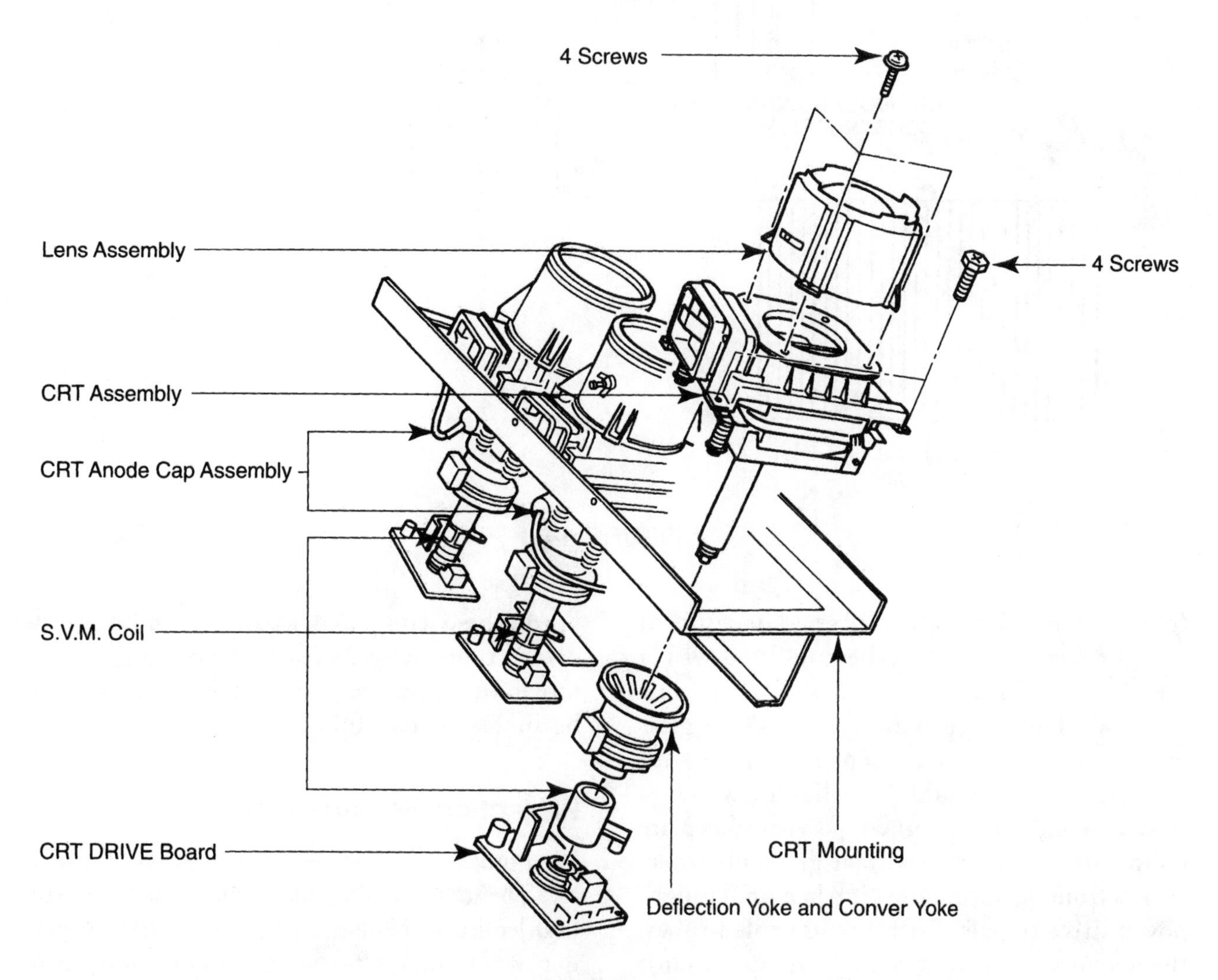

FIGURE 4-5

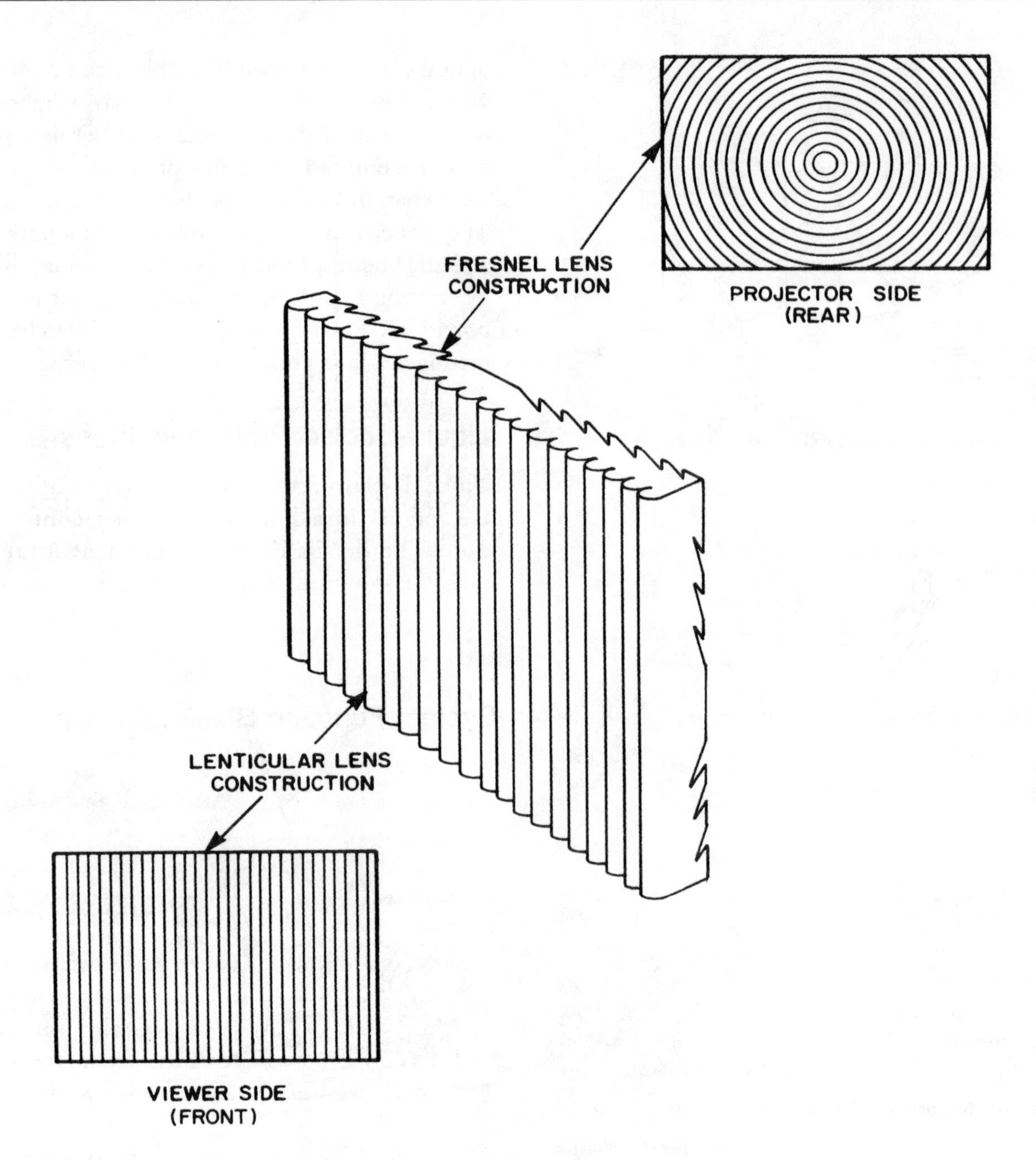

FIGURE 4-6

(red and blue) slant-face CRTs and one (green) straight-face CRT. The tubes are fitted with a metal jacket housing that has a clear glass window. The space between the clear glass window and the tube's faceplate is filled with an optical, clear liquid. The liquid, which is insulated and self contained, prevents faceplate temperature rise and thermal gradient differentials from forming across it when under high-power drive signals. With liquid-cooled tubes, the actual safe power driving level can almost be doubled compared to that of non-liquid-cooled tubes. This increases the overall system's screen brightness as the drive level wattage can be increased two-fold.

Projection Screen Construction

Most projection TV screens are of a two piece assembly. The front (viewing side) section will have a vertical lenticular, black-striped section. The rear portion is a vertical, off-centered Fresnel construction.

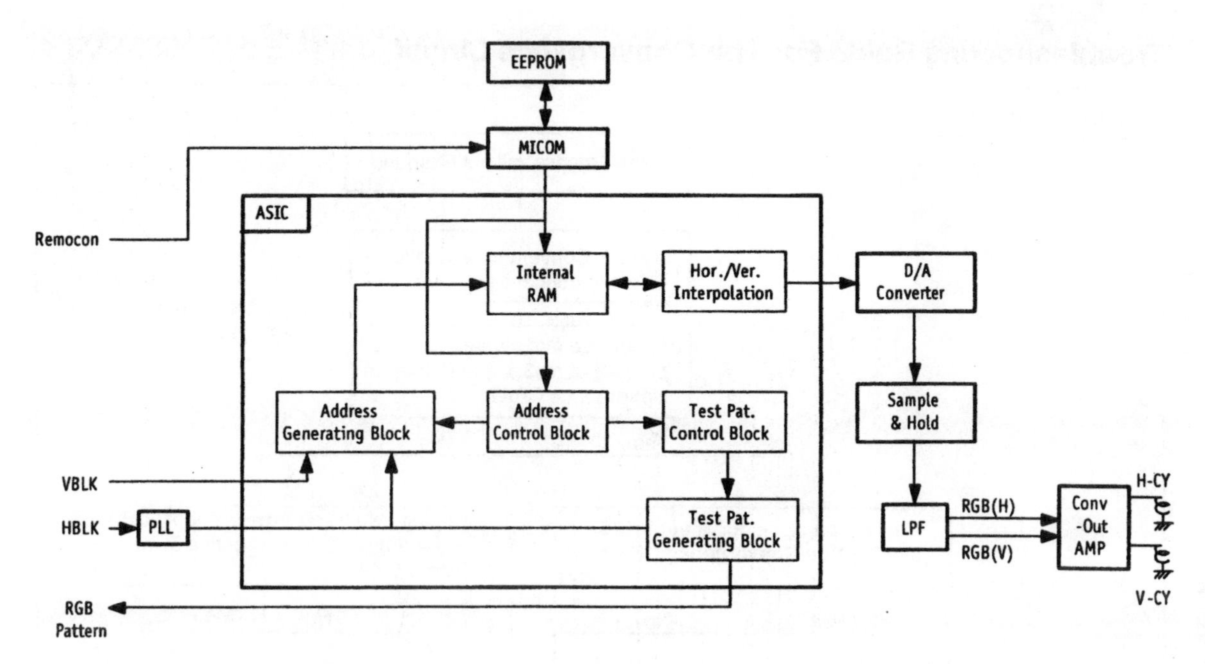

FIGURE 4-7

The black striping not only improves initial contrast but also enhances picture brightness and quality for more viewing pleasure under typical room ambient conditions.

The Fresnel lens consists of many concentric rings as shown in the (**Fig.** 4-6) drawing. Each ring is made to reflect light rays by the desired amount, resulting in a lens that can be formed in thin sheets.

If the surface of the sheet is divided into a large number of rings, each ring face may be flat, and tilted at a slightly different angle. The resulting cross section of the lens resembles a series of trapezoids.

As noted in **Figure 4-6**, the Fresnel lens is incorporated on the back (projection side) of the projection screen.

Digital Convergence Operation

Rear-screen projection TV sets require a convergence circuit to compensate for any misconvergence caused by any difference of the Red, Green, and Blue beam's mechanical alignment. The digital convergence circuit can adjust the convergence accurately by generating a crosshatch pattern for adjusting and moving the cursor, displaying the point of adjustment.

System Function Block Diagram

The digital convergence circuit block diagram is shown in **Figure** 4-7. The outline operation of each block section is as follows:

The EEPROM has the convergence data for the adjusting points. The average points for most models is 45.

Micom controls the convergence data to send from an EEPROM to an ASIC (application specific integrated circuit) when powering ON and OFF after adjustment. The PLL generates the main clock for the system by synchronizing to the horizontal blanking signal.

The Address Generating Block generates the number (position) of scanning lines by synchronizing to vertical and horizontal blanking (BLK) signals.

The Horizontal/Vertical Interpolation Block calculates convergence interpolation data of the actual scanning position in real time and then reconstructs it to fit the D/A converter, and then sends it onto the D/A converter.

The test pattern control and test pattern generating blocks generate the test pattern and cursor during the convergence adjustment mode.

The D/A converter converts digital convergence adjustment data from ASIC into analog data. It uses a 16 bit D/A converter.

Troubleshooting Guide For The Convergence Circuit

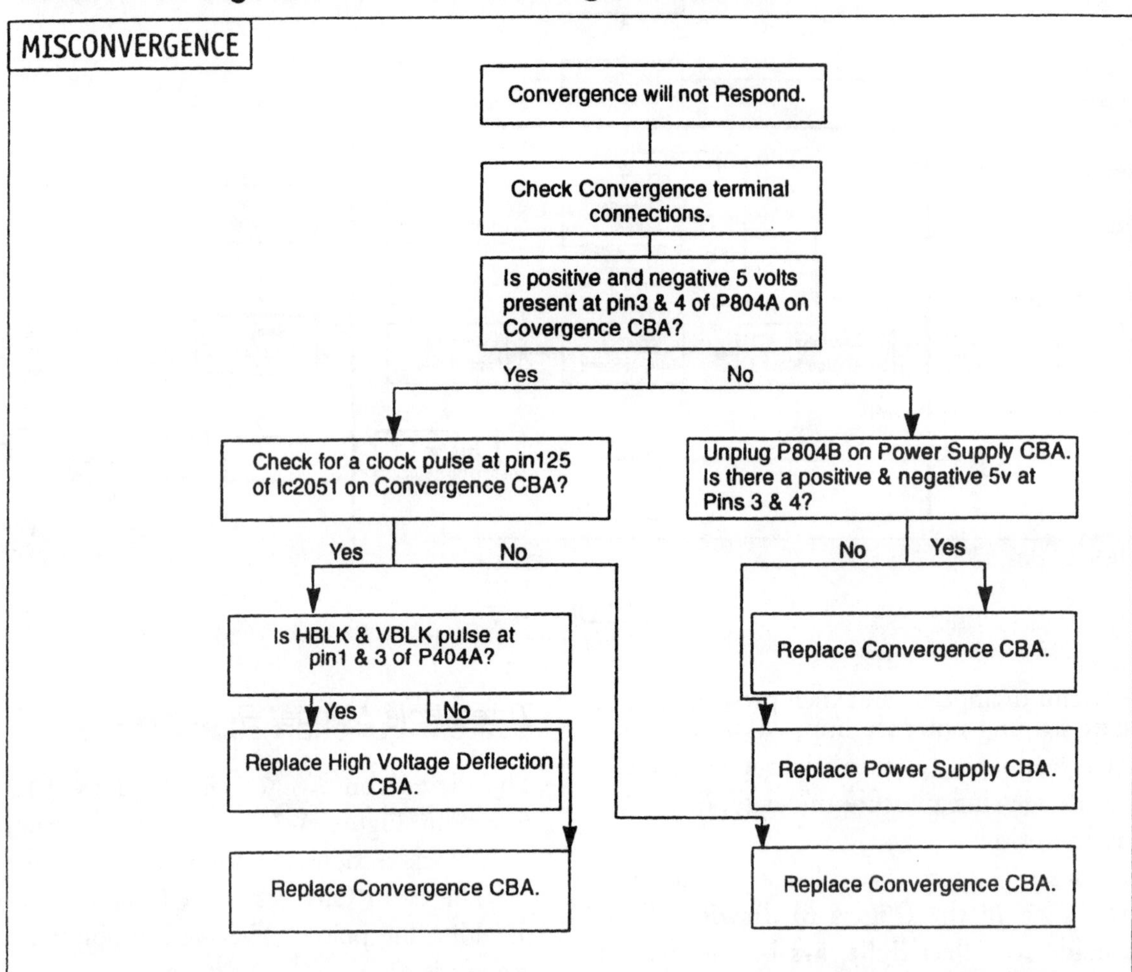

FIGURE 4-8

Sample and hold block demultiplexes convergence data from the D/A converter into horizontal/vertical values. In addition, to avoid glitches by setting the time of the D/A converter, this block reduces glitches by sampling stabilized output from the D/A converter after a constant time.

The LPF block interpolates among adjusting points horizontally. That is, it connects adjusting points smoothly from stairs-like output data by filtering.

Finally, there is compensation by the magnetic field by a flowing amplitude convergence compensation waveform through coil CY generated by the successive operation that compensates the misconvergence. The troubleshooting guide for the digital convergence system is shown in **Figure 4-8**.

Projection Receiver Preliminary Setup

G2 Screen Control Setup — The following test equipment is required:

1. Oscilloscope
2. 100/1 Oscilloscope test probe

Screen Adjustment Procedure

1. Set the receiver to an in-house digital signal. Convert the screen format to 16:9 ratio by pressing the RATIO button.
2. Adjust picture data-CONTRAST to 20, and then BRIGHTNESS to 20, and COLOR to 50.
3. Enter the factory adjustment mode change DRIVE (VP3-8) Data to #32.

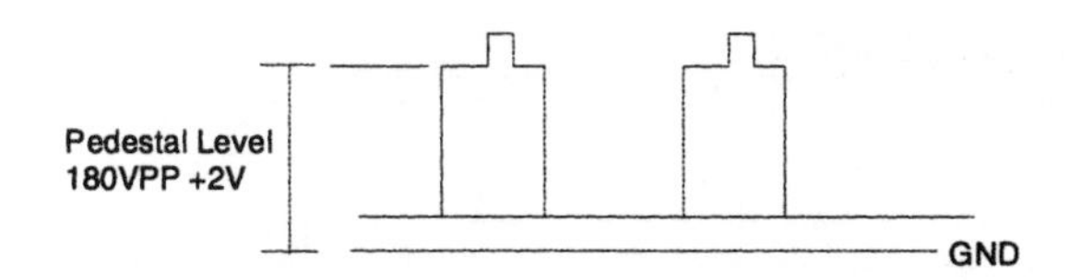

FIGURE 4-9

4. Connect scope probe between R924G and Ground of G CPT/CRT Board.
5. Adjust Screen Control so that the Black level Pedestal voltage is at 180 ± 2 Volts. Refer to **Figure 4-9**.
6. Adjust Red and Green to 180 ± 2 Volts using this method.

Yoke Tilt Alignment Procedure

1. Enter the Convergence Alignment mode by pressing the remote Menu button until the Menu disappears, and then press 9, 8, 7, 5, Enter. The HDTV receiver will display an internally generated Crosshatch Pattern.
2. Mute the Convergence Alignment Data by pressing 5 and Enter.

Adjustment Procedures

1. Mute Green CRT by pressing 7 on the remote.
2. Mute Red CRT by pressing 6 on the remote.
3. Mute Blue CRT by pressing 8 on the remote.
4. Select Green CRT only by pressing 6 and 8 to mute the Red and Blue CRTS.
5. Loosen and rotate the Green Yoke to adjust picture tilt so that the picture is level. The difference in Ll and L2 should be less than 2mm as shown in **Figure 4-10**.

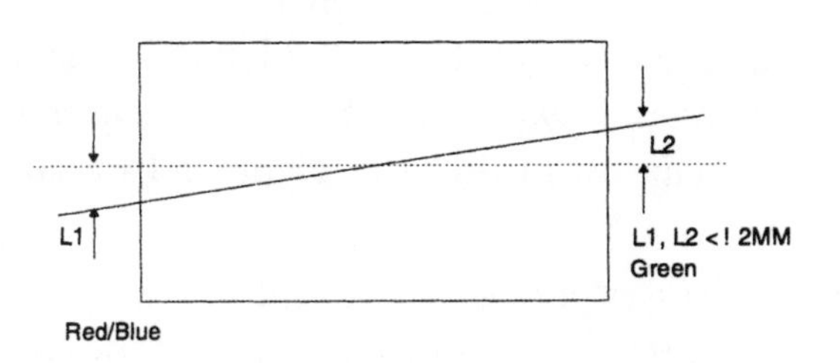

FIGURE 4-10

6. Select the Red and Blue raster by pressing 6 and 8 buttons again on the remote.
7. Adjust the slope of the Red and Blue Raster so that it will match the Green Raster.

NOTE: To adjust raster slope, loosen the yoke clamping screw; do not forget to retighten clamp screw after adjustment is completed.

The Two-Pole Magnet Adjustment

1. Enter the Convergence Alignment mode by pressing and holding the Menu button until the Menu disappears, and then press 9, 8, 7, 5, and then Enter. The HDTV receiver will display an internally generated Crosshatch Pattern.
2. Mute the Convergence Alignment Data by pressing 5 and then Enter.

Three-Pole Magnet Preparation

1. Mute Green CRT by pressing 7 on the remote control unit.
2. Mute Red CRT by pressing 6 on the remote control unit.
3. Mute Blue CRT by pressing 8 on the remote control unit.

The Adjustment Produres

1. Mute Red and Blue CRTs by pressing 6 and 8 on the remote control unit to show the Green Raster.
2. Use the centering rings on the deflection yoke to center the Green raster on the screen. Refer to **Figure 4-11** illustration.

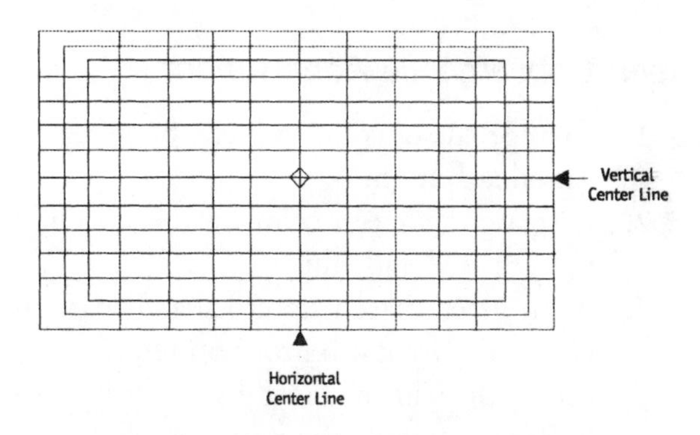

FIGURE 4-11

3. Adjust Red and Blue raster by using the same method.
4. After adjusting, store original convergence data.
5. To restore the Convergence Data, select 0 and then press Enter on remote unit.

Stigmator Alignment Procedure Alignment

1. Operate the TV receiver for approximately 30 minutes prior to making this adjustment.
2. Preadjust CRT alignment, Raster position, and lens focus.
3. Make sure Magnet Astigmator is approximately 45 mm from the end of CRT socket. See drawing in **Figure 4-12.**
4. Enter the Convergence Alignment mode by pressing and holding the Menu button until the Menu disappears, and then press 9, 8, 7, 5, and then Enter. The HDTV receiver will display an internally generated crosshatch pattern.
5. Press 3 on the remote control unit to switch back to the dot pattern.

Pattern Position Adjustment

1. Select pattern shift mode by pressing 9 and 4 buttons.
2. Adjust screen centering by pressing the 5, 6, 3, and 4 buttons.
3. To quit the pattern shift mode, press ENTER button.
4. You can save adjusted phase/pattern position adjustments by pressing 9, 1, and then ENTER.

Green Convergence Admustment

1. Change cursor color to green by using the Antenna button.
2. Display only green color on the screen by muting Red and Blue CRTs.
3. Align green pattern to crosshatch pattern. Use 5, 6, 3, and 4 buttons on remote unit. Move the cursor to adjust convergence. Follow numeric sequence as shown in **Figure 4-14.**

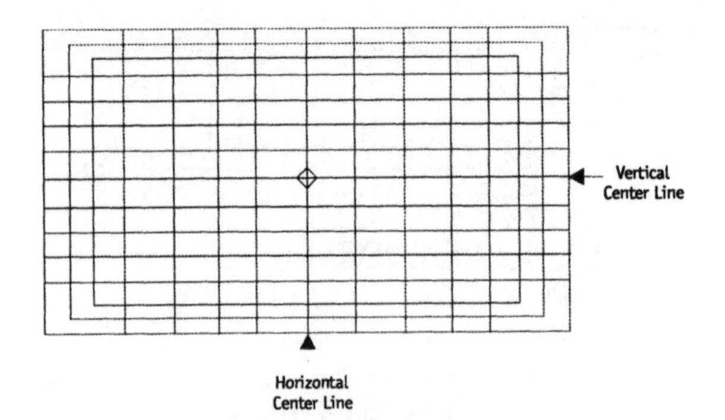

FIGURE 4-12

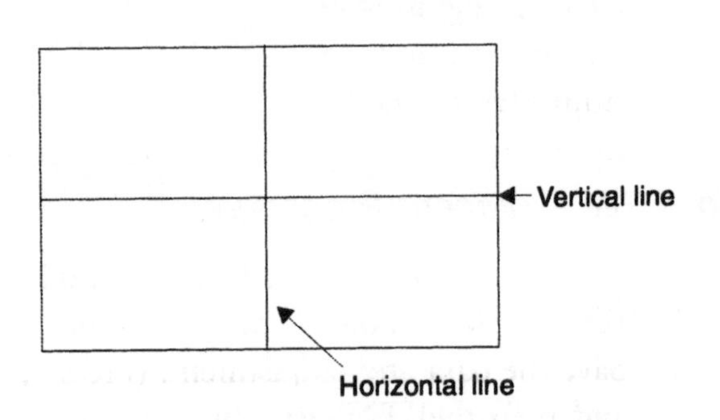

FIGURE 4-13

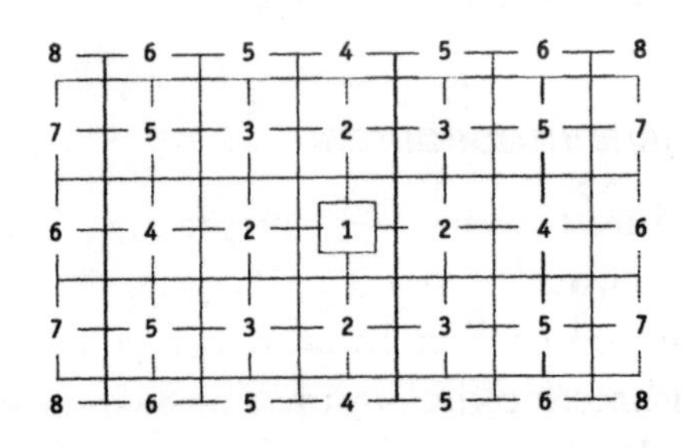

FIGURE 4-14

Red Convergence Adjustment

1. Restore Red color on the screen (Green color is being displayed; Blue is muted).
2. Select Red cursor. Align to Blue; press "antenna" button on remote.
3. Match the Red screen with the Green screen in same way with that of green convergence adjustment. Refer to **Figure 4-14** again.

Blue Convergence Adjustment

1. Restore Blue color on the screen (Green color displaying; Red is muted).

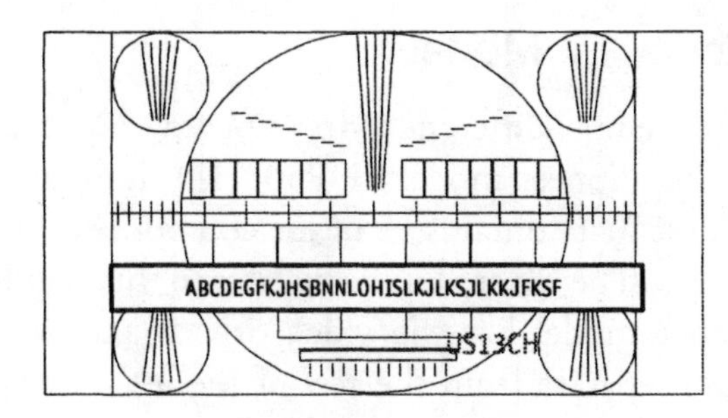

FIGURE 4-15

2. Change cursor color to Blue with antenna button.
3. Adjust the Blue raster to Green screen in same way as with red convergence adjustment.

Saving the Adjusted Data

1. Save the data after adjustment. (Press 9, 1, and then the "ENTER" button.
2. Quit the convergence adjustment mode. Press the "QUIT" button on the remote unit.

OSD Position/Adjustment (If Needed)

1. Convert the screen format into 4:3 mode and check caption OSD position, and adjust only when initial screen of basic establishment differs from that shown in **Figure 4-15** drawing.
2. Change into Factory adjustment mode (9, 8, 7, 6) and select IIVP18 OSD H-POSITION, and adjust caption OSD position to be like that shown in **Figure 4-15**.

White Balance Adustment

NOTE: The white balance adjustment requires a Brightness meter.

Preliminary Adjustments

1. This white balance adjustment must be performed in a dark room.
2. Make these adjustments after (G2) cutoff and Focus adjustments have been performed.
3. The Brightness meter must be located in 3M distance from the center of the screen. Adjust the focus of the Brightness meter to optical screen and set Field degree to 1 C.
4. Convert the screen format into 16:9 mode. (This is performed with the MENU Special Mode.)
5. Inject 100% white signal (White Screen Pattern).

Adjustments Procedures

1. Set Contrast to 1110011, Set Brightness to 116011, and Color Level to 115011.
2. Change into factory adjustment mode and check if the data of 3-8 (R-DRIVE — B-CUTOFF) is 32. 3 3 — Select Medium with "MUTE" button and adjust IIVP3 R-DRIVE" and IIVP4 G-DRIVE" so the color coordinates to be X=296, Y=303 (80OOcK-5MPCD). Fix the data of "B-DRIVE" 32 and do not adjust.
4. Select COOL with "MUTE" button and repeat 2 and 3 so the coordinate to X=282, Y=268 (100OOcK-5MPCD).
5. Select WORM with "MUTE" button and repeat 3-5 so the color coordinate to be X=313, Y=320 (65OOcK-5MPCD).
6. If the data changes, repeat c-e.

Sub Bright Tint and Color Adjustment

Perform Sub Bright adjustment as follows:

- Inject color bar pattern and select VP9 S-BRIGHT in the service menu and adjust to menu number 2. Refer to the pattern shown in **Figure 4-16**.

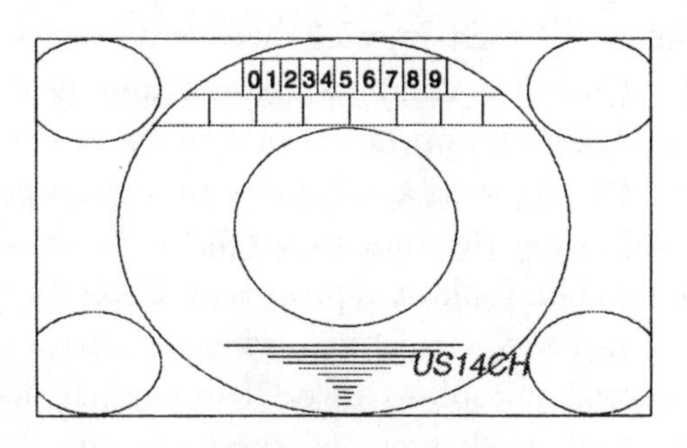

FIGURE 4-16

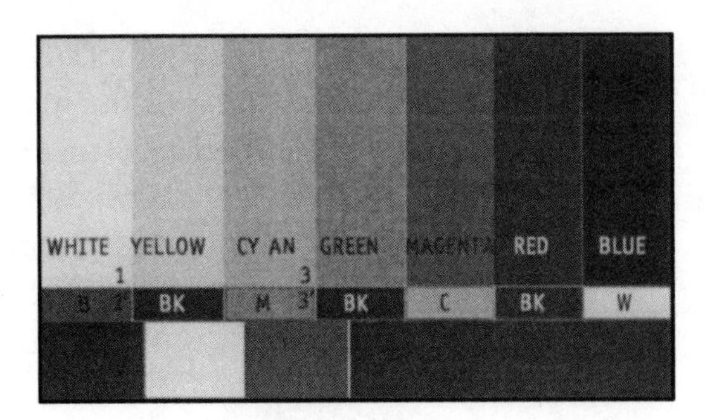

FIGURE 4-17

Sub Color and Tint Adjustment

1. Receive in-house 2CH (SMPTE pattern) and cover Red and Green lenses.
2. Select "VP SUB COLOR" in adjustment mode and adjust not to distinguish 1 with 111.
3. Select "VP10 SUB HUE" and adjust not to distinguish 3 with 31. Attach 1) adjustment Item and Initial data in adjustment mode. Refer to the Sub Color and Adjustment Standard shown in **Figure** 4-17 illustration.

On-Screen Display

The drawing in **Figure 4-18** shows a projection HDTV receiver with the "call-outs" for the on-screen operation modes.

The Basics of Plasma Panel Displays

Plasma technology was invented by Bitzer and Slattow in 1964 and was later developed by Owens-Illinois Glass for military and aerospace applications. The basic flat-panel structure incorporates a series of red, green, and blue cells sandwiched between two glass plates or substrates. Neon and xenon gas is then injected between these sealed plates to form a complete plasma display panel. The fundamental operation of a plasma display is activated by feeding a voltage to these transparent and addressing electrodes, causing the internal gas to ionize. This ionization process produces an emission of ultraviolet radiation, which in turn excites the colored phosphors in each individual plasma display cell. Each type cell produces one illuminated pixel.

Plasma TV Models

There are now different styles of plasma TV receivers. The Pioneer model PDP505HD is an all-in-one unit that contains all of the components. This model will be covered in detail later in this chapter. Another model has the video drivers and power supply separate from the rest of the TV receiver.

Another manufacture of a flat-screen unit is Revox's model E-542, uses remote video drivers and a power supply. This remote system allows the separate plasma monitor to be operated at least 60 feet away. The remote system allows the monitor to be slimmed down to only 2-k inches thick, and has the size of a large wall picture. This model can be upgraded to HDTV when the HDTV digital signal is being received.

The Plasma HDTV Advantage

Besides the slim wall picture, the main advantages of the plasma TV screen is very sharp focus and good illumination qualities. Because each cell is activated separately, there is no loss of focus. In addition, dark or hot spots have also been eliminated because of the plasmal's inherently even illumination. Another benefit of the plasma screen is that it is not susceptible to magnetic distortion. Speakers or other magnetically active components have no affect on the quality of the picture. The screen can also be moved or turned in different directions without having any color distortions.

Considering the high performance video output of the plasma monitor, it will interface very well with DVDs, laserdiscs, and even the personal computer should be very good for classroom education demonstrations.

Conventional CRT, Projection Sets, and Plasma Screen Comparisons

The conventional TV viewing screen is a vacuum tube with a faceplate and neck (gun) assembly that is referred to as a CRT. The CRT has three electron beams (for red, green, blue) that sweep across the phosphor strips or dots to produce a color picture. The CRT creates pixels by illuminating these RGB phosphors. As the beams sweep they are controlled or modulated for various brightness

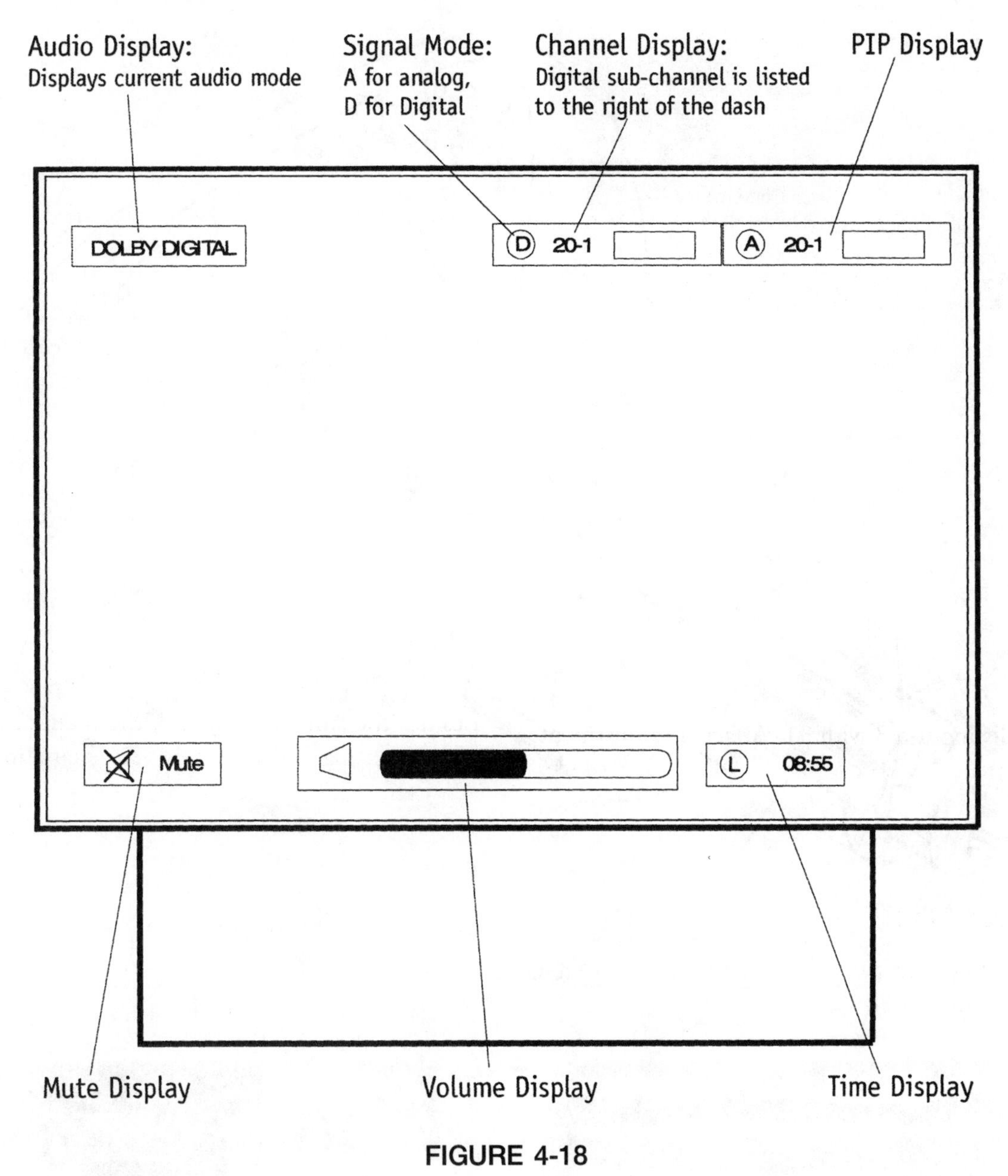

FIGURE 4-18

levels to produce a viewable color picture. This CRT operation is illustrated in the **Figure 4-19** drawing.

Rear-screen HDTV projection receivers use three smaller, very high brightness CRTs to project an image on the back of the viewing screen. And of course, the three projection set CRTs scan continuously across the screen columns. As shown in the **Figure 4-20** drawing, the projection CRTs expand smoothly along rows of phosphor dots. In this same drawing you will note that the fixed pixel displays sample the image in both directions, thus response capabilities are much faster. The drawings in **Figure 4-21** compare a high-definition monitor VGA to XGA with 4X3 resolution and 16X9 resolution pictures.

Let's now look at the benefits of the PLASMA PANEL for a high definition monitor as used for TV viewing:

- Unlimited installation possibilities
- A flat, thin screen that is cool and convenient
- Can be easily hung on the wall
- Swing it out from the wall for viewer comfort
- Extremely high performance

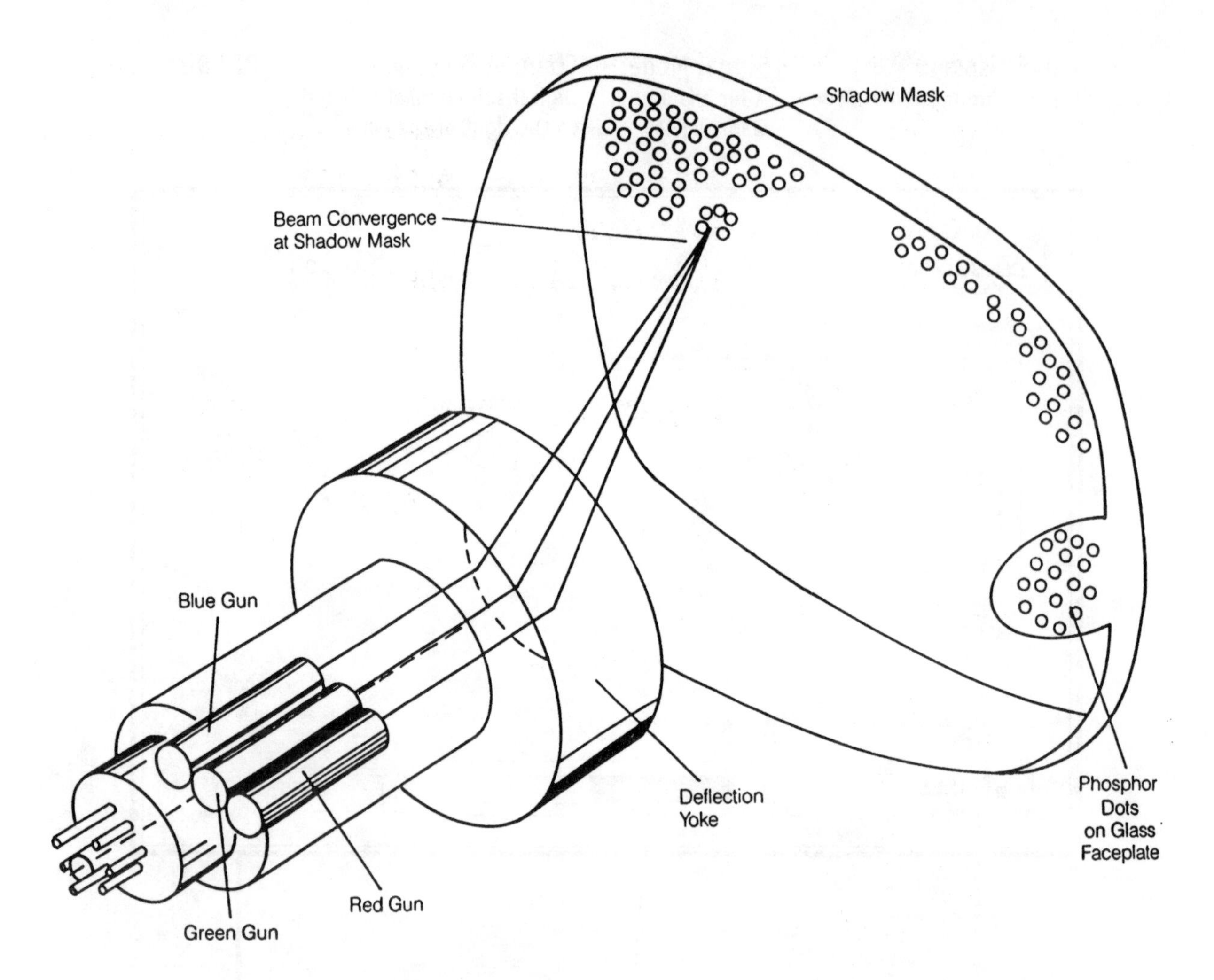

FIGURE 4-19

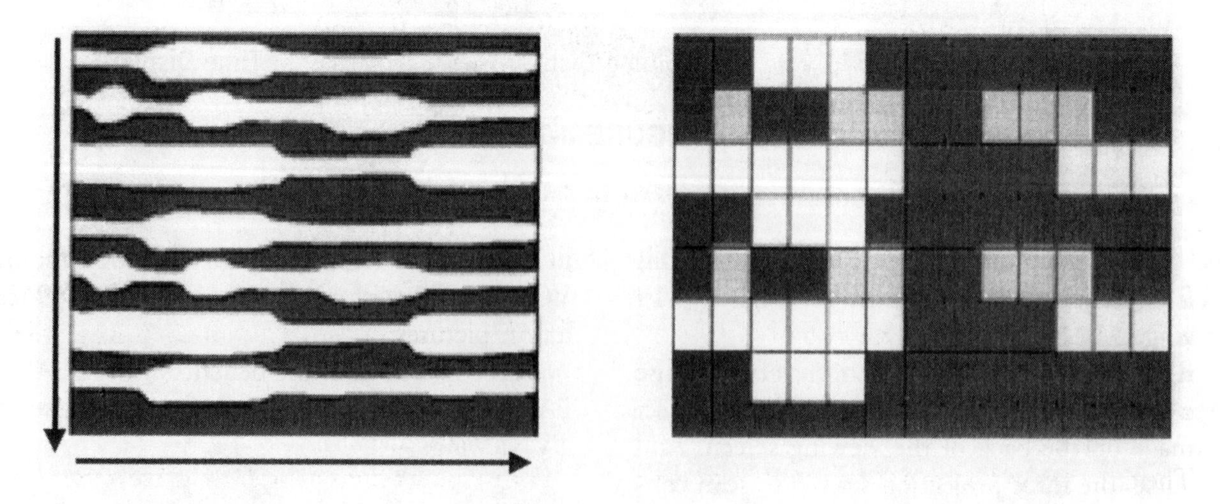

FIGURE 4-20

High Definition Monitor - Comparing VGA to XGA

•By the Numbers

Video Standard	*Resolution* 4x3	Resolution 16x9
VGA	640 by 480	852 by 480
SVGA	800 by 600	
XGA	1,024 by 768	1280 by 768

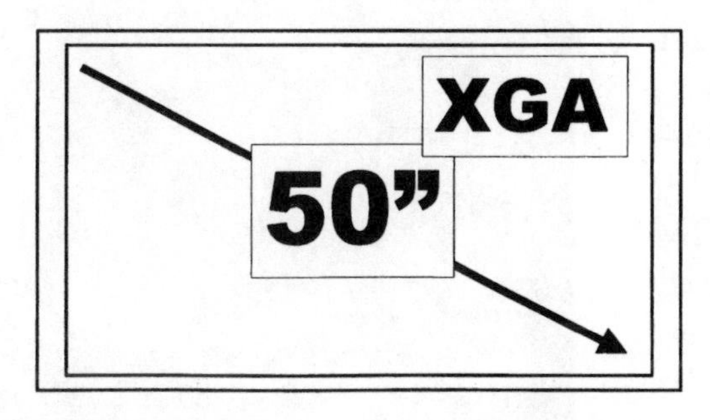

FIGURE 4-21

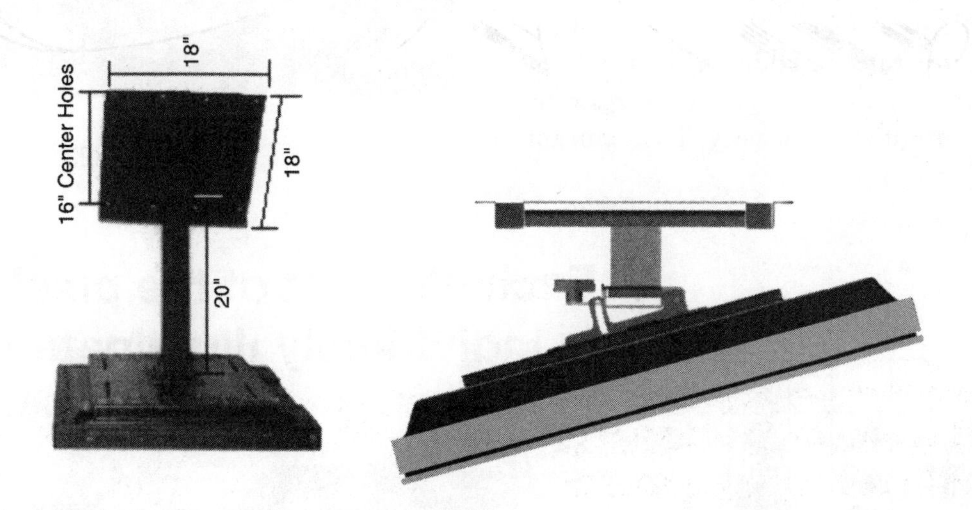

FIGURE 4-22

- Excellent HDTV monitor
- XGA level computer monitor
- Perfect picture through matrix technology
- Matrix-style display has a perfect geometry

The drawings in **Figure** 4-22 illustrate two views of the HDTV Plasma Monitor wall installations arrangements.

The features of the Pioneer PDP-505HD HDTV monitor product information are as follows:

- Encased cell structure
- Black striping
- Pioneer automatic format converter
- 8X processing
- PureCinema
- HD progressive processing
- New pixel driving sequence
- New menu system
- New remote control features

Current Plasma Panel Technology

The current Pioneer Electronics Company generation of plasma display panels use a "ROW" type configuration for the construction of the elements of each cell. Refer to the illustration in **Figure 4-23.** As you will note in the (**Fig.** 4-24) drawing, each element of the pixel is individually illuminated. The current design problem with all other types of plasma panels is that they suffer from light leakage from element to element within the vertical color column. This light leakage is illustrated in the (**Fig.** 4-25) drawing.

In the PIONEER model PDP-505HD HDTV MONITOR they use an exclusive new encased cell structure that prevents light from leaking from cell to cell. This patent-pending technology is used exclusively by the Pioneer plasma screen HDTV receiver monitors. This encased cell structure technique is illustrated in **Figure** 4-26. The drawing in **Figure** 4-27 is a close-up view of this encased cell structure. In addition, Pioneer's encased cell structure has been able to increase the overall light output and efficiency. To obtain this added light output, the additional top and bottom walls are now coated with phosphor and also emit light. This technique is shown in the (**Figure** 4-28) drawing.

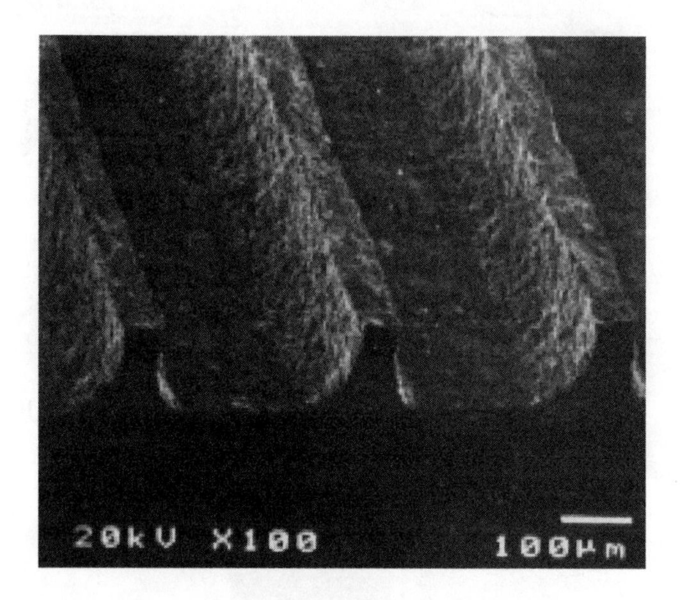

FIGURE 4-23

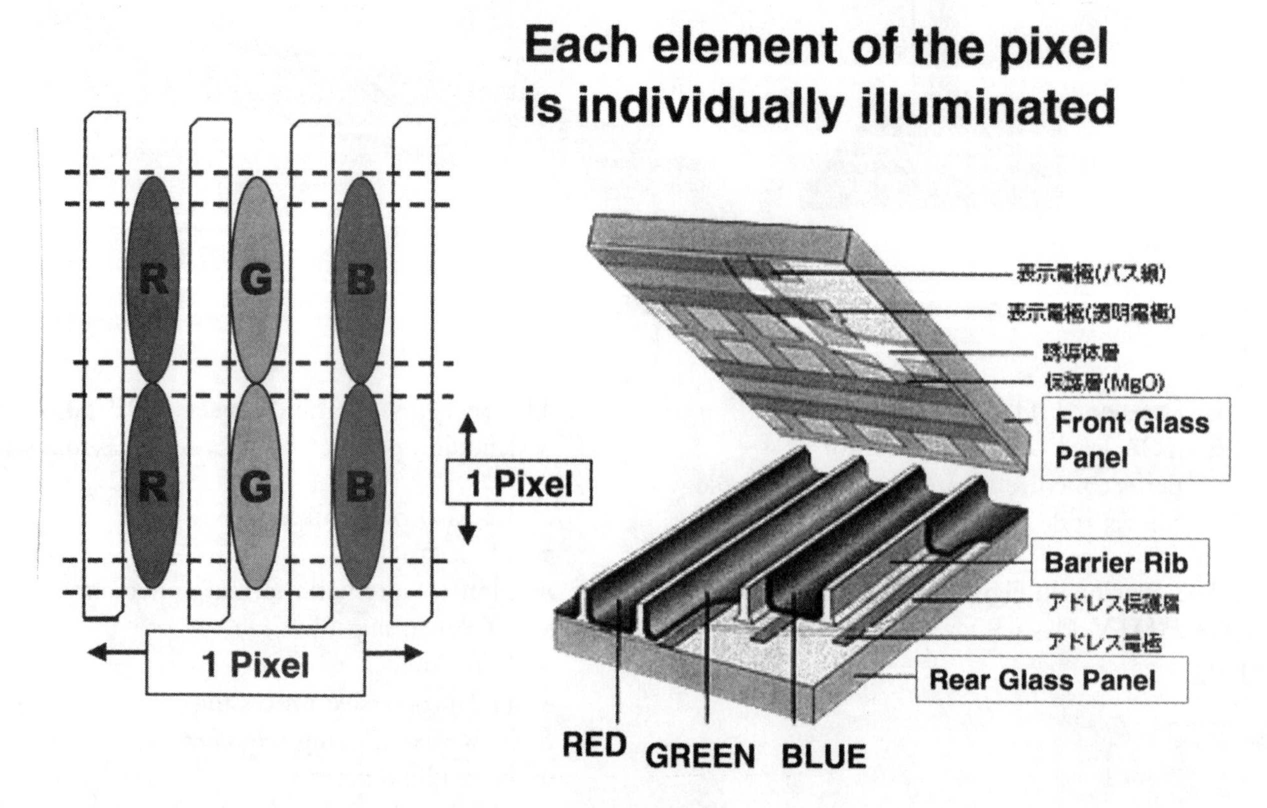

FIGURE 4-24

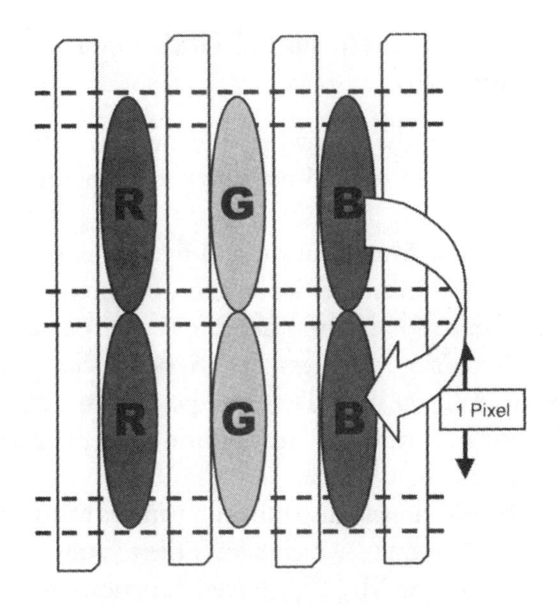

FIGURE 4-25

FIGURE 4-26

FIGURE 4-27

FIGURE 4-28

Black Stripe Coating

Pioneer's HDTV monitor plasma screen has a new black stripe coating technique. This BLACK STRIPE coating helps produce deep solid blacks by absorbing external light and reducing light reflections. Producing black striping at XGA resolution requires extreme precision during plasma screen production. The black stripe coating is illustrated in **Figure 4-29**.

DIGITAL SIGNAL VIDEO PROCESSING

The Screen-Saver Mode

The PIONEER Plasma HDTV MONITOR has features that include the screen-saver mode. These monitors will detect an extended lack of motion on the plasma screen and automatically dim it down to a safe level. When using the monitor screen for computer read-out, this would be an advantageous feature. This feature will usually be

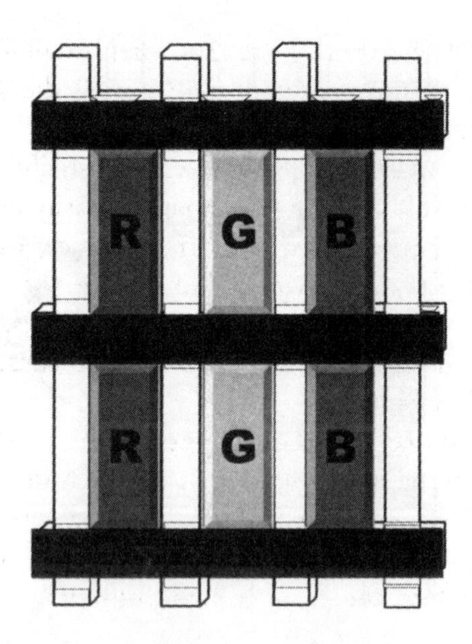

FIGURE 4-29

enabled after about five minutes of power-up operation.

You can program the flat screen's adjustments in three different ways: at the monitor's right-hand control panel, with the remote control hand unit, or with a PC using the RS232C controller. Now, for a brief run-through of some of these adjustments.

Pioneer Plasma Panel Monitor Adjustments

The plasma unit has four major operating modes: Normal, which allows setting of the screen-size switching and full auto-zoom; Menu, which is used for setting the picture quality and image positioning; Intergrator, which mainly adjusts white balance; RS232C, which enables adjustments by using a PC. Although the plasma unit is preset at the factory, ambient conditions where the plasma panel is to be used may require some fine-tuning changes. Should this adjustment process not go as you plan, you may perform an initialization (reboot) and the system will return to the factory preset condition.

Remote Control Hand Unit Initialization

You can use the following information with the remote control unit to perform the simple initialization.

Press the Menu button and five items will appear on the screen: White balance, Picture, Screen, setup, and Total Initialize. Move the cursor to total initialize and press Set. A perfunctory will appear: Are you sure? Then you position the cursor to "yes" and press the "set" button, and the menu screen will then return.

The additional setup function handles baud rates and mirror modes (which allows the picture to basically be reversed). The screen functions set the parameters for Horiz. Position, Vert. Position, Clock Freq., and Clock PHS. The white balance sets the parameters for Red High, Green High, Blue High, Red Low, Green Low, and Blue Low. The picture function on the menu sets the parameters for Contrast, Bright, Color, Tint, Sharp, and Detail. Again, pressing Total Initialize at any point in the programming will set the plasma unit to the factory preset condition. Of course, there are many more options than this, especially if you are using a PC to do the programming. This start-up will be all you need to know to put the picture into operation until you can make the fine-tuning adjustments you may require for the sharpest picture quality.

Tips on Plasma Screen Installation

When you need to install the plasma panel screen on a wall or ceiling you must make some structural inspections. For wall or ceiling mounting, there needs to be some sturdy frame or studding behind the mounting surface. Stud finders, bubble levels, and some careful measurement calculations can take the guesswork out and make a professional-looking installation.

These plasma units have variable-speed cooling fans located at the panel top-side. And there are many air vents located all around the sides of the panel frame. During installation of this unit make sure that these fans and vents have a large enough area clearance for good airflow.

Some plasma monitor screens may have a problem referred to as "pseudo-contour." This is a pattern of striped shadows that may accompany a moving image that contains certain colors or different levels of brightness. In later models the system designers used different video drivers and tweeked the various circuits, and have kept this effect to a minimum.

Plasma HDTV Maintenance

Let's now take a brief look at some maintenance tips to be performed for plasma monitors. Some of the same precautions that you have used for conventional CRT sets can be used for the plasma display panels as well. Still images affect the plasma screen the same way they affect a CRT screen since phosphors are used in both applications. Blue and green phosphors degrade faster than red, so it is a good idea to adjust the white balance every 1000 hours of operation. Note that when programming certain plasma monitors with a PC, the screen-saver function is not normally activated.

The plasma display screen is usually coated with an anti-glare material that can easily be damaged. When cleaning the screen surface, use caution while gently wiping the surface with a soft cloth. In most cases, a cleaning solution will discolor the monitor screen surface, or cause it to become opaque.

As noted previously, some plasma models require forced-air cooling because of a restricted air space of the cabinet enclosure. Dirt and dust should be removed from the vents to keep the internal temperature cool. For dirt and dust removal you should use a low-suction vacuum cleaner with a soft brush attachment. Also, check make sure all of the fans are operating properly.

Portions of this chapter's technical information and drawings are courtesy of the Zenith Electronics Corp. and Pioneer Electronics Corp.

CHAPTER 5

HDTV Scanning Process Circuits

INTRODUCTION

This chapter contains information on the various sweep systems found in High-Definition color TV receivers. This coverage will include the horizontal and vertical sweep formats and how these circuits operate. We will also cover vertical and horizontal oscillator/AFC operation in addition to a review of the aspect ratio for standard color TV as compared to the HDTV picture format.

We will present some actual HDTV horizontal sweep deflection circuits, vertical sweep circuits, and digital convergence circuits.

After you understand how these circuits operate you can then move on into some actual troubleshooting techniques of how to repair these circuits. This will include circuit diagnosis, circuit diagrams, test equipment service techniques, and test point connections.

HDTV PICTURE FORMAT

To have a high definition picture you may think 1000 or more sweep lines are required. This is referred to as the 1080i or computer resolution picture format. However, this is not actually true in the "real world." If we use the same number of lines as used with interlaced scanning, the progressively-scanned picture will offer a higher resolution picture. Let's now look at these two HDTV formats in the ATSC DTV standard.

One format has 1080 lines x 1920 pixels per line and the other has 720 lines x 1280 pixels per line. These two pixel arrays are used with three different frame rates: 60 frames per second for live video, and 24 or 30 frames per second for material produced on film. In all, there are six HDTV formats in the standard mode. They all use progressive scan except for the one specified at 1080 x 1920, which operates at 60 frames per second. This particular format requires the most bits. The only way to fit this format into a 6 MHz standard TV broadcast channel is to use interlaced scanning. This method actually compresses the data further and reduces the bit rate by only reproducing half of the picture at one time.

HDTV ASPECT RATIO

High-Definition television has a resolution approximately twice that of a conventional TV receiver in both the horizontal (H) and vertical (V) dimensions. The NTSC standard picture has an aspect ratio (the ratio of picture width to picture height) of 4:3, while the HDTV system has an aspect ratio of 16:9. Because of this, the set viewer can receive almost six times more picture information. Plus, the 16:9 HDTV aspect ratio is the same format in which most motion pictures are filmed. This enables the films to be broadcast directly in HDTV. It also eliminates much of the editing required to fit these motion pictures into the NTSC screen format.

PROGRESSIVE SCAN FORMAT

Progressive scan is a noninterlaced scanning process. The complete video frames are transmitted at one time. This eliminates scan lines and flicker that is also used in some computer monitors. The progressive scan technology is necessary to reproduce a bright, sharp picture on a large screen TV set.

A direct view, or projection TV, creates an image by illuminating pixels on a CRT. When an electronic beam touches a pixel, the pixel is illuminated. The pixels are arranged in lines called scanning lines. There are a total of 525 scanning lines. In a conventional, interlaced TV set, the electron beam scans every other line. First, the beam scans the odd-numbered lines. Then, it scans the even-numbered lines. At a normal viewing distance of 15 to 20 feet, the average viewer should not see the scanning lines in a TV. The number of scanning lines does not change with the size of the TV set's screen. Therefore, the larger the viewing screen, the more visible the scanning lines become. If a 71-inch television was a standard NTSC-interlaced scanned TV, the viewer could easily notice the scanning lines, and might notice a flicker in the picture. By using progressive scanning in the larger picture tubes and projection TV sets, virtually all flicker is removed, making the scanning lines virtually invisible.

A video picture is made up of many still shots shown in sequence and at a speed that prohibits the eye from detecting the movement from picture to picture. An interlaced television set develops a picture image by scanning fields. Each field contains 262.5 scanning lines. Two fields make up one frame (525 scanning lines). A frame is one still picture. An interlaced television screen flashes 30 frames (or 60 fields) per second. The two fields that make up the frame are the odd field and the even field. As the name implies, the odd field scans the odd-numbered scanning lines and the even field scans the even-numbered scanning lines.

The progressively-scanned television has the same number of scanning lines as a conventional TV set. It is different, however, from the conventional TV because it doubles the horizontal frequency (vertical remains the same) and scans the lines in sequential order. The increased horizontal frequency allows the TV to scan the lines at twice the speed. Thus, in the time a conventional TV scans one field, the progressively-scanned TV scans one complete frame. The TV receiver flashes 60 complete frames per second.

The progressively-scanned TV system uses the same NTSC video signal as the conventional television. However, since the TV is scanning twice as many lines as a conventional TV, it requires twice as much video information to complete a frame. This problem can be solved in two ways. If the information is moving video, the technique is to use line doubling. If the video is stationary, a frame doubling process is used.

Line doubling is a process of doubling the line information. The progressive-scan television takes one field of information and formulates the missing field to complete one frame. For example, if the television has the odd field of information, it converts that information to digital information. It examines each line of information and fills in the even scanning lines. In other words, it examines lines 1 and 3 to create line 2. This process is used for motion video only. Should the video be motionless, the television receiver uses Frame Doubling to fill in the missing information on line 2.

With still video, or graphic displays, the format for progressive-scan television delays one field of information and combines it with the second field. Both fields are scanned at the same time to create one frame. It then copies the combined fields and scans them into a second frame. This allows the TV to scan two full frames in the same amount of time a conventional television scans one frame. The name of this process is called frame doubling. Both frame doubling and line doubling are essential to reproduce an exceptionally bright and sharp picture on a large picture tube screen or a "home theater" system's projection screen.

HDTV Horizontal Deflection Circuit Operation

For most color TV sets, the same output circuit develops horizontal deflection and the high voltage. For HDTV progressively-scanned TV sets, these functions are divided into separate circuits. This division improves high-voltage regulation, thus improving overall focus and brightness.

High-Voltage Circuit Notes

The operation of the high-voltage circuit is the same as in a conventional TV set. The high-voltage circuits in this chapter, for an HDTV progressively-scanned TV, include a regulation circuit that maintains a constant 31 kilovolts.

High-Voltage Regulation Operation

The high-voltage output of the flyback feeds a distribution block (CR Block Z450) as shown in **Figure 5-1**. The CR block distributes the high voltage to the anodes of all three CRTS. The internal voltage divider supplies a feedback to Q437. The first of two amplifiers internal to Q437 is a buffer amplifier. The feedback is applied to the first amplifier to pin 12 and sent out on pin 14. Pin 14 connects to a voltage divider that uses a variable resistor (R466). R466 is the high-voltage control adjustment. The output of the first amplifier is applied to the second amplifier at pin 2 of Q437.

The input to the second amplifier is the inverting side of the amplifier. Therefore, the output is 180 degrees out of phase with the input. The output of the second amplifier is pin 1, and is applied to Q436 and Q435. Q436 and Q435 regulate the current flow through the primary side of the winding, thus controlling the high-voltage output from

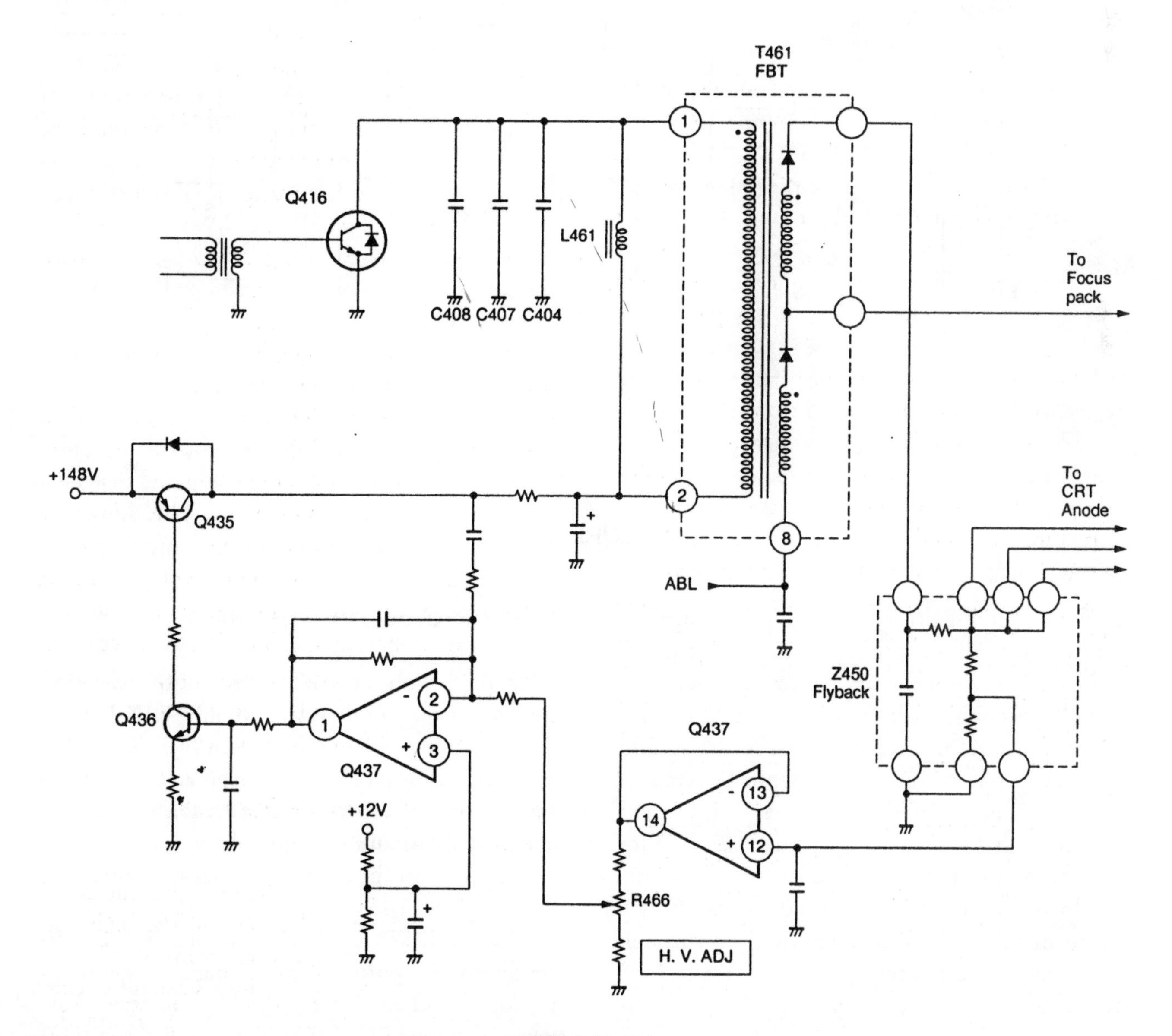

FIGURE 5-1

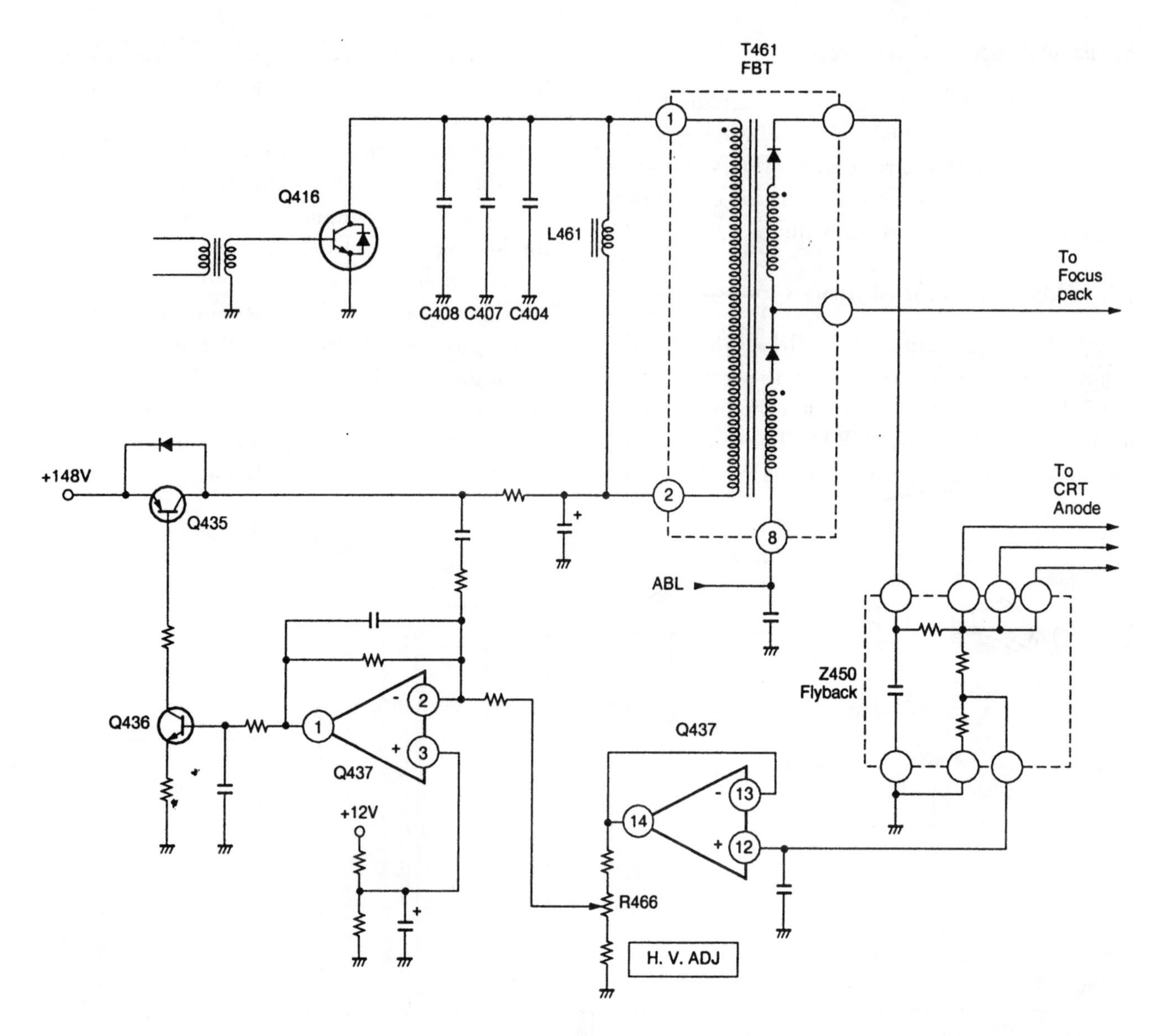

FIGURE 5-2

the flyback transformer. If the high voltage increases, the feedback applied to pin 12 of Q437 increases. When the voltage on pin 14 increases, the voltage going back into pin 2 of Q437 increases. Pin 2 is an inverting input to the second amplifier; therefore, the output on pin 1 decreases. As the voltage on pin 1 decreases, the current flow through Q435 and Q436 decreases. With less current through Q435 and Q436, less current flows through the primary side of the flyback transformer. The decrease in current flow through the primary windings results in a decrease in high voltage on the secondary winding. The circuit operates in the opposite fashion when the high voltage decreases.

X-Ray Protection Operation

X-Ray protection is needed to prevent the television from producing an excessive amount of X-Ray radiation. An increase in the amount of X-Ray radiation is caused by an increase in high voltage. The X-Ray protect circuit shuts the television set down if an excessive amount of high voltage is detected.

Referring to the circuit drawing in **Figure 5-2**, note that the T461 secondary winding at pin 9, which feeds D471 and C471, produces a DC voltage directly proportional to the high voltage that's developed. When the high voltage increases, the voltage developed across C471 increases proportionally. The voltage developed at C471 is applied to a

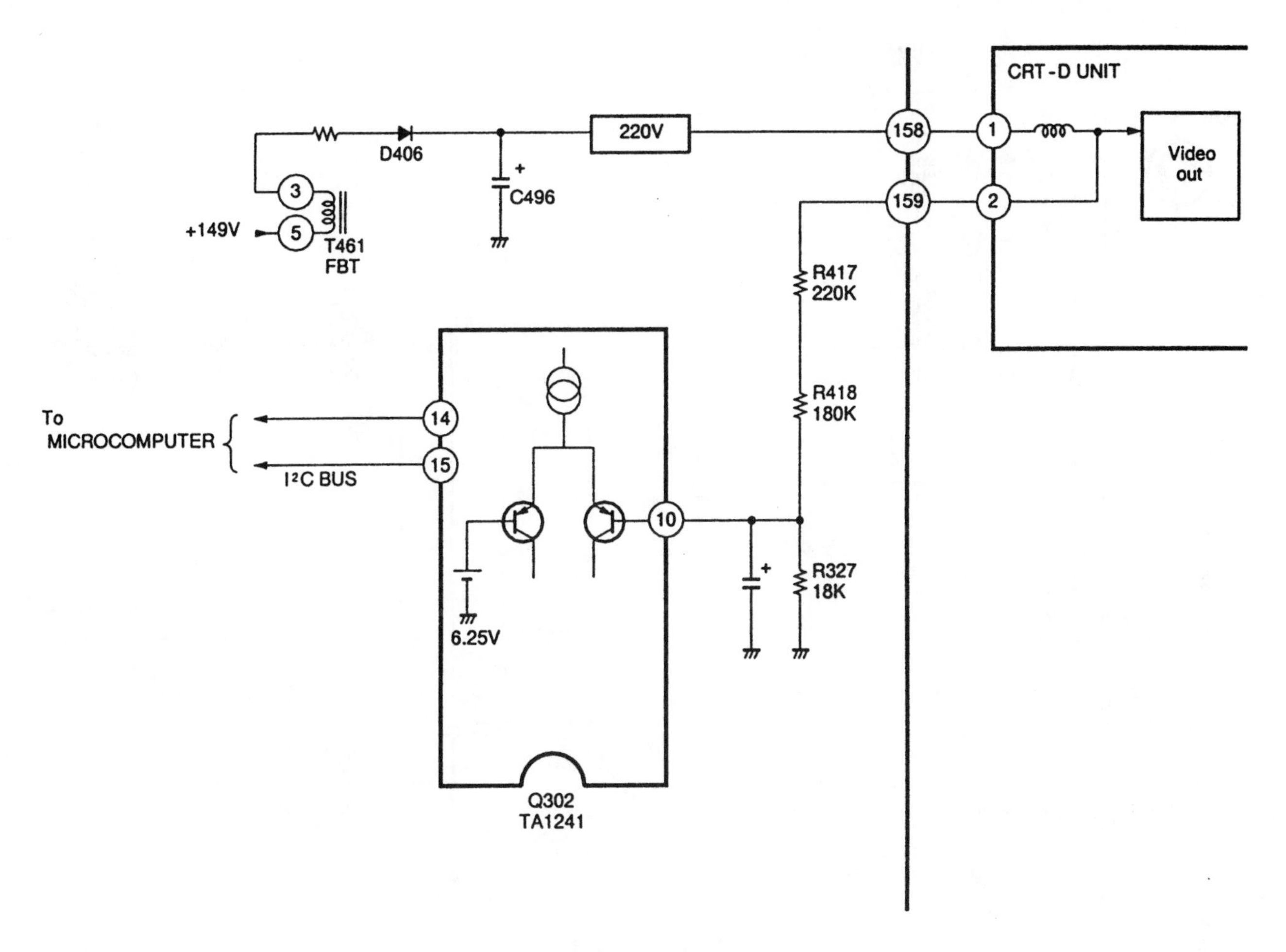

FIGURE 5-3

resistor divider consisting of R451, R452, and R453. The emitter of Q430 connects to the voltage divider. As the high voltage increases, the voltages at C471, and on the emitter of Q430, increase. Connected to the base of Q430 is Zener diode D472. If the voltage on the emitter is high enough, the Zener diode conducts, turning Q430 ON. The Q430 conduction allows a voltage to be developed on the collector. The collector voltage of Q340 turns Q429 ON. When Q429 turns ON, current is conducted between its emitter and collector, and a voltage appears on the emitter. The emitter voltage of Q429 is applied to the SCR (D846), putting the television into a shutdown mode condition.

The 220-Volt Low-Voltage Protect Circuit

Referring to **Figure 5-3**, note that pins 3 and 5 are either side of the winding of the flyback transformer that provides 220 volts to the CRT drive circuit. Diode D406 and C496 rectify the 220 volts. A loss of this voltage results in excessive cathode current in the CRT. The excessive current would damage the CRT by burning the phosphor.

The low-voltage protection circuit turns the television OFF if a loss of this 220 volts occurs. If the 220 volts DC drops, pin 10 of Q302 senses the drop. Internal to Q302 is a reference voltage of 6.25 volts. Q302 communicates via the microcomputer bus. The microprocessor then turns OFF the television and flashes the power indicator. Unlike the shutdown mode, the television can be turned back ON by the power button on the remote. The TV set does not have to be unplugged first.

Horizontal Stop Protection Circuit

Because the horizontal deflection and the high-voltage circuit are separate circuits, it is possible for

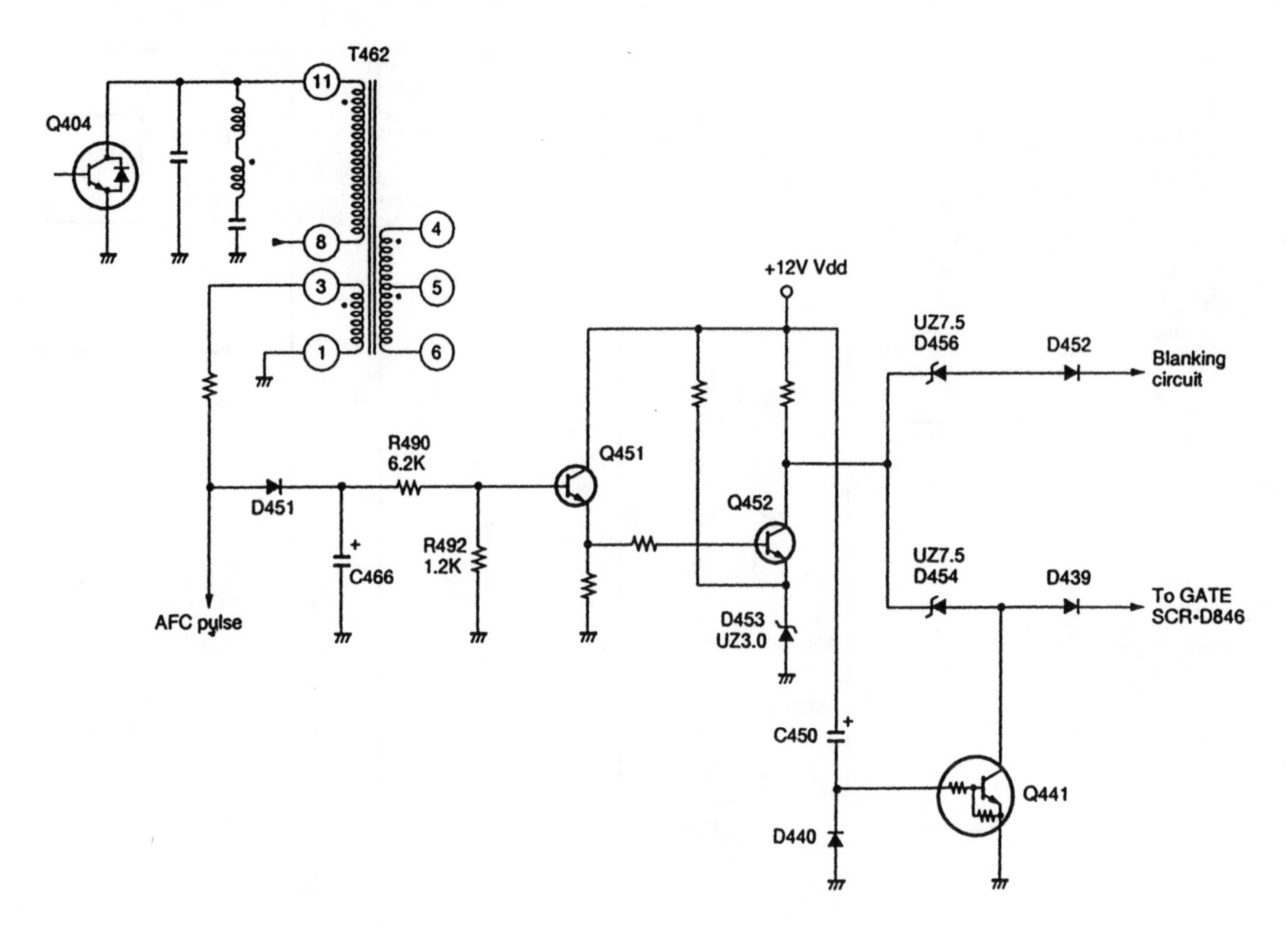

FIGURE 5-4

the deflection circuit to fail and still produce high voltage. In this scenario, one bright vertical line would appear on the screen. This vertical line would burn the phosphors of all three CRTs. To prevent damaging the CRTs, the horizontal stop protection circuit engages the shutdown circuit if a loss of the deflection occurs.

Circuit Operation

T462 is the deflection transformer shown in the **Figure** 5-4 protection circuit. During normal operation, current is induced into the secondary windings between pins 3 and 1. The current is rectified and filtered to produce a DC voltage across C466. This voltage is applied to the base of Q451. Q451's emitter connects to the base of Q452. During normal operation, both of these transistors are ON, and the voltage on the collector of Q452 is approximately 3.6 volts. Should horizontal deflection be lost, the voltage applied to the base of Q451 drops and both transistors turn OFF. The voltage on the collector of Q452 increases to 10.8 volts. The increased collector voltage is applied to the blanking circuit to black out the picture and to the gate of D846 to engage the shutdown circuit.

Protect Delay Circuit

Transistors Q441, C450, and D440 prevent the shutdown circuit from engaging when the TV set is first turned ON. At turn ON, the 12-volt Vdd appears before the horizontal deflection starts. At this time, Q452 is OFF and 10.8 Volts is on the collector. This 10.8 Volts blanks the video by engaging the blanking circuit, and normally would engage the shutdown. But, as C450 charges, Q441 is ON, preventing the 10.8 Volt application to the SCR. Once C450 is fully charged, Q441 turns OFF. By this time, horizontal deflection has started and normal operation begins.

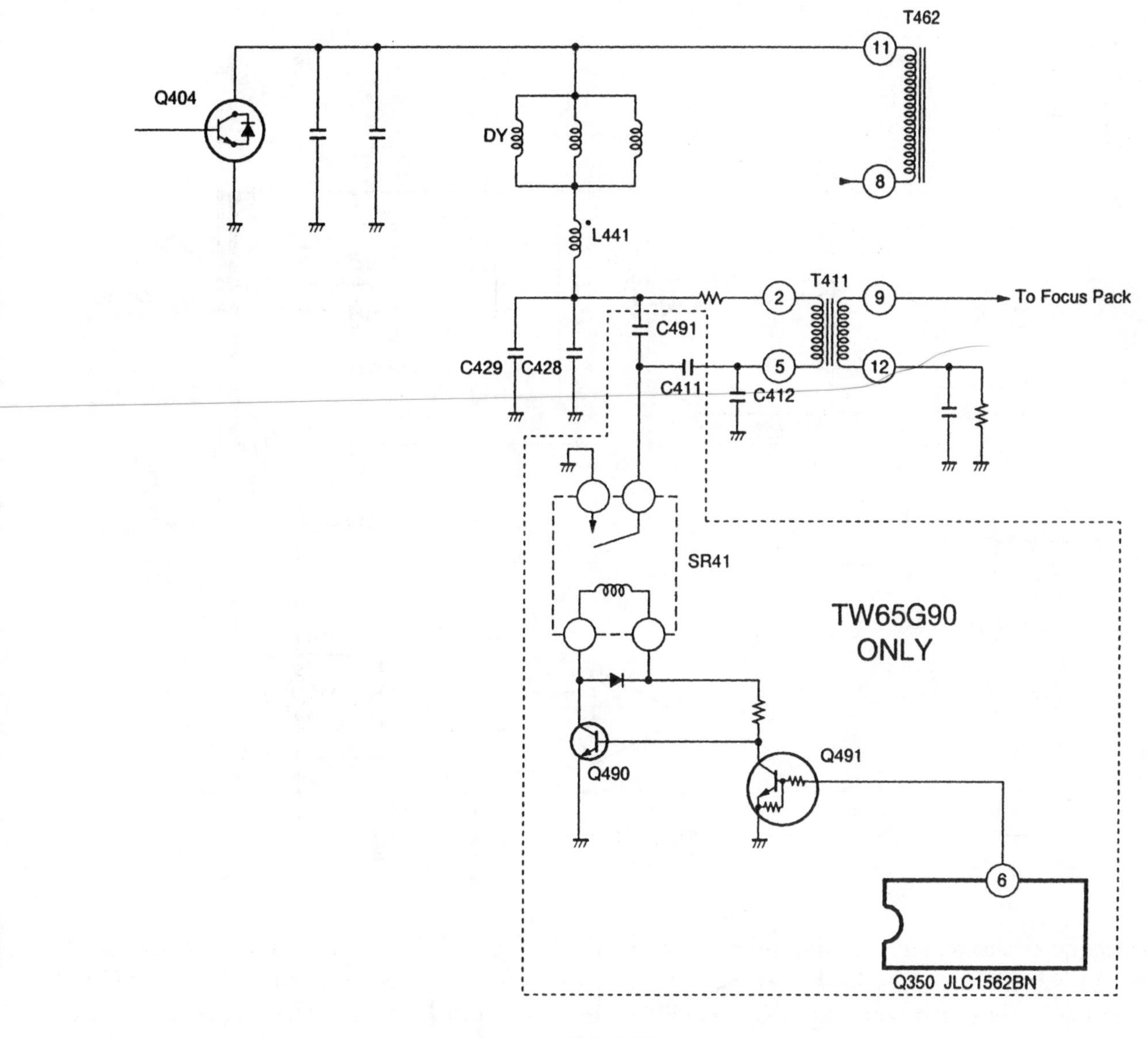

FIGURE 5-5

S-Character Capacitor Switching

An S-Curve correction capacitor (S Character Capacitor) prevents non-linear expansion on the sides of the picture caused by the angle of deflection. C429 and C428 are the S-character capacitors of all of the progressively-scanned TV receivers. For theater-wide sets, an additional circuit is added as shown in **Figure 5-5**. This circuitry adds additional capacitance to the S-character correction circuit when operating in the Theater-Wide mode. When SR41 closes, C491 is added in parallel to C429 and C428. When the Theater-Wide mode is selected, Q350 pin 6 goes to a logic HIGH, turning Q491 and Q490 ON. SR41 closes adding additional capacitance. C411 and C412 are dynamic focus capacitors.

Horizontal Deflection Circuits

Generally, the deflection circuits in progressively-scanned TV sets operate on the same basic principle as conventional TV receivers. Let's now check out these differences.

Dynamic Pincushion Circuit Operation

The only difference between the conventional TV sets deflection circuit and the progressively scanned units is the addition of a regulation circuit for

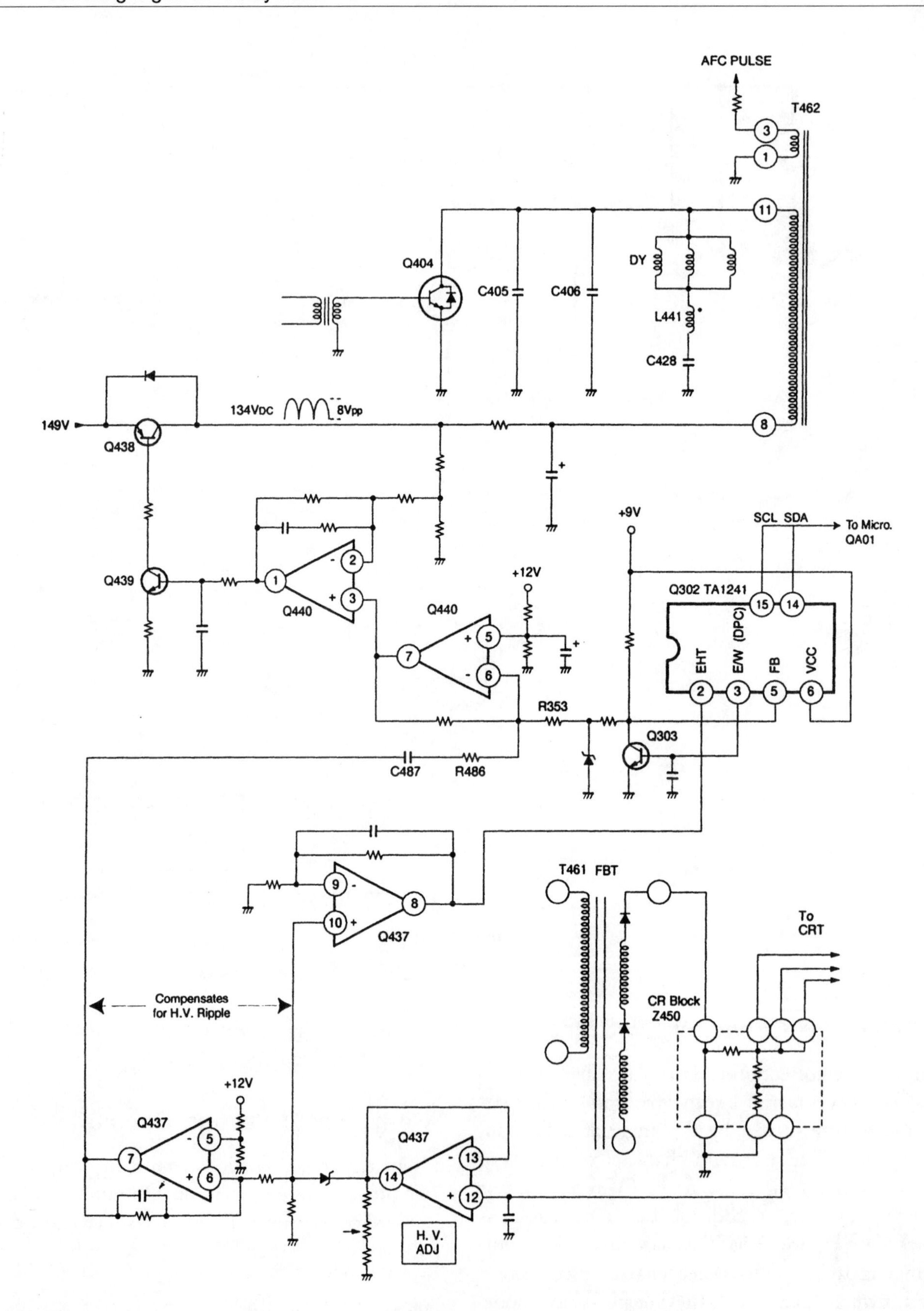

AFC PULSE
T462
Q404
C405
C406
DY
L441
C428
149V
Q438
134VDC
8Vpp
Q439
Q440
+12V
+9V
SCL SDA
To Micro. QA01
Q302 TA1241
EHT
E/W (DPC)
FB
VCC
R353
Q303
C487
R486
Q437
T461 FBT
To CRT
CR Block Z450
Compensates for H.V. Ripple
H. V. ADJ

FIGURE 5-6

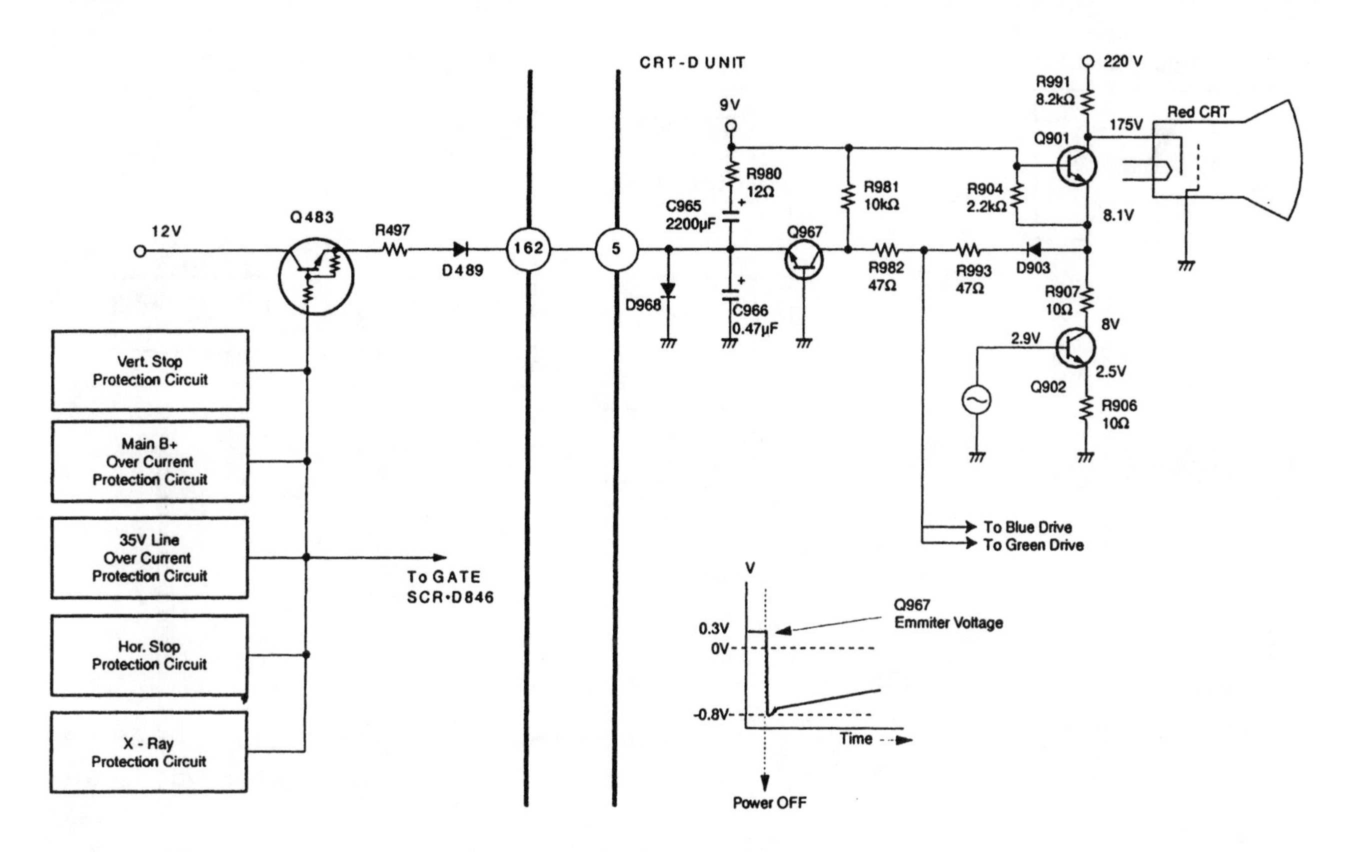

FIGURE 5-7

Dynamic Pincushion Correction (DPC). In the **Figure 5-6** circuit, the DPC is processed in Q302. As in conventional models, the DPC modulates the horizontal output signal for the pincushion adjustment. In the progressively-scanned television, the DPC modulates the 149 volt DC Main B+ with a parabolic waveform.

Signal Flow

The DPC parabolic waveform is obtained from pin 3 of Q302, applied to Q303, and then to Q440 at pin 6. The inverted output from pin 7 is applied to Q440, pin 3. The output of Q440 pin 1 is applied to Q439 and Q436, which modulate the Main B+ going to Q404.

Adjustments for DPC

On the collector of Q404 is 134 volts DC, modulated with an 8-volt peak-to-peak parabolic waveform. The width (WID) and dynamic pincushion (PARA) controls are adjusted in the service mode. The WID adjustment controls Q438 collector voltage, which controls the picture. The PARA adjustment controls the peak-to-peak value of the parabolic waveform on the collector of Q438. This adjustment affects the East-West (E-W) bow or pincushion effect in the picture.

High-Voltage Ripple Compensation

Also applied to pin 6 of Q440 is a feedback voltage from the CR block. This circuit compensates for any high-voltage ripple. The signal coming off of the CD Block is amplified and fed to the DPC circuit on pin 6 of Q440 to compensate the width of the picture. This signal is also fed to pin 2 (EHT) of Q302 to compensate for picture height.

CRT Spot Killer for CRT Protection

The CRTs hold a charge after the TV set is turned OFF. Thus, electron emissions are still present inside the CRTs. Because there is no deflection, a bright spot would appear in the CRT's center. This spot would burn the phosphor, damaging the CRTs. The spot-killer circuits, shown in **Figure 5-7**, eliminate this problem by providing a way to discharge the CRT before all deflection sweeps stop. When

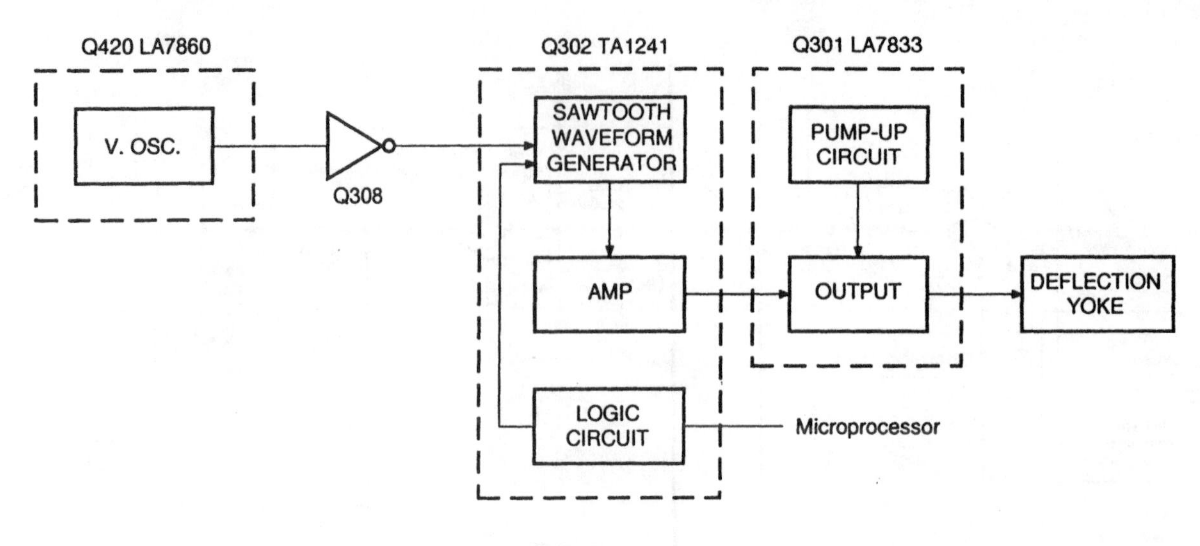

FIGURE 5-8

the television in ON, and Q967 is OFF, C966 and C965 are fully charged. The charge on C966 keeps a positive voltage on the emitter of Q967, maintaining a reverse bias. D968 is a clamp for the emitter voltage, keeping it at 0.3 Volt. When the 9-Volt Vdd drops at turn OFF, C965 discharges and applies -0.8 Volt on the emitter of Q967. The -0.8 Volt on the emitter turns Q967 ON. When Q967 turns ON, the CRTs discharge to ground through the CRT drive transistor and Q967.

Because the cathode voltage decreases when the spot killer engages, the electron beam intensifies. If vertical or horizontal deflection were not present at the time, the CRT's phosphors would burn. For this reason, Q483 is added. Q483 inhibits the spot killer when the shutdown circuit engages. The shutdown circuit monitors the horizontal and vertical deflection. If a loss of deflection occurs, the shutdown engages. Q483 then turns ON and applies a positive voltage to the emitter of Q967, keeping it OFF. At this time, the CRTs are protected by the vertical and horizontal stop protection circuits. The shutdown also monitors several power supply sources.

Vertical Sweep and Output Circuitry

The vertical deflection block diagram is shown in **Figure 5-8**. The vertical oscillator circuit is located within Q420. The sawtooth waveform generator, logic circuit, and an amplifier are located internal to Q302 (TA1241). Transistor Q301 (LA7833) contains the pump-up circuit and the final output.

Theory of Operation

The purpose of the vertical output circuit is to provide sawtooth current with good linearity through the deflection yoke.

Refer to **Figure 5-9** for vertical output discussion. When switch "S" closes, C2 charges to VP voltage. This is the first part of the sawtooth waveform as shown in **Figure 5-9**a. When the switch opens, C2 discharges through resistor Rl.

The value of Cl and R1 determine the discharge rate. The discharge rate, being slower than the charging rate, creates the second part of the sawtooth waveform.

The differential amplifier amplifies the signal and applies it to the deflection yoke. The negative feedback through resistor R2 controls the gain of the amplifier. The actual vertical deflection circuit is shown in **Figure 5-10**.

Sawtooth Waveform Generation

The block diagram for the ramp generator is shown in **Figure 5-11**. The vertical deflection pulses from Q420 are applied to pin 21 of IC 302. The waveform shape block is a differentiator (high-pass filter). It ignores the pulse width of the incoming signal to prevent the width of the pulse from being a factor.

The signal is then applied to a trigger detection block and goes on to the pulse width block. The signal is applied to the vertical ramp generator, which creates the vertical ramp. The microproces-

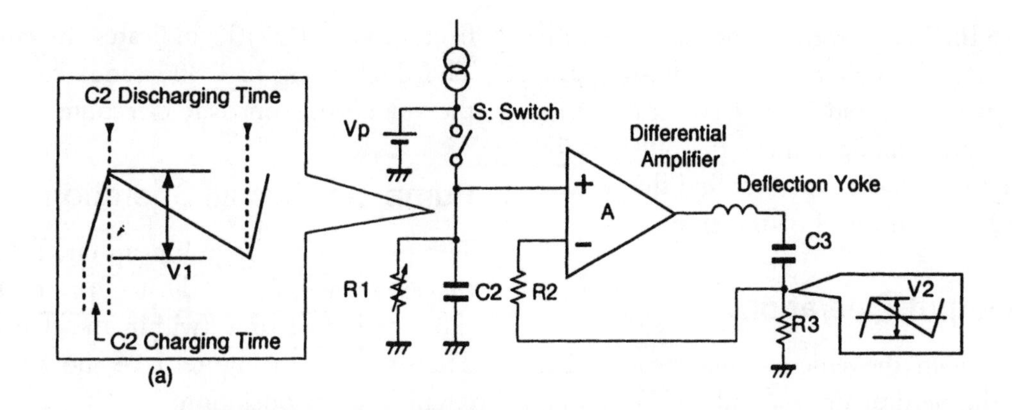

FIGURE 5-9

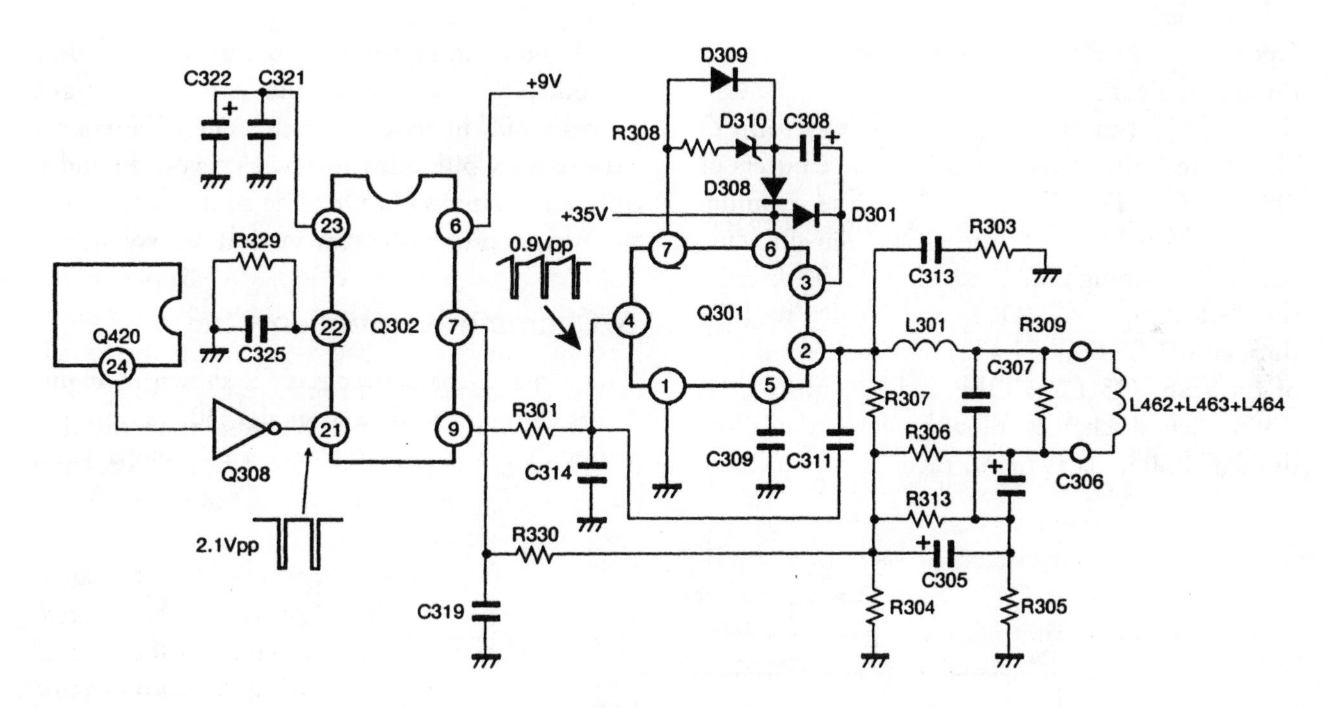

FIGURE 5-10

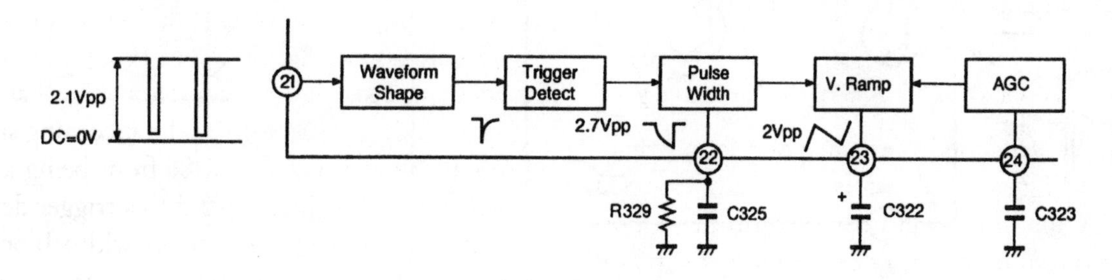

FIGURE 5-11

sor controls the pulse width block and vertical ramp generator through the bus. Adjusting the HIT (vertical height) and VLTN (vertical linearity), in the service mode, adjusts the pulse width and vertical ramp blocks. You will find the vertical output of Q302 on pin 9 of the chip.

Vertical Output Operation

Chip Q301 contains the vertical output circuit, which consists of the vertical drivers and the pump-up circuit. Refer to **Figure 5-12** for the internal block diagram of Q301. Q2 amplifies the input from pin 4 and applies it to Q3. Q3 and Q4 amplify the current through the vertical deflection yoke.

The basic circuits and waveforms are shown in **Figure 5-13**. As diagram b shows, the total current through the yoke waveform d is a total of the ON times of Q3 (waveform a) and Q4 (waveform c) Waveform e shows the voltages at the emitters of Q3 and Q4. For the first part of the scanning period, Q4 is OFF. Q3 turns ON, allowing current to flow through R305, to charge C306 through the deflection yoke (DY), and finally to flow through Q3 (indicated by I1). For the second part of the scan, Q3 turns OFF and Q4 turns ON. C306 then discharges allowing current to flow through R305, via Q4, and then through the deflection yoke (DY)(I2 indicates the current flow). The discharging of C306 sends current through the yoke in the opposite direction.

Pump-Up Circuit Operation

The purpose of the pump-up circuit is to return the electron beam to the top of the screen in a short period of time (within the flyback period). The drawings in **Figure 5-14** show the pump-up circuit and its operation.

During the scanning period, pin 7 switches to ground potential. As shown in **Figure 5-14a**, this allows current to flow through C308, building a charge across the capacitor. The voltage on pin 3 is at the 35-volt source voltage.

As noted in **Figure 5-14 b**, during the flyback period, pin 7 switches to the Vcc potential and places C308 in series with the Vcc. The voltage charge on C308 adds to the Vcc voltage and is applied to pin 3 of Q301. The combined voltages applied to pin 3 is approximately 60 volts.

S-Character Correction Circuit

The actual S-character circuit is shown in **Figure 5-15**. Using the feedback signal, Q302 performs S-correction. A parabolic waveform develops across

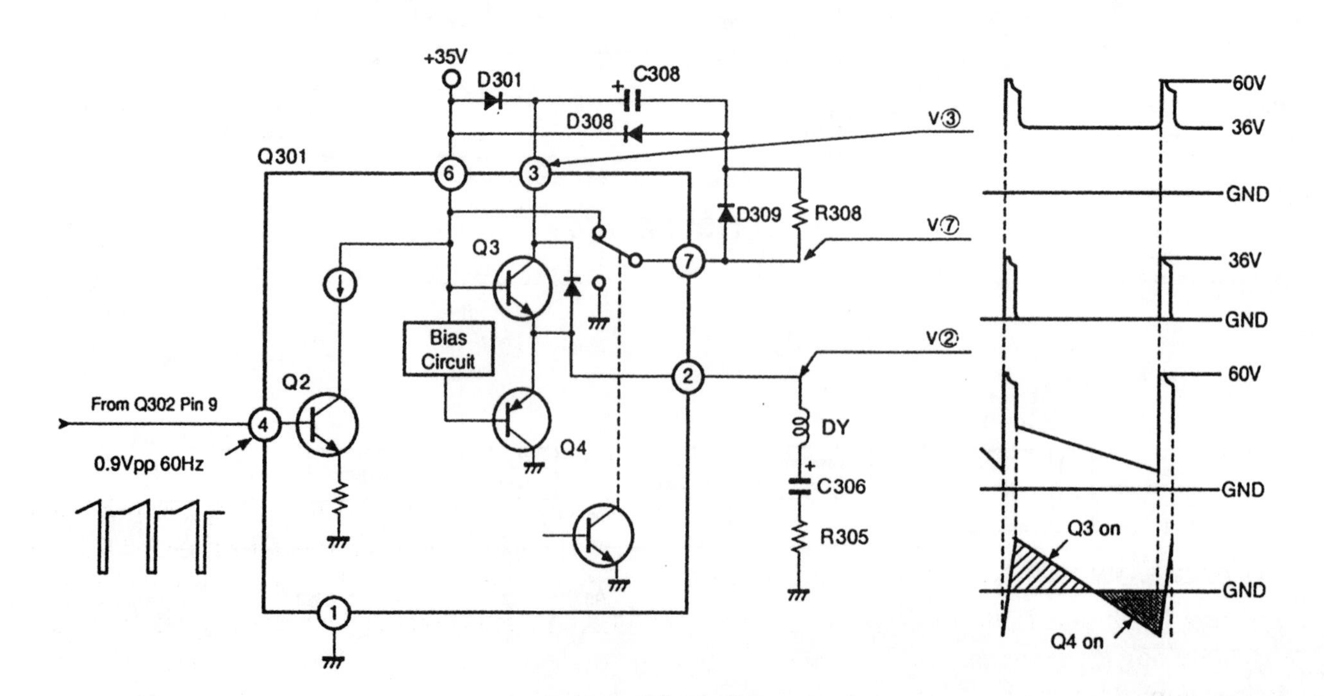

FIGURE 5-12

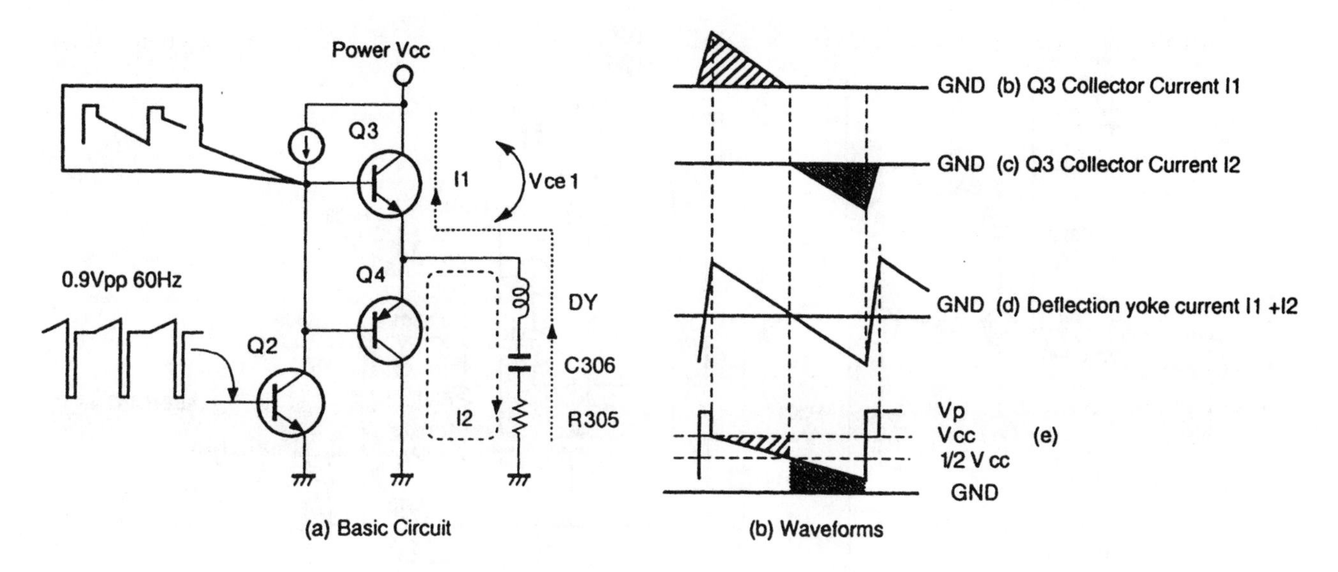

FIGURE 5-13 A and B

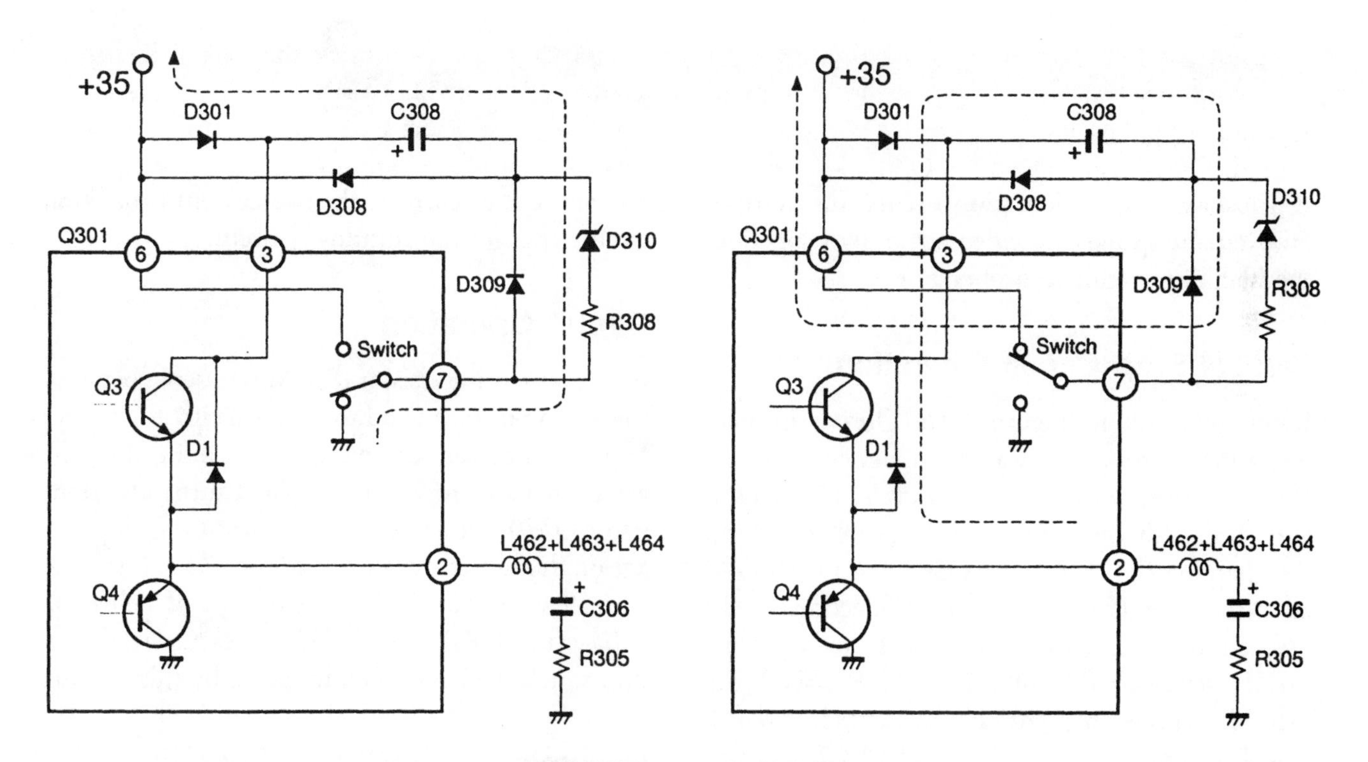

FIGURE 5-14 A and B

C306, and is integrated by R306 and C305, which is then applied to pin 7 of Q302.

Up and Downward Linearity Balance

Resistors R308 and R307 form a voltage divider that develops a voltage at pin 7 of Q301. This voltage improves linearity balance characteristics of the output stage.

Vertical Stop

When a loss of vertical deflection occurs, one bright horizontal line appears across the screen. This condition is a serious problem, especially with projection TV sets. Because the projection television picture tubes have no shadow masks, the bright horizontal line, caused by a loss of vertical sweep, burns the fluorescent screen and damages the CRT.

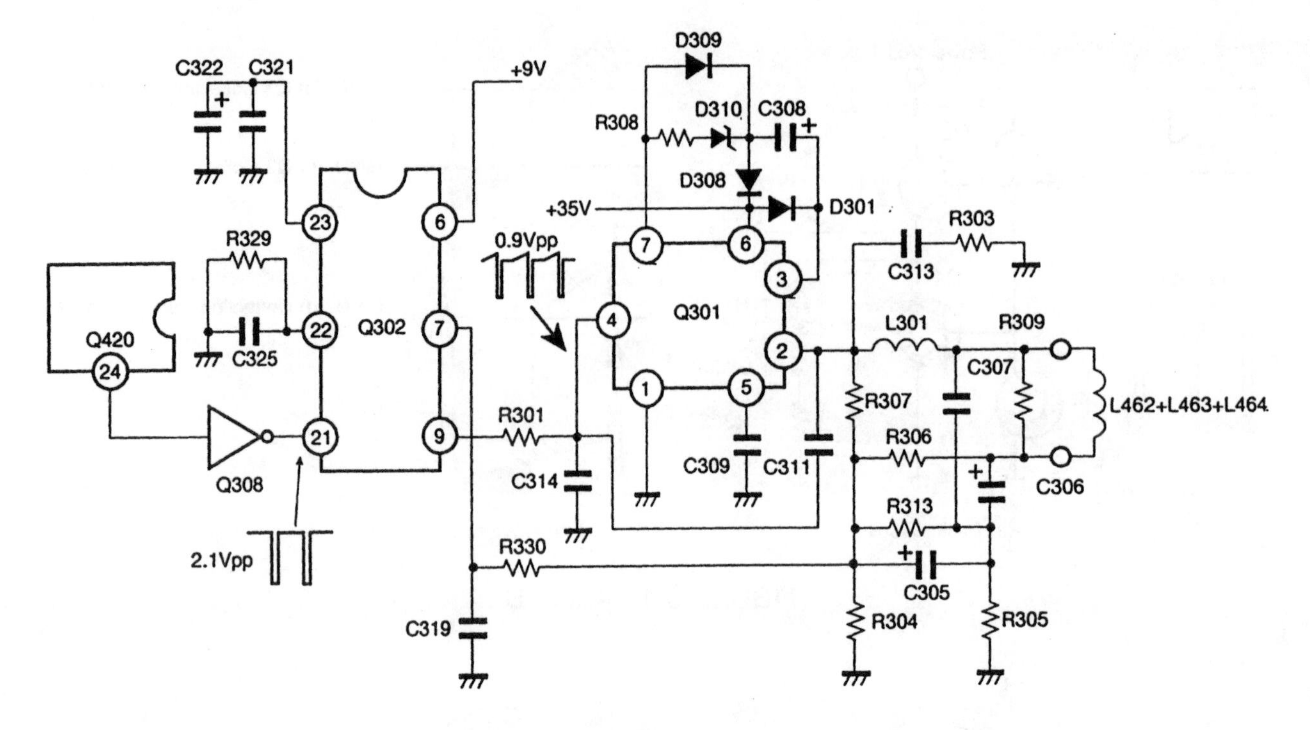

FIGURE 5-15

If a loss of vertical deflection occurs, the vertical Stop circuit blanks the video signal by triggering the blanking circuit into operation.

Vertical Stop Circuit Operation

Refer to circuit in **Figure 5-16**. During normal operation, a sawtooth waveform develops across R305. As the sawtooth cycles, it cycles Q384 ON and OFF. Although C384 is cycling ON and OFF, the charging of C388 keeps Q385 and Q386 ON all of the time. If a loss of vertical deflection develops, all three transistors turn OFF and a logic HIGH develops on collector Q386. This logic HIGH is applied to Q505 in the blanking circuit. The blanking circuit prevents any RGB output from Q501.

+35-Volt Overcurrent Protection Circuit

The +35-volt overcurrent protection circuit is shown in **Figure 5-17**. The vertical output is powered by a +35 volt supply from the main power supply. This supply is monitored by an overcurrent protection circuit. If the vertical output pulls an excessive amount of current, the overcurrent protection circuit engages the shutdown circuit.

Circuit Operation

A current sensing resistor (R370) is in series between the +35 Volt source and the vertical circuit IC chip. If the load current exceeds a certain limit, the voltage drop across R370 increases and turns ON transistor Q370. Otherwise, Q370 turns ON, the voltage on the collector rises toward the +35 volt supply. The Zener diode (D370) conducts and delivers a voltage to SCR D846. D846 places the television into shutdown code when triggered by this voltage.

Dynamic Focus Circuit Operation

The parabolic waveform that is applied to the Focus Pack (Z410) is developed from two separate parabolic waveforms: the vertical parabolic and horizontal parabolic.

Refer to **Figure 5-18** for focus circuit operation. The first parabolic waveform develops from the negative feedback from the vertical output. The signal at R305 is applied to pin 6 of Q306. The first operational amplifier, internal to Q306,

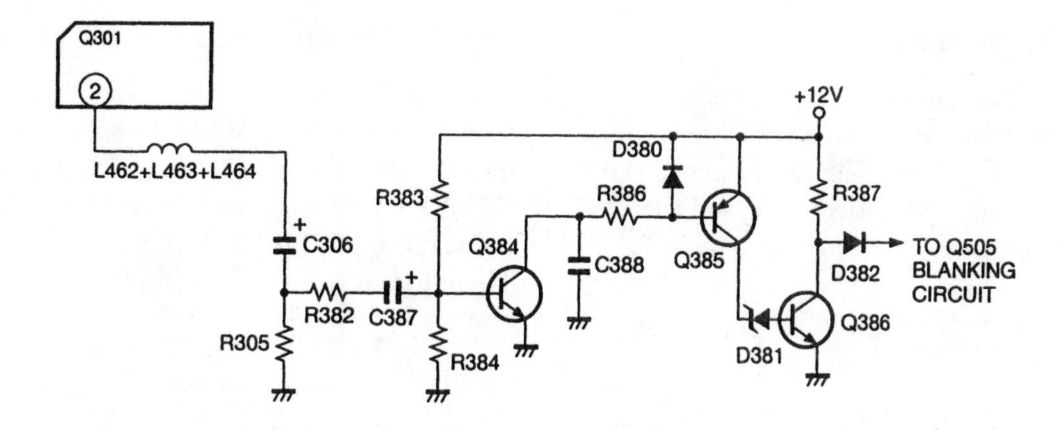

FIGURE 5-16

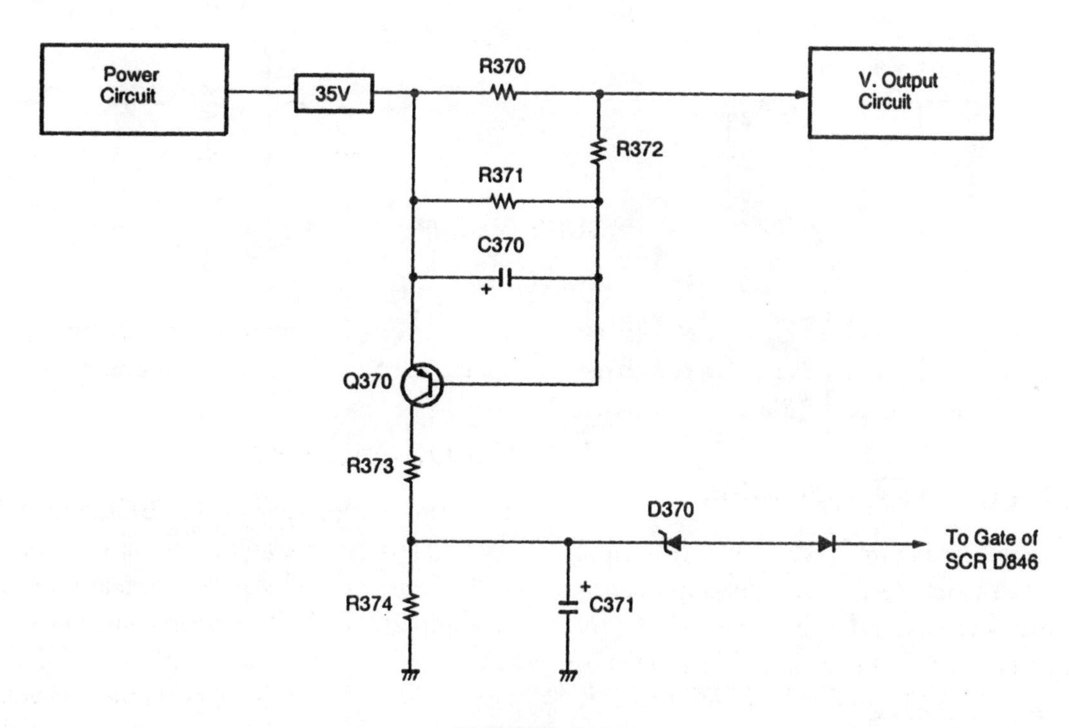

FIGURE 5-17

amplifies and wave-shapes the signal, and then outputs a parabolic signal on pin 7. The signal is applied to a second amplifier in Q306 on pin 2. This amplifier inverts the signal and outputs it on pin 1. The signal at pin 1 is then amplified by Q307 and integrated into the horizontal parabolic waveform in T411. Z470 is a voltage rectifier and filter that produces a DC bias voltage for Q370.

The horizontal parabolic wavefcrm is taken from the S-character capacitors (C411 and C412). The signal is coupled and integrated to the vertical-parabolic signal by the step-up transformer T411. When in the Theater-Wide mode, SR41 closes and introduces additional capacitance into the circuit. This capacitance reduces the peak-to-peak value of the horizontal parabolic waveform.

The combined parabolic waveform is applied to each of the CRTs through the focus pack (Z410). Each focus adjustment is coupled to the parabolic waveform by a 330 pf capacitor. The focus adjustments are voltage dividers that control the amplitude of the waveform applied to the focus controls.

Digital Convergence

The convergence circuit in a projection TV receiver function aligns all three CRTs to produce one sharp color picture. The digital convergence system is more accurate, easier to use, and takes less space than the analog convergence system.

Digital Convergence Circuit Operation

Refer to the digital convergence block diagram shown in **Figure 5-18**. When the television is turned on, RH009 and CH010 provide the reset to QH001. Horizontal (31.5 kHz) and vertical sync pulses are applied to QH00L for synchronization. The horizontal sync is also applied to the PLL to create a basic clock for QH001.

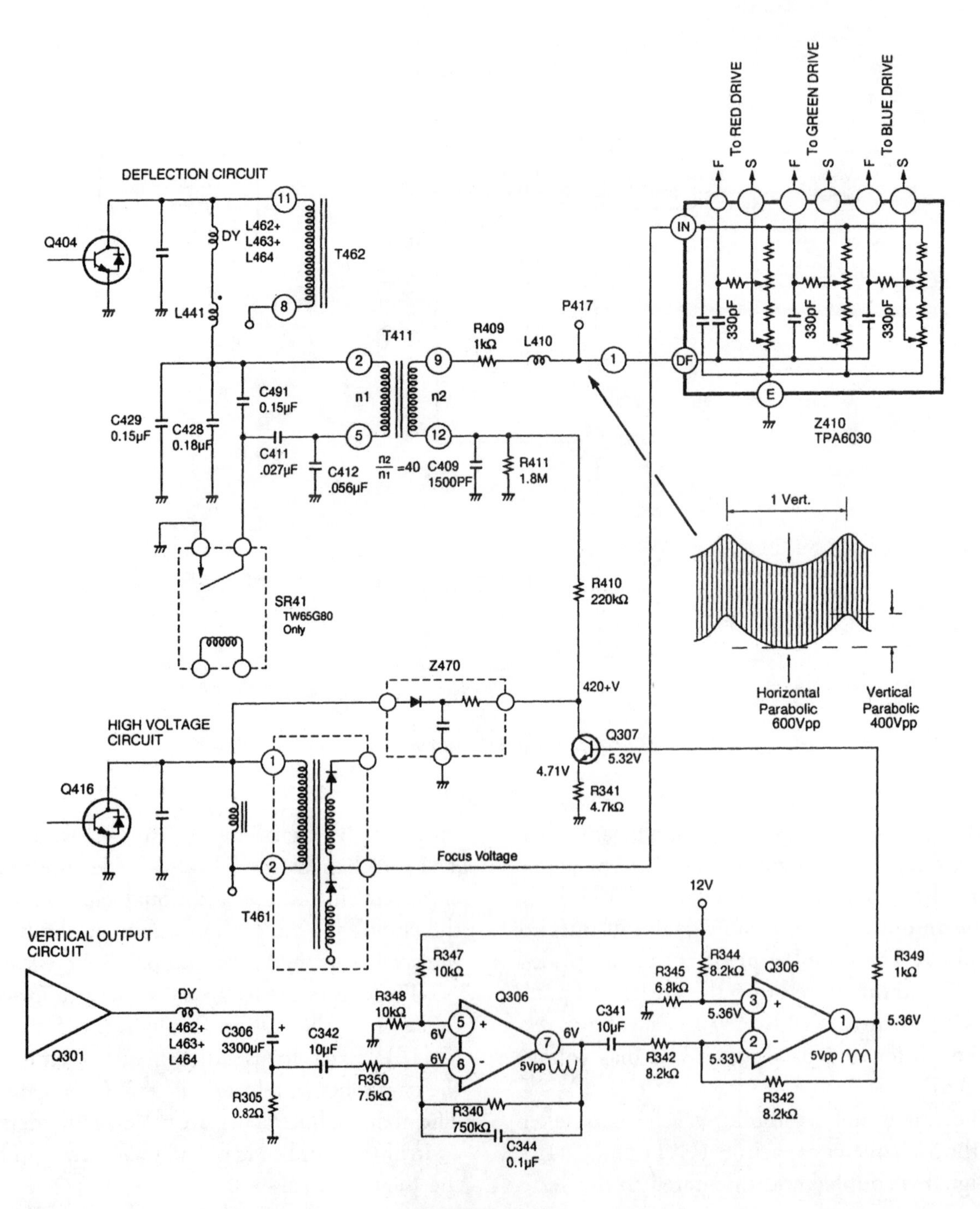

FIGURE 5-18

Controlled by the microprocessor, QH00L reads or writes information from the memory in QH174, develops a digital signal, and performs serial data transfer to the D/A converters. The D/A converters convert the digital information into analog signals. The analog signals are wave-shaped and amplified by QH170, QH171, and QH172. Lastly, the analog signals are smoothed out by filters and outputted to the convergence output drivers.

CONVERGENCE CONTROLS

The convergence adjustments operate when the set is placed in the service mode. The flow chart for entering the convergence service mode is illustrated in **Figure 5-20**.

Once in the service mode, pushing the No. 7 button on the remote control brings up the convergence crosshatch, the register information disappears, and a blinking cursor appears. The cursor

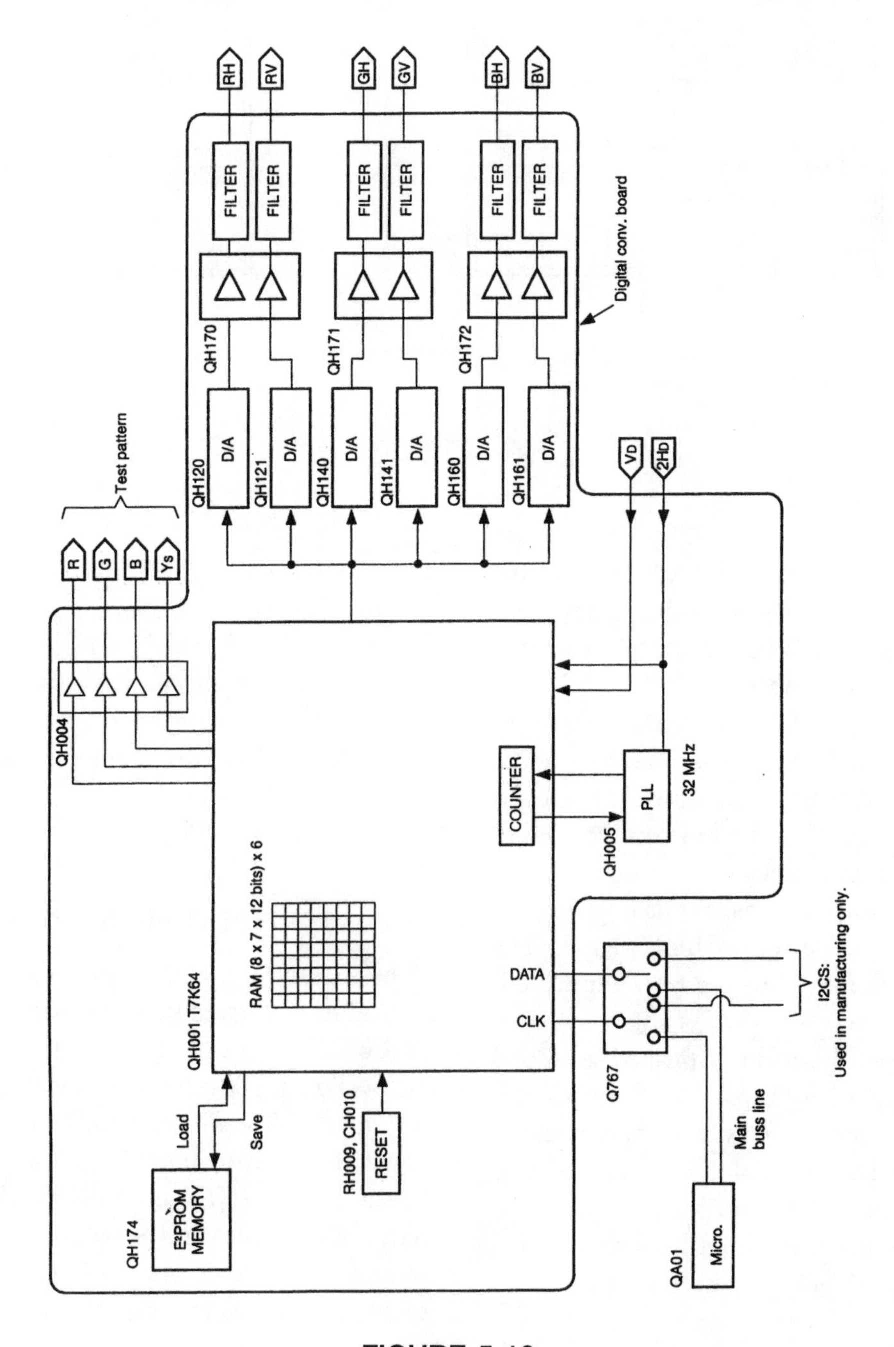

FIGURE 5-19

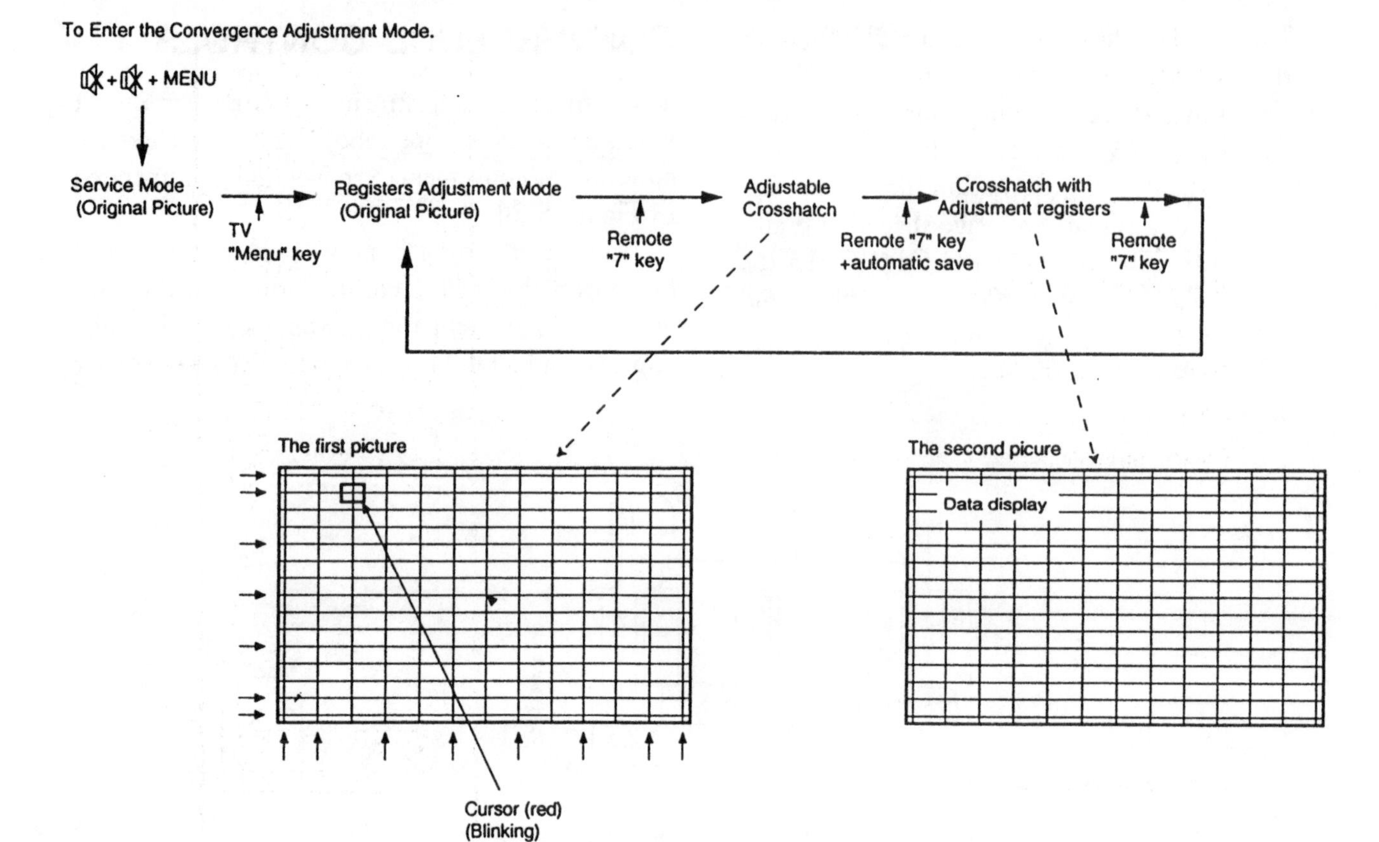

FIGURE 5-20

selects the adjustment point. The color of the cursor indicates the color of the CRT being adjusted. The color of the cursor changes by pressing the No. 3 button on the remote.

The cursor moves the crosshatch pattern by pushing the No. 2, 6, 8, and 4 buttons on the remote control unit. Once the adjustment point is selected, the No. 2, 6, 8, and 4 buttons are used to adjust the crosshatch pattern.

After the adjustment is made, pushing the No. 5 button causes the cursor to blink again. The cursor can then be moved to the next adjustment point.

When the No. 7 button is pushed a second time, the adjustment information is stored into memory, and the service register reappears; however, the crosshatch remains on the screen. Refer to the flowchart outline in **Figure 5-20**. Pushing the No. 7 button a third time brings back the normal picture, with the TV set in the service mode.

The remote control functions for digital convergence are shown in **Figure 5-21**. In addition to the function previously covered, the crosshatch patterns may be turned OFF independently. The No. 100 button toggles the red crosshatch OFF and ON; the 0 button toggles the green crosshatch OFF and ON; and the ENT button toggles the blue crosshatch OFF and ON.

Convergence Adjustment Procedure

When converging all three CRTs, set all the customer's convergence controls — for both red and blue — to the center of their ranges. Using an overlay for a reference, set the overall geometry of the Green crosshatch with the distortion correction adjustments in the service mode. Refer to table 5-1. Then, converge the green crosshatch to the overlay. Start in the middle of the picture and work outward. Once the green is converged, converge the other colors to the green

(1) 100 key

Red test pattern ON/OFF

(2) 0 key

Green test pattern ON/OFF

(3) ENT key

Blue test pattern ON/OFF

(4) 7 key

Mode picture change-over

(5) 5 key

Cursor shift/data change mode change over

(6) 8 key

Cursor down/adjusting point down

(7) 2 key

Cursor up/adjustment point up

(8) 6 key

Cursor right/adjustment point right

(9) 4 key

Cursor left/adjustment point left

(10) 3 key

Cursor color change

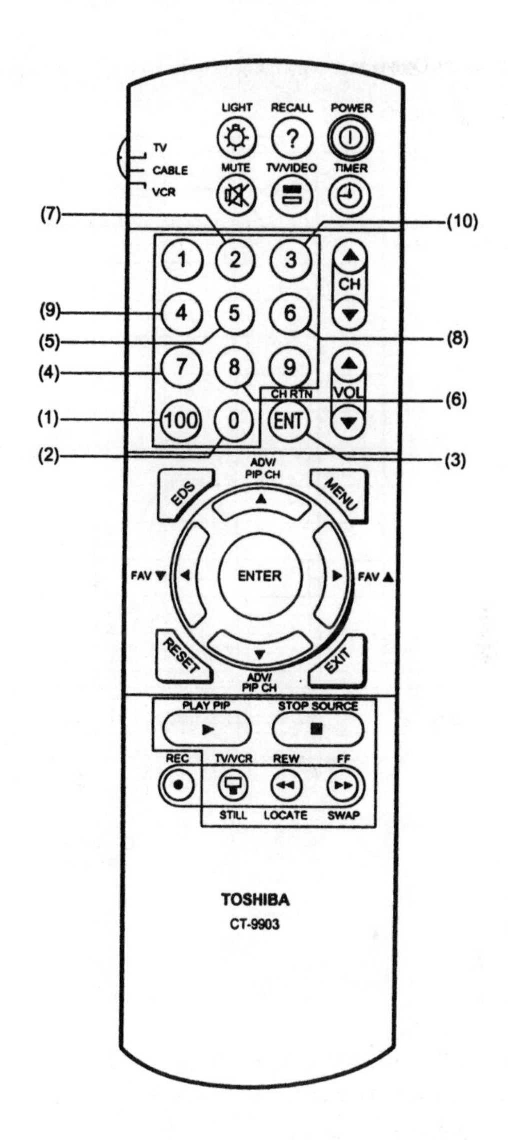

FIGURE 5-21

crosshatch. When aligning a crosshatch, turn off the unnecessary crosshatch(es). For example, when converging the red crosshatch to the green crosshatch, turn off the blue crosshatch. The screen geometry pattern is shown in the (**Fig. 5-22**) drawing.

HDTV Wide-Screen Convergence

The convergence for HDTV wide screen is very similar to the standard 4:3 television. The only difference is that there are three additional picture sizes, and each picture size is converged separately. The order of this convergence is illustrated in **Figure 5-23**. As shown in **Figures 5-24** and **5-25**, the full picture is converged first, wide 1 second, wide 2 third, and wide 3 is coverged last. Pressing the picture-size button toggles through the different screen sizes. The screen geometry patterns are shown in **Figures 5-26** and **5-27**.

Convergence Output Circuit

The convergence output circuit amplifies the convergence correction signal and drives the convergence yokes.

The output circuit consists of two main driver ICs: Q751 and Q752. Q751 amplifies the vertical correction for all three CRTs (V-R/V-G/V-B). Q752 amplifies the horizontal convergence correction for the CRTs (H-R/H-G/H-B).

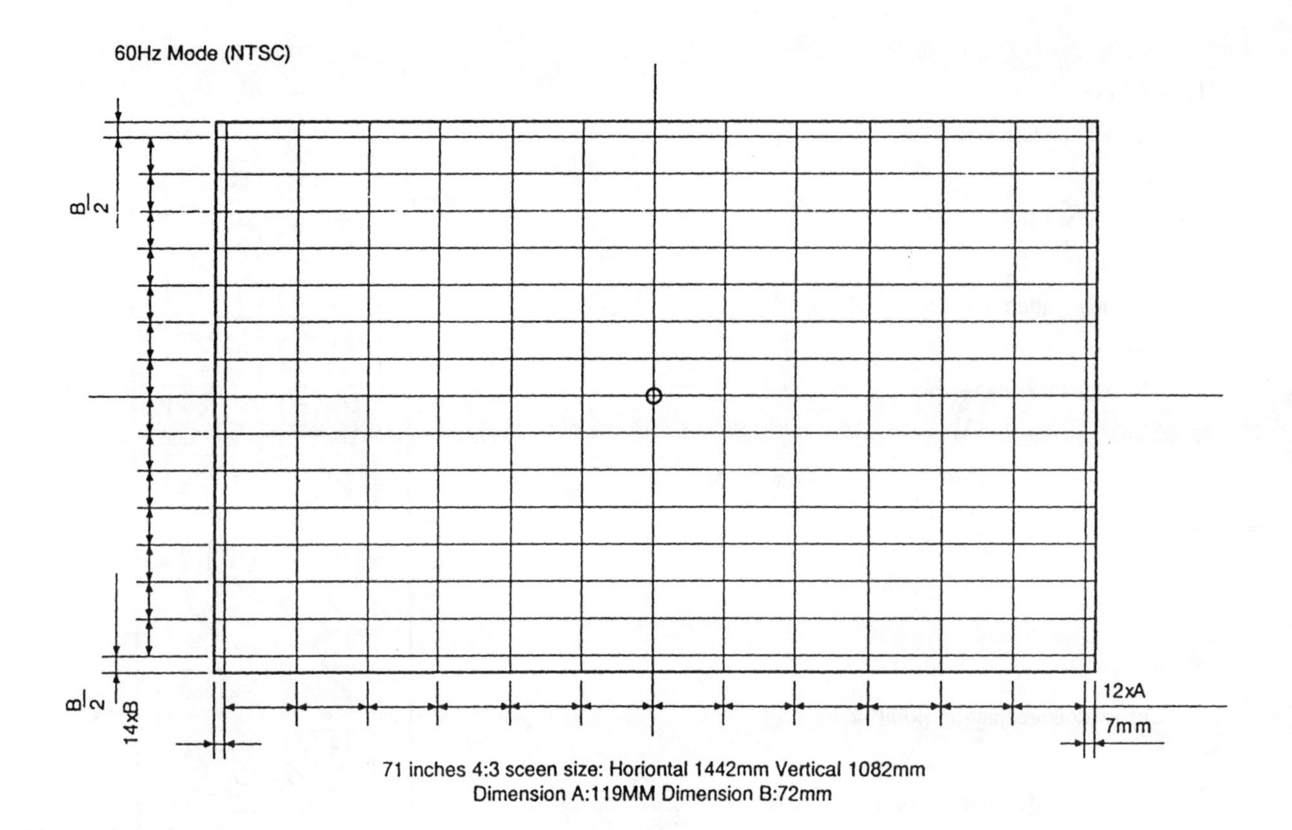

FIGURE 5-22

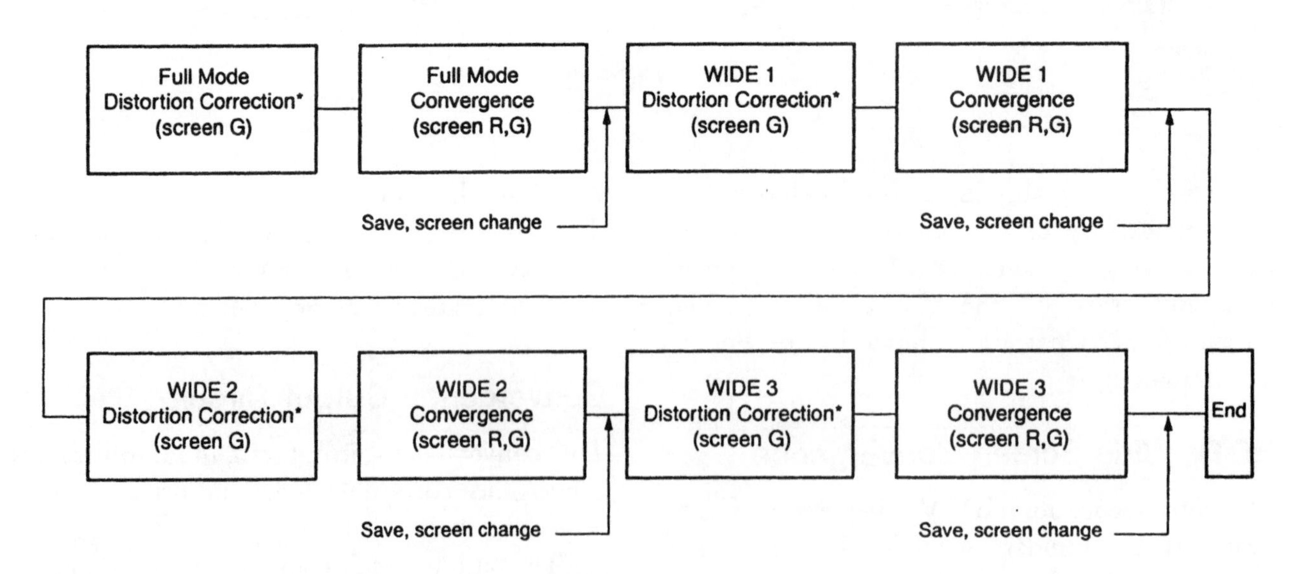

FIGURE 5-23

Dimensions of Each Picture Size for the 16:9 Television

65 inches 16:9 screen size:

- Horizontal 1442 mm
- Vertical 810 mm

(1) FULL

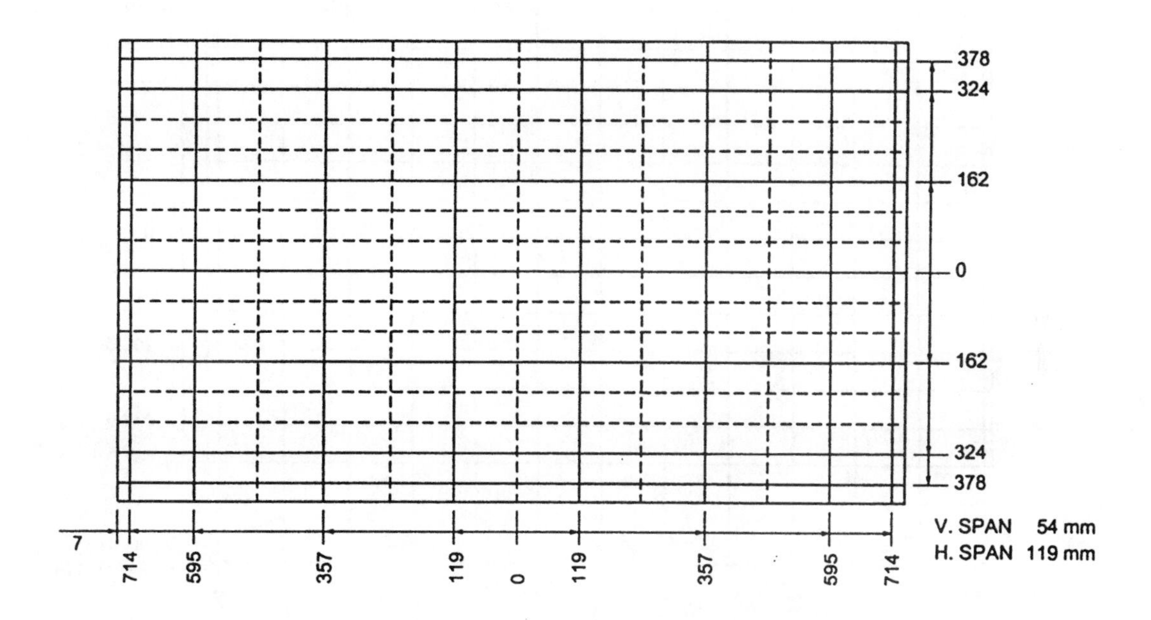

FIGURE 5-24

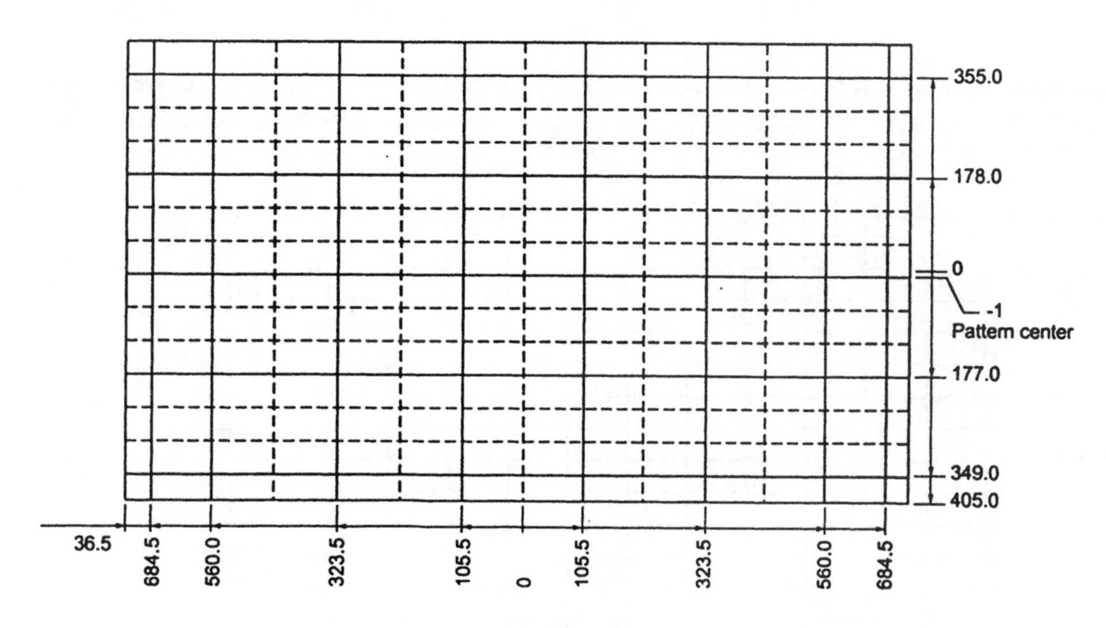

FIGURE 5-25

Note:

- In this mode, the cursor may be located out of the viewing area. Therefore, adjust with care.

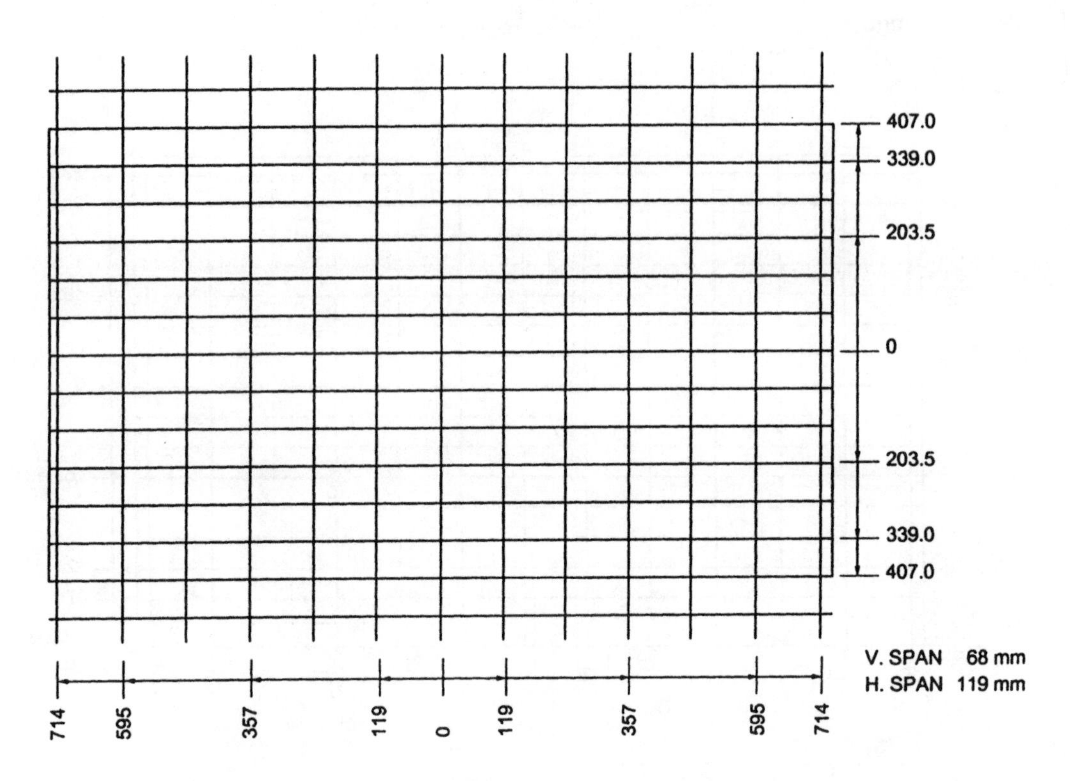

FIGURE 5-26

Note:

- In this mode, the cursor may be located out of the viewing area. Therefore, adjust with care.
- On this screen, convergence pattern center is displayed on screen center, but picture center is indicated at 22 mm up on screen.

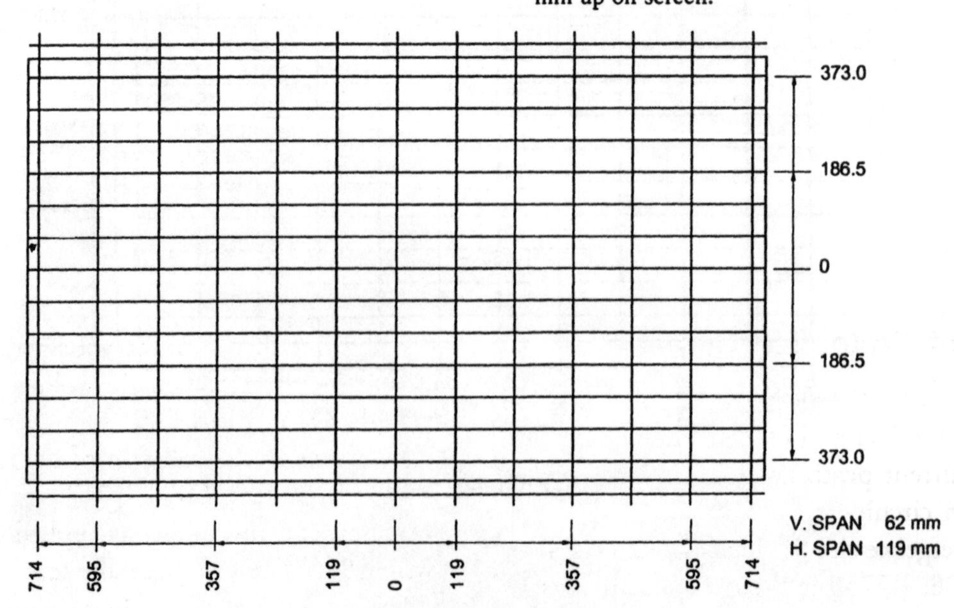

FIGURE 5-27

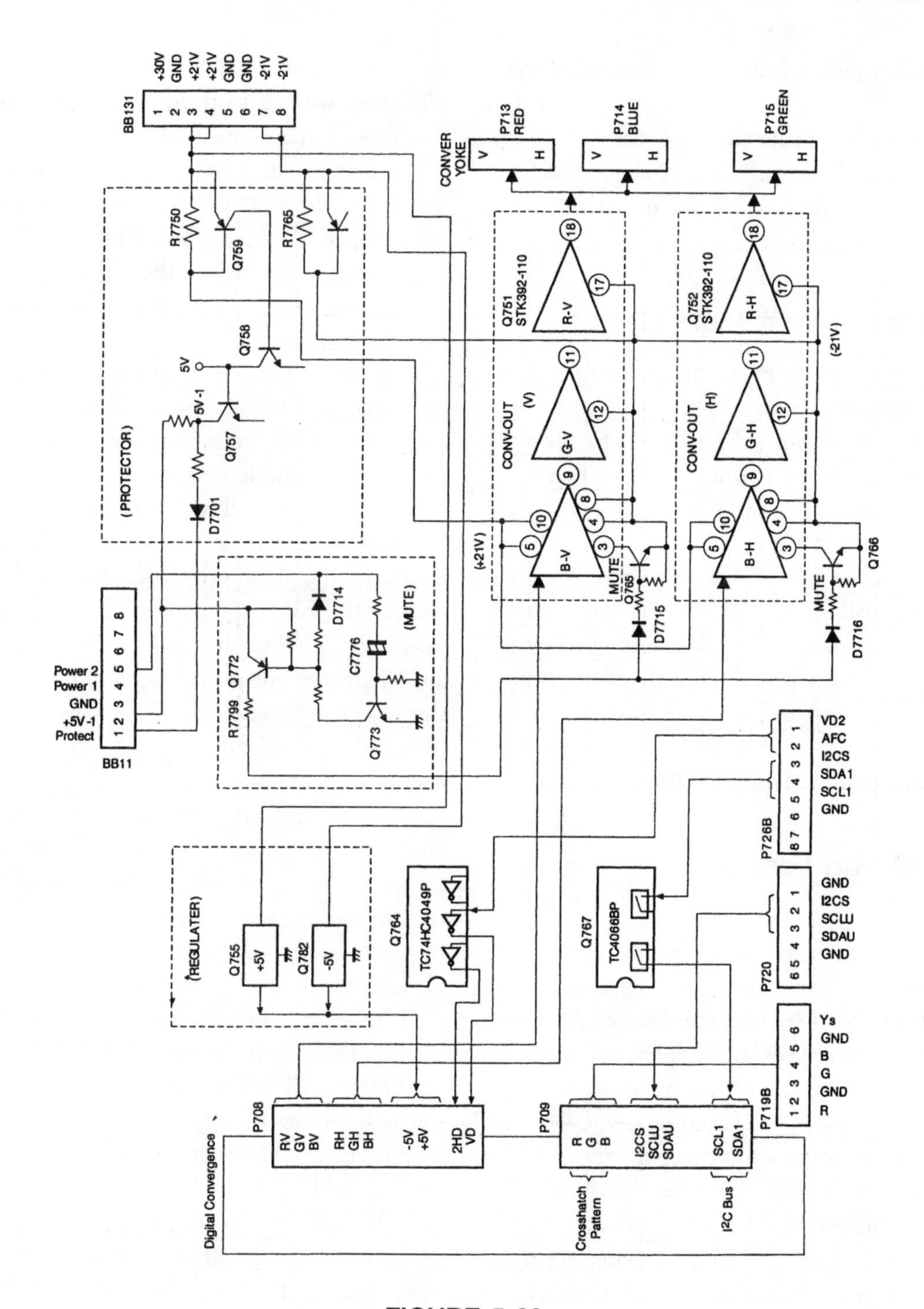

FIGURE 5-28

Overcurrent Protection Circuit

The -21 volt and +21 volt power supplies provide voltage to both Q751 and Q-752. Each power line has an overcurrent protection circuit that triggers the shutdown circuit.

The convergence circuit block diagram is shown in **Figure 5-28**. R7750 is the overcurrent sensing resistor for the +21 volt line. If the current on the +21 volt line increases, the voltage drop across the resistor increases. When the voltage drop is large enough, Q759 turns ON. This turns ON Q758. The voltage on the collector of Q758 drops, causing Q757 to turn OFF. Q757's collector voltage increases, and the voltage is applied to the gate of SCR D846. The voltage on the gate of D846 causes D846 to conduct, thus disengaging relays SR81 and SR83. For more information on shutdown circuit operation refer to chapter eleven on power supplies.

The overcurrent protection for the -21 volt line operates in the same way. The components for the -21 volt line are resistor R7765 and transistors Q762, Q761, and Q757.

Convergence Output Mute Circuit

The convergence output mute circuit disables the cutputs when the television is OFF, during power-up, and during shutdown. The +5 volts are always present when the TV set is plugged in. When the television is OFF, Q772, Q766, and Q765 are turned ON, Q773 is OFF, and the mute is engaged. When power is first applied, there are 4.9 volts at pin 5 of BB11, which turns on Q77 maintaining the on state of Q772, while charging C7776. Once C7776 is fully charged, all four of the transistors turn OFF and the mute disengages. The convergence circuit troubleshooting flow chart drawings are shown in **Figure 5-29**.

Troubleshooting Vertical Sweep Circuits

Television vertical sweep circuits appear to be the most difficult ones to troubleshoot. It seems that even a small change in component values can cause vertical deflection problems such as picture foldover, reduced deflection, and non-linear sweep. And, the same symptoms can be caused by an expensive deflection yoke or a low-cost component.

In this section we will examine the theory of vertical deflection and find out why they are tough to troubleshoot and repair. The following vertical sweep circuit operation can be used for conventional and HDTV receiver service techniques.

VERTICAL YOKE AND CRT BEAM DEFLECTION

For you to see how the vertical sweep stages operate, an understanding of the CRT (picture tube) beam deflection is required. Inside the CRT, an electron gun emits a stream of electrons. These electrons travel to the CRT face and strike the phosphor surface to produce light.

Should the stream of electrons travel to the CRT without any magnetic or electrostatic field influence, the electrons will only strike the center of the CRT and produce a white dot. To cause the dot to make lines across the CRT screen, the electron beam must be influenced by an electrostatic or magnetic beam.

For television sets, a magnetic field is produced by the coils of a yoke mounted around the neck of the CRT. The yoke is made with coils wound around a magnetic core material.

When current flows in the vertical yoke coils, a magnetic field is produced. The yoke's core concentrates the magnetic field inward through the neck of the CRT. As the electrons pass through the magnetic field on the way to the CRT face, they are "deflected" or pulled upward or downward by the yoke's magnetic field. This causes the electrons to strike the CRT face at points above or below the center.

To see how electrons are "deflected," let's review the interaction of magnetic fields as you refer to the drawings in **Figure 5-30**. You may recall that an individual electron in motion is surrounded by a magnetic field. The magnetic field is in a circular motion surrounding the electron. As electrons travel through the magnetic field of the yoke, the magnetic fields interact. Magnetic lines of force in the same direction create a stronger field while magnetic lines in opposite directions produce a weaker field. The electron is pulled toward the weaker field.

The direction of the current in the yoke coil determines the polarity of the yoke's magnetic field and whether the electron beam is deflected upward or downward. Current flow in the direction shown in **Figure 5-30a** causes the electron beam to deflect upward. Current flow in the opposite direction through the yoke coils (**Figure 5-30b**) reverses the magnetic field, causing the beam to deflect downward.

How far the electrons are repelled when passing through the yoke's magnetic field is determined by the yoke and the level of current flowing in the vertical coils. The higher the current, the stronger the magnetic field and resulting electron deflection.

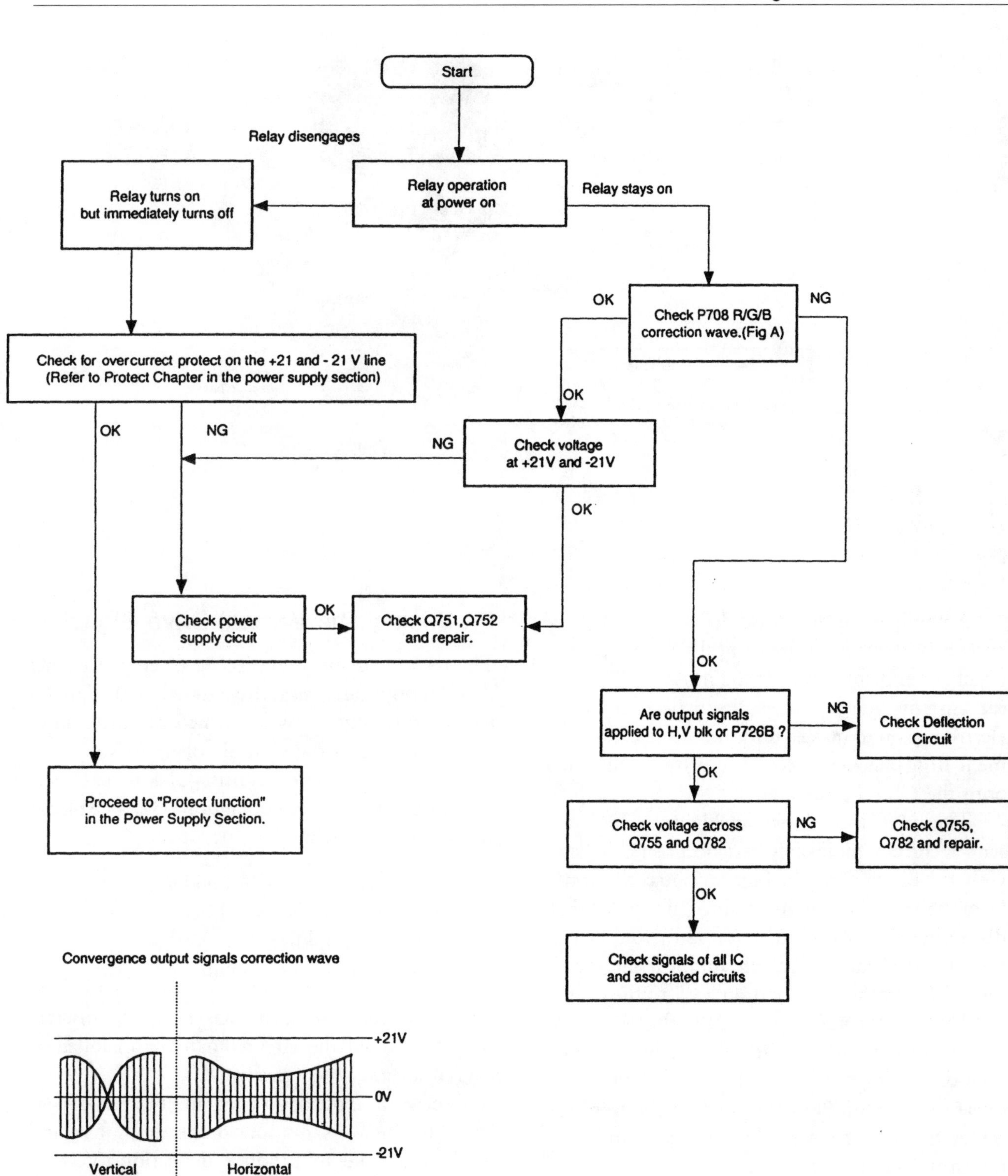

Fig A

FIGURE 5-29

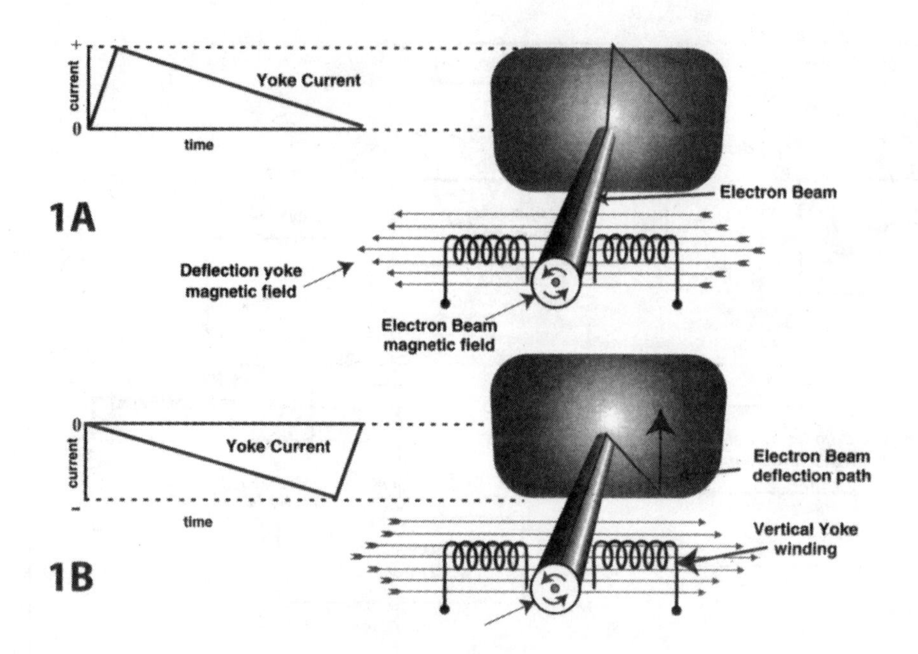

FIGURE 5-30

A requirement of vertical deflection is that the current in the coils of the vertical yoke increase an equal amount for specific time intervals. This linear current change causes the deflection of the electron beam to make a uniform or smooth movement from the top to center and from the center onto the CRT bottom.

The waveforms shown in **Figure 5-30** represent a current increasing and decreasing in level with respect to time. In **Figure 5-30a**, the waveform shows the current increasing quickly and then decreasing slowly back to zero. This would cause the electron beam to quickly jump to the top of the CRT and then slowly drop back to the center.

The waveform in **Figure 5-30b**, shows the current increasing slowly in the opposite direction and then decreasing quickly back to zero. This would cause the electron beam to slowly move from the center to the bottom of the CRT face and then return quickly to the center.

During normal operation, the yoke current increases and then decreases as illustrated in the (**Fig. 5-31** a and b) waveforms. The current changes directions, alternating between approximately 60 times per second (**Figure 5-30 a and b**.) The alternating current moves the electron beam from the top of the CRT face to the bottom and then quickly back to the top.

Producing The Vertical Drive Signal

The vertical circuits in the TV receiver are the ones that develop the vertical drive signal. This signal is fed to an output amplifier, which produces alternating current in the vertical yoke.

The vertical section consists of four basic circuits or blocks as shown in **Figure 5-31**. The blocks contain the following circuits:

1. Oscillator or digital divider
2. Buffer/predriver amplifier
3. Driver amplifier
4. Vertical output amplifier

The circuitry for these stages may be discrete components on the circuit board or included as part of an integrated circuit (IC).

The vertical oscillator generates the vertical signal. The signal is output to the amplifiers and drives the yoke to produce deflection. Vertical oscillators may be freerunning circuit oscillators or more modern digital divider generators.

The output of a vertical oscillator must be a sawtooth-shaped waveform. A ramp generator is often used to shape the output waveform of a freerunning oscillator or digital divider. A ramp generator switches a transistor on and off, alternately charging and discharging a capacitor. When the transistor is off, the capacitor charges to the

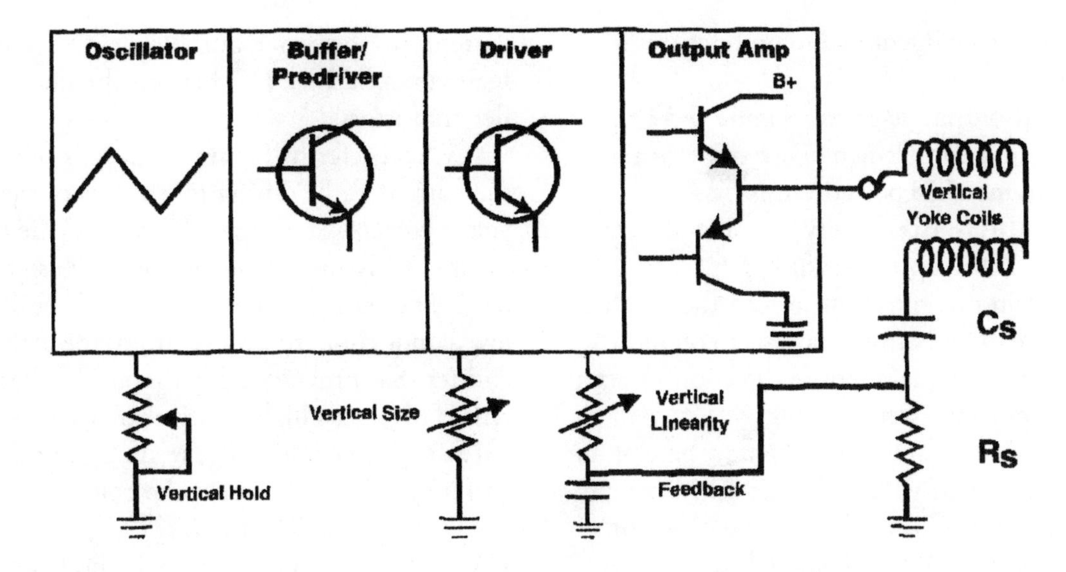

FIGURE 5-31

supply voltage via a resistor. When the transistor is switched on, the capacitor is discharged.

VERTICAL SYNC AND LOCK

The vertical oscillator must be synchronized with the video signal so a stationary picture can be viewed on the CRT screen. The oscillator frequency is controlled in two ways. A vertical hold control may be used to set the freerunning oscillator close to the vertical frequency. Vertical sync pulses, removed from the video signal, are applied to the vertical oscillator locking it to the proper frequency and phase. If the oscillator is not synched, the picture will roll vertically. The picture rolls upward when the oscillator frequency is too low, and downward when the frequency is too high.

There are several intermediate amplifier stages between the output of the vertical oscillator and output amplifier stage. Some common stages are the "buffer," "predriver," and/or "driver." The purpose of the buffer amplifier stage is to prevent loading of the oscillator, which may cause frequency instability or waveshape changes.

The predriver and/or driver stages shape and amplify the signal to provide sufficient base drive current to the output amplifier stage. The output stage transistor then produces the yoke current required for full and linear deflection.

The predriver and/or driver amplifier stages are usually DC coupled and use AC and DC feedback much like audio amplifier stages. Feedback maintains the proper DC bias and waveshape to insure the current drive to the yoke remains constant as components, temperature, and power supply voltages drift.

AC feedback in most vertical circuits is obtained by a voltage waveform derived from a resistor placed in series with the yoke. The feedback is often adjusted with gain-shaping controls called the Vertical Height, or Size, and Vertical Linearity controls.

DC feedback is used to stabilize the DC voltages in the vertical output amplifiers. DC voltage from the output amplifier stage is used as feedback to a previous amplifier stage. Any slight increase or decrease in the balance of the output amplifiers is offset by slightly changing the bias. Since the amplifiers are direct coupled, adjusting the bias on the output transistors brings the stage back into balance.

A lot of the difficulty in troubleshooting vertical stages is due to the feedback and DC coupling between stages. A problem in any amplifier stage, yoke, or its series component alters all the waveforms and/or DC voltages, making it difficult to trace the problem.

Developing Vertical Yoke Current

The vertical yoke may require up to 500 mA of alternating current to produce full deflection. A power output stage is required to produce this level of current.

A vertical output stage, see **Figure 5-32**, usually consists of a complementary, symmetrical circuit with two matched power transistors. The transistors conduct alternately in a push-pull arrangement. The top transistor conducts to produce current in one direction to scan the top half of the picture. The bottom transistor conducts to produce current in the opposite direction to scan the bottom portion of the picture.

Most vertical output stages are now part of an IC and are powered with a single positive supply voltage. The voltage is applied to the collector of the top transistor. In this balanced arrangement, the emitter junction of the transistors should measure approximately one-half of the supply voltage. In series with the vertical yoke coils is a large electrolytic capacitor. This capacitor passes the AC current to the yoke but blocks DC current to maintain a balance DC bias on the output amplifier transistors.

As you refer to **Figure 5-32**, let's now analyze the current paths during four cycles of the vertical yoke's current. Starting with time "A" the top transistor, Qt, is turned on by the drive signal at its base. The transistor is biased "ON" resulting in a low conduction resistance from the collector to emitter that provides a high level of collector current. This puts a high positive voltage (+ V) potential at the top of the yoke, resulting in a fast rising current in the deflection yoke coils.

During time "A" capacitor Cs charges toward + V and current flows through the yoke and the top transistor, Qt. This pulls the CRT's electron beam from the center of the CRT quickly to the top. During time "A" an oscilloscope connected at the emitter junction displays a voltage peak as shown

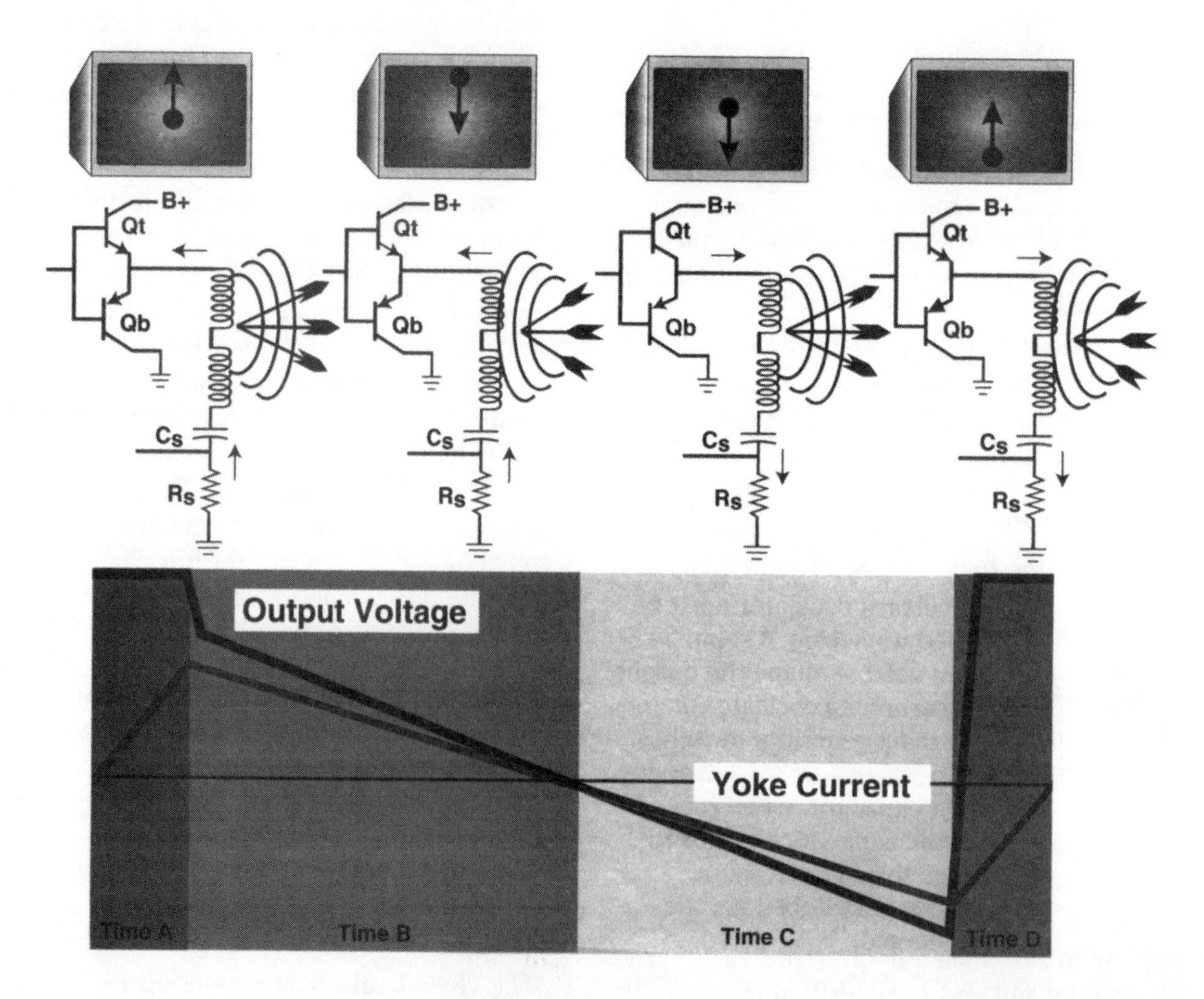

FIGURE 5-32

in the V output waveform. The inductive voltage from the fast-changing current in the yoke along with retrace "speedup" components cause the voltage peak to be higher than + V.

The current flowing in the yoke during time "A" produces a waveform as viewed from the bottom of the yoke to ground. This is the voltage drop across Rs that is representative of the current flowing through the yoke.

During time "B," the drive signal to Qt is reduced, slowly increasing the transistor's "emitter-to- collector resistance." Current in the yoke steadily decreases as the transistor's emitter-to-collector resistance increases, thus reducing its collector current. The voltage at the emitter junction falls during this time and capacitor Cs discharges. A decreasing current through the yoke causes the CRT's electron beam to move from the top to the center of the screen.

To produce a linear fall in current through the yoke during time "B," an accurate drive waveform is required at the base of Qt to meet the transistor's linear operating characteristics. The drive waveform must decrease the transistor's base current at a constant rate. Hence, the transistor must operate with linear base-to-collector current characteristics. Reductions in base current must result in proportional changes in collector current. The linear decrease in yoke current is shown by the I Yoke or Vrs waveform.

At the end of time "B," transistor Qt's emitter-to-collector resistance is high and the transistor is approaching the same emitter-to-collector resistance as the bottom transistor Qb. The capacitor Cs has been slowly discharging to the falling voltage at the emitter junction of the output transistors. Just as the voltage at the emitter junction approaches 1/2 + V, the bottom transistor begins to be biased "ON" to start time "C". This transition requires that the conduction of Qt and Qb, at this point, be balanced to avoid distortion in the center of the CRT.

During time "C," the resistance from the collector to emitter of transistor Qb is slowly decreased by the base drive signal and the collector current increased. Capacitor Cs begins to discharge, which produces current through the yoke and through Qb. As Qbls resistance decreases and its collector current increases, the voltage at the emitter junction decreases. This can be seen on the "V" output waveform as it goes from 1/2+v toward ground during time "C." Then the current will increase at a linear rate through the yoke.

The resistance decrease of Qb must be the mirror opposite of transistor Qt's during time "B." If not, the yoke current would be different in amplitude and/or rate rise time causing a difference in CRT beam deflection between the top trace and bottom trace. At the end of time "C," the emitter-to-collector resistance of Qb is low and the current in the yoke approaches a maximum level. At the start of time "D," the emitter-to-collector resistance of Qb is increased quickly and collector current decreased. This quickly slows the discharging current from capacitor Cs through the yoke and transistor. As the current is reduced, the trace is pulled quickly from the CRT bottom and back to center. Time "A" begins once again, and the cycle repeats.

Vertical IC Operations

The following vertical circuit chips are used in many conventional and HDTV receivers. The deflection yoke and other components external to the IC are required to work in conjunction with these vertical chips. Let's now see how one of these typical vertical ICs operate.

This vertical oscillator is a vertical countdown stage located within the set's signal processor IC as shown in the **Figure 5-33** circuit. The vertical countdown block receives an input from the horizontal oscillator, divides it by 262, and is gated by vertical sync. The output is a synchronized vertical pulse on pin 27. The vertical signal is buffered by a transistor and applied simultaneously to the microprocessor and character generator to synchronize on-screen graphics. The signal is also routed to pin 3 of IC550. IC550 contains a ramp generator, buffer amplifier, and the vertical amplifier stage.

The vertical signal to the ramp generator charges and discharges C558 producing a sawtooth or ramp signal at pin 7. The amplitude of the ramp signal is adjustable with the vertical size control. The lack of an input to the ramp generator would result in no output and no vertical yoke deflection.

The output of the ramp generator feeds the buffer amplifier and then the output amplifier at pin 9. DC voltage at pin 9, developed by the voltage divider networks, sets the bias to the amplifier and provides DC feedback to stabilize the gain of the amplifier and center the deflection.

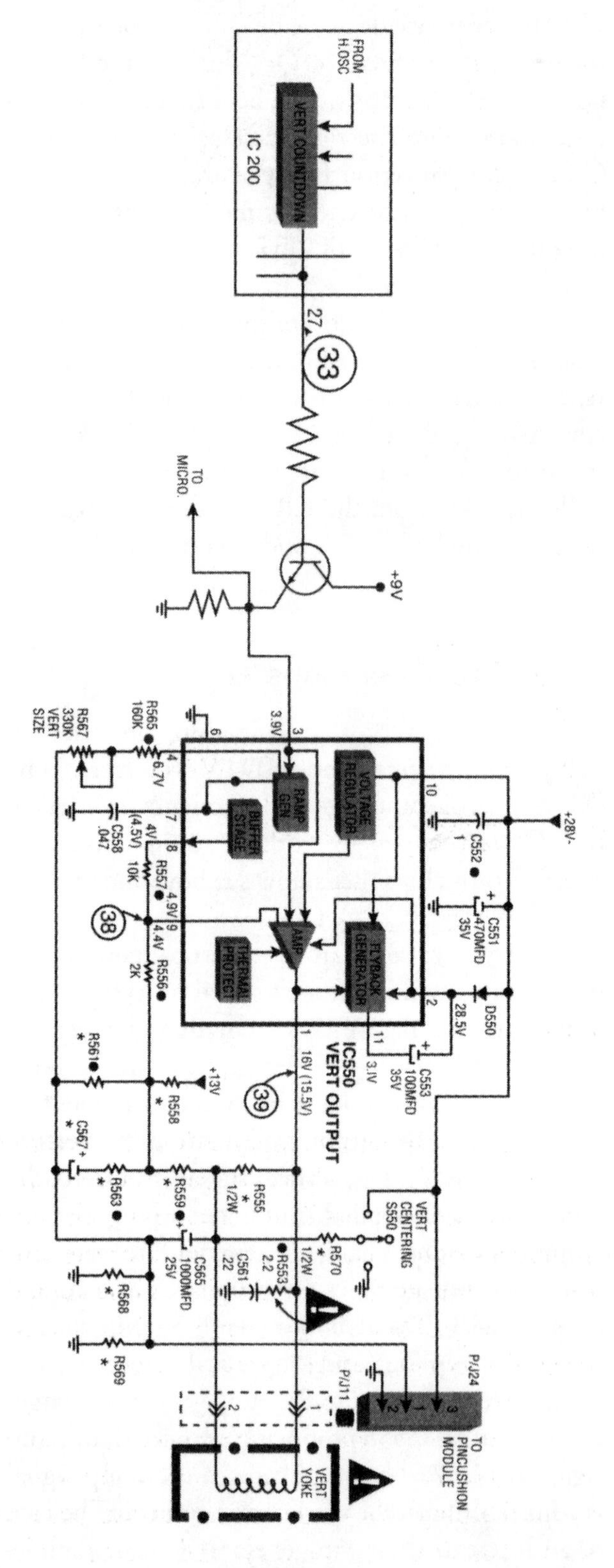

FIGURE 5-33

The output amp of IC550 contains the complementary output amplifier transistor stage. The output amplifier causes current to alternate through the yoke by conduction through the parallel resistors R568 and R569 charging and discharging capacitor C565. During retrace time, the flyback generator block switches in capacitor C553, which was charged to 28 volts during trace time. The capacitor and D550 conduct, doubling the supply voltage to the output, and produces a waveform amplitude of 58 volts peak during retrace at pin 1. The added voltage works to increase the rate of current and speed up retrace time.

A portion of the signal across the yoke series capacitor, C565, is shaped by R563/C567 and added to the DC at pin 9. This AC feedback shapes the drive signal and maintains proper drive linearity. Problems associated with IC550, or components surrounding it, can all cause reduced or improper deflection due to the DC and AC signal feedback. A quick method to accurately test the yoke at full operational current indicates that the circuit has a relatively inexpensive component causing the trouble.

Vertical Yoke Testing

Sencore's TVA92 Video Analyzer dynamically tests the vertical yoke by subbing for the drive signal that normally drives the yoke. This lets you analyze the yoke's ability to produce a full, linear deflection pattern. The TVA92 Sencore Video Analyzer is shown in **Figure 5-34**.

The analyzer's vertical yoke drive output signal is produced by circuits similar to the vertical stages of a TV receiver. A ramp generator produces a sawtooth waveform. Vertical amplifiers then shape and amplify the signal to the analyzer's vertical yoke drive output jack as shown in the **(Fig. 5-35)** drawing.

Vertical sync pulses originating in the analyzer,s video generator are input to the TVA92's vertical ramp generator to sync lock the yoke drive. Because the vertical yoke drive signal is locked to the video, you can simply watch the set's CRT to determine if the vertical yoke coils produce proper vertical deflection.

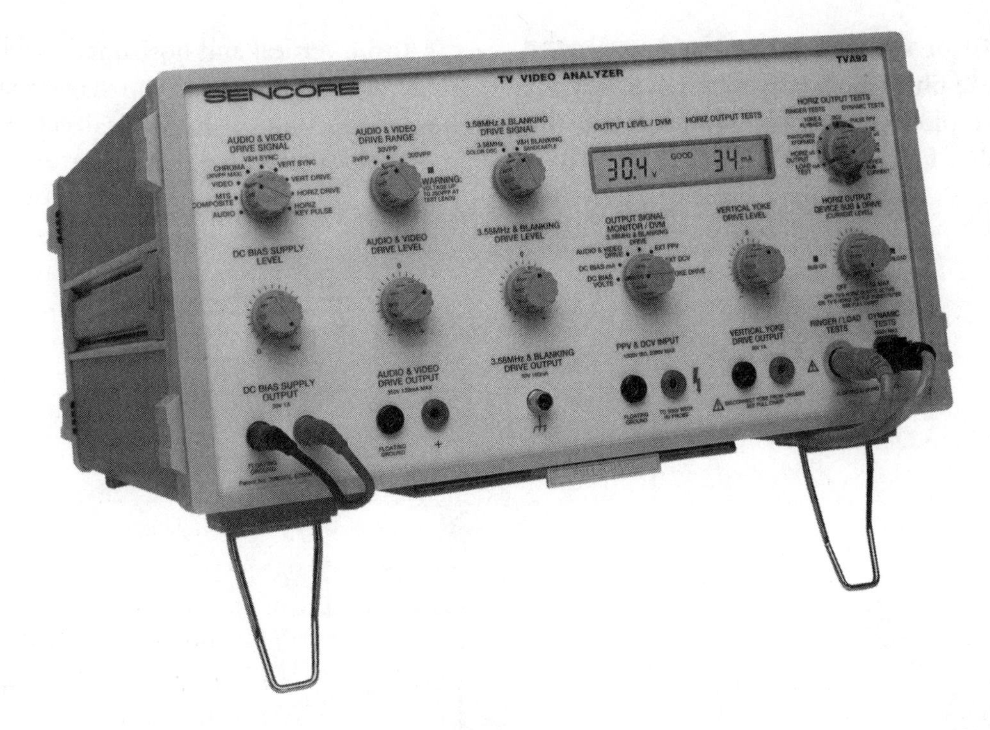

FIGURE 5-34

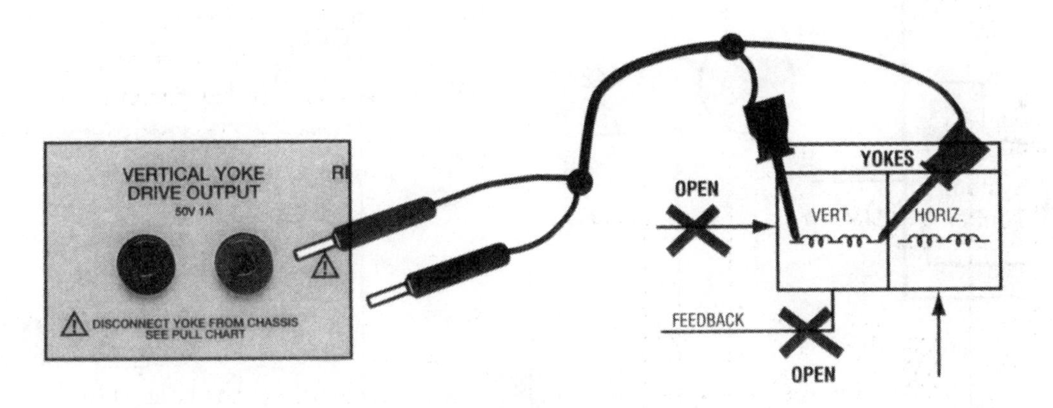

FIGURE 5-35

The vertical yoke drive level control lets you increase the output signal (AC current) to the yoke being tested. The control varies the output from approximately 0 to 40 Volts P-P and deflection current from 0 to 1.5 amps peak. When the analyzer's output signal monitor DVM switch is set to "Yoke Drive," the digital LCD display indicates the output peak-to-peak signal level.

Connecting the Analyzer for Yoke Tests

The analyzer's vertical yoke drive test signal is on a different jack than other substitution signals since it is applied to the circuit under test differently.

The vertical output amplifier in a TV receiver is a power output amplifier. It drives current into a low impedance yoke load. "Swamping out" the amplifier's output current drive, while at the same time substituting for it, would likely cause component damage. Thus, the yoke must be disconnected from the circuit before applying a test vertical yoke drive.

There are two ways to disconnect the vertical yoke from the circuit. The easiest is just to remove the plug that connects the vertical yoke to the circuit board. You may find a single plug connecting the vertical and horizontal yoke to the circuit board or separate vertical and horizontal plugs. If these plugs are separate, remove only the vertical yoke plug connection. Note the illustration shown in **Figure 5-35.**

If the vertical and horizontal yoke connections share a common plug, you have two choices on how to disconnect the vertical yoke. You can remove the yoke plug from the circuit board or

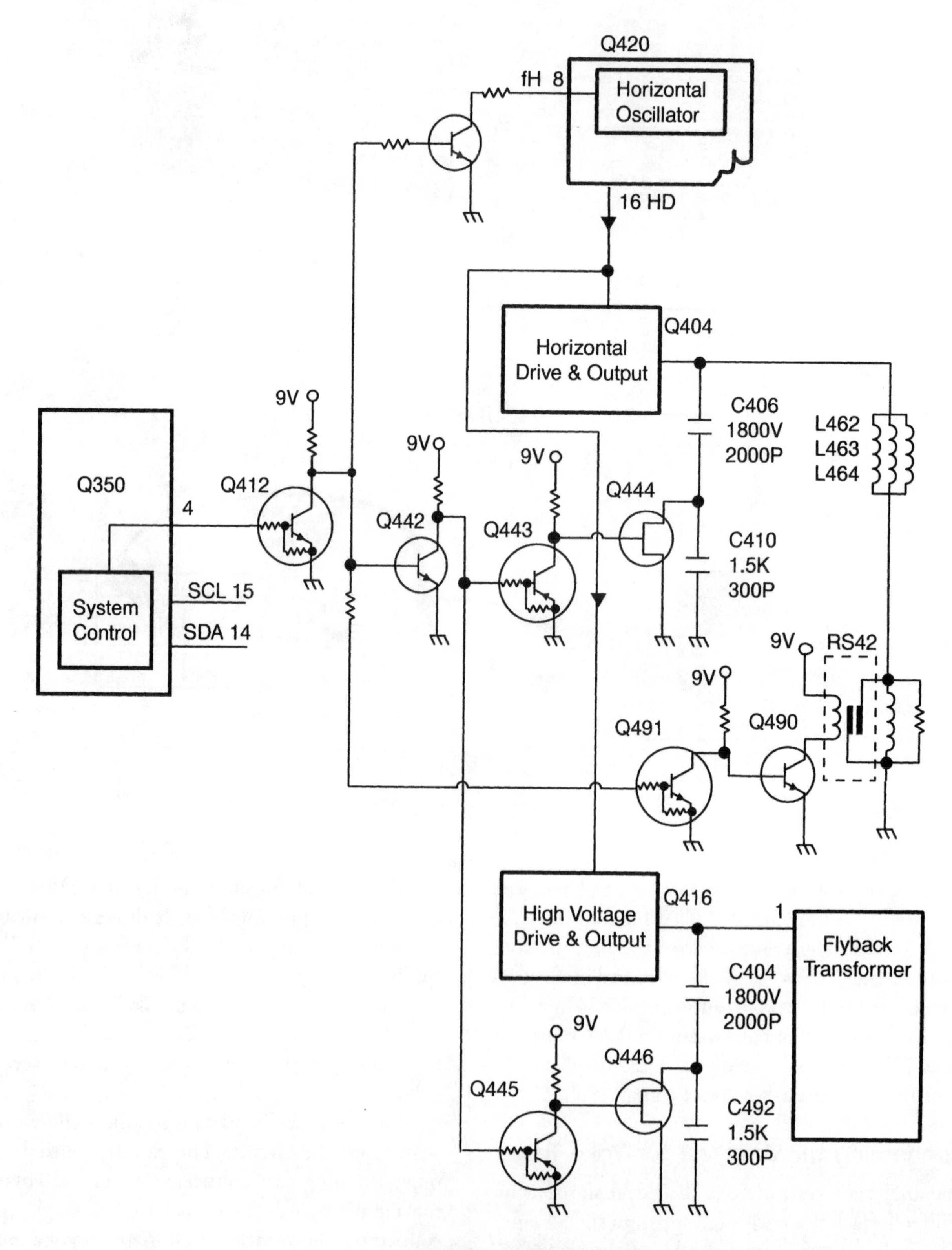

FIGURE 5-36

unsolder the wires at the terminals of the yoke. If you remove the plug, connect jumper wires from the horizontal yoke leads on the plug to the corresponding horizontal connections on the circuit board. This connects the horizontal yoke to the TV set's circuit board.

Another method of disconnecting the yoke is unsoldering the wires from the yoke windings. Most yokes have mounted terminals on the yoke assembly that you can access by removing a plastic cover. These terminals connect the ends of the yoke windings to the wires leading to the circuit board. The vertical coils are the outside windings of the yoke. Connecting wires (usually green and yellow) run from the vertical yoke terminals to the set's circuit board.

To connect the vertical yoke drive output test of the TVA92 to the vertical yoke, use the direct test leads. With the TV receiver unplugged from AC power, insert the banana plugs into the vertical yoke drive output test jacks. Connect the test lead clips to the vertical yoke terminals or to the yoke plug. You do not have to observe test lead polarity.

Testing the Vertical Yoke

To test the vertical yoke, start with the vertical yoke drive control set to "0" and the output signal monitor/DVM switch set to "Yoke Drive." Then feed in an RF signal with a "crosshatch" video pattern at 1000 microvolts.

Now apply power to the TV set and turn the unit "ON." Select the TV channel to match the test video crosshatch pattern frequency. Check the receiver CRT for a bright horizontal line across the center of the screen.

Increase the vertical yoke drive level control while observing the CRT. Adjust the control in either the positive or negative direction until the crosshatch pattern nears the top and bottom of the display area. If the vertical yoke is good, you will see a full and near-linear deflection. The crosshatch pattern should display near-perfect squares from the top to the bottom of the screen if the yoke is good.

If the vertical yoke is bad, the results will be similar to the problem seen when the chassis vertical circuits are driving the yoke. Achieving full deflection may require an increase in the drive signal level control, resulting in foldovers or nonlinearities in the crosshatch video pattern seen on the display.

HORIZONTAL OUTPUT CIRCUIT OPERATION

How the horizontal output system of TV receivers works can be a mystery. The next chapter section will try to unravel that mystery for you.

The basic concept of these circuits has not changed much over the years, but scan-derived supplies, and start-up and shutdown circuits, have made the horizontal output stage a tough one to troubleshoot.

Key Horizontal Circuit Components

All horizontal output stages operate virtually the same way regardless of make, model, or type of HDTV set. A block diagram of a typical HDTV horizontal output stage is shown in the (**Fig. 5-36**) drawing. All output stages drive a sawtooth current into the primary winding of the flyback transformer and receive power from the main B+ supply. The B+ supply can deliver peak currents of several amps while retaining a regulated voltage of about 130 volts DC. The peak-to-peak current required by the output stage depends on the CRT size and the number of scan derived supplies.

In **Figure** 5-37 you will find a simplified horizontal output stage circuit. It consists of six key components:

1. Horizontal output transistor (Q1)
2. Sweep or flyback transformer
3. Retrace timing capacitor or "safety cap" (Ct)
4. Damper diode (D1)
5. Horizontal yoke
6. Yoke series capacitor (Cs)

FUNCTION OF OUTPUT TRANSISTOR

The horizontal output transistor (HOT) functions as a switch. It provides a path for current to flow through the flyback's primary winding and hori-

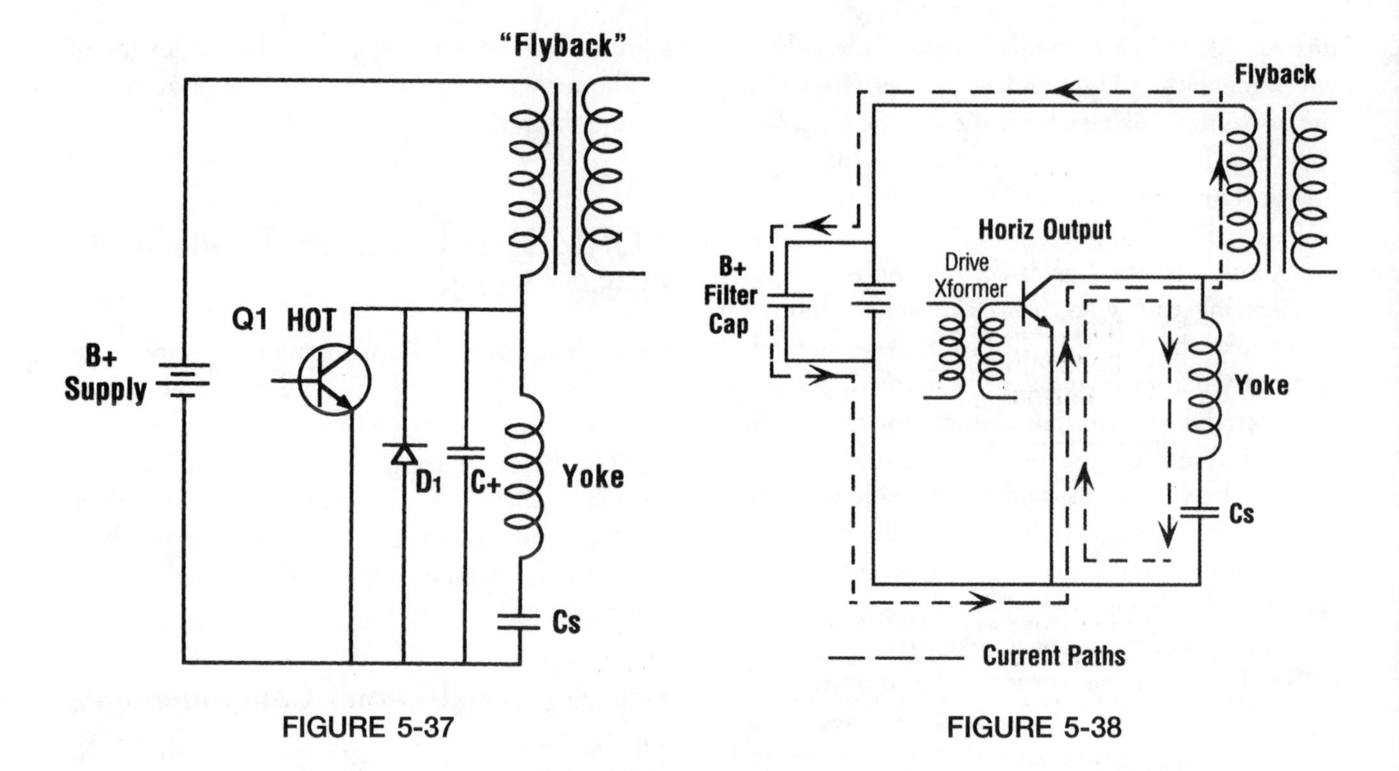

FIGURE 5-37

FIGURE 5-38

zontal yoke as shown in **Figure** 5-38. The HOT is switched ON and OFF by a signal applied to the base. Because this is a power transistor, a current drive signal is required. This drive current is supplied by the horizontal driver and driver transformer. In addition to current step-up, the driver transformer provides impedance matching.

The horizontal output transistor passes current levels ranging from 200 mA to 1.5 amps in large-screen TV sets that have multiple scan-derived power supplies. These are average current ranges, while some peak current may be 4 amps.

With transistors, the collector current equals the base current multiplied by the current gain (beta). Thus, the base drive current must be sufficient to produce the required collector current and may be as much as 100 to 300 mA. If the base drive is insufficient, the emitter-to-collector resistance of the conducting HOT will be too high and the transistor will become too hot. Sufficient drive is also important for fast switching.

The horizontal output transistor is switched ON and OFF at the horizontal frequency of 15,734 Hz. The horizontal oscillator (which controls the driver stage) begins to turn the horizontal output transistor on approximately 30 to 35 microseconds before horizontal as shown in **Figure** 5-39. The HOT conducts until the start of horizontal sync and then is abruptly turned OFF.

The time it takes to switch the HOT from ON to OFF is important. As the transistor is switched, the emitter-tocollector resistance changes from approximately 5 ohms (ON) to 10 megohms (OFF). The current flowing through the transistor during the on/off transition produces heat, and the longer the transition, the greater the heat buildup, and the greater the chance of a thermal failure.

The drive current produces a voltage waveform at the base of the HOT that is similar to a squarewave. These waveforms have a peak-to-peak range from 5 to 30 volts; however, much of the amplitude is due to voltage spikes caused by the switching action. The waveform only confirms the presence of drive to the HOT. It cannot confirm if the base current drive is adequate for normal operation. Reduced or improper drive results in transistor heating, reduced deflection (width), picture foldover, and shortened HOT life.

Output HV Transformer Operation

The sweep output transformer is also referred to as a flyback or IHVT. (An IHVT is a flyback transformer that includes the high voltage multiplier.)

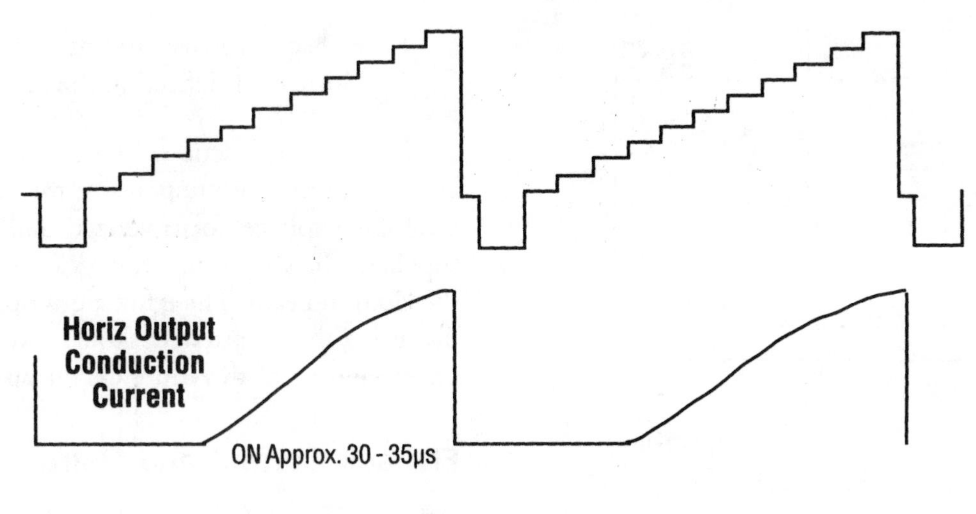

FIGURE 5-39

The flyback is primarily responsible for developing high voltage. It is constructed with a powdered iron core or ceramic core to work efficiently at high frequencies.

The flyback includes one primary winding and many secondary windings. The main secondary winding supplies voltage pulses to the voltage multiplier. Other secondary windings supply CRT filament power, keying pulses, and scan derived power supplies, as shown in **Figure 5-40**.

Let's now see how these pulses are produced. The current in the flyback primary rises at a linear rate when the HOT is conducting. This produces a constant amount of induced voltage in the flyback windings. However, when the HOT is abruptly turned OFF, the magnetic field in the flyback core rapidly collapses and induces a high voltage into the flyback's primary and secondary windings, as illustrated in **Figure 5-41**.

The rate that the magnetic field collapses is controlled with timing components. If they were not, induced voltage spikes of several thousand volts would be produced across the flyback primary. These spikes would exceed the breakdown rating of the horizontal output transistor and flyback, and produce excessively high voltage to all flyback windings.

Retrace Timing Capacitor

The retrace timing capacitor slows down the rate of the flyback's collapsing magnetic field. This is a very critical function. If the retrace capacitor's value decreases, or if it becomes open, the amplitude of the flyback pulse will increase several thousand volts. Also, safety shut-down circuits are added to disable the output stage should the high voltage increase to unsafe levels. The retrace timing capacitor is also referred to as the "safety capacitor."

Damper Diode

The damper diode completes the resonant current path for the flyback primary and deflection yoke by turning on during the time when the current through

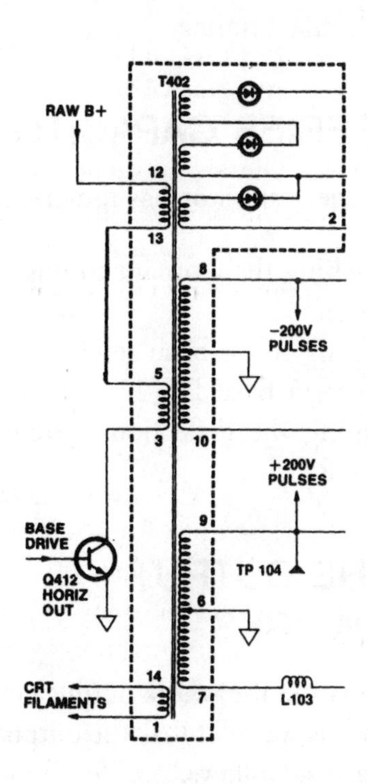

FIGURE 5-40

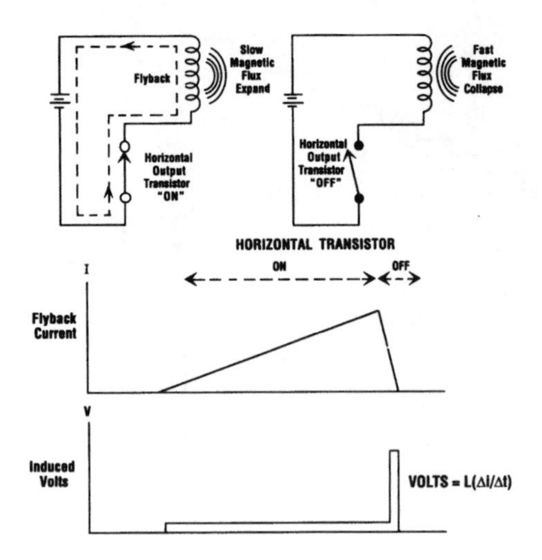

FIGURE 5-41

the HOT reverses. If the damper diode opens, the HOT is forced to operate in reverse breakdown, which will cause the transistor to fail. The damper is a fast-switching, high-current diode.

HORIZONTAL DEFLECTION YOKE

The rising and falling sawtooth current flowing in the yoke produces horizontal electron beam deflection. Because it is part of the output stage, the yoke also affects retrace timing.

YOKE SERIES CAPACITOR

The yoke series capacitor has four functions:

1. Matching the resonant timing of the yoke current
2. Helping to establish retrace time
3. Prevent a fixed DC bias on the yoke
4. Shaping the deflection current to match the CRT

HOW THE OUTPUT STAGE WORKS

Let's now analyze the whole output circuit operation. To do this we will break the output operation into two parts as follows:

1. Flyback primary current and retrace time
2. Horizontal deflection operation

In the first function, flyback primary current and retrace time are responsible for producing the CRT high voltage, focus voltage, and scan derived supplies. The second function deals with deflecting the electron beam. These functions do interact, but discussing them separately should give you a better understanding of the entire circuit operation.

Flyback Current and Retrace Time

Refer to **Figure** 5-42 for the flyback action and current paths at four time intervals during one output cycle. It begins with the horizontal output transistor turning ON. When the HOT is turned ON, current flows into the flyback's primary from the B+ power supply. All of the power required by the output stage, including the secondaries, is delivered to the circuit from the B+ supply during this time. The current and magnetic field in the flyback's core continue to build until the transistor is turned OFF.

During the next three periods, the magnetic energy expands and collapses in the flyback. This produces current that charges and discharges the retrace timing capacitor. The current flow through the flyback primary winding transfers power to the secondary and its load.

The magnetic field that was stored in the flyback's core begins to collapse immediately after the HOT is turned OFF. This is the beginning of retrace time and corresponds with the start of horizontal sync. With the HOT switched off, the retrace timing capacitor is effectively placed in parallel with the flyback primary. Thus, a resonant circuit is formed as shown in **Figures** 5-42 b and c. The time constant of the resonant circuit is determined mainly by the value of the retrace capacitor and the inductance of the primary winding. The yoke components in parallel with Ct (the yoke and Cs), also have an effect on retrace timing.

The collapsing magnetic field causes current to flow through the low impedance of the B+ supply's filter capacitors and into Ct. This current charges Ct and produces the large pulse at the HOT collector.

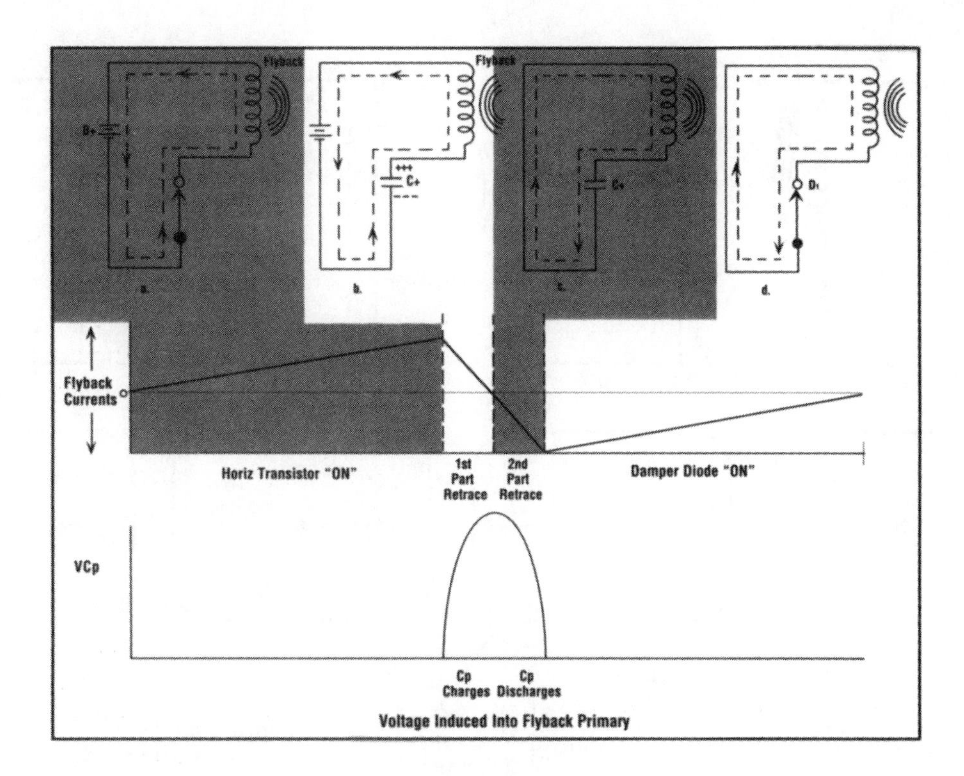

FIGURE 5-42

After the magnetic field has completely collapsed, Ct begins discharging, which causes current flow back into the primary in the opposite direction. A magnetic field rebuilds with the opposite polarity. This action completes the second part of retrace and corresponds to the falling portion of the HOT collector pulse. A properly operating output stage has a retrace time (flyback pulse duration) of 11.3 to 15.9 microseconds.

When Ct has completely discharged, the magnetic field begins to collapse. The collapsing field induces a voltage with a polarity that forward biases the damper diode. The damper diode serves as a switch and allows the magnetic energy (current) in the flyback and yoke to decay at a controlled rate. (The damper diode would not be needed if the horizontal output transistor could conduct current in both directions.)

With the damper diode turned ON the circuit is highly inductive, so the current in the flyback primary once again increases slowly. Approximately 18 microseconds later, the horizontal output transistor is once again turned ON and the cycle is repeated.

The flyback transformer works like any other transformer in that the energy in the primary is transferred to the secondaries. If all of the secondary loads were open, most of the energy stored in the magnetic field would return back to the primary circuit. However, the secondary circuits draw power from the primary. Thus, as the load on the secondary windings increases, more current flows in the primary and more current is drawn from the B+ supply. Some problems, such as a shorted secondary load circuit or a shorted flyback winding, may cause such a great load that the circuit cannot compensate for the power demand. This will cause the horizontal output transistor (HOT) to overheat and short out, causing the flyback primary to open up or the B+ supply to fail.

HORIZONTAL YOKE DEFLECTION

Another function of the horizontal output stage is to provide deflection current. The HOT's collector current is split between the flyback and horizontal yoke. Both paths share the damper diode and retrace timing capacitor.

Capacitor Cs, which is in series with the yoke, has four functions:

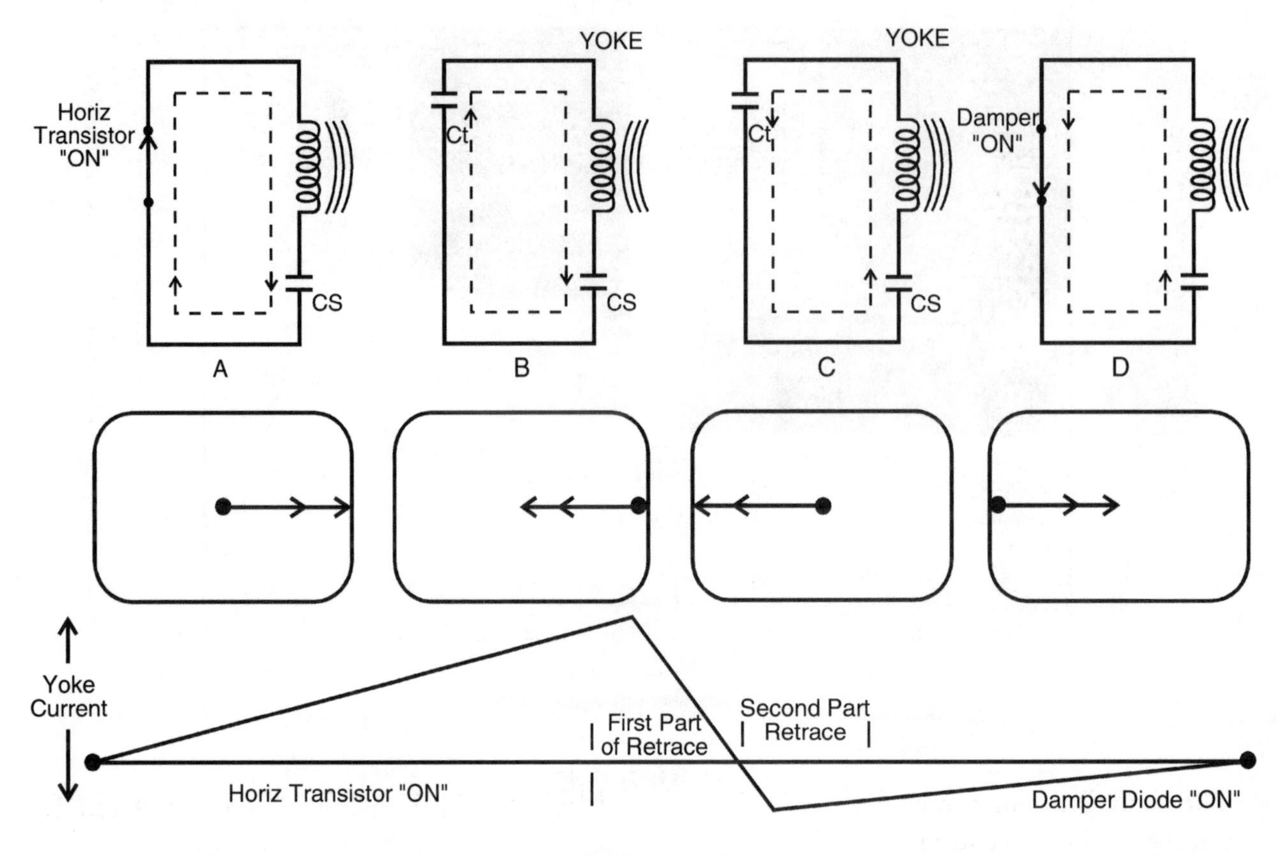

FIGURE 5-43

1. Cs is primarily responsible for determining the time constant of the deflection portion of the horizontal output stage.
2. Affects the retrace time.
3. Prevents a fixed DC bias from developing on the yoke that would cause improper picture centering.
4. Shapes the sawtooth rise in deflection current to match the slight curvature of the CRT.

Figure 5-43 drawing shows the deflection current at the same time intervals as illustrated in Figure 5-42. When the HOT is turned ON, the bottom side of the yoke series capacitor, Cs, is connected to the top of the yoke. Because Cs is fully charged, it begins discharging through the horizontal output transistor. The resulting current produces an expanding magnetic field in the yoke that moves the electron beam from the center of the screen toward the right side of the screen.

When the horizontal output transistor opens, the retrace timing capacitor is added to the circuit, as shown in the (**Fig. 5-43b**) drawing. This increases the resonant frequency and causes the yoke's magnetic field to rapidly collapse. This is the beginning of retrace during which the beam is snapped from the right side of the screen back to the center. The induced voltage causes current to flow and returns the energy that was stored in the yoke's magnetic field to capacitors Ct and Cs. The retrace timing capacitor is replenished with charging current from the flyback transformer and becomes the current source for the yoke current.

During the second part of retrace, Ct and Cs discharge, forcing the current to flow in the opposite direction as shown in **Figure 5-43c.** The timing is identical to the first part of retrace and the beam moves quickly from the center to the left side of the screen.

When capacitors Ct and Cs are fully discharged, the yoke's magnetic field begins to collapse, as shown in **Figure 5-43d**. The induced voltage forward biases the damper diode into conduction. Note that this occurs at the same time as it did in **Figure 5-42.**

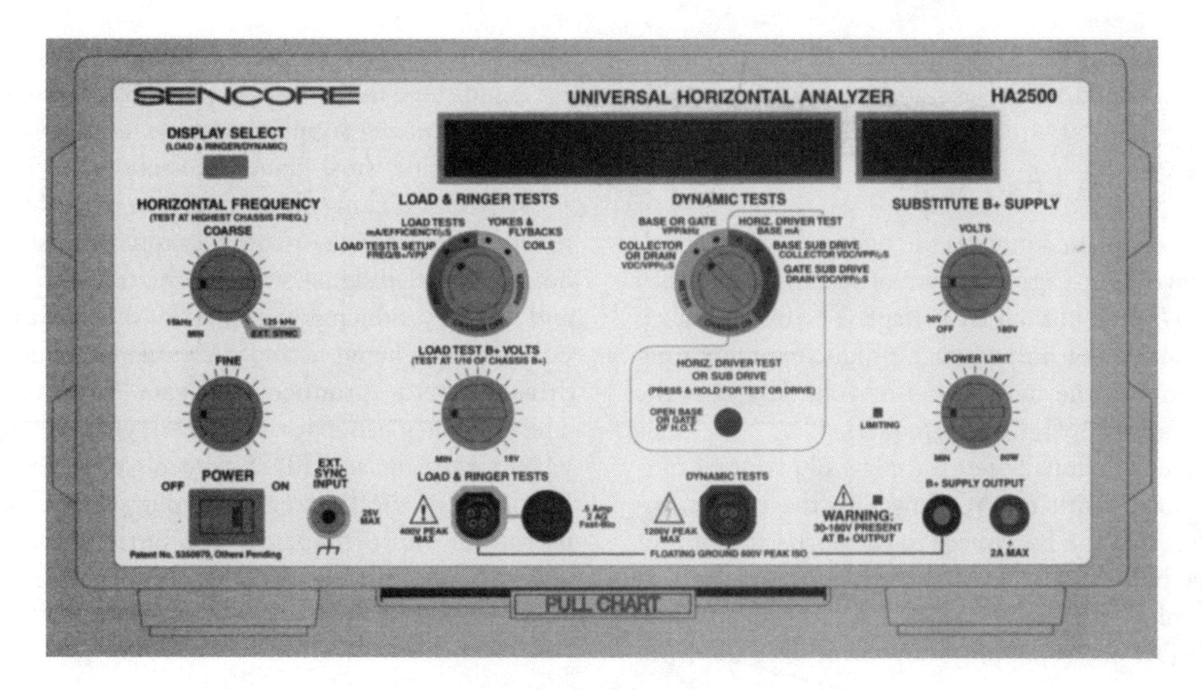

FIGURE 5-44

The circuit's timing is now determined by the yoke and capacitor Cs and mimicks the timing of the trace as it moved right. The yoke's collapsing magnetic field produces current through the damper diode, which returns energy to the circuit and charges capacitor Cs.

When the yoke's magnetic field is collapsed, the damper diode stops conducting. The horizontal output transistor must immediately begin to conduct, otherwise there will be horizontal nonlinearities in the center of the raster.

Please note that these circuits do not operate separately as this simplified explanation may appear to suggest. These circuits *DO NOT* function independently. The flyback current is transferred to the yoke by retrace timing capacitor Ct. The yoke and flyback currents share the conduction time of the horizontal output transistor, damper diode, and retrace timing capacitor. Because of this interaction, most problems in the horizontal output circuits affect both the flyback and yoke current.

USING THE SENCORE HA2500 HORIZONTAL SYSTEM ANALYZER

The Sencore HA2500 features exclusive analysis tests and substitution capabilities to localize horizontal circuit defects faster than using conventional troubleshooting techniques. Its exclusive tests isolate defects that previously required many hours of extensive troubleshooting or component swapping. The HA2500 may be synchronized to a video or RGB signal generator to become an integrated part of your service bench, or it may be used for field service. The front panel of the HA2500 is shown in **Figure** 5-44.

HA2500 User Information Guide

These unique analyzing tests of the HA2500 provide effective troubleshooting capabilities to isolate horizontal and related chassis symptoms. Some tests are performed with power to the chassis off, while other tests are performed with the power to the chassis on. The HA2500 "chassis-off" tests include the Load Tests and Ringer Tests. The HA2500 "chassis-on" Dynamic Tests, include the Meter Tests, Horiz. Driver Tests and Sub Drive Tests. The Substitute B+ Supply adds additional chassis-off or chassis-on analyzing capabilities. Guidelines for using each of the HA2500 tests are shown in **Chart 5-1.**

Doing The Load Test

The Load Test simulates the operation of the horizontal output stage at its normal operating frequency and 1/10 its normal B+ voltage. The test energizes the horizontal output stage producing alternating currents in the flyback and/or yoke just as if the chassis was operating normally.

To functionally test the chassis horizontal output stage, with no AC power to the chassis, requires that the B+ voltage to the horizontal output stage be supplied. A variable DC power supply to 18 volts provides the voltage for the Load Test. The DC power supply is adjustable with the front panel LOAD TEST B+ VOLTS control. The voltage is indicated in the display during the Load Test Setup. The power supply current is limited to 250 mA to provide short circuit protection.

A MOSFET switching transistor provides the function of the sets horizontal output transistor. A horizontal drive signal switches the transistor ON and OFF, producing currents in the horizontal output stage being tested. A variable frequency drive generator produces the gate drive signal. The frequency generator is adjusted with the front panel COARSE and FINE controls to match the frequency of the horizontal output stage. This is useful for multifrequency computer monitors. The drive may also be decoded from an Ext. Sync Input signal.

HA2500 TEST	WHEN TO USE	WHAT IT TELLS YOU
Load Tests functions.	• Full AC cannot be applied: B+ supply dead or bad. H.O.T heats or fails. B+ supply squeals, burns-up components or blows fuses. • X-ray shutdown symptom.	• Output stage normal. • Short/Load on B+ supply. • Output stage pulse timing. • Output stage efficiency.
Ringer Test	• Load Tests is abnormal. • Suspect flyback, yoke, or horizontal coil defect.	• If coil or transformer is bad from shorted turn(s).
Collector or Drain Meter Tests	• Load Test normal but suspect no HV. or defl.	• B+ Supply volts. • Horiz. Output pulse VPP. and time.
Base or Gate Meter Tests	• Load Test normal but suspect no HV or defl.	• Input Horiz. Drive VPP. • Input Horiz. Drive Freq.
Horiz. Driver Test	• H.O.T runs hot but OK with SUB & DRIVE. • Reduced HV and/or Defl. but B+ voltage normal. • Repeat H.O.T. failures.	• Base Drive Current output from Horiz. Driver. • Drive mA low or reduced.
Base or Gate Sub Drive	• When horizontal drive is missing or suspect. • Test horizontal output with the chassis off using the Substitue B+ Supply.	• If horiz. output stage OK. • Isolates symptom to driver or horiz. output stage.
Substitute B+ Supply	• B+ Supply dead, damaged, suspect of noise or intermittent. • B+ HV/Defl. regulator suspect • Suspect HV breakdown in horizontal output stage. • X-ray shutdown symptom.	• If horiz. output stage OK. • Isolates symptom to B+ supply or horiz. output. • Isolates symptom to HV/ defl. regulator or output. • Adj. B+ to test X-ray latch.

CHART 5-1

Measuring circuits indicate parameters during the Load Tests Setup and Load Tests functions. A DC voltmeter, frequency counter, and P-P voltmeter provide readouts of the HA2500 front panel control settings and horizontal output stage peak-to-peak voltage during the Load Tests Setup. A DC supply current meter, microsecond timebase meter, and efficiency calculator provide measurements during the Load Tests procedures.

Setting Up the Load Test

The Load Test Setup displays guide you in setting the proper horizontal test frequency and B+ volts to ensure the most accurate Load Tests results. The Load Tests Setup is used to confirm proper connections and to set the horizontal output stage, under test, as close as possible to the required 1/10 of normal level. Load Test Setup requires selecting the proper horizontal frequency and a B+ voltage that is 1/10 of the TV chassis normal voltage level.

NOTE: You should always perform the Load Tests Setup before analyzing the Load Tests mA, %, and timebase readouts. An improper horizontal frequency and/or B+ voltage can cause Load Test results that are beyond the typical normal range for the chassis type. If you change the horizontal test frequency during the Load Tests, repeat the Load Test Setup.

Occasionally, horizontal output stage problems, improper circuit connections, or an improper test frequency or B+ voltage may produce inaccurate Load Test Setup results. **Chart 5-2** indicates possible readings and their most likely causes.

Ringer Testing Techniques

The Ringer Test identifies one of the three common failures of a flyback transformer. The three common flyback failures include the following:

1. Shorted turn(s) in a winding
2. Leakage between separated winding
3. 3-IHVT multiplier defect

The most common flyback failure is a shorted turn or turns. This failure occurs when individual wire turns of one of the windings short together. This is a different type of failure than a leakage or short circuit path between separate windings of the flyback transformer. A shorted turn(s) defect can only be detected using the Ringer Test.

To perform a Ringer Test on a flyback or IHVT, connect the Ringer test leads to the primary winding. The drawing in **Figure 5-45** shows connections for Ringer testing. A shorted turn in any other winding will reflect back to the primary and cause it to ring incorrectly. You do not need to

Load Tests	Setup Readouts		
Frequency (kHz)	VDC	VPP	Most Likely Causes
Display's Highest Horiz. Freq.	1/10 normal B+ (Highest Display Freq.)	90 -110VPP (Bipolar type)	Proper Load Test Setup
	Can't increase DCV	0.0 VPP Low VPP	Load Test mA ≅ 250 mA "Current Limiting" Severe B+ Load.
Display's Highest Horiz. Freq.	1/10 normal B+	0.0 VPP	Improper connections Open inductor/flyback circuit path.
Display's Highest Horiz. Freq.	1/10 normal B+	Low VPP	Load Test B+ too low. Load Test Freq. too high.
Display's Highest Horiz. Freq.	1/10 normal B+	High VPP	Load Test B+ too high. Load Test Freq. too low.

CHART 5-2

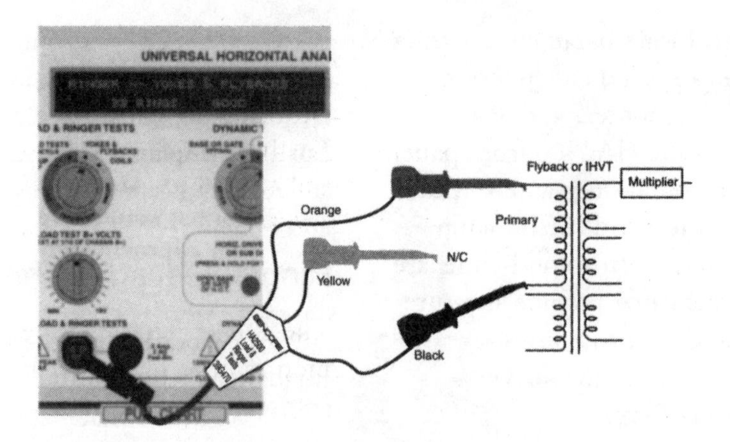

FIGURE 5-45

ring each winding separately. A flyback or IHVT usually has several windings that only have a few turns and do not normally ring greater than a count of 10.

You can ring a flyback or IHVT while it is connected to the circuit. However, other components and circuits often cause good flybacks to ring incorrectly in circuit. If a flyback rings less than 10, when tested in the circuit, disconnect or open the following paths:

1. Horizontal output transistor
2. B+ input to the flyback primary
3. Deflection yoke
4. CRT filament - unplug the CRT socket
5. Flyback scan-derived secondaries

Retest the flyback after disconnecting each circuit or component. If the flyback rings "good," it does not have a shorted turn.

If the flyback tests "bad" after disconnecting each of the above circuits or components, unsolder it and completely remove it from the circuit. If the flyback primary still rings less than 10, ring the rest of the windings. If one of the other windings rings above 10, the flyback is good. If not, the flyback has a shorted turn and must be replaced.

A flyback or IHVT may develop a leakage path between two separate windings or between a winding and the transformer's core or mounting bracket. This failure is usually not detected by the Ringer Test. A high resistance leakage path often pulls down the chassis B+ supply, even with the horizontal output transistor removed. This may occur at reduced voltages and be identified by the Load Test or may go unnoticed until approaching normal flyback operating voltages. You can use a high potential leakage tester to isolate leakage between windings, core, or mounting frame.

Another failure, which is common, occurs in flybacks that contain voltage multipliers. These flybacks are referred to as integrated high voltage transformers, or IHVTs. If the multiplier portion of an IHVT fails, the high voltage and/or focus voltages are low or missing. To test this failure, activate the horizontal output stage and IHVT with the Load Test and perform Load Test Setup. Now measure the high voltage and focus output voltages. These voltages should measure approximately 1/100 of the normal value. Low or missing voltages indicate a defect problem.

Powered Up — Horizontal Output Stage Checks

The "METER" section of the HA2500 Dynamic Tests switch provides automatic measurements to quickly confirm normal operation of the horizontal output stage. When abnormal, the measurements indicate if input voltages or drive signals are improper or if a fault exists in the horizontal output stage.

COLLECTOR AND DRAIN METER INTERPRETATION

The Collector or Drain DC voltage readout measures the DC voltage at the collector or drain of the horizontal output transistor with respect to ground. The DC voltage is the B+ power supply voltage applied to the horizontal output stage. The DC voltage measurement is designed to measure the full range of B+ voltages found in video displays ranging from 0 to 400 volts.

The DC voltage readout should closely agree with the schematic or the B+ known to be normal for the chassis being tested. DC voltage readings that vary considerably from normal indicate a change in the normal load current to the horizontal output stage. It may also be an indication of a problem with the B+ power supply.

If you do not have a schematic, you can estimate the normal B+ voltage. While the normal B+ voltage varies among TV receivers, there are some common norms in B+ voltages depending on the display's size and resolution capabilities. Color televisions with CRT sizes of 13 inches or more will usually have B+ voltages ranging from 115 to 140 volts. A VGA-only monitor commonly uses a B+ supply voltage of 75 to 95 volts. A monitor operating at a frequency of 31.5 KHz may have a B+ voltage of 80 volts, and increase to 130 volts as frequencies near 60 KHz. As you use the Collector or Drain Dynamic Tests you will gain further insight as to the normal B+ voltages found in various displays.

A missing B+ voltage results in a very low or zero voltage reading. Loss of B+ voltage prevents the horizontal output stage from producing alternating currents in the inductor, flyback transformer, or yoke. Missing B+ voltage usually results from an open circuit defect within the B+ supply or high voltage/deflection regulator. Voltage readings near zero may also be caused by a short in the horizontal output stage.

A lower-than-normal B+ voltage causes reduced currents in the horizontal output stage and less high voltage and/or deflection. Reduced B+ voltage may be an indication of an excessive load or current demand from the horizontal output stage. Perform the Load Test to identify and isolate horizontal output stage shorts or loading problems. If the Load Test indicates normal operation, you should suspect the B+ power supply or high voltage/deflection regulator. The Substitute B+ Supply can also be used to further isolate the defect.

A higher-than-normal B+ voltage causes increased currents in the horizontal output stage and increased high voltage and/or deflection. If the voltage is considerably higher than normal, safety circuits will "shut down" the horizontal output stage. Higher-than-normal B+ voltage indicates a B+ power supply regulation problem, or a high voltage/deflection regulator problem.

Higher-than-normal B+ voltage to the horizontal output stage is often normal when the horizontal output stage is not operational or is not drawing current from the B+ supply or high voltage/deflection regulator. This commonly occurs when the horizontal drive is missing to the horizontal output stage. Missing drive results in no current being drawn from the B+ supply.

The collector or drain peak-to-peak voltage readout provides additional verification in interpreting whether the B+ voltage is normal, low, or high. Refer to **Figure 5-46** for the Collector or Drain dynamic analysis test waveform. If the P-P voltage range of the horizontal output stage or flyback pulse is near normal, the B+ voltage is most likely correct. A common cause of a high horizontal output stage P-P voltage pulse is a higher-than-normal B+ voltage. A common cause of a lower-than-normal horizontal output stage P-P voltage pulse is a lower-than-normal B+ voltage. Use the VPP readout to further confirm normal, high, or low B+ voltage.

HIGH VOLTAGE/DEFLECTION REGULATOR PROBLEM TROUBLESHOOTING

Many TV set symptoms can be caused by defective B+ power supply, regulators, or faulty interaction between these stages and the horizontal output stage(s). These trouble symptoms can best be isolated by using a substitute B+ supply. For example, when the B+ power supply is "dead" or defective, subbing the chassis B+ supply permits testing of the horizontal output stage and high

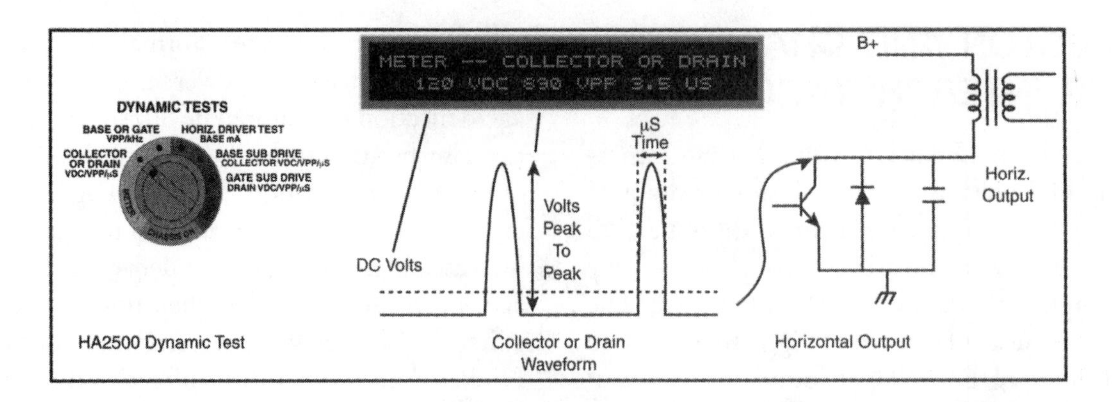

FIGURE 5-46

voltage/deflection regulators to full operating voltages. This lets you identify high voltage component problems before investing in parts and time to repair the power supply. This will also help you keep out of trouble when estimating repair costs.

The substitution check for the B+ supply is especially helpful in troubleshooting breakdown failures of horizontal output components that result in a high current load on the sets B+ supply, at or near full operating voltages. These failures at times cannot be detected by the reduced voltage Load Tests. Often these breakdown failures can quickly damage the horizontal output transistor and power supply components leaving little time to isolate the cause. By subbing the B+ Supply to the horizontal output stage and gradually increasing the B+ supply voltage and/or the output power limit, these problems can be isolated.

B+ Voltage Substitution Techniques

Before substituting the sets B+ supply, the chassis B+ power supply to the horizontal output stage or HV/deflection regulator must be removed by disconnecting the B+ voltage path. This prevents the sets B+ voltage from reaching the input of the horizontal output stage or the input to the high voltage/deflection regulator. Once the B+ voltage path is opened, the substitute B+ test unit can be connected and used to substitute for the sets B+ voltage to these circuit points. Failure to open the B+ supply will result in output currents between supplies and the potential of damaging the TV sets power supply and/or the subber test unit.

The typical B+ voltage path of a TV receiver is shown in the (**Fig. 5-47**) circuit diagram. The B+ voltage is output from the switch mode or linear supply. The B+ voltage is regulated and stepped down to produce normal high voltage and/or deflection by a high voltage or deflecting regulator. The regulator is usually a switching "buck converter" type as shown, or a linear pass regulator. The B+ voltage is input to the transistor, and the reduced output B+ voltage is applied to the input of the horizontal output stage.

Substitution of the chassis B+ voltage isolates problems to the horizontal output stage, the high voltage/deflection regulator, or the main B+ power supply. Two key circuit locations for subbing the chassis B+ supply are the input to the horizontal output stage and the input of the high voltage/deflection regulator. The input of the high voltage/deflection regulator corresponds to the output of the main B+ supply.

Components at the B+ input of the horizontal output stage coil or flyback transformer typically make opening the B+ path quite easy. Locate the B+ input on the schematic or circuit board by tracing from the collector of the horizontal output transistor to the coil or output transformer. Use an ohmmeter to identify windings with continuity to the collector and note the components to these points. The B+ voltage input to the horizontal output stage usually has a bypass capacitor of 1 to 100 microfarads, at 150 volts rating, to ground. A diode, small value resistor, coil, or trace jumper may be found on the path for the B+ input voltage to this pin. If in doubt, use the Load Test to

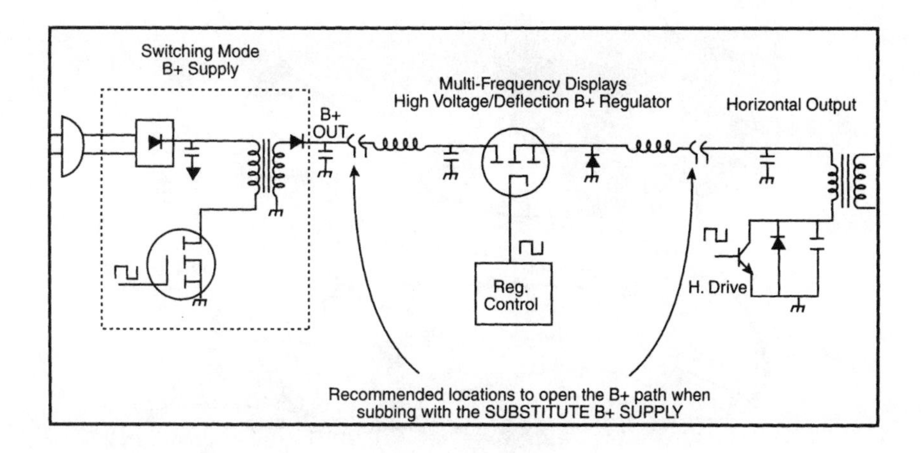

FIGURE 5-47

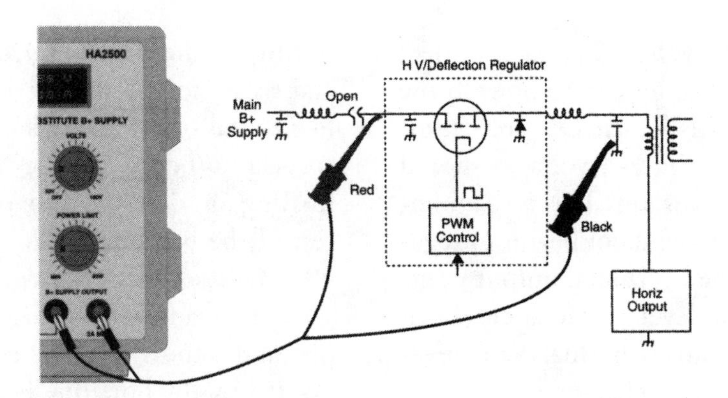

FIGURE 5-48

confirm you have properly identified the B+ input. Open the diode, coil, resistor, or wire trace jumper to open the B+ voltage path to the input of the coil or flyback transformer.

The B+ voltage output from the main power supply can be identified on the schematic or PC board by finding the largest DC output voltage or highest DC voltage rated filter capacitor on the output side of the switch mode power supply transformer. Trace the B+ supply output voltage to the high voltage/deflection regulator or horizontal output stage. A coil or jumper may be unsoldered and lifted from the PC board to open the B+ output voltage path. When opening the B+ voltage path near the switching power supply circuitry, be careful not to open the voltage feedback path for the regulator control of the switch mode power supply.

Testing the HV/Deflection Regulator

Substituting the B+ supply input to the high voltage and/or deflection regulator, is a quick and easy test that lets you isolate problems affecting the regulator. High voltage/deflection regulator defects can cause shutdown symptoms or high-current draw from the main power supply causing component damage. Subbing a reduced voltage to the input of the HV/deflection regulator lets you analyze the regulator and horizontal output stage avoiding high currents and set shutdown.

When substituting B+ to the HV/deflection regulator, start at a reduced B+ substitute voltage and slowly increase the voltage. With the Dynamic Test leads connected to the horizontal output transistor, monitor the collector or drain

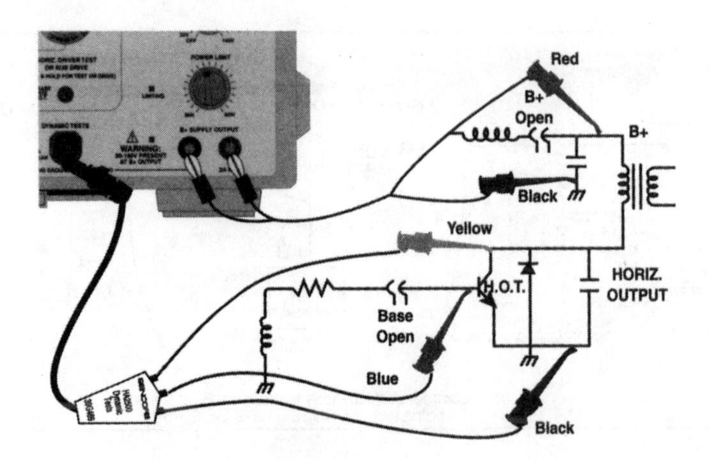

FIGURE 5-49

parameters for volts DC, volts P-P, and microseconds readings. Note test setup connections in the (**Fig. 5-48**) drawing. Initially, the DC volts readout should increase and the P-P voltages should begin to hold relatively constant. This is the point where the HV/deflection regulator begins to regulate the B+ voltage to the horizontal output stage. This should also be the level at which the horizontal output stage produces normal VPP readings, high voltage, and/or deflection.

As you increase the substitute B+ voltage, monitor the VDC and VPP readouts. If VPP readings rise above 1200 VPP (bipolar output) or 900 VPP (MOSFET output), the HV/deflection regulator, or its controlling circuits, are defective. A common defect is a shorted MOSFET switching transistor in the HV/deflection regulator. If the B+ voltage to the horizontal output stage begins to regulate too early, it will result in low DC and P-P voltages to the horizontal output stage. This HV/deflection regulator or control defect causes reduced high voltage and/or deflection.

Using the Sub B+ Supply and Subdriver

The Substitution B+ supply and base or gate subdriver may be used simulataneously to analyze a horizontal output stage. The subdriver permits testing of the horizontal output stage when the chassis is not producing horizontal drive to the horizontal output transistor, or when the drive is suspect. When using the substitute B+ supply, a missing or defective drive prevents an accurate test of the horizontal output stage. See **Figure 5-49** for these test setup connections.

An advantage of using the substitute B+ supply and subdrive together is that both inputs, needed by the horizontal output stage to operate, are substituted. This permits analysis of the horizontal output stage for defects without applying AC voltage to the chassis.

To simultaneously substitute for the chassis B+ voltage and horizontal drive to the base or gate of the horizontal output transistor, connect the test leads as shown in the (**Fig. 5-49**) test setup. Note that the substitute B+ supply and base sub drive both require that circuit paths be opened in either a chassis ON or chassis OFF test condition. Open the B+ voltage path and connect the substitute B+ supply output to the horizontal output stage side of the opened path. The base or gate subdrive requires that the gate or base path for the horizontal drive be opened. The base or gate must be opened regardless of whether the TV set's drive is missing or is present. The circuitry preceding the base, or gate, alters the operation of the subdrive output circuit.

Information and circuit drawings used in portions of this chapter were supplied courtesy of Toshiba Electronics Corporation and Sencore, Inc.

CHAPTER 6

Chroma and Video HDTV Circuits

INTRODUCTION

A block diagram of a typical HDTV receiver's video circuit operation introduces this chapter to the electronics TV service technician. These blocks include the comb filter, color demodulator, up-converter, video processor/color matrix, and the R, B, G, CRT circuits.

This chapter covers; information video switching circuitry and composite "S" Video switching systems as well as details of the internal blocks that make up a video processing IC.

The chapter continues with the video switching and processing circuit operations found in Thomson multimedia (RCA-GE) MM101 HDTV chassis. We will review some information you will find in Zenith Corporation's HDTV receiver video and color circuits. This will include the overall block diagram, color stream switch, CRT output circuits, CRT drive circuit, and video signal flow chart.

The chapter concludes with an overview of the digital video signal processing found in the Pioneer model PDP-505HD plasma, flat-screen HDTV monitor.

HDTV CHROMA CIRCUITRY OVERVIEW

The video flow block diagram in **Figure 6-1** shows the video input points and basic signal path for the video stages of a typical HDTV receiver. Depending on the input signal type, the input may bypass one or more of the video stages. As an example, the RF signal must pass through all the video stages, but the colorstream input (R-Y, B-Y, and Y) bypasses most of the video stages.

HDTV-ready sets accept NTSC (480I), 480P, and 1080I signals. However, the progressively scanned TV sets only display 480P or 1080I. These televisions convert the NTSC signal to 480P for display. All other formats must be converted externally to one of the previously mentioned formats before being applied to one of these receivers.

Note that in the industry, 480I and 480P are also referred to as 525I and 525P. 480 is the number of active scanning lines or lines containing video information; 525 is the number of total scanning lines.

The tuner and IF blocks are in the first video stages. This is often referred to as the front end of a TV receiver.

The next stage contains the comb filter. It separates the composite video signal into a chrominance (color C signal) and a luminance (detail or black and white, Y signal). After the Y and C separation, the color demodulator stage divides the chrominance signal into R-Y and B-Y signals.

Next, the up-converter doubles the horizontal frequency and changes video signals from an interlaced (480I) signal to a progressive signal (480P).

In the last stage, the signal enters video chroma deflection (VCD) IC, Q501. This is the video processing and color matrix IC. Q501's video

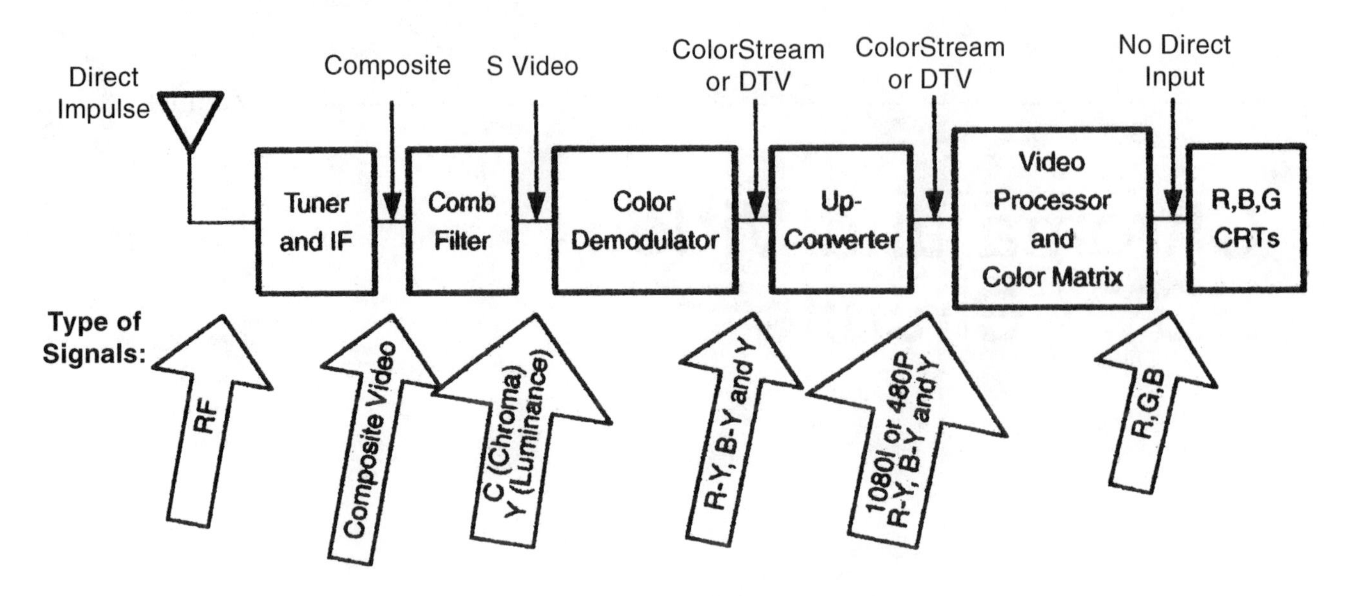

FIGURE 6-1

processing portion applies all the customer and service controls to the video signal, e.g. color, tint, R-cut, V-cut, etc. The color matrix combines the R-Y and B-Y to develop RGB signals that, with the luminance signal, are sent to the CRT's for display. Q501 also applies the on screen display (OSD), convergence grid, and closed captioning to the video signal.

VIDEO SWITCHING

To fully understand the video switching circuitry, the following explanations will break down the video switching into two parts. The first is for the composite and S-video switching. The second one explains the HDTV and Colorstream inputs.

Composite and S-VIDEO Switching

Figures 6-2 and 6-3 illustrate the basic operation of the video switch IC QVO1. Using the Video 1 input as an example, these diagrams demonstrate the QVO1's function. The other inputs operate in the same manner and are represented in the drawings by a notation of "Other."

Figure 6-2 shows the signal flow for the S-video (Y and C) input. The Y and C signals enter QVO1 on pins 11 and 9. Internal switch SW1 selects the proper Y and C outputs, and pins 44 and 42 of QVO1 apply the signals to the video processing IC, Q501. An adder also combines the Y and C signals into a composite signal that the TV receiver uses as a monitor output, located on the TV set's back panel, and as a sync source for the horizontal and vertical circuits. Internal switch SW2 selects the composite video signal and applies it to pin 46 as the output of QVO1.

The final output of QVO1 must be a luminance (Y) and a chrominance (C) signal. If the Video 1 input is a composite signal, it must be separated into Y and C signals. As shown in Figure 6-3, the composite video signal enters QVO1 on pin 7. SW2 selects the input, outputs it on pin 46, and applies it to the comb filter for Y and C separation. The composite signal on pin 46 is also used for the monitor output and for vertical and horizontal sync. After the comb filter separates the composite signal into Y and C signals, it reapplies them to QVO1 on pins 40 and 48. SW1 selects these signals as the source and applies them to pins 44 and 42.

Microprocessor QAO1 controls the internal switches through a 12C bus (on pins 27 and 26) and QVO1's internal D/A converter. An internal switch, placed in the S-video input jack, determines the type of input operation. The switch places a logic HIGH or LOW on the composite input of QVO1. The information passes through QVO1's internal D/A converter to the microprocessor. The switch defaults to open, which then indicates the use of a composite input. When the

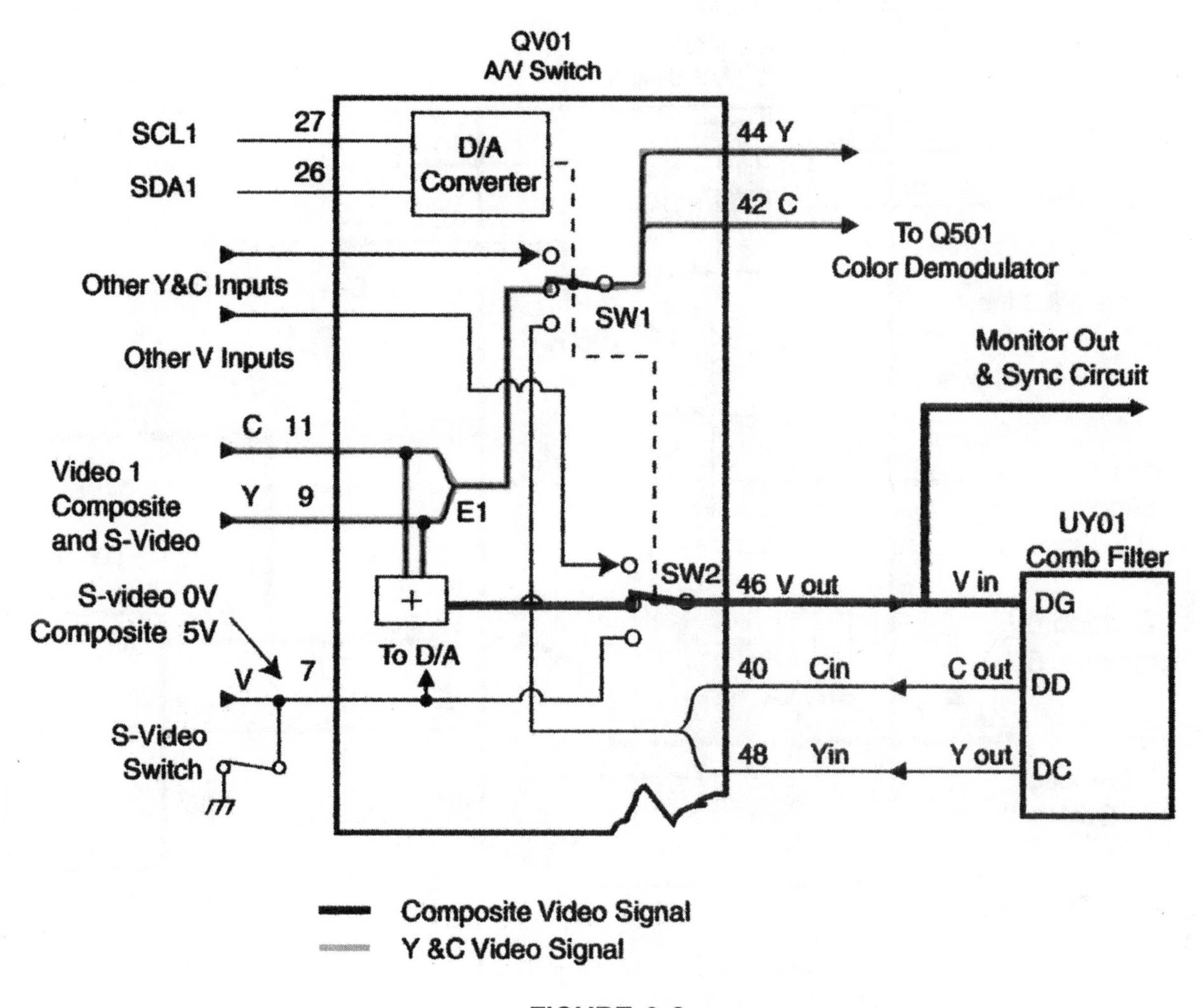

FIGURE 6-2

S-video cable is connected, the switch closes and places a logic LOW on the composite input.

The drawing in **Figure 6-4** shows the internal schematic for the composite and S-video switching IC, QVO1.

The picture in picture/picture out of picture (PIP/POP) switching operates in the same manner as the main picture; however, the composite signal is not separated into Y and C signals and reapplied to QVO1. The PIP/POP module itself separates the composite video signal and makes the selection between the separated composite signal and S-video signal.

VIDEO PROCESSING IC CHIP

The block diagram of the QVO1 IC with pinouts of video signal paths is shown in **Figure 6-5**. The selected composite video (V) signal is taken out of pin 46 of QVO1 and applied to the Video Output terminal on rear of the TV receiver. The same composite signal is applied to the 3D YCS comb filter. The 3D YCS comb filter separates the luminance (Y) and chrominance (C) signals and returns them to pins 40 and 48 of QVO1. The final outputs of QVO1 are pins 42 and 46. The Y and C signals are applied to Q501 (V/C/D) for video processing.

The video signal for the PIP/POP is taken from pins 32 and 34. Then, the signals are applied to the dual circuit.

THE S-VIDEO SIGNAL

When a cable is connected to the S-terminal input, an internal switch in the S-terminal shorts to ground

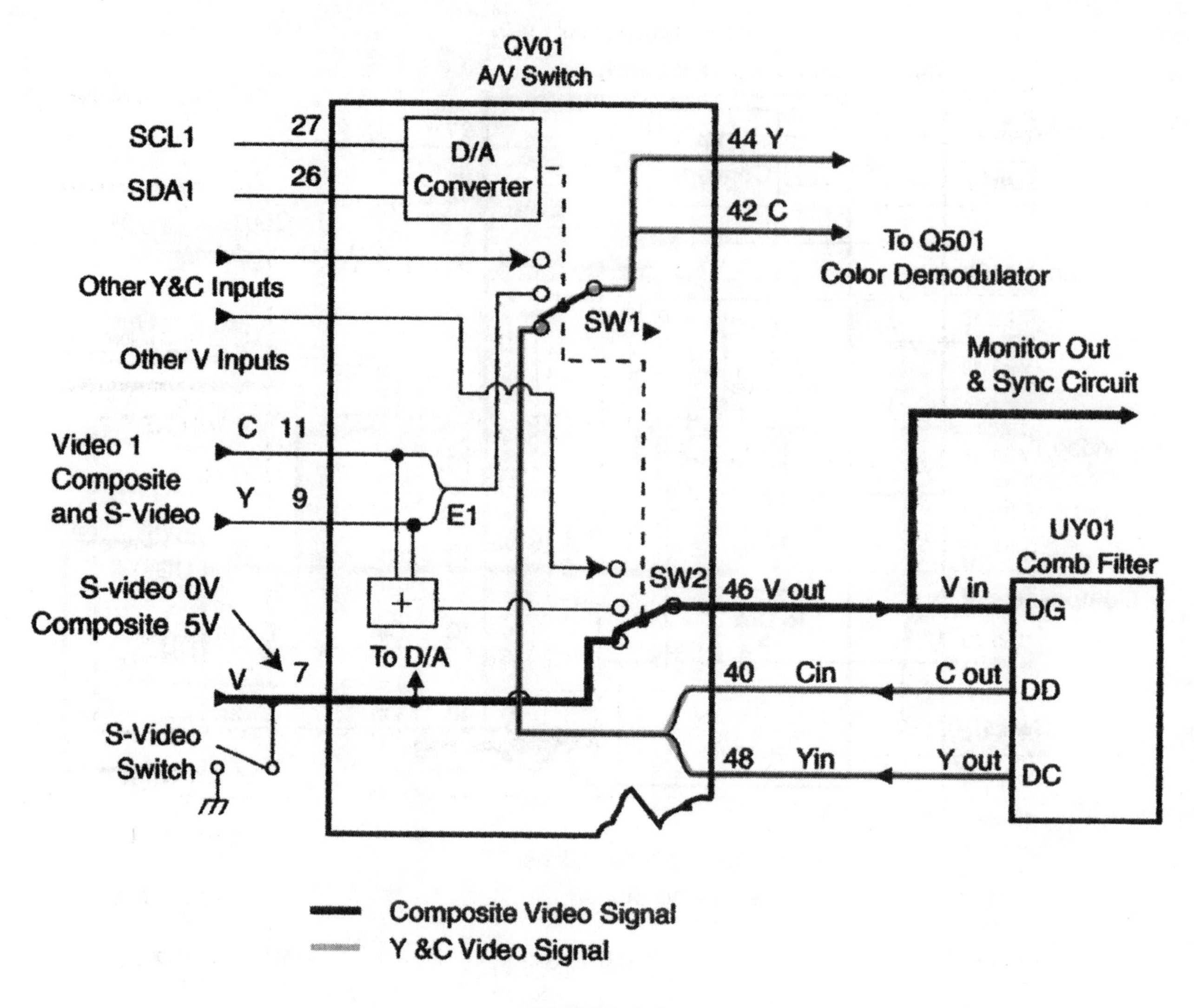

FIGURE 6-3

GND. The V-terminal (composite video terminal on QVO1), of the corresponding input, drops in bias level. When the S-video is selected, QVO1 sends the Y/C signals of the selected channel directly to the final outputs (pins 42 and 44) without passing through the comb filter.

REVIEW OF COLOR TV VIDEO PROCESSING

The MM101 RCA chassis uses YUV (Y black and white) with sync, U (R-Y), V (B-Y) component video for most of the video processing. Component video is generally used by the broadcast industry due to its higher bandwidth and the fact that fewer artifacts are created when component signals are combined, and then separated. In most cases, modern cameras output an RGB component signal from the pick-up device. The RGB mix is "tweaked" to provide a proper gray scale with available phosphors. The agreed mix to obtain a standard white was 30 percent Red, 59 percent Green and 11 percent Blue. When pointing a camera at a standard white card, the RGB output voltages will track these percentages. The RGB signals are then transferred from the camera to a video mixer where special effects and graphics are applied. The signal is then converted to composite video and modulated into an RF carrier for terrestrial or cable broadcast. The TV receiver captures the RF carrier, separates the composite signal, and then processes it back into its original RGB components to drive the CRT.

Internal Diagram of QV01

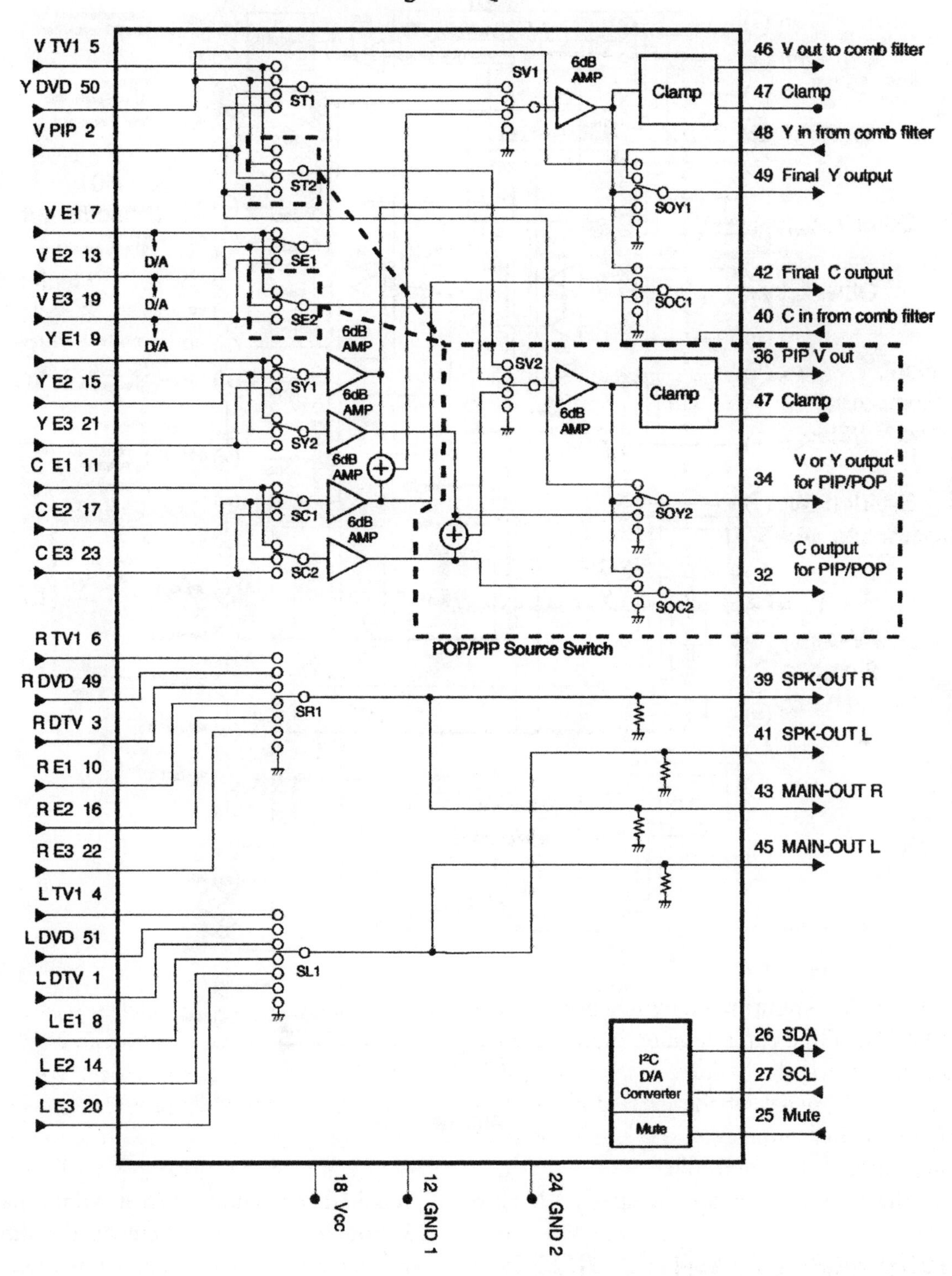

L = Left Audio
R = Right Audio
V = Composite Video
C = Chrominance Signal
L = Luminance Signal
E1 = Video 1 Input
E2 = Video 2 Input
E3 = Video 3 Input
DVD = Colorstream Input
DTV = Digital TV Input
SDA = Serial Data
SCL = Serial Clock

FIGURE 6-4

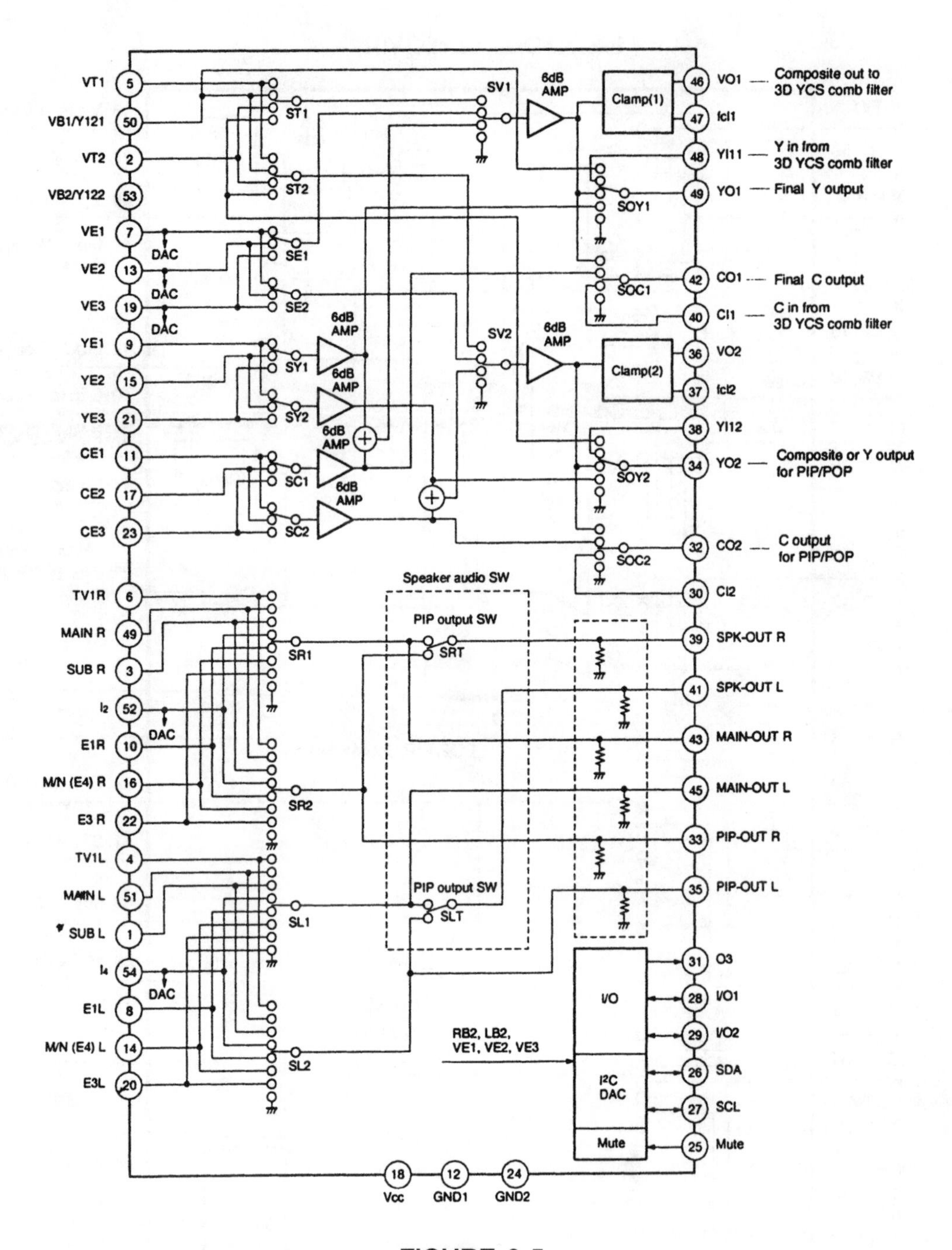

FIGURE 6-5

In the NTSC system, approximately 4 MHz of bandwidth is required to produce the 330 lines of resolution specified by the original NTSC committee who determined the criteria for acceptable video. In the days of black and white TV, this was not technically difficult. However, with color, to achieve identical resolution, a 4 MHz bandwidth for each color channel would be required. If the original RGB signals from the cameras existed, 12 MHz (3 x 4 MHz) of bandwidth would be required to produce a color signal. This would have required twice the existing (6 MHz) channel bandwidth. To solve this difficulty and make certain the new color system was compatible with the black and white system, composite video was introduced. RBG channels from the camera were combined in the proper percentages to produce a standard 4 MHz black and white luma signal that would be backward compatible with all existing TV receivers. To reproduce the original three-color channels, a decision was made to take advantage of the human eye's inability to resolve fine color detail,

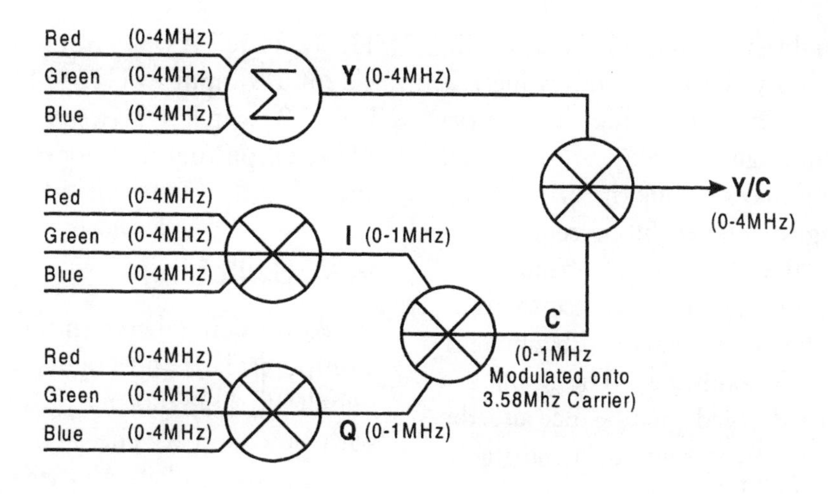

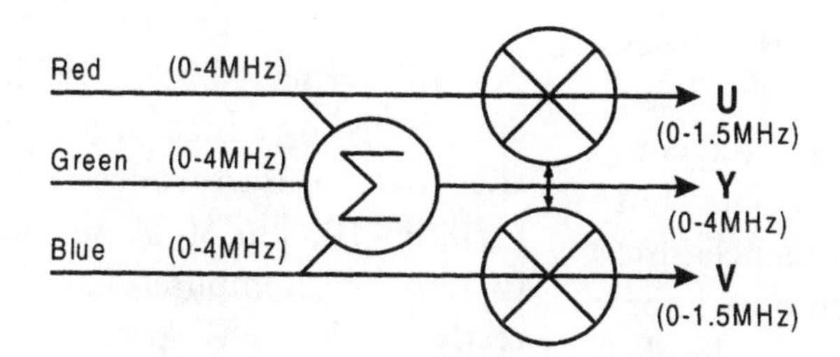

FIGURE 6-6

and reduce the bandwidth of the chroma signal to 1 MHz. This still meant it would take three additional MHz of bandwidth to reproduce the RGB color signal.

By combing the color signals into two difference signals (I & Q) created from the combination of the original RGB information, and then further combining these two signals together using phase modulation, a resulting signal, about 1 MHz wide, was created. The I signal consisted of 60 percent Red, 28 percent Green, and 32 percent Blue. The Q signal consisted of 21 percent Red, -52 percent Green and +31 percent Blue. This signal was then placed on a sub-carrier, which was in turn interlaced into the black and white signal. The combined signal fit into the original 4 MHz bandwidth as illustrated in **Figure 6-6**. This process is called encoding. The resulting signal, containing both black and white (B&W), along with color (chroma) information, is called "composite" or Y/C video. There has always been unavoidable crosstalk between the luma and chroma signal. When luma interferes with chroma, it is called "cross color," which produces the familiar spurious colors on newscaster jackets, or "dot crawl," during graphics overlayed on screen information. Crosstalk from chroma to luma is called "cross luminance" and shows up as a herringbone pattern on B&W TV receivers. Since B&W TV sets have almost disappeared, cross luminance is not a problem for viewers today.

VIDEO PROCESSING (RCA MM101 CHASSIS)

A YUV signal consists of three components generated by mixing the original RGB signals from the source. Again refer to Figure 6-6. They are the luminance signal (Y with sync), the Red color signal less the luminance signal U, (R-Y), and the Blue color signal less the luminance signal V, (B-Y). Because there is no encoding of the color signals to a separate color carrier, crosstalk is not an issue. The color bandwidth is increased to 1.5 MHz, allowing more fine color picture detail. Even though the human eye has trouble resolving fine color detail, it is desirable to keep video resolution as high as possible through the video processing chain in order to maintain the original quality of the signal. The YIQ processing may degrade the original signal from processing as much as 50 percent, resulting in less than 500 KHz of usable bandwidth. YUV processing generally degrades less than 25 percent, keeping the usable bandwidth above 1 MHz under the worst conditions. The result is a much cleaner, more detailed picture. Because the color and B&W signals are not mixed, laying graphics over a component YUV signal is much easier and results in less crosstalk than composite video.

YUV, in analog terms, may be called YPrPb. In digital video terminology it's known as YCrCb, but the mixed signals still adhere to the basic YUV format. The RCA MM101 chassis refers to YUV as YPrPb for external input signals, but internal processing uses the YUV nomenclature.

The intermingling of luma and chroma information in composite systems complicates the decoding process in receivers. Thomson multimedia uses complex filtering to ensure the minimum amount of composite signal crosstalk possible. The MM101 can process YUV to avoid the occurrence of signal crosstalk. The fine color detail is also greatly improved.

Video Input Switching

The MM101 is capable of accepting virtually any type of standard NTSC video input. These include a YUV (YPrPB) input (for DVD), computer VGA/SVGA inputs, VGA video to accept digital TV set top box inputs, three composite video inputs (with S-video (Y/C) capability), and standard NTSC broadcast video for the main and PIP tuners. The video input switching network consists of U22401, U22402, U38200, U16501 and U16500. Note that all video inputs are converted to YUV format prior to switching. Refer to block diagram in **Figure 6-7**.

YUV (YPrPb) Input

The YUV (Y, R-Y, B-Y) or YPrPb input is handled by U22402. The control signal for U22402 is provided by U38300 and is applied to pins 9, 10, and 11. "Selected YUV" from the Video Processor U22300 is input to U22402 at pins 3, 1, and 13. U22402 switches between external and internal YUV, outputting the proper signal on pins 4, 15, and 14.

VGA Switching

VGA switching between the two VGA inputs, Computer/Text and Digital TV, is done by video switch IC U38200 via pins 2, 3, and 4, along with 6, 7, and 8. The VGA Computer/Text inputs are most likely VGA/SVGA signals from a computer and will always be 2 H or greater. Digital TV inputs may be from many different sources such as a set top decoder box or satellite receiver, but will normally be 2 x H video. The two sources are selected by a high or low signal on pin 5 from the main microprocessor via bus expander U38300-10. The VGA output from U38200 has two paths. Computer/Text mode is routed directly to the video processor RGB inputs as it does not require tint, color, or autocolor processing. Contrast (pix level) and brightness (black level) are still available. The VGA Computer/Text mode is selectable via front panel controls or the remote. VGA Digital TV video is first converted to YUV, and then sent to the main YUV switch, U22401, pins 13, 1, and 3.

Baseband Switching

Video switches U16500 and U16501 provide all baseband (composite), S-video, and tuner video switching. Composite video inputs 1, 2, and 3 are applied to U16501 pins 5, 6, and 8. PIP video from the PIP tuner and main video from the main tuner are applied to pins 1 and 20, respectively, for selection. The front panel baseband video input is applied to U16501-3. The MM101 chassis has three (3) S-video inputs. All three are applied to video switch IC, U16500, across pins 5/6, 8/3, and 1/10. The two Y/C outputs from U16500 are sent to pins 15/16 and 17/18. The S-video inputs may be selected as the main or PIP display. IC's U16500 and U16501 are controlled via the IIC bus and the system control microprocessor.

All composite and S-video inputs from U16500 and U16501 are sent to the F2PIP IC. Composite

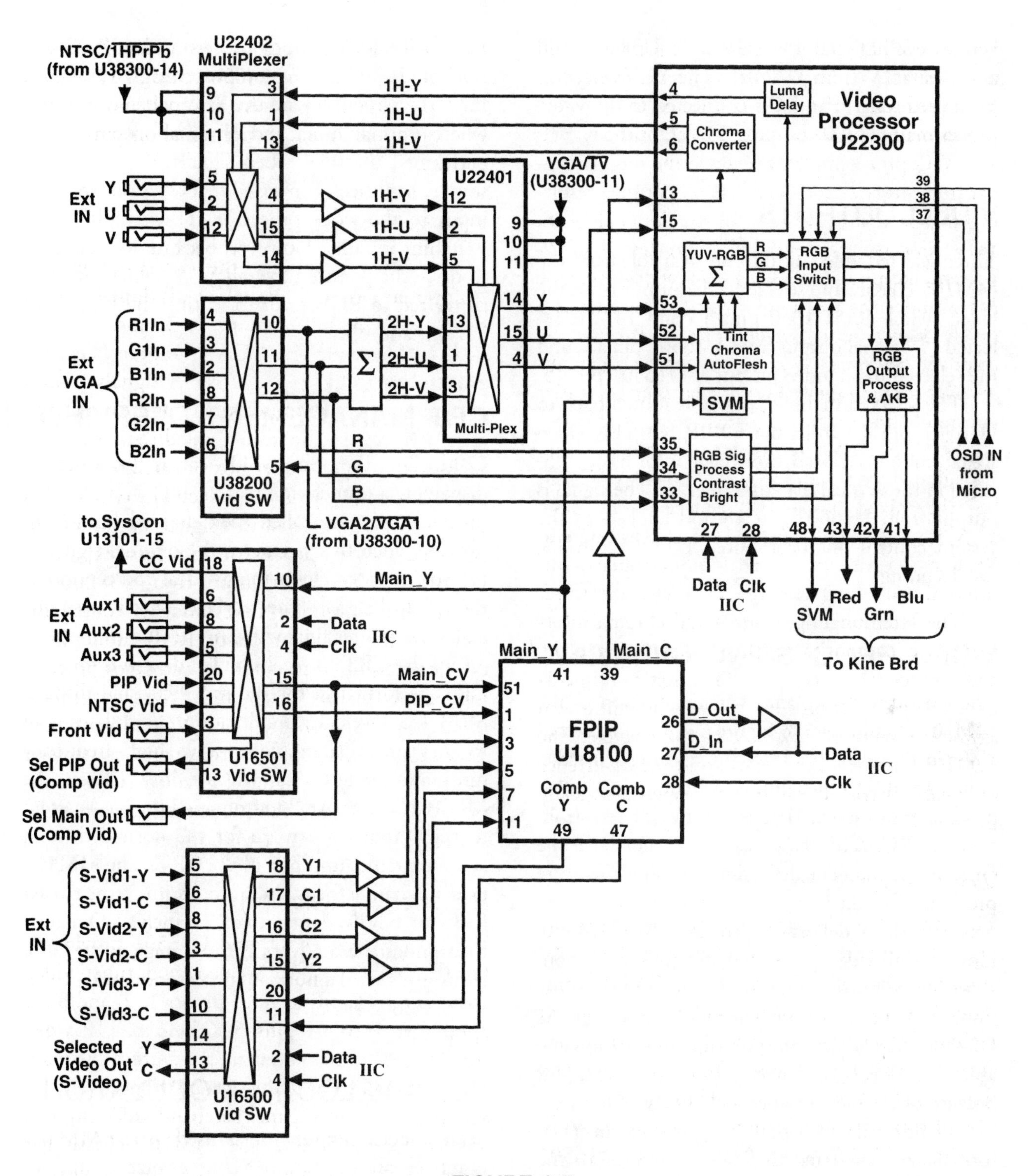

FIGURE 6-7

video is separated into Y/C (S-video) and sent to the video processor, U22300 pins 13 and 15. S-Video does not require any separation. The video processor converts the incoming Y/C signals from the F2PIP, U18100 pins 41 and 39, into YUV, and then passes the YUV signal to the video switching network.

Main YUV Switching

U22401 is the main YUV switch and is the primary YUV signal selection path. All incoming video, with the exception of VGA Computer/Text, is either already available in YUV or has been converted to YUV format. U22401 chooses between the converted VGA Digital TV YUV from U38200, and all other YUV from U22401. The YUV selection from pins 14, 15, and 4 is connected to the video processor, U22300, on pins 53, 52, and 51.

VIDEO OUTPUTS

Three selected video outputs are available from the MM101 chassis: the selected Y/C output from the video switch IC, U1650 pins 13 and 14, the selected PIP signal output from U16501 pin 13, and the selected main video output at U16501 pin 15.

The selected main "Y" signal is routed back to U16501 pin 10 from the F2PIP IC. This video signal contains closed captioned information located in the vertical blanking interval. The signal is output from U16501 pin 18 and applied to the system control microcomputer, U13101 pin 15, for decoding.

VIDEO PROCESSING OVERVIEW

The video processing and switching section of the MM101 chassis, refer to Figure 6-7, is composed of a Video Processor IC, U18100, and five (5) video switching ICs. Video input switching and video processing in the MM101 is slightly different from previous RCA/GE chassis as it does not have the familiar T-Chip. The deflection functions that were previously handled by the T-Chip are performed separately in a deflection processor IC, U14350. However, all video processing and AKB operation, as well as color, tint, brightness, and contrast functions, still occur within the video processor IC U22300. Signal switching remains much the same as earlier chassis, but may appear more complex because of a larger number and variety of inputs.

All internal video processing is done in YUV format and converted to RGB for output to the CRT drivers. All composite inputs (baseband video) are separated into Y/C by the F2PIP IC. The video processor converts the Y/C signal to YUV for further use.

There is a separate processing path for the RGB inputs. RGB comes in on pin 33, 34, and 35 for VGA "test operation." VGA text processing provides only contrast and brightness before application to the RGB output processing and AKB section of the IC. No color processing is available. Pins 51, 52, and 53 are the YUV video input pins where normal luma and chroma operations are performed. Normal video signals, no matter the source, have been converted to YUV prior to being input to the video processor. This simplifies the switching circuits. However, because RGB is used to drive the picture tube, all YUV signals must be converted to RGB prior to exiting the IC at pins 41, 42, and 43.

CRT MANAGEMENT OVERVIEW

Controlling the beam width scan is one aspect of displaying a proper video on a CRT. Beam current must also be controlled in a defined fashion to provide video that is true to the original signal or to provide video closer to an "ideal" perception of the original signal. The MM101 chassis uses several circuits after final video processing to properly set up the CRT to receive video data and for peak CRT performance for the specific visual display. SVM (scan velocity modulation) modulates scan to increase apparent contrast of high frequency luminance video. As with previous chassis the MM101 uses an AKB (automatic kine bias) system to track and compensate for the normal drift in beam current cutoff bias of the CRT. The MM101 uses a dynamic focus circuit to optimize the corner focus of CRTs larger than 27 inches. Dynamic focus modulates or "varies" the voltage to the CRT focus grids with a horizontal sawtooth and a vertical parabolic signal.

SCAN VELOCITY MODULATION

Scan velocity modulation is used in the MM101 family chassis to produce a sharper picture without enhancing noise. Note the block diagram in **Figure 6-8**. Large amplitude transitions (black and white) with very fast rise and fall times are hard for most CRTs to display as the tube must go from cutoff, or very nearly cutoff, to high beam current in a very short amount of time. To assist the CRT in the MM101 chassis, a separate yoke coil is placed around the neck of the tube in the vicinity of the electron

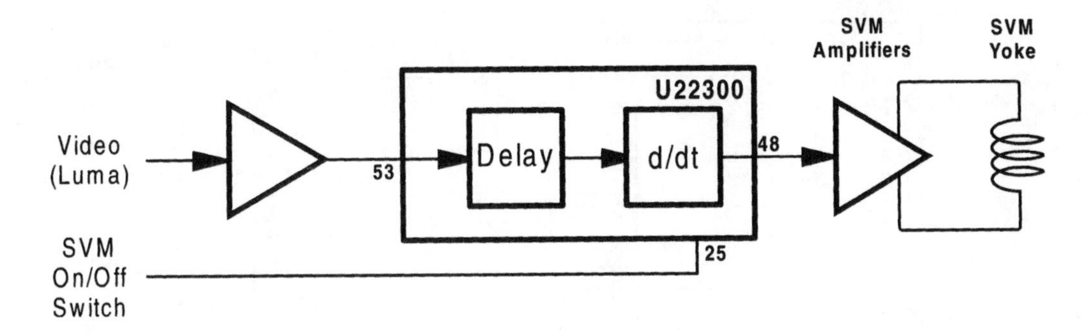

FIGURE 6-8

guns. As the electron beam travels across the screen, the SVM effectively reduces the size of the illuminated phosphor area while increasing the contrast. When the beam accelerates, fewer electrons will hit the phosphor, making the dark edges on the screen image appear even darker. When the beam decelerates, more electrons hit the phosphor, making bright edges on the screen image appear brighter. SVM circuitry is so effective, it can momentarily either double or completely stop beam scanning velocity as needed. SVM uses the first derivative of the luma video signal. The first derivative results in a slope-detected output that peaks in amplitude during any high frequency black-to-white or white-to-black transition. During low level, or low frequency video amplitudes, the SVM signal is zero.

SVM is relatively unchanged with the exception of added IIC bus control from the microprocessor to shut off SVM effects during OSD portions of the scan. The entire horizontal line is shut off, not just the OSD portion of the line.

ZENITH HDTV VIDEO CIRCUIT OPERATION

Refer to the block diagram of Zenith's QVO1 video IC and composite video signal operation system, shown in **Figure 6-9**.

The selected composite video (V) signal is taken off of pin 46 of QVO1 and applied to the VIDEO OUTPUT terminal on the HDTV receiver's back panel. The same composite signal is applied to the 3D YCS comb filter. The 3D YCS comb filter separates the luminance (Y) and chrominance (C) signals and returns them to pins 40 and 48 of QVO1. The final outputs of QVO1 are pins 42 and 46. The Y and C signals are applied to Q501 (V/C/D) for video processing.

The video signal for the PIP/POP is taken from pins 32 and 34. Then, the signals are applied to the dual circuit.

S-Video Signal

When a cable is connected to an S-terminal input, an internal switch in the connector shorts to ground. The V-terminal (composite video terminal on QVO1) of the corresponding input then drops in bias level. When the S-video is selected, QVO1 sends the Y/C signals of the selected channel directly to the final outputs (pins 42 and 44) without passing through the comb filter.

VIDEO PROCESSING CIRCUITS

The video processing block diagram drawing is shown in **Figure 6-10**. The Y and C video signals from the A/V selector circuit are applied to pins 15 and 13 of Q501. The "Y" signal passes through QVO1 unchanged. The C-signal is demodulated within Q501 and converted to R-Y (I) and B-Y (Q) signals. Q501 conveys the three signals on pins 4, 5, and 6.

The I, Q, and Y signals are horizontally compressed and are superimposed with the PIP/POP video signal in the dual circuit, and applied to the UP-converter (UP CON). The UP CON doubles the horizontal scanning frequency and converts the signal to a progressively scanned signal. The converted I, Q, and Y signals are returned to pins 53, 52, and 51 of the Q501 chip.

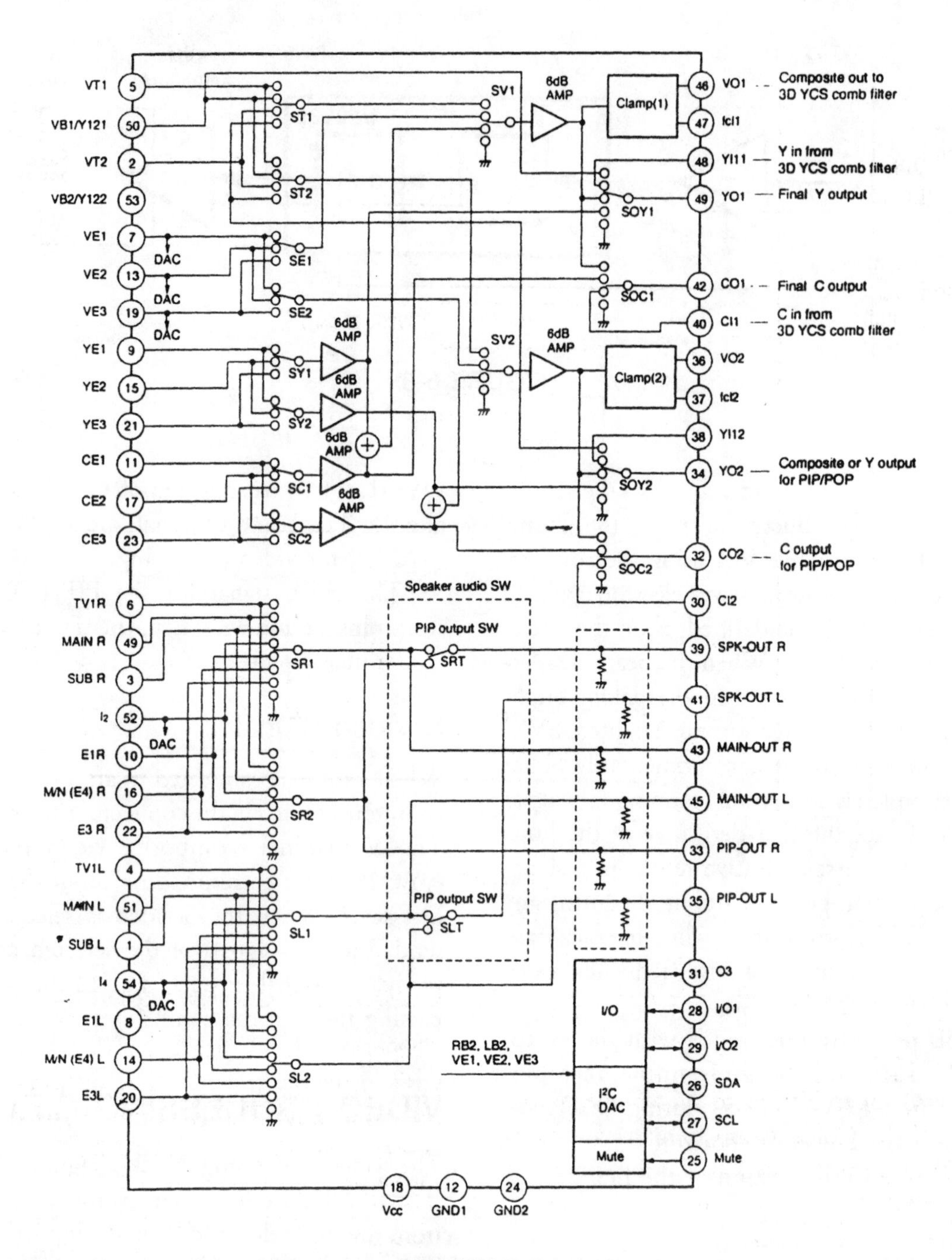

FIGURE 6-9

The Q501 chip controls brightness, contrast (unicolor), color intensity, and tint. It also corrects overall picture quality. The last stages in Q501 convert the I, Q, and Y signals to RGB signals and combine the on-screen display (OSD), closed captioning, and internal test patterns to RGB signals. The final RGB signals are sent out on pins 43, 42, and 41 and applied to the CRT drive circuit. The chroma IC block diagram is shown in **Figure 6-11**.

Colorstream Switch

The luminance (Y) signal input for the ColorStream uses the VIDEO 2 input terminal; the Cr (R-Y) and Cb (B-Y) use exclusive terminals.

As shown in **Figure 6-12**, the input identification for ColorStream is activated by setting pin 54 of QVO1 from LOW to HIGH when a cable is connected to the Cb input. Refer to **Figure 6-13** and note QVO1 communicates the activation of

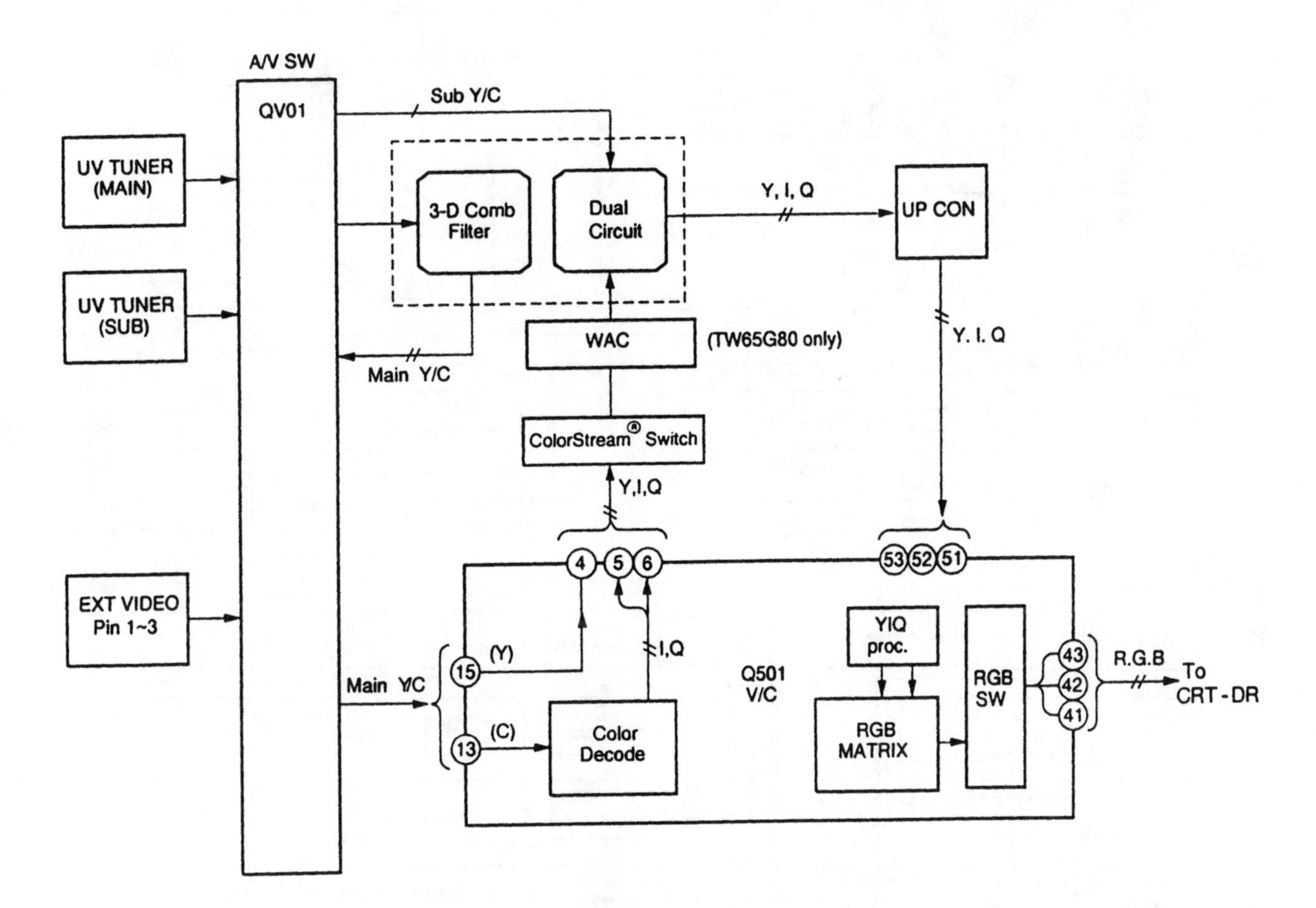

FIGURE 6-10

the ColorStream inputs to the microprocessor via the I square C bus.

When VIDEO 2 is selected, pin 14 of the microprocessor goes HIGH to switch the output of QW01. If the ColorStream input is not being used, VIDEO 2 can be used as a composite video input. Then, the switching is preformed internal by the QVO1 switch block.

CRT Output Circuit

The CRT output circuit is composed of a cascode amplifier (a cascode amplifier has a common-emitter and a common-base amplifier connected in series), fine white improvement circuit, and a blue extension circuit. As shown in **Figure 6-14**, the fine white improvement circuit and DC bias circuit are included in the R drive, and a blue extension circuit is included in the "B" drive unit.

Output Circuit Operation

The circuit in **Figure 6-15** shows the Red (R) CRT drive output. All three outputs operate in the same way. The R-output is used as an example. The output transistor Q901 connects to Q913 in a cascode configuration. The input signal is applied to the base of Q913. The collector of Q901 develops the output signal. The DC bias circuit controls the voltage level at the emitter of Q913. This voltage determines the gain of the amplifier and places Q913 into cutoff. R916, R914, R931, R932, and R936 establish the cutoff voltage. C914, L912, and L913 correct the frequency response of the amplifiers and operate as a high frequency peaking filter. To reduce blooming, a fine white circuit controls the DC bias during white-to-black transitions.

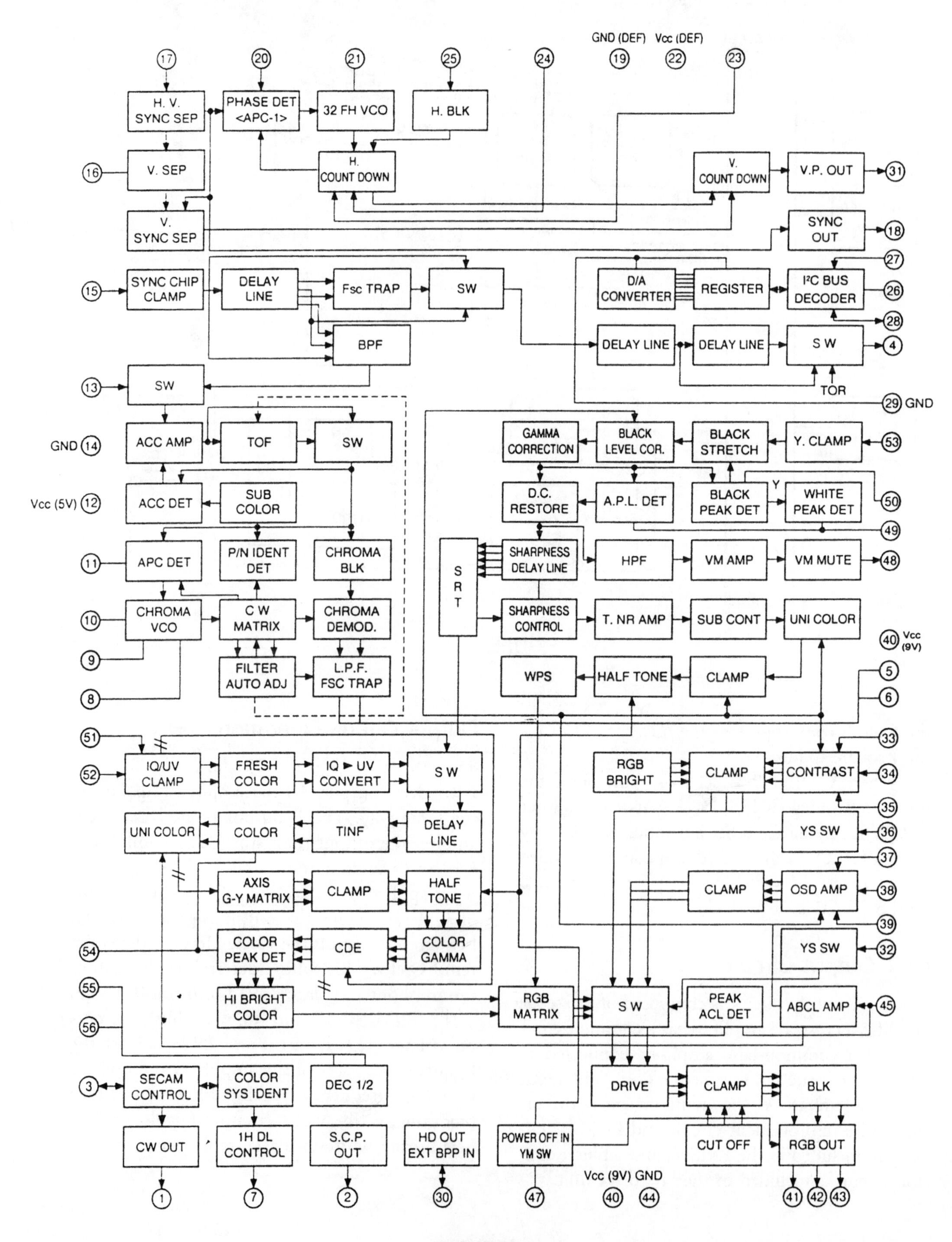
GND (DEF)
Vcc (DEF)
H. V. SYNC SEP
PHASE DET <APC-1>
32 FH VCO
H. BLK
V. SEP
H. COUNT DOWN
V. COUNT DOWN
V.P. OUT
V. SYNC SEP
SYNC OUT
SYNC CHIP CLAMP
DELAY LINE
Fsc TRAP
SW
D/A CONVERTER
REGISTER
I²C BUS DECODER
BPF
DELAY LINE
DELAY LINE
S W
TOR
SW
GND
GND
ACC AMP
TOF
SW
GAMMA CORRECTION
BLACK LEVEL COR.
BLACK STRETCH
Y. CLAMP
Vcc (5V)
ACC DET
SUB COLOR
D.C. RESTORE
A.P.L. DET
BLACK PEAK DET
Y
WHITE PEAK DET
APC DET
P/N IDENT DET
CHROMA BLK
S R T
SHARPNESS DELAY LINE
HPF
VM AMP
VM MUTE
CHROMA VCO
C W MATRIX
CHROMA DEMOD.
SHARPNESS CONTROL
T. NR AMP
SUB CONT
UNI COLOR
Vcc (9V)
FILTER AUTO ADJ
L.P.F. FSC TRAP
WPS
HALF TONE
CLAMP
IQ/UV CLAMP
FRESH COLOR
IQ ► UV CONVERT
S W
RGB BRIGHT
CLAMP
CONTRAST
UNI COLOR
COLOR
TINF
DELAY LINE
YS SW
AXIS G-Y MATRIX
CLAMP
HALF TONE
CLAMP
OSD AMP
COLOR PEAK DET
CDE
COLOR GAMMA
YS SW
HI BRIGHT COLOR
RGB MATRIX
S W
PEAK ACL DET
ABCL AMP
SECAM CONTROL
COLOR SYS IDENT
DEC 1/2
DRIVE
CLAMP
BLK
CW OUT
1H DL CONTROL
S.C.P. OUT
HD OUT EXT BPP IN
POWER OFF IN YM SW
CUT OFF
RGB OUT
Vcc (9V) GND

FIGURE 6-11

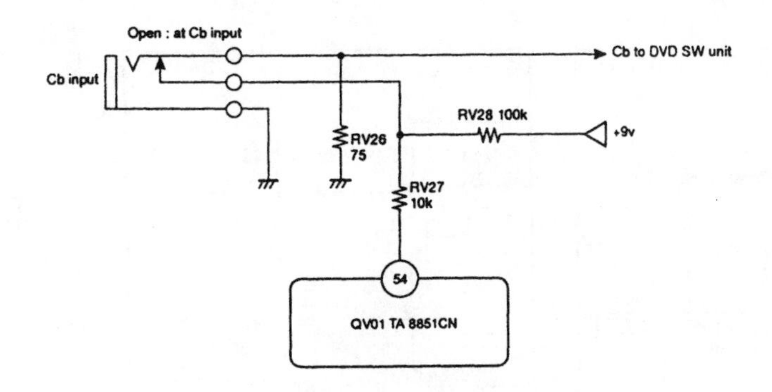

FIGURE 6-12

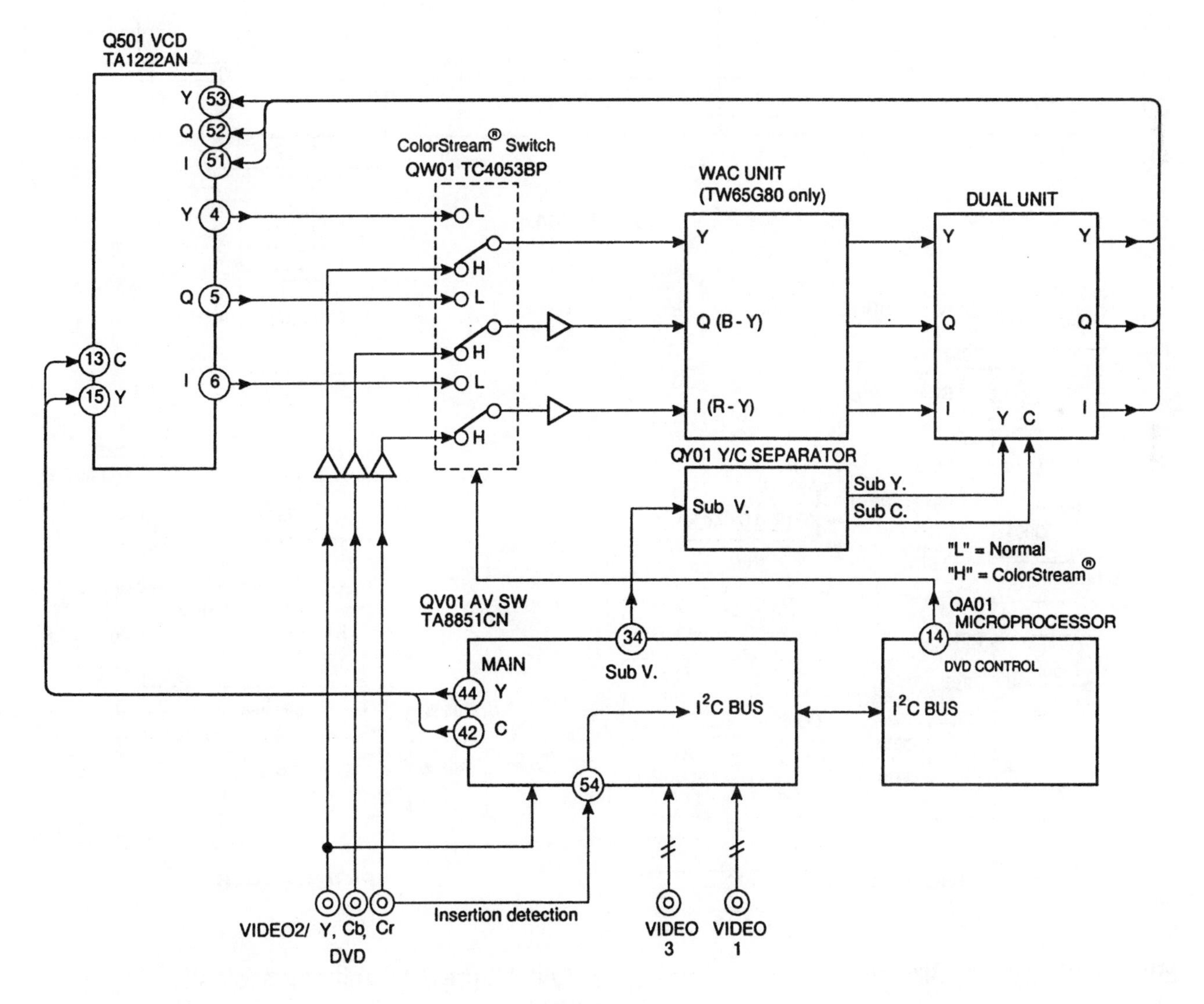

FIGURE 6-13

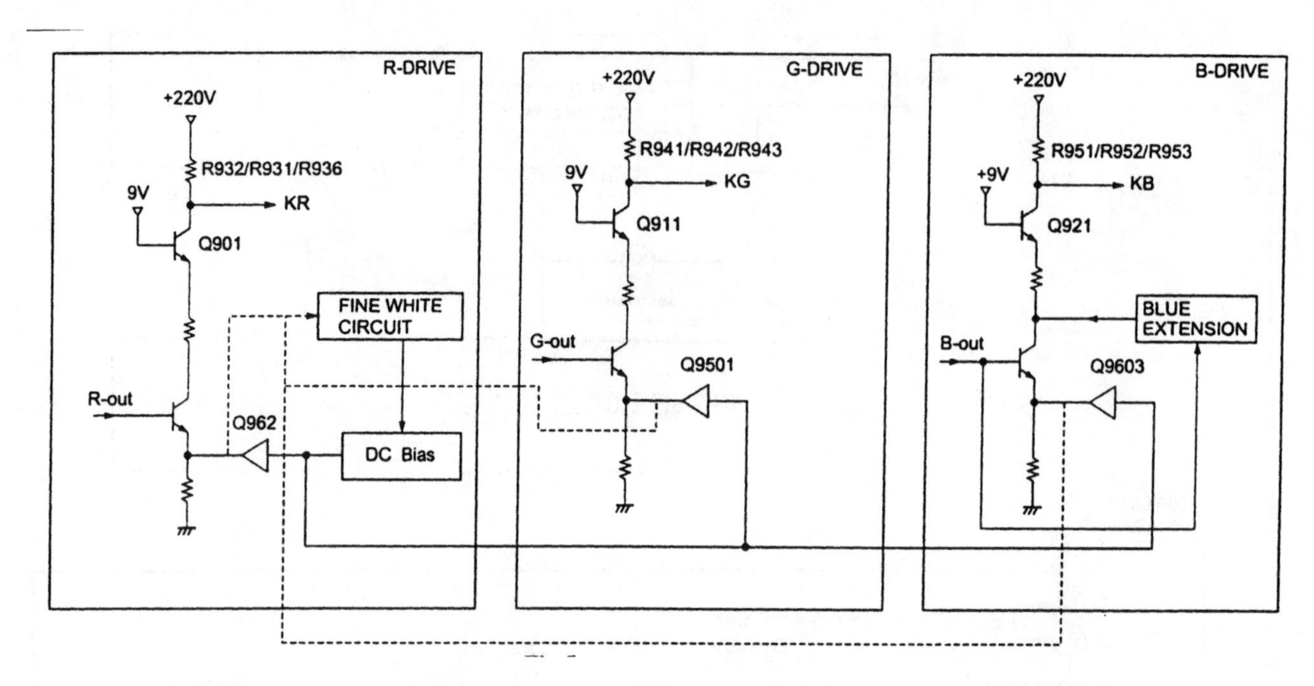

FIGURE 6-14

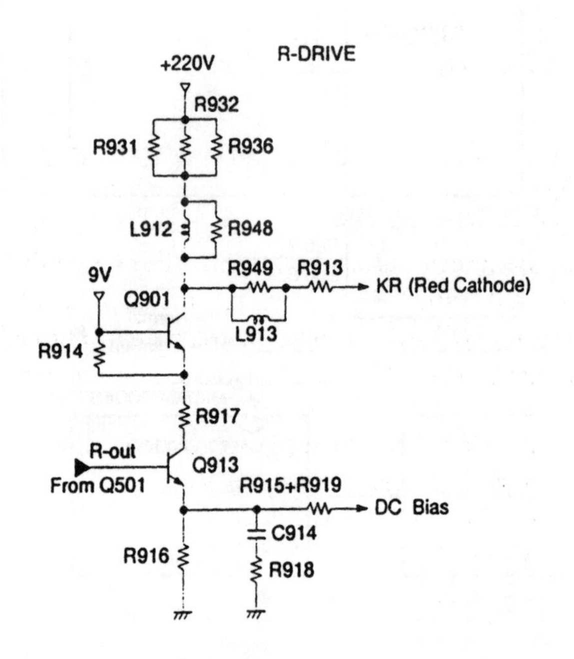

FIGURE 6-15

FIGURE 6-16

Blue Extension Circuit

Depending on the B-out level, the blue extension circuit controls Q921's emitter current and corrects the B-luminance signal to obtain an even white level.

When the B-out level rises, the base voltage of Q9605 exceeds the cut-off level specified by the emitter of Q9601. Then, Q9605 turns on and the gain of Q921 rises. If the level continues to rise, Q9602 turns ON and mutes Q9605. This segment of the waveform is labeled "C" in **Figure 6-17.**

Combined Operation

The block diagram in **Figure 6-18** illustrates the 3D-YCS comb filter being used as a combination for both 3-dimensional and 2-dimensional Y/C separation. The composite video signal is applied to

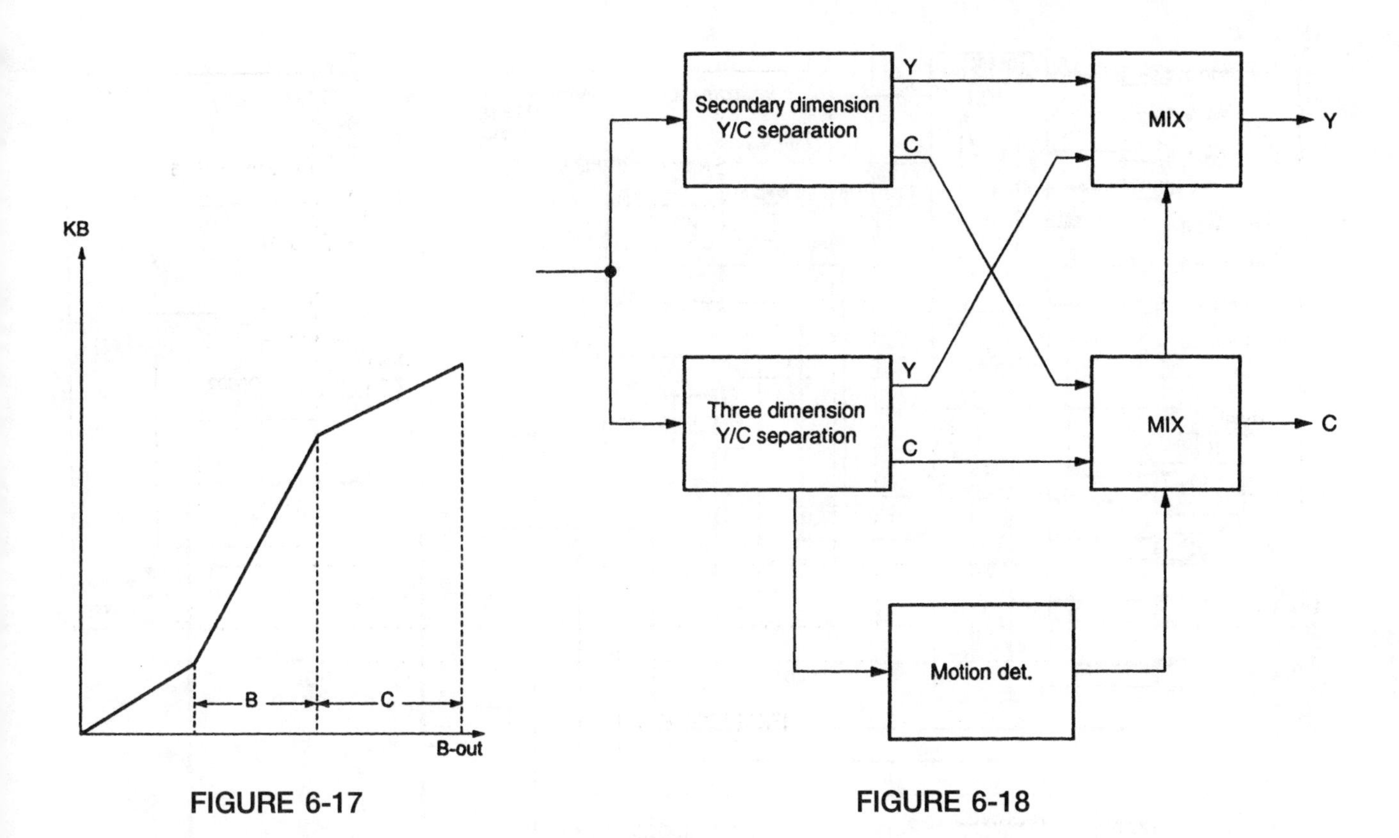

FIGURE 6-17

FIGURE 6-18

QZO1 and converted to digital information. The first frame of the video signal is delayed and stored in memory. If an area of a picture is stationary, 3-dimensional Y/C separation engages. If the area of the picture has movement, 2-dimensional Y/C separation is used. By using both the 2D-YCS and 3D-YCS, the defects of each system are compensated for and optimum Y/C separation is obtained.

VIDEO INPUT SWITCHING OVERVIEW

The RCA/GE MM101 chassis is capable of accepting virtually any type of standard NTSC video input and several other common video inputs. These include a YUY (YPrPb) input (for DVD), VGA/SVGA computer inputs, and VGA video to accept compatible digital TV set top box inputs. They also include 3 composite video inputs (with S-video (Y/C) capability), and standard NTSC broadcast video for the main and PIP tuners.

The composite video input switching network includes U16501 shown in the block diagram in **Figure 6-19**. It also supports three rear jack panel inputs, a front panel input and two composite signals to be displayed as a main or insert (PIP) picture. Any of the inputs, or either of the tuners, can be driven out of the SELECTED COMPOSITE OUTPUT jack. The IC switch is a TEA-6415C, 8x6 bus-controlled matrix switch. The eight inputs to the CV Matrix switch are as follows:

- Main Tuner Composite Video
- PIP Tuner Composite Video
- Front Composite Video
- YPrPb Y
- AUX-1
- AUX-2
- AUX-3
- Main Y from S-video

The outputs are as follows:

- Main Tuner Composite Video
- PIP Tuner Composite Video
- External Composite Video Note: Only incoming composite video may be switched to this output.
- Composite Video signal to the data slicer in the microprocessor.

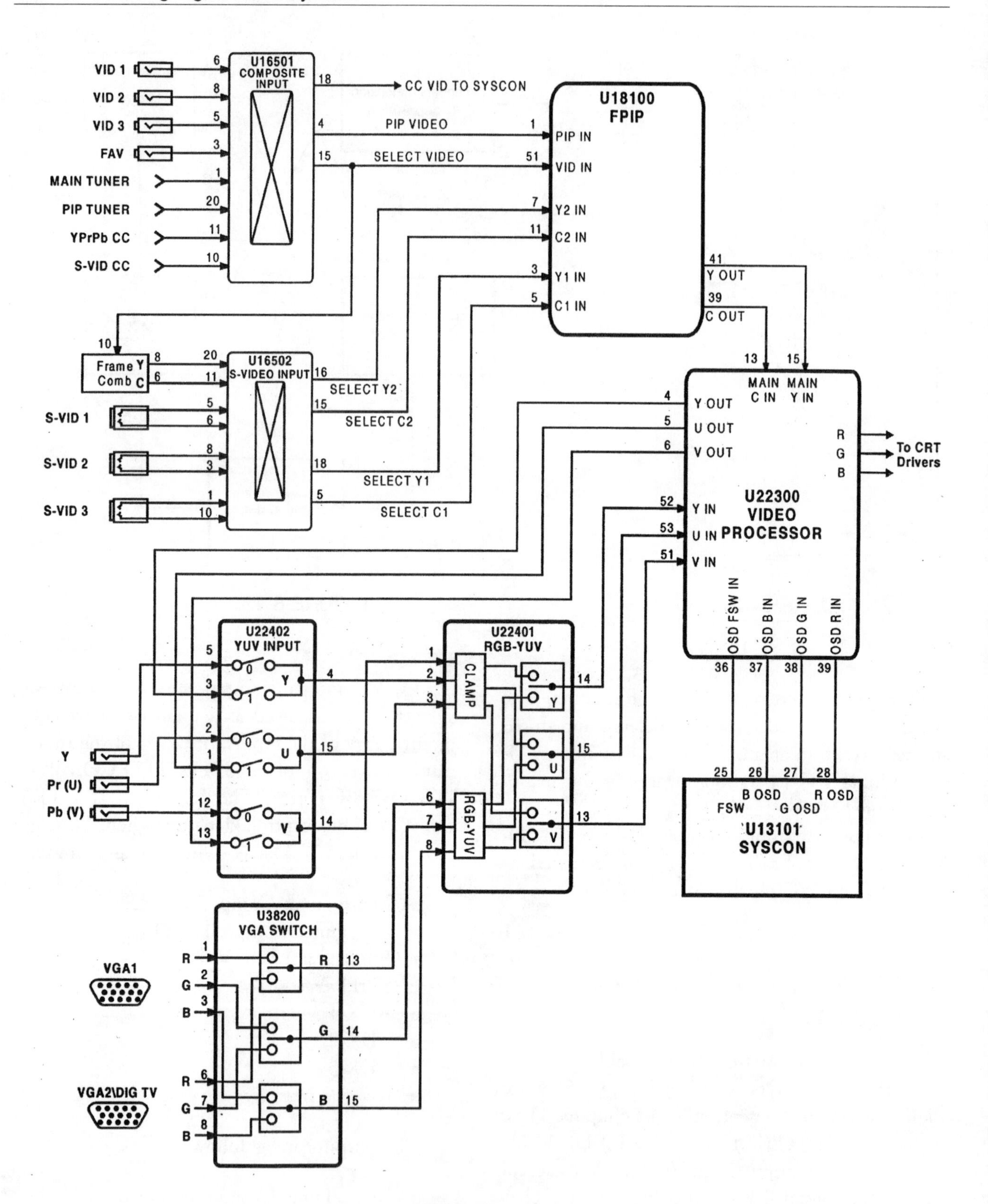

VID 1
VID 2
VID 3
FAV
MAIN TUNER
PIP TUNER
YPrPb CC
S-VID CC
U16501 COMPOSITE INPUT
CC VID TO SYSCON
PIP VIDEO
SELECT VIDEO
U18100 FPIP
PIP IN
VID IN
Y2 IN
C2 IN
Y1 IN
C1 IN
Y OUT
C OUT
Frame Comb
U16502 S-VIDEO INPUT
SELECT Y2
SELECT C2
SELECT Y1
SELECT C1
S-VID 1
S-VID 2
S-VID 3
MAIN C IN
MAIN Y IN
Y OUT
U OUT
V OUT
Y IN
U IN
V IN
U22300 VIDEO PROCESSOR
To CRT Drivers
OSD FSW IN
OSD B IN
OSD G IN
OSD R IN
U22402 YUV INPUT
U22401 RGB-YUV
CLAMP
RGB-YUV
Y
Pr (U)
Pb (V)
B OSD
R OSD
FSW
G OSD
U13101 SYSCON
U38200 VGA SWITCH
VGA1
VGA2\DIG TV

FIGURE 6-19

The CC Video output can be any of the available composite video inputs or the "Y" from the displayed S-video. No provisions are made to output S-video or any of the various VGA inputs.

The main and PIP tuners have mute ability to reduce jacks, sharing the audio inputs with their respective composite video inputs. Combed Y/C from the optional frame comb is also input to this switch. An auto detection scheme will select S-video over composite video if sync is detected on the corresponding Y/C input. Registers are provided to report the status of the sync detectors present on the two S-video inputs routed to the U18100 chip.

U22402 switches between composite and S-video signals that have been converted to YUV format and the rear jack panel YPrPb inputs.

U38200 switches between the two VGA/SVGA inputs. U22401 switches between incoming RGB from the VGA/SVGA input switch and incoming YUV from the YUV input switch. U22401 also converts the RGB from U38200 to YUV so that only YUV exits. The video processor now accepts the selected YUV for video processing.

IC SWITCHING AND VIDEO PROCESSING

Video switching in the RCA/GE MM101 chassis is complex. It may be helpful to note that the video switching takes place inside the two main video ICs, the video processor, U22300, and the F2PIP IC, U18100, that is shown in the **Figure 6-20** circuit drawing.

Baseband and S-video signals are routed directly to the F2PIP IC. Inside the F2PIP IC, baseband video is digitized and then separated to Y/C (S-video). Once in Y/C format, if a frame comb filter is not available, the video is output to an analog comb filter from pins 47 and 49. It then re-enters on pins 43 and 45 and is available at the main video switch inside the F2PIP IC chip.

Y/C video is fed directly to the main video switch for selection. A composite signal generated from the Y/C input is fed back to the main and PIP analog input switches. The composite signal for the PIP switch is used for the PIP picture when an S-video input is selected as the PIP input. There are two S-video inputs to allow display of the two S-video sources during PIP operation. (In normal operation, the second S-video input is polling the other S-video inputs, looking for incoming sync in the auto-select process.) The PIP analog switch output is digitized and processed, adding brightness, color, and tint control. It is then separated to its Y/C components and fed to the PIP overlay switch where it is placed on top of the selected main video input.

The main video switch selects the video for the main picture output. The output is fed to the PIP overlay switch. During normal video it is the full video display. When PIP is selected, the reduced PIP display (about 1/9th full screen) is superimposed on top of the main display.

COMB FILTERING

Two methods of comb filtering are available in the MM101 chassis: an optional frame comb filter and traditional analog comb filtering used with the F2PIP IC.

Frame Comb Filter

The frame comb filter receives composite video from the composite video input switch and, following a sync tip clamp, the signal is digitized to 8 bits. An external low pass filter in front of the A/D converter minimizes aliasing components from the sampling process. The digital comb, with two 1fH delay lines, adoptively separates the luma and chroma signals, which are then converted back to analog. The Y/C signals are low-pass filtered externally and routed back into the S-video switch, U16502. If composite video is selected, the S-video that results from the frame comb filter is input to the F2PIP IC, U18100 pins 3 and 5. This allows the frame comb module to be used instead of the F2PIP comb filter. The frame comb filter is auto-detected at startup. If system control detects the frame comb IC at power up it will configure the switches for frame comb operation. Otherwise, the analog comb filter in the F2PIP is utilized.

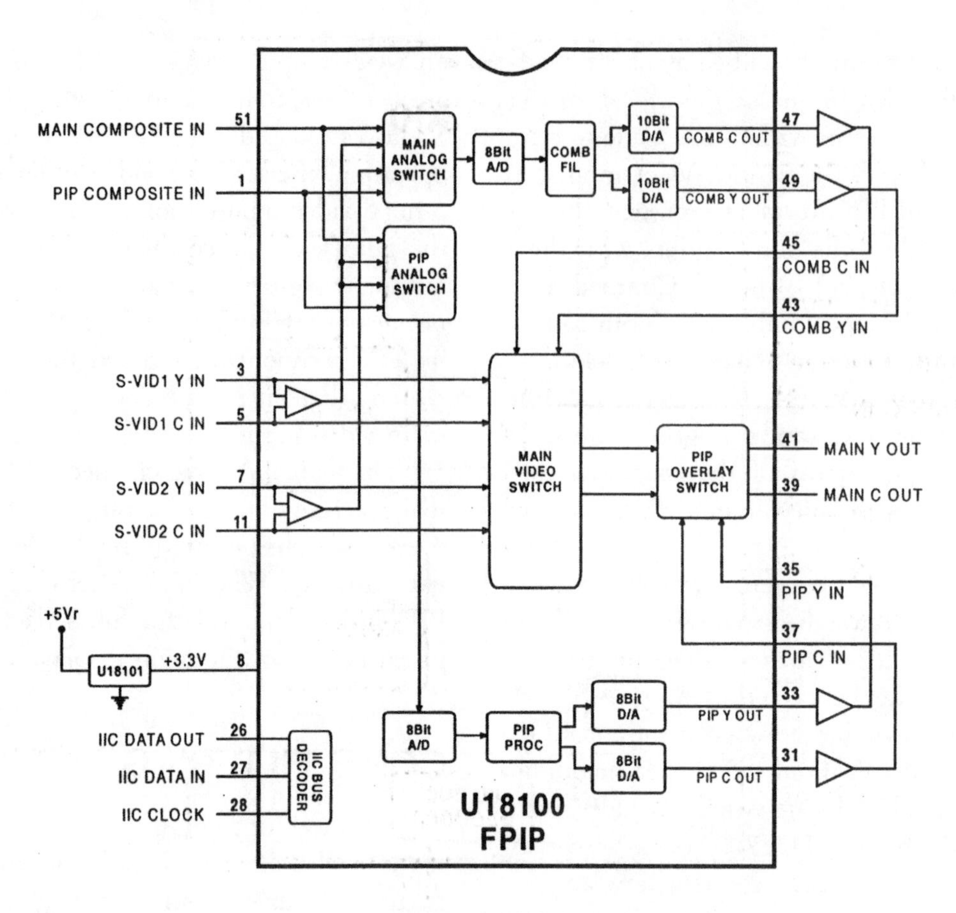

FIGURE 6-20

Analog Comb Filter

U18100, the F2PIP IC, may perform both comb filter and PIP functions. The two main inputs, Y1/C1 and Y2/C2 are fed from the S-video switch, U16502. The third Y/C input is from the external analog comb filter. A Y/C output from the main internal composite input switch is comb filtered, and then re-enters the F2PIP IC on pins 43 and 45. This comb-filtered Y/C signal is connected to the internal main video switch for access by the F2PIP.

Video Processing

The main video processor, U22300, in **Figure 6-21**, accepts a YUV input adding overall brightness, contrast, color, and tint control of the selected signal. A luma derivative is separated prior to any video processing to drive SVM (scan velocity modulation). The YUV signal is then converted to RGB and placed on the RGB IN switch. RGB IN switches between the main RGB video and OSD (generated in RGB format) coming from system control. The resulting signal, either main video or main video with OSD overlayed, is output in RGB format to the CRT drivers.

Incoming VGA is handled in two ways. Either input VGA1 is always expected to be a computer video adapter input, or user control of the brightness and contrast is all that is provided.

Input VGA2 can be switched between a computer video adapter and television signal video as might be expected from a digital television converter device or some other form of compatible entertainment video device input. When switched to digital TV mode, user control of color, tint, and sharpness returned.

Incoming Y/C video from the F2PIP IC pins 13 and 15 is converted to YUV format and then goes back out of pins 4, 5, and 6 for further switching.

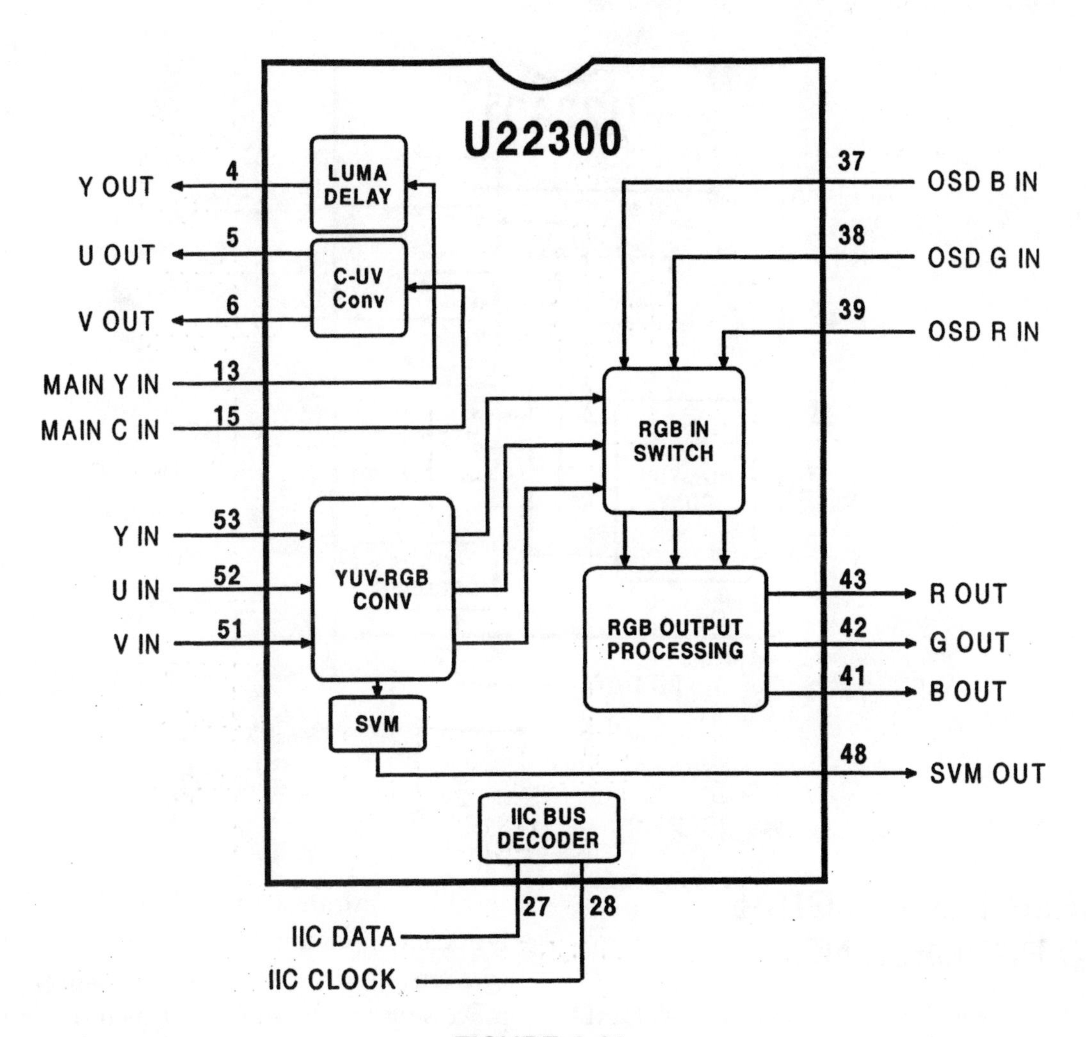

FIGURE 6-21

The back-end of the video processor IC accepts YUV format video and enhances it with the following functions:

1. User controls for pix, black level, sharpness (with noise coloring), color saturation, and tint.
2. Black stretch with auto-pedestal to expand blacks to full dynamic range.
3. Edge replacement to sharpen the luma edges.
4. SVM processing to enhance edges.
5. Noise reduction to improve appearance of noisy sources.
6. Auto-flesh is available, for NTSC signals only, to minimize scene to scene variations in tint.
7. Beam current limiting.

All incoming video signals eventually are converted to YUV before entering the main video processor IC for final processing. The U22401 chip, shown in the **Figure 6-22** drawing, is a video switch with unique capabilities that enable it to switch between YUV signals, RGB signals, or a combination of the two. In this case, RGB from the VGA switch is placed on pins 6, 7, and 8. The RGB is converted internally by the IC to YUV format and then made available to the main YUV switches.

The other video signals, previously converted to YUV, enter on pins 1, 2, and 3 and are routed to the main YUV switches. The main switches then select the inputs via IIC commands from system control.

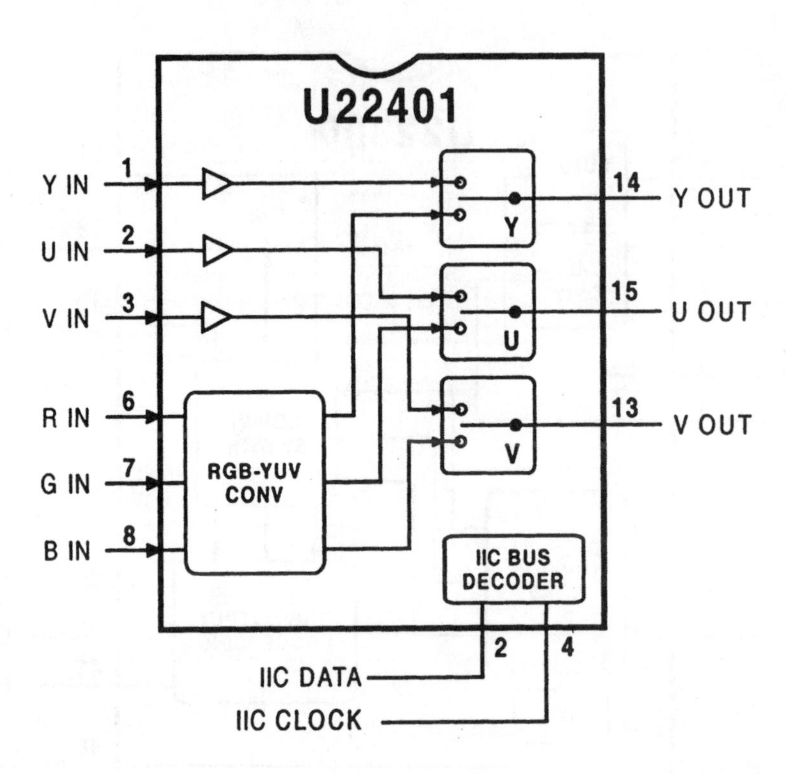

FIGURE 6-22

PIONEER HDTV DIGITAL VIDEO PROCESSING

In this section we will cover the model PDP-505HD Pioneer flat screen plasma digital signal processing for the video portion of this receiver.

Automatic Format Converter (AFC)

Pioneer's automatic format converter achieves 8 times the normal NTSC signal density for the ultimate in picture quality from conventional sources. This 8X, high density conversion process is illustrated in the **Figure 6-23** drawing.

This digital video signal processing, using the automatic format converter, achieves 8 times the normal NTSC signal density for the ultimate in high picture quality from conventional sources. The drawing, in **Figure 6-24** illustrate this density progressive format. The drawings in **Figure 6-25** show, for comparison, the original (1 field), the double density progressive format, the quadruple density progressive format, and pioneer's 8-fold density progressive format.

The model PDP-505HD AFC with PureCinema circuit detects a film-based source and converts it to a progressive format with precise processing for the smoothest picture presentation. Refer to drawings in **Figure 6-26**. Each original still film frame is recreated and displayed alternately, 2 or 3 times, for an incredibly pure image.

This digital signal video processing AFC converts 1080i HDTV signals to 1080P for more efficient processing and a sharper picture. The 2X progressive conversion is shown in the **Figure 6-27** drawing.

Pioneer's "New Clear," pixel driving technologies nearly eliminate the traditional "reset-discharge" cycle to eliminate unwanted light emissions and maintain deep rich blacks.

Pixel Driving Technologies Benefits

The benefits of pixel driving technologies for flat screen plasma HDTV receivers are as follows:

- Solid blacks on the screen
- Clean and clear detail in dark screen areas
- Lower energy usage resulting in less fan noise

Some of the technical information and drawings in this chapter are courtesy of Thomson multimedia, Zenith Electronics Company, Pioneer Electronics of America, and Toshiba of America Electronics Company.

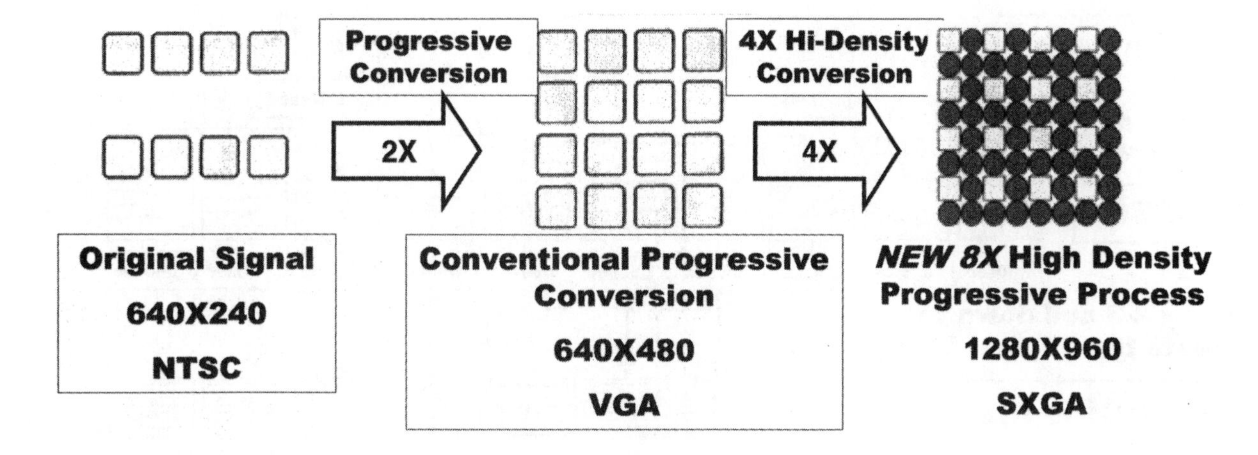

FIGURE 6-23

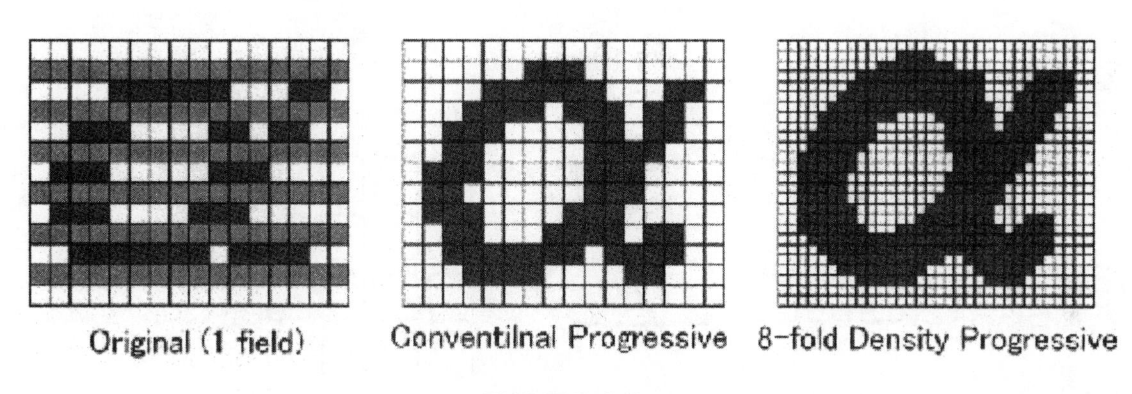

FIGURE 6-24

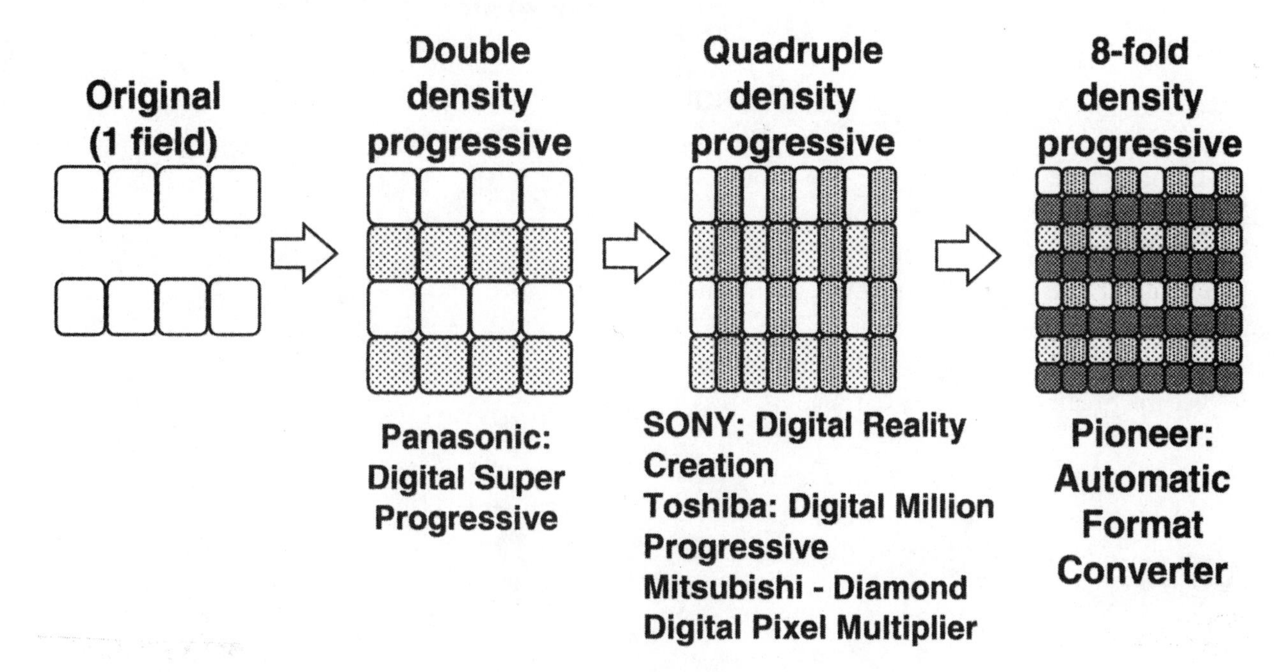

FIGURE 6-25

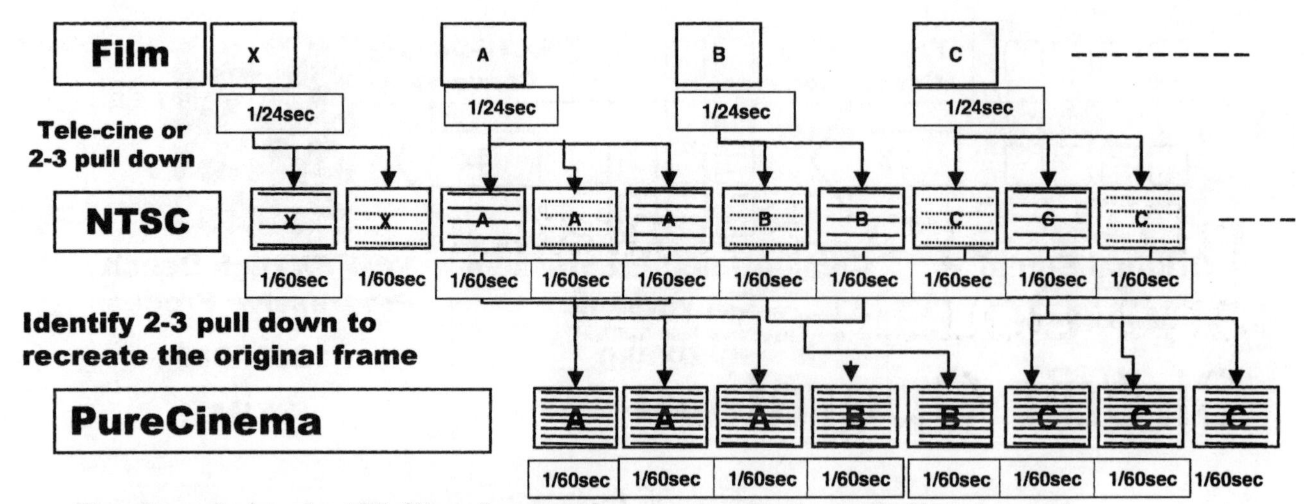

FIGURE 6-26

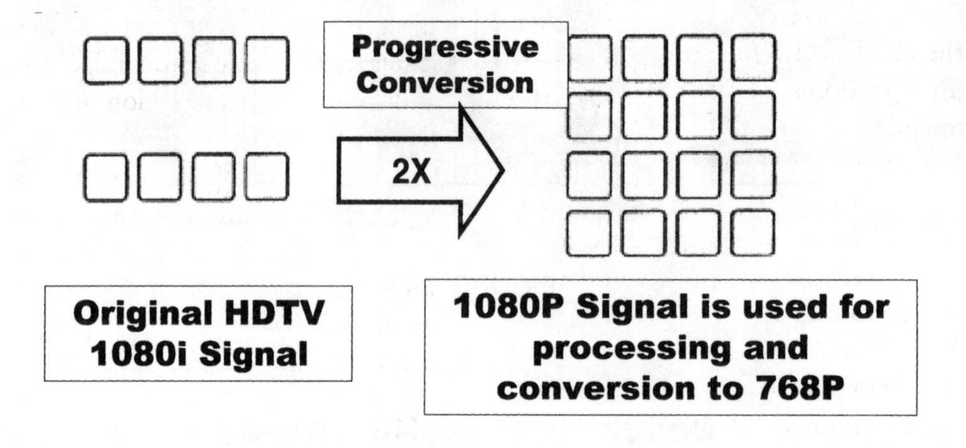

FIGURE 6-27

CHAPTER 7

HDTV Sync Separation and Double-Speed Circuit Operation

INTRODUCTION

This chapter begins with an overview of the HDTV evolution and non-interlaced scanning systems. This will include the HDTV 1125/60 standard, MPEG-2 video system operation and, other HDTV sweep and sync standards.

Then, we'll move on to a review of analog video sync signal waveforms and vertical interval test signals (VITS). Next, we'll cover information on horizontal and vertical sync pulse standards.

The chapter continues with the sync signal path for a typical HDTV receiver that has provisions for 480i and 1080i horizontal sweep speeds. Next, we'll review information on digital logic control and DTV sync detection and separation circuit operation. Also, circuit operation of the double-speed (up-con) circuits will be discussed.

PROBLEMS WITH INTERLACED SCANNING

Let's now review the problems that non-interlaced TV pictures have. With very high resolution receivers, a small amount of flicker is very noticeable when the screen phosphors fadeout. With each individual phosphor dot the rate of decay will vary the light output. And, of course, the interlaced scanning lines will be very noticeable. As the viewer comes closer to the TV screen the lines are more prevalent and distracting to the viewer. This is how the viewing ratio distance was developed for 4.5 to 6 times the height of the TV screen. From this distance, the scanning lines appear to merge and the eye thinks a complete picture image is being displayed. Of course, on large 36-inch CRT sets and large-screen projection screens the scanning lines will be seen.

To eliminate the flicker and scanning lines being viewed, the following fix was developed. This technique now used on computer monitors and HDTV receivers uses a non-interlaced refresh or progressive scanning. With progressive scanning, every line of picture information displayed is scanned by the beam at each sweep across the panel. This technique will enhance the vertical resolution of the set's display and the viewer can be much closer without seeing any sweep lines. With the progressive scanning technique the viewing distance ratio can now be cut down to 2.5 times the height of the set's display screen.

DEVELOPING THE HDTV STANDARD

The TV system you are now viewing, and the NTSC standard, were approved by the FCC in 1940. This conventional system has 525 horizontal scanning lines and interlaced scanning. Also, transmission is separate for the chrominance and luminance signals, and the system has a frame rate of 60Hz. Of these 525 lines, you will only see 483. The rest are for interval timing, sync, and other functions. All of these signals take up a bandwidth of 4.2 MHz. Other countries use the NTSC system. However, Europe uses pal system that has a 50 Hz frame rate, which is dictated by their power line frequency.

Another TV standard, developed by the French, is called SECAM, or sequential coleur avec memoire, and is used in eastern European and Russian countries. With these different TV systems there is also an incompatibility factor. As an example, the SECAM system cannot display a PAL TV signal image because of differences in the broadcast transmitter and station equipment. And the NTSC systems cannot reproduce the PAL station broadcasts as the frame frequencies are different.

In the late 1980s, at the request of TV broadcasters, the FCC convened a committe to act as the HDTV advisory service. This committee was named ACATS for Advisory Committee on Advanced Television Service. The FCC decided that the new HDTV signals would be broadcast on currently unusable channels and that the TV stations would be temporarily assigned a second channel for the digital HDTV channel.

Three design teams that started work on the HDTV standards combined all their work together in the early 1990s to produce a high-definition system design. Many consumer TV electronics product companies and engineering colleges, including MIT, formed the Digital HDTV Grand Alliance.

These high-definition television standards produced by the Grand Alliance set the technological standards for the merging of broadcast TV, cable TV, telecommunications, and computer technologies. And of course, this new HDTV digital system standards affects both the TV signal transmission and the color TV receiver designs.

The amount of screen luminance for HDTV doubles both the horizontal and vertical range for the viewer. The HDTV system provides about 4 times as many pixels as the conventional NTSC system. The wider screen aspect ratio produced by the HDTV systems gives the viewer much more visual information. The HDTV system increases vertical deflection with the use of 1125 lines for scanning the screen. There is more video picture detail because of a 5-fold increase of the video bandwidth.

The NTSC system has an aspect ratio, which is the ratio of picture width to picture height, of 4:3. However, the increased picture width for HDTV produces an aspect ratio of 16:9. For this reason the viewer gains a screen area of almost 6 times as much information. HDTV will not only be useful for home video viewing, but for industrial, information capture, storage, and retrieval applications. It will also be valuable for educational, medical, and various cultural applications. With all of these applications, HDTV will provide the picture quality for training, teleconferencing, and product features for sales promotions.

EXPLANATION OF THE HDTV 1125/60 STANDARD

The HDTV design team chose the 1125 scanning lines with a picture refresh rate of 60 Hz. This type of standard compares with the type of resolution given by projecting a 35mm formatted film slide onto a large screen and having 1035 active scanning lines in the display picture. Also, with this standard, the 1125/60 system addresses the need to convert from conventional systems that have 525 and 625 scanning lines.

The 1125/60 format allows existing TV signal transmissions to convert from the NTSC 525/59.4 standard via available large-scale IC chips and supply the global market of video information. During the HDTV conversion period the new HDTV digital system stations will share existing channels and unused channels. During this period some HDTV broadcast stations may be temporarily assigned a second channel for this transition period.

A Review of Other HDTV Standards

Along with the 16:9 aspect ratio and the 1125/60 scanning refresh standard, the HDTV standards system will have the following:

- A new sync waveform.
- A luminance bandwidth of 30Hz.
- 2:1 interlaced scanning combined with non-interlaced scanning.
- Two-color difference signals with bandwidths of 15MHz.

- A horizontal blanking duration of 3.77 microseconds.
- An active horizontal picture duration of 29.63 microseconds.

These HDTV standards developed by the Grand Alliance take advantage of the interlaced scanning used for TV signal transmission and reception, as well as the non-interlaced scanning found on computer monitors.

With non-interlaced, or progressive, scanning, the HDTV system provides a choice of 24, 30, and 60 frames-per-second scanning with a 1920 x 720 pixel dot resolution and a 24 and 30 frames-per-second scan with a 1920 x 1080 pixel dot resolution.

By using the formats developed by the Grand Alliance, the HDTV system provides direct compatibility with computer systems. Another plus to the non-interlaced scanning formats is that the system also offers 60 frame-per-second interlaced scanning at a resolution of 1920 x 1080. The use of interlaced scanning becomes necessary for the two 1920 x 1080 x 60 formats because there's no method for compressing the formats into a 6 MHz channel width. Each of the formats feature square pixels, a 16:9 aspect ratio, and 4:2:0 chrominance sampling.

HDTV SYNC WAVEFORM OBSERVATIONS

The HDTV horizontal blanking interval accommodates the new sync waveform. With this improved sync waveform, the HDTV system ensures compatibility across all systems, achieving precise sync, and has a sync structure that will continue to have noise immunity into the future. This HDTV sync signal eliminates jitters by placing the horizontal timing edge at the center of the video signal dynamic range. Thus, the timing edge has a defined midpoint centered on the video blanking level.

In addition to improving the sync waveform, these design changes have improved the capability of the HDTV system to reproduce colors through the use of 4:2:0 chrominance sampling. When you compare HDTV to the standard NTSC system, you will find it provides a broader choice of colors that aligns with newer film technologies and print media. With this greater capability of the HDTV system standard to produce a broader spectrum of colors, it will affect both camera and receiver display technology.

Many of the improvements in resolution and color reproduction are obtained by a technique that has a 30 MHz luminance bandwidth and two color difference signals with a bandwidth of 15 MHz each. In effect, the use of 30 MHz and 15 MHz bandwidths was a result of the decision to use 1125 scanning lines. This called for a system that required a bandwidth of at least 25 MHz.

The team's decisions to combine increased horizontal and vertical resolution with wider luminance and chrominance bandwidth, resulted in a larger number of pixels. With the 1920 horizontal pixels, the HDTV system becomes a platform for many applications of computer display technologies for medical imaging techniques, computer aided design programs, and various manufacturing methods.

With the HDTV broadband, a 20 Mbit-per-second digital transmission system will enable the marriage of industrial, medical, entertainment, and educational technologies that utilize packetized data transport structure based on the MPEG-2 compression format. Each data packet is 188 bytes in length with 4 bytes designated as the header or descriptor and 184 bytes designated as an information payload. With this type of high compression data transportation, the HDTV system can deliver a wide variety of audio, voice, video, data, or multimedia services and can inter-operate with various other imaging and delivery systems.

The plan at this time is for "over-the-air," broadcasts of HDTV signals to be carried via 8-VSB vestigial sideband station transmissions. The HDTV signals for cable will use the 16-VSB vestigial sideband system. This system will minimize any potential interference between the HDTV broadcasts and conventional NTSC transmissions. Each of these standards use digital technology to provide a high data rate of information, but with no increase of spectrum space. This higher data rate signal, used over the HDTV cable system, enables it to carry two full HDTV signals in a single 6 MHz cable channel.

The MPEG-2 Video Format

The digital HDTV system format uses a video compression technique based on the MPEG-2 video compression standard while the audio system relies on the Dolby AC-3 five-channel sound system. The system was developed by the Motion Picture Experts Group, hence MPEG. The MPEG-2 pictures consist of a luminance matrix and two chrominance matrices. It ensures synchronization between the audio and video playback. In the 4:2:0 format used in HDTV, the chrominance matrices are one-half the size of the luminance matrixes in both the vertical and horizontal planes of the picture. While the bidirectional frame motion compensation, or B-frame, used in MPEG-2, improves picture quality, the MPEG format supports interlaced and progressive scanning.

HDTV Sound System

The HDTV system uses the Dolby AC-3 sound format that encodes multiple channels as a single channel. As a result, this format can operate at data rates as low as 320 kbps. The Dolby AC-3 algorithm represents five full bandwidth channels represented as follows:

1. Left audio channel
2. Center audio channel
3. Right audio channel
4. Left-surround sound channel
5. Right-surround sound channel
6. A limited bandwidth low-frequency subwoofer channel.

The Dolby AC-3 format was designed to take full advantage of the characteristics of the human ear and to permit the noise-free reproduction of the transmitted sound. Refer to chapter nine for more information on the audio signal found in the HDTV system.

As has been pointed out previously, HDTV requires a different set of sync signals and these standards take advantage of the MPEG video compression and Dolby audio compression formats. And in many cases, the HDTV standard takes advantage of standards and technologies that were used only for computer displays.

HDTV SYNC AND TIMING NOTES

The techniques to "sync-in" digital HDTV systems are very different than those found in the analog NTSC system. In the NTSC system, the receiver can use the sync information that it has received to drive a clock that provides timing for locking in the picture and audio information. But with a digital compressed system, digital HDTV, the amount of data generated for each picture is variable, depending on the picture-coding method and the complexity of the data. Thus, there is no exact concept of synchronism between program transmission and the display of the program in a compressed digital HDTV bit stream, compared to the synchronism that is found in a conventional analog NTSC system.

However, synchronization for the digital television is very important. The received data needs to be processed at a certain rate. The rate must match the rate at which the data is generated and transmitted. The loss of synchronization will lead to serious problems, such as buffer overflow or underflow at the decoder, which will culminate in a loss of synchronization of display presentations.

Let's now take a brief look at how the system prevents a loss of synchronization. The sync is kept in check when the transmitting station sends timing information to the adaptation headers of selected packets to serve as a reference for timing comparisons at the decoder. This is accomplished by transmitting a sample of a 27 MHz clock in the program clock reference (PRC) field. This sample indicates the expected time at the completion of the reading of that field from the bit stream at the transport decoder. The phase of the local clock running at the decoder is compared to the PCR value in the bit stream, the instant at which it is obtained, to determine whether the decoding process is synchronized.

Actually, the PCR from the bit stream does not directly change the phase of the clock in the digital HDTV receiver, but only serves as an input to change the clock rate. However, this does not occur during channel change nor during insertion of local programming. Keep in mind that the nominal clock rate in the decoder system is 27 MHz.

Note that these digital HDTV standard specifications are for transmitting synchronizing data to the TV receiver, but do not specify how the sync is recovered in the TV set.

Another detail to remember is that the audio and video samples are locked into the TV receiver decoder system clock that is derived from the PCR values. This makes the design engineer's job easier because fewer numbers of local oscillators will be required to drive the complete decoding process. This locked system technique will also cause the HDTV sets to have a rapid sync acquisition.

COMPRESSED BIT STREAM CONSIDERATIONS

The digital HDTV system has to have the TV receiver randomly enter the application bit stream, for audio and video, so that the receiver can acquire the signal when the set is first turned ON or when a program change occurs. It will only be possible to enter into the elementary bit stream when the coding for the elementary bit stream for the application supports it directly. One such example is for the video bit stream to have provisions for random entry through the concept of interframes (I-frames) that are coded without any prediction. Thus, it can only be decoded with no other information available. A good place for the receiver to begin receiving the video bit stream would be at the beginning of the video sequence header information that comes before the data in an I-frame. Usually any time a digital HDTV receiver begins to receive a program, as when it is first turned ON, the channel is changed, or a signal is disrupted (random entry by the ATV standard), it should coincide with the start of PES packets when they are used for audio and video.

The digital transmitted signal contains the information in the transport layer that the receiver requires to be able to latch onto the data stream. There is a flag in the adaption header of each packet that indicates whether that packet contains a random access point. The data payload starts with the data that forms the random access point of entry into the elementary bit stream itself.

The digital TV system works this way so that when the set is turned ON, channels are changed, or a signal disruption occurs, the receiver need only look for a sync point in the transport bit stream, and can just ignore packets until it sees one with a flag. This technique also makes it easier for the receiving unit to find an entry point in the elementary bit stream once the transport level synchronization has taken place.

The basic reason for the frequent random entry points into the digital HDTV transmitted programs is to have quick channel changes at all times.

ANALOG TV SYNC SIGNALS REVIEW

On an analog TV set the scan line will start on the left side of the screen and also at the left side of the scope trace. The video information line on the TV screen and the scope trace display will end on the right side of both displays. **Figure 7-1** shows one horizontal line across the TV screen of a 10-line color test pattern as viewed on a scope. The whiter areas of the picture are at the top of the trace and the darker parts will be at the bottom of the scope waveform.

In normal TV picture views, each line will be a little different from the one before and after each one. In this picture the beam goes from light to dark as it scans across the color bar pattern. The

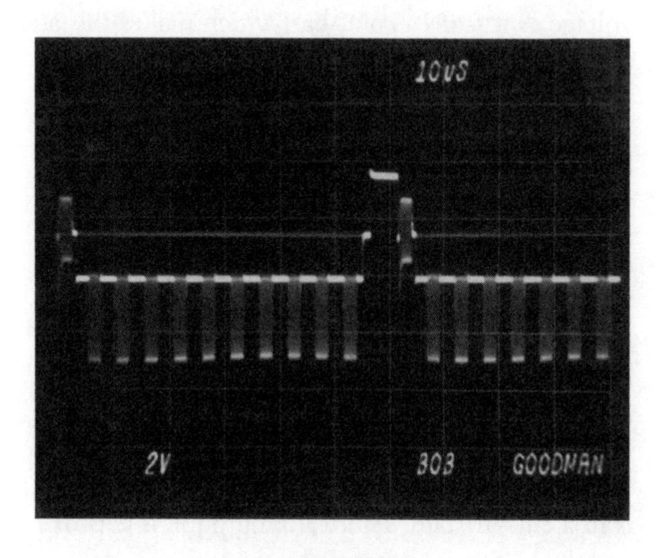

FIGURE 7-1

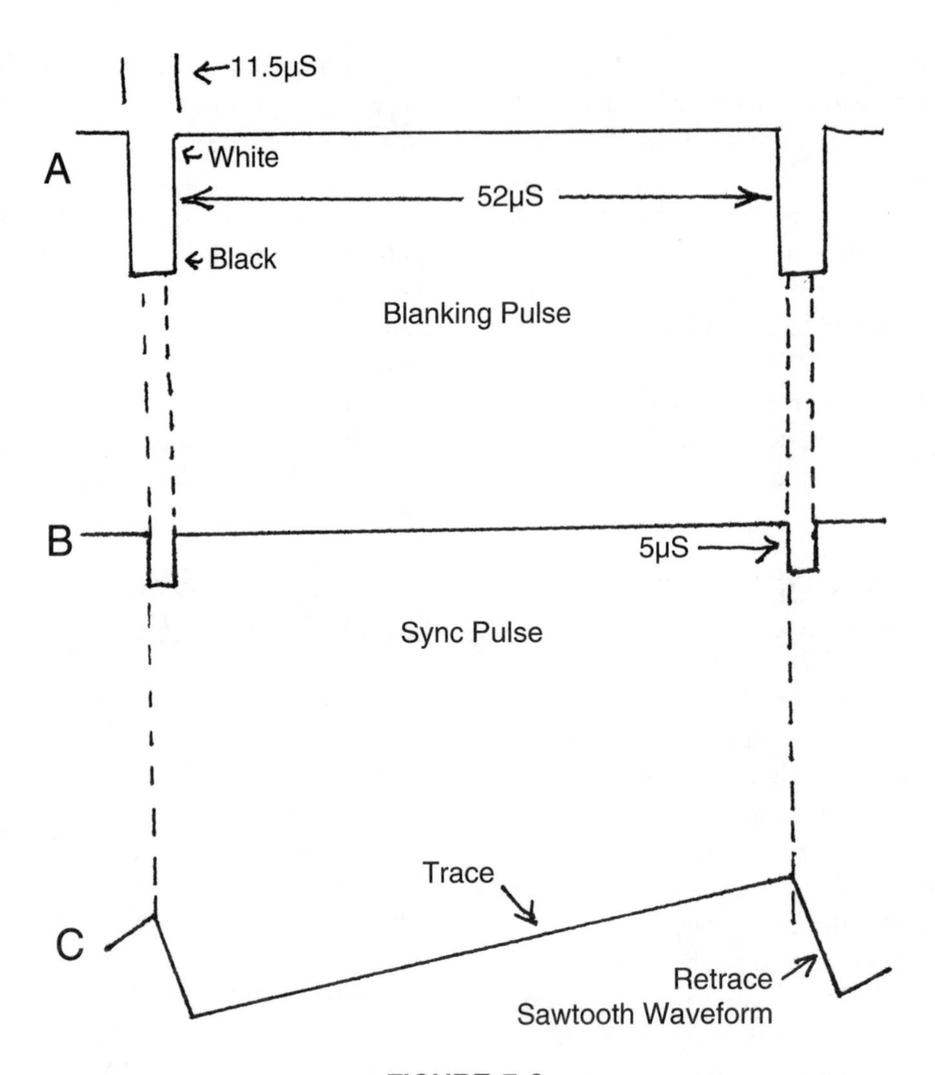

FIGURE 7-2

signal level will go near maximum as it moves between bars (white area) and then to a lower voltage as it scans a color bar, which makes this an alternating light and dark pattern waveform.

HORIZONTAL BLANKING PULSE

When the video line ends, the CRT must be shut off so that as the beam goes back to the left (retrace) this line does not show on the screen. Thus, there is a need for a blanking pulse, shown in **Figure** 7-2A, going in a negative direction beyond a black level, referred to as the "blacker-than-black" level. Simply put, if a 12-volt pulse turns off the CRT, then a 15 volt pulse will make sure the CRT is in a cut-off state. The blanking pulse is approximately 11.5 microseconds in duration.

HORIZONTAL SYNC PULSE

Syncing in the horizontal sweep oscillator in the receiver takes place during the blanking interval. The horizontal sync pulse drives the video signal deeper into the blacker-than-black area, holding the CRT in cutoff. The amplitude of the sync pulse is about one-third of the blanking pulse's P-P voltage, or 25 percent of the complete video signal amplitude.

The horizontal sync pulse as shown in the **Figure** 7-2**B** drawing, starts 1.5 microseconds after the leading edge of the blanking pulse and lasts for 5 microseconds. The sync pulse ends about 5 microseconds before the end of the blanking interval.

The sweep waveform, shown in **Figure** 7-2**C**, tells you the sweep retrace period begins at the leading edge of the sync pulse and must be completed before the end of the blanking pulse time. The leading edge of the sync pulse tells the horizontal oscillator to begin the next cycle. The horizontal oscillator is triggered on each and every cycle.

THE COMPOSITE VIDEO SIGNAL

The composite video signal refers to all of the signals that make up the color picture video information. This includes the video picture information, horizontal sync pulses, the blanking pulses, vertical blanking pulses, equalizing pulses, and VITS and VIRS test information. A complete horizontal line of video is shown in the **Figure** 7-3 scope trace photo.

As the video signal is amplified and processed throughout the TV circuits, the signal will be inverted by 180 degrees. As an example, sync pulses will now be at the top of the trace and white video signal level will be at the bottom of the wave trace. Always keep this in mind while using the scope to troubleshoot these circuits. These different signal polarity changes at different circuit locations may cause you to make an incorrect diagnosis.

The inverted video signal in Figure 7-3 also shows the color burst on the "backporch" of the horizontal blanking pedestal. These details will be discussed next.

3.58 MHZ COLOR BURST SIGNAL

As noted, the color sync (burst) signal is located next to the back porch on top of the blanking pulse. The color sync signal is referred to as the "color burst" which is a sine wave that is used to maintain the phase and frequency of the 3.58 MHz color oscillator to produce correct color picture reproduction. The horizontal blanking/sync and 3.58 MHz color burst scope waveform with callouts is shown in **Figure** 7-4. The 3.58 MHz burst signal from the TV station keeps the color oscillator "in-step" for proper demodulation of the color signals and is referred to as the "color subcarrier." The color subcarrier actually has a frequency of 3.579545 MHz. The color burst must contain 8 to 10 cycles of the subcarrier signal. The space between the decay of the horizontal sync pulse and the start of the burst signal is sometimes referred to as the breezeway.

The front porch time length is 1.5 microseconds and the horizontal sync pulse time is 5 microseconds. The rest of the time period is called the back porch, which contains 8 to 10 cycles of the 3.58 MHz color subcarrier signal. The TV transmitters maintain these precise standards, while the TV receiver has no control. Thus, as the technician doing the troubleshooting, you only need to verify

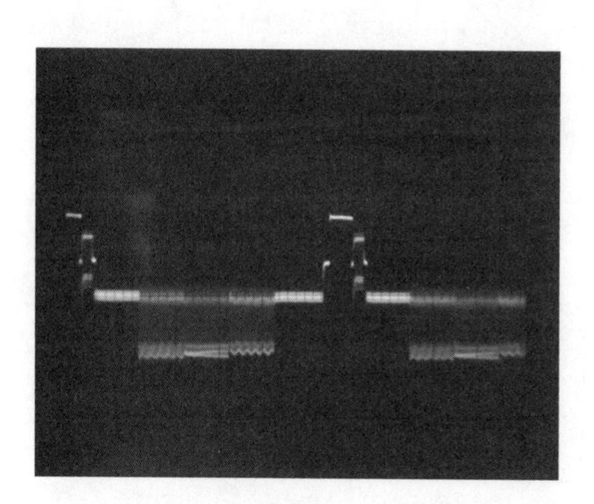

FIGURE 7-3

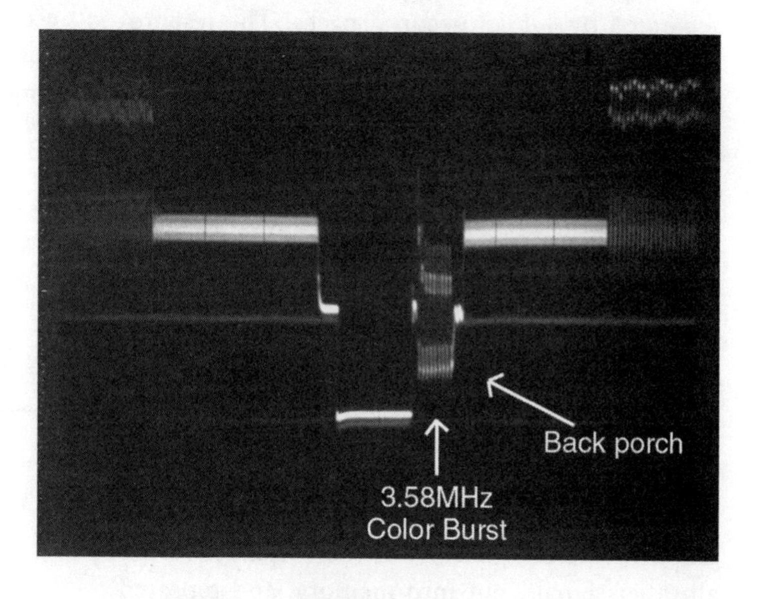

FIGURE 7-4

that the signal has the correct parts in the right proportions.

VERTICAL INTERVAL TEST SIGNALS (VITS)

The following is a thumbnail review of the VITS and VIR test signals found on the back porch of the vertical blanking signal. The vertical blanking back porch is used for a family of signals on lines 17 through 21, and usually vary from the first and second fields. To view these test signals, specialized equipment is required and is primarily used by TV stations, networks, and cable TV engineers. It should be of interest to a top-notch technician to know what these signals are, how they are used, and where they are found in the transmitted signal.

MULTIBURST TEST SIGNAL

The multiburst test signal is transmitted on line 17 of the first field as a frequency response test of the signal after many stages of amplification and other receiver steps of demodulation within the receiver. It starts with an 8 microsecond level of 100 IRE and then drops to the 40-unit level where 6 bursts of sine-wave signals are added. Each sine-wave signal is on for 6 microseconds, and then separated by 1 microsecond spaces. The test frequencies are 0.5, 1.5, 2.0, 3.0, 3.58, and 4.2 MHz. The test frequency amplitudes are ± 30 units (from +10 to +70 IRE units). Good frequency response is obtained when all six test signals have the same amplitude at the test point. Poor frequency is shown by a reduced amplitude at the higher frequencies.

LINE 18 VITS TEST SIGNAL

The second pulse on line 18 of the first field, shown in the **Figure** 7-5 drawing, is a window signal used to measure overshoot and ringing in amplifier circuits. Line 18 is then separated from all other signals, put into memory, and repeated for the full screen display. The screen is dark for the first half of the trace, and then white for one fourth of the display, and then black again. Amplifiers that ring show several cycles of damped oscillation, which appear as darkened lines in the window.

The stairstep pattern on line 18 of the second field can have any number of steps up to 10. In **Figure** 7-6 the stairstep pattern has 10 levels. On each of the steps is a burst of 3.85 MHz color subcarrier signal at zero phase that extends ± 10 IRE units above and below the step level. This signal is used for amplitude distortion evaluation over the range from full white to full black at two frequencies, 115 KHz of the step voltage and at 3.58 MHz of the subcarrier voltage.

VERTICAL INTERVAL REFERENCE (VIR)

The VIR line is used in the receiver for automatic color alignment and automatic brightness circuits with sets so designed. The signal rises to +70 units, where a 24 microsecond run of 3.58 MHz subcarrier at zero phase reference signal can be used for automatic color control. The subcarrier amplitude changes from +50 to +90 IRE units. The level then drops to +50 units for 12 microseconds, and then to the black reference level for another 12 microseconds. The difference between these last two levels is used to automatically compensate the brightness level for the best viewing under various conditions.

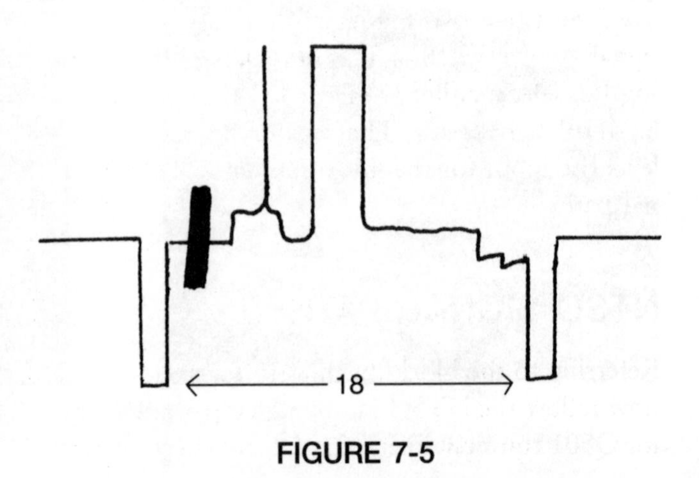

FIGURE 7-5

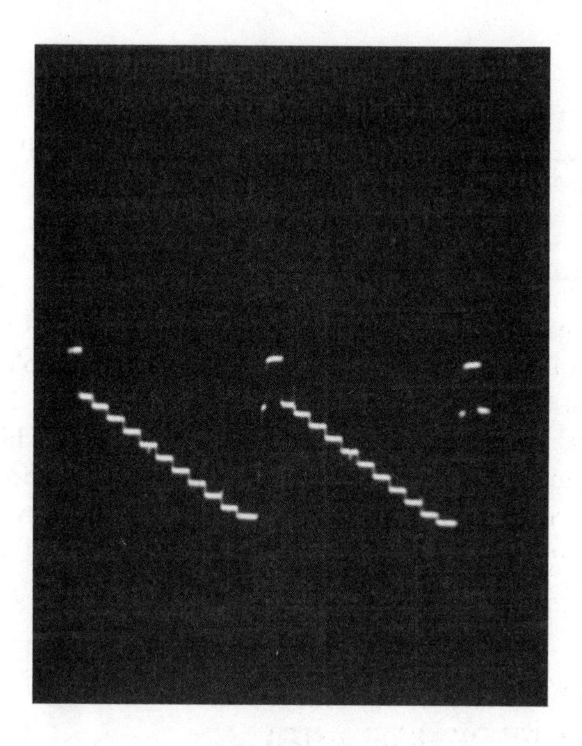

FIGURE 7-6

Line 20 of the first field is reserved for TV station use when the channel number, time of day, or other data can be transmitted. It is set up for 48 bits on line 20 for about 1440 bits-per-second (bps). The end user can make use of this signal only if it transmitted.

HDTV Logic Control Circuits

The Colorstream (DVD) and DTV input signal paths are shown in **Figure** 7-7, while the logic commands for the switching ICs are found in **Figure 7-8.** These two inputs apply R-Y, B-Y, and Y signals to two switching ICs. QV30, the first switching IC, selects either DVD or DTV for the 1080i or 480P signal path. The second IC, QVO7, selects the input for the 480i (standard NTSC) signal path.

NTSC SIGNAL PATH

Referring to the block diagram in Figure 7-7, let's now follow the NTSC (480i) signal path. Transistor Q501 converts the luminance and chrominance, NTSC signals, to Y, I and Q signals and applies them to switching IC QW01. QW01 selects between the Q, Y, and I signals from Q501, or the Colorstream and DTV (480i) inputs selected by QVO7, and applies its selected output to the wide aspect converter (WAC).

The WAC installs the gray side panels (mattes) to a 4:3 aspect ratio picture signal to fit the 16:9 ratio of the HDTV screen. This module is used in the large screen HDTV receivers. The 480i signal then passes through the dual PC module. This module gets its name because it has two functions. One function is a comb filter and the other is the picture-in-picture (PIP) or picture-on-picture (POP). The PIP/POP portion of this module is for viewing a second video source with the original picture. Large screen (16:9) receivers use the POP to add a second picture to the screen. It reduces the width of the main picture and places the second picture alongside the first. The PIP, used in the 4:3 TV sets, places the second smaller picture in a corner of the main picture. Both the PIP/POP also have other features related to the second picture.

After the PIP/POP, the 480i video signal goes through the up-converter. This module converts the 480i video signal into a 480P progressive scan signal. The up-converter doubles the horizontal frequency from 15.734 KHz to 31.5 KHz and, by using frame and line doubling, doubles the number of scanning lines. Although the up-converter doubles the horizontal frequency, the vertical frequency remains the same.

In the next step, the 480i (now converted to 480P) signal runs through switching IC Q2201, which selects either the up-converted 480i or the input from switching IC QV30. Then, Q2201 applies its output to the video processing transistor, IC Q501.

The last stage the 480i signal passes through is Q501. Q501 detects whether the signal is Colorstream (R-Y,R-D, and Y) or I, Q, and Y (converted from 480i). The IC converts its type of signal into RGB, adds the on-screen display (OSD) and closed caption, and applies the set-up and customer adjustments. Q501's RGB output is applied directly to each CRT in the projection receiver.

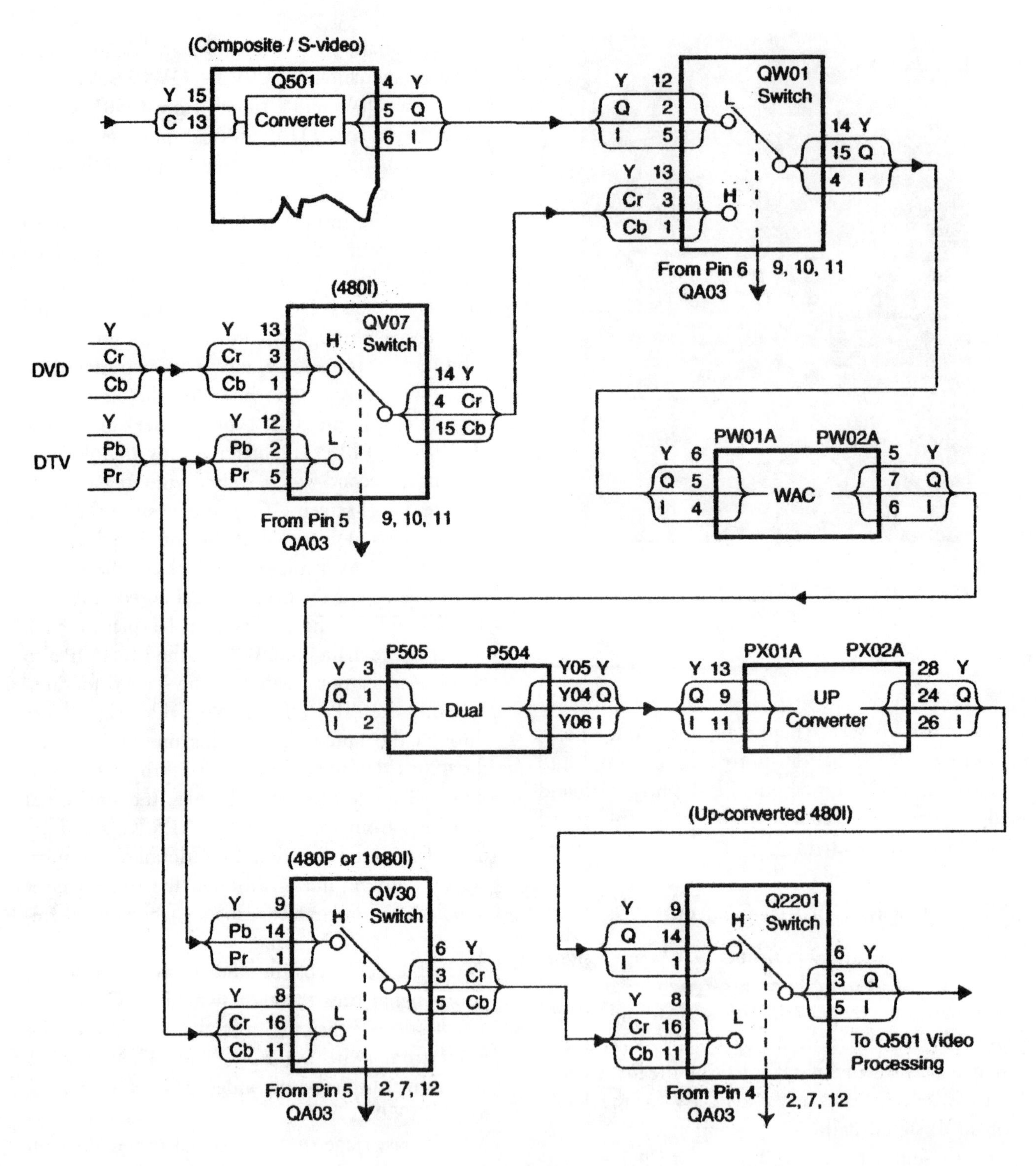
(Composite / S-video)
Q501
Converter
Y 15
C 13
4 Y
5 Q
6 I
QW01
Switch
Y 12
Q 2
I 5
L
H
Y 13
Cr 3
Cb 1
14 Y
15 Q
4 I
From Pin 6
QA03
9, 10, 11
(480I)
QV07
Switch
DVD
Y
Cr
Cb
DTV
Y
Pb
Pr
Y 13
Cr 3
Cb 1
Y 12
Pb 2
Pr 5
H
L
14 Y
4 Cr
15 Cb
From Pin 5
QA03
9, 10, 11
PW01A
PW02A
WAC
Y 6
Q 5
I 4
5 Y
7 Q
6 I
P505
P504
Dual
Y 3
Q 1
I 2
Y05 Y
Y04 Q
Y06 I
PX01A
PX02A
UP
Converter
Y 13
Q 9
I 11
28 Y
24 Q
26 I
(Up-converted 480I)
(480P or 1080I)
QV30
Switch
Y 9
Pb 14
Pr 1
Y 8
Cr 16
Cb 11
H
L
6 Y
3 Cr
5 Cb
From Pin 5
QA03
2, 7, 12
Q2201
Switch
Y 9
Q 14
I 1
Y 8
Cr 16
Cb 11
H
L
6 Y
3 Q
5 I
To Q501 Video
Processing
From Pin 4
QA03
2, 7, 12

FIGURE 7-7

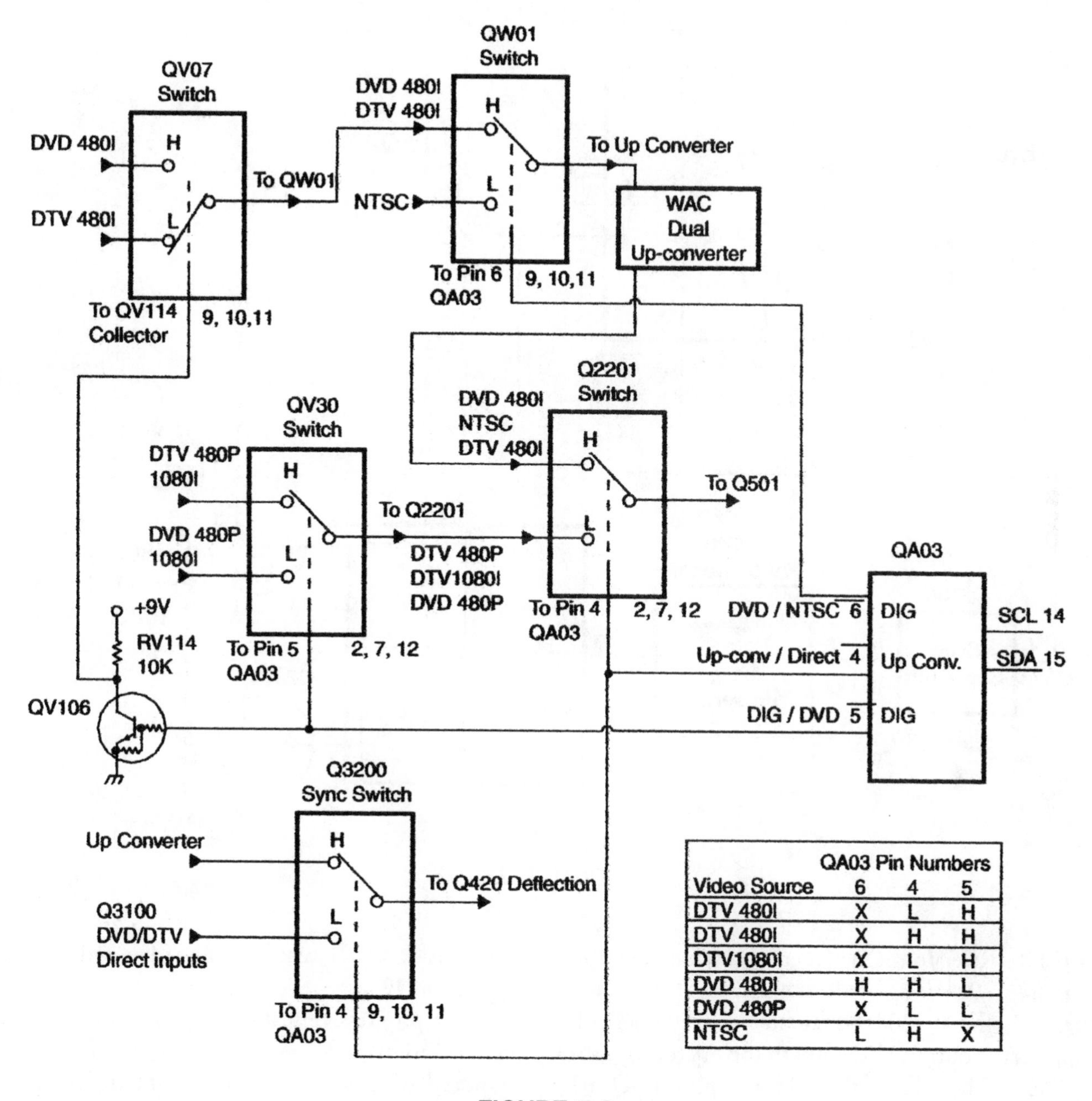

Video Source	QA03 Pin Numbers 6	4	5
DTV 480I	X	L	H
DTV 480I	X	H	H
DTV1080I	X	L	H
DVD 480I	H	H	L
DVD 480P	X	L	L
NTSC	L	H	X

FIGURE 7-8

SIGNAL PATH FOR 1080I AND 480P

The 1080i and 480P use the same video path and bypass several switches and modules. QV30 selects the DTV and DVD inputs and applies its output directly to switching IC Q2201. After passing through this switch, the 1080i or 480P follow the same circuit path as the up-converted 480i signal discussed in the preceding paragraphs.

SYNC DETECT CIRCUITS FOR DTV/HDTV

There is no difference in the video path between the 1080i and 480P circuitry. However, the horizontal circuit must be changed to accommodate the proper horizontal frequency for each signal. In order to detect the type of signal being used, the luminance output of QV30 is applied to sync separator Q3100 and frequency-to-voltage converter

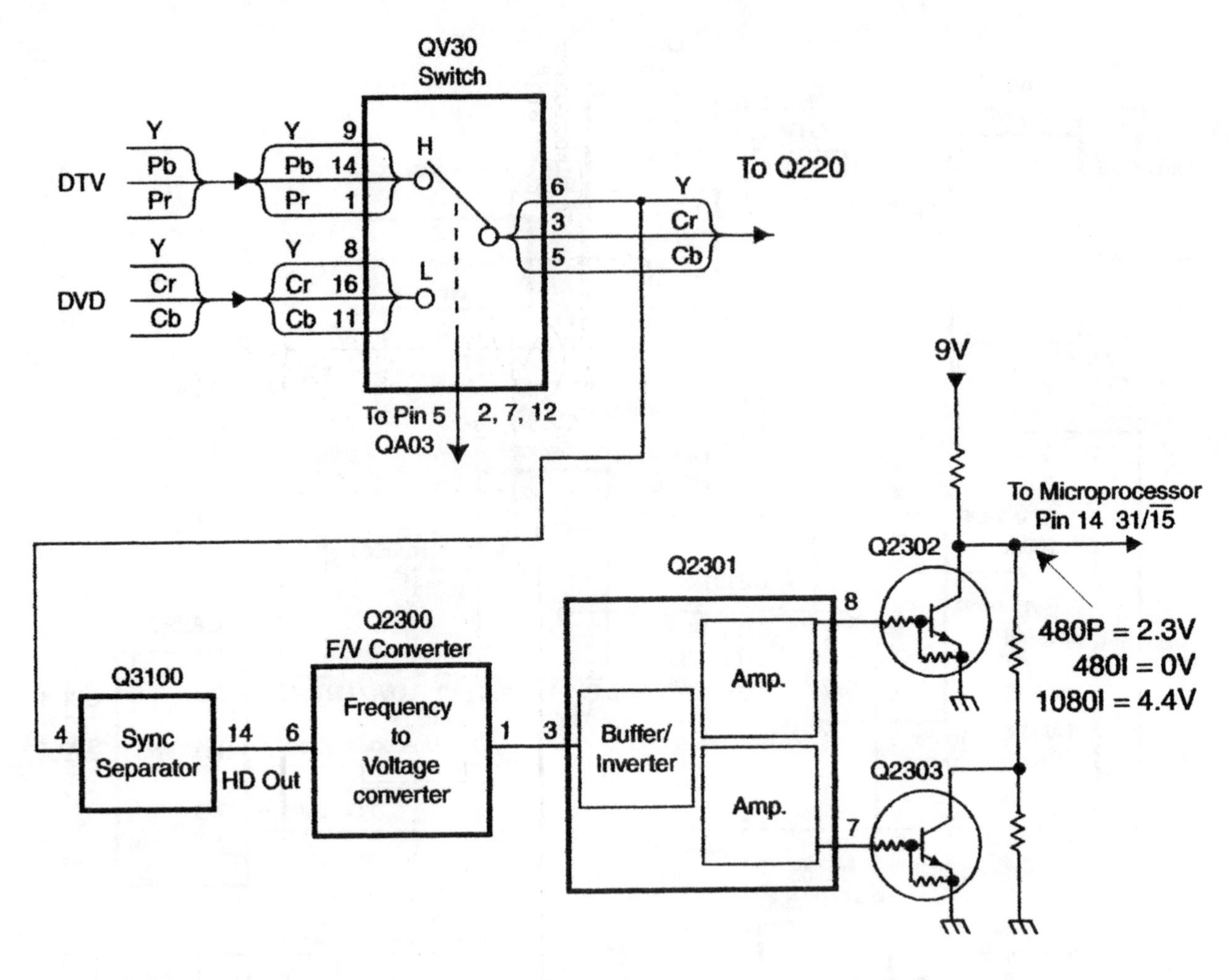

FIGURE 7-9

Q2300. Refer to the sync detect block diagram in **Figure 7-9**. Along with transistors Q2302 and Q2303, the Q2300 buffer and amplifier circuit applies the output of Q2300 to the microprocessor (QA01). The value of the voltage applied to QA01 indicates the type of input signal. The microprocessor controls the input/output expander, QA02, which makes the proper changes to the horizontal circuit for displaying the input. The DTV/HDTV sync separator block diagram is shown in **Figure 7-10**. Refer to block diagram in **Figure 7-11** for the sync switch. The horizontal switch is illustrated in the **Figure** 7-12 block diagram.

TOSHIBA PROJECTION HDTV SYNC CIRCUITRY

In Toshiba's conventional NTSC receivers you will find a sync separator, oscillator, and AFC circuits built inside Q501 which is the video and chroma deflection IC.

The progressive scanned TV sets use a similar video chroma deflection IC. However, because the standard IC chip cannot process the double horizontal frequency, a second IC has been added. This second IC, Q420, contains a horizontal and vertical oscillator, a vertical ramp generator, and the AFC circuits.

SYNCRONIZATION SIGNAL FLOW

The block diagram in **Figure 7-13** shows the flow of the sync signal. The luminance signal enters Q501 on pin 17. The sync separator located in Q501 extracts the horizontal and vertical sync pulses from the luminance signal. The vertical sync pulses are sent out of pin 31 and applied to the up-converter

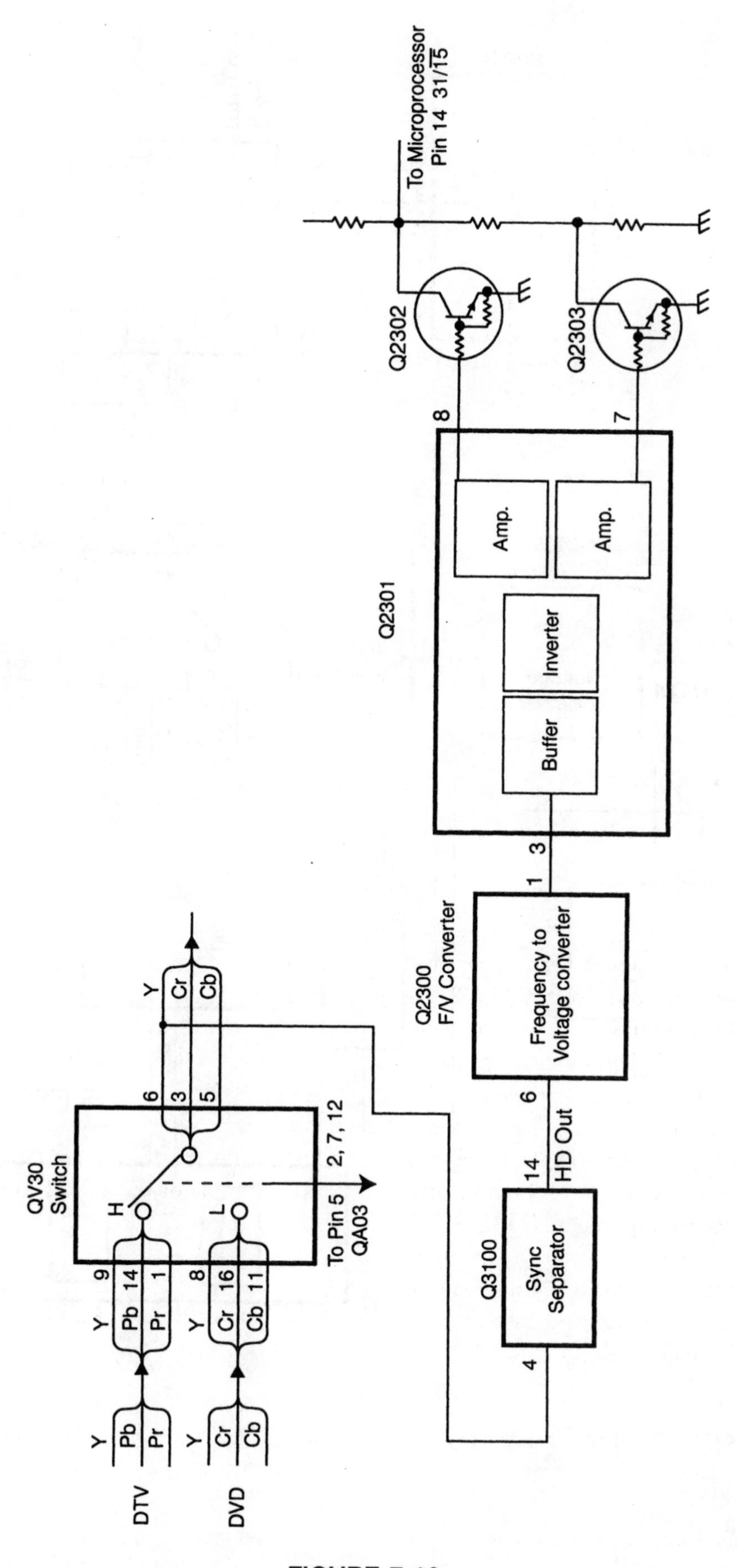
DTV
Y
Pb
Pr
DVD
Y
Cr
Cb
Y
Pb 14
Pr 1
9
Y
Cr 16
Cb 11
8
QV30
Switch
H
L
6
3
5
Y
Cr
Cb
To Pin 5
QA03
2, 7, 12
Q3100
Sync
Separator
4
14
HD Out
6
Q2300
F/V Converter
Frequency to
Voltage converter
1
3
Q2301
Buffer
Inverter
Amp.
Amp.
8
7
Q2302
Q2303
To Microprocessor
Pin 14 31/$\overline{15}$

FIGURE 7-10

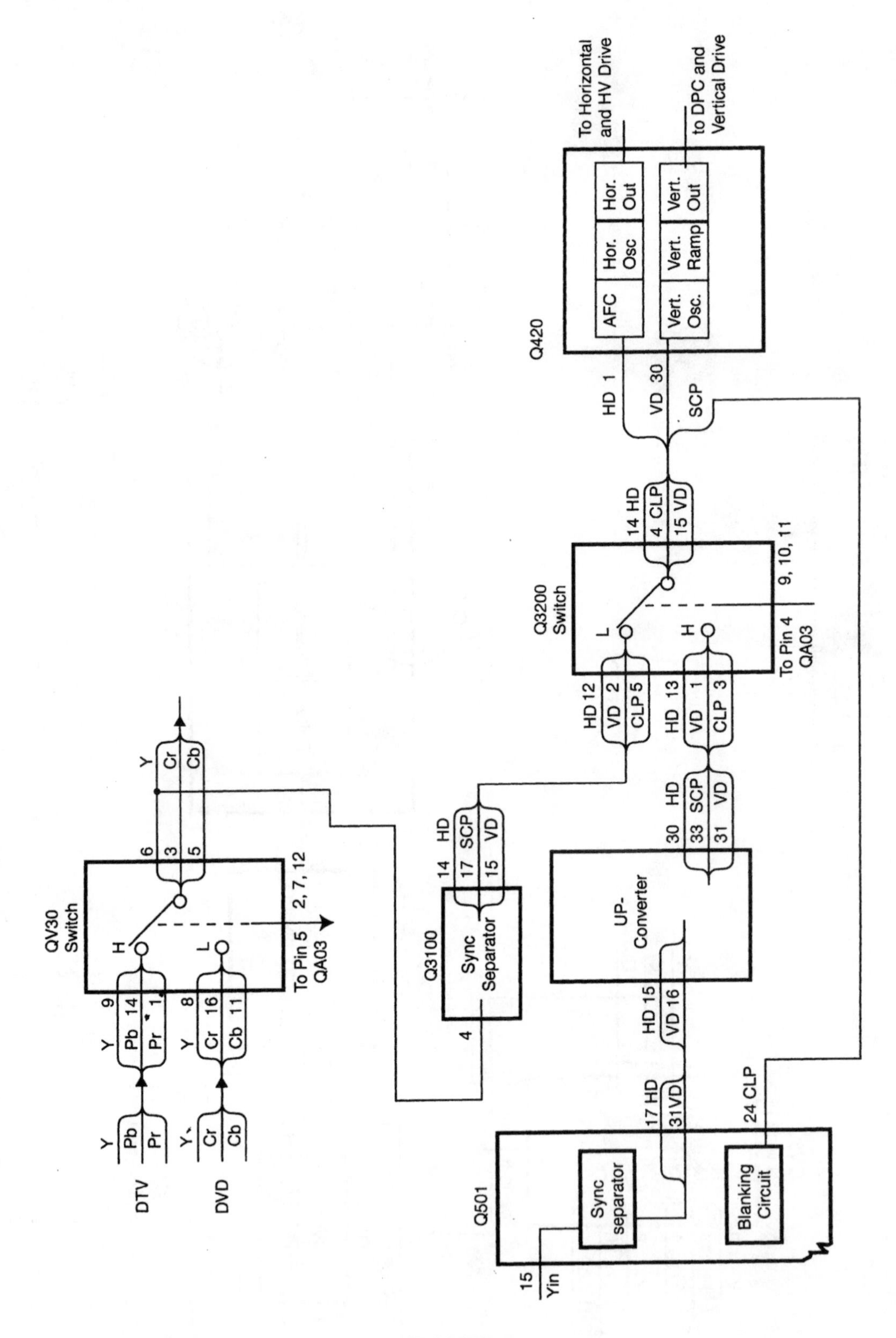
To Horizontal and HV Drive
to DPC and Vertical Drive
Q420
AFC
Hor. Osc
Hor. Out
Vert. Osc.
Vert. Ramp
Vert. Out
HD 1
VD 30
SCP
14 HD
4 CLP
15 VD
Q3200 Switch
9, 10, 11
To Pin 4 QA03
HD 12
VD 2
CLP 5
HD 13
VD 1
CLP 3
QV30 Switch
2, 7, 12
To Pin 5 QA03
DTV
DVD
Y
Pb
Pr
Cr
Cb
Q3100
Sync Separator
14 HD
17 SCP
15 VD
UP-Converter
30 HD
33 SCP
31 VD
HD 15
VD 16
17 HD
31VD
24 CLP
Q501
Sync separator
Blanking Circuit
15
Yin

FIGURE 7-11

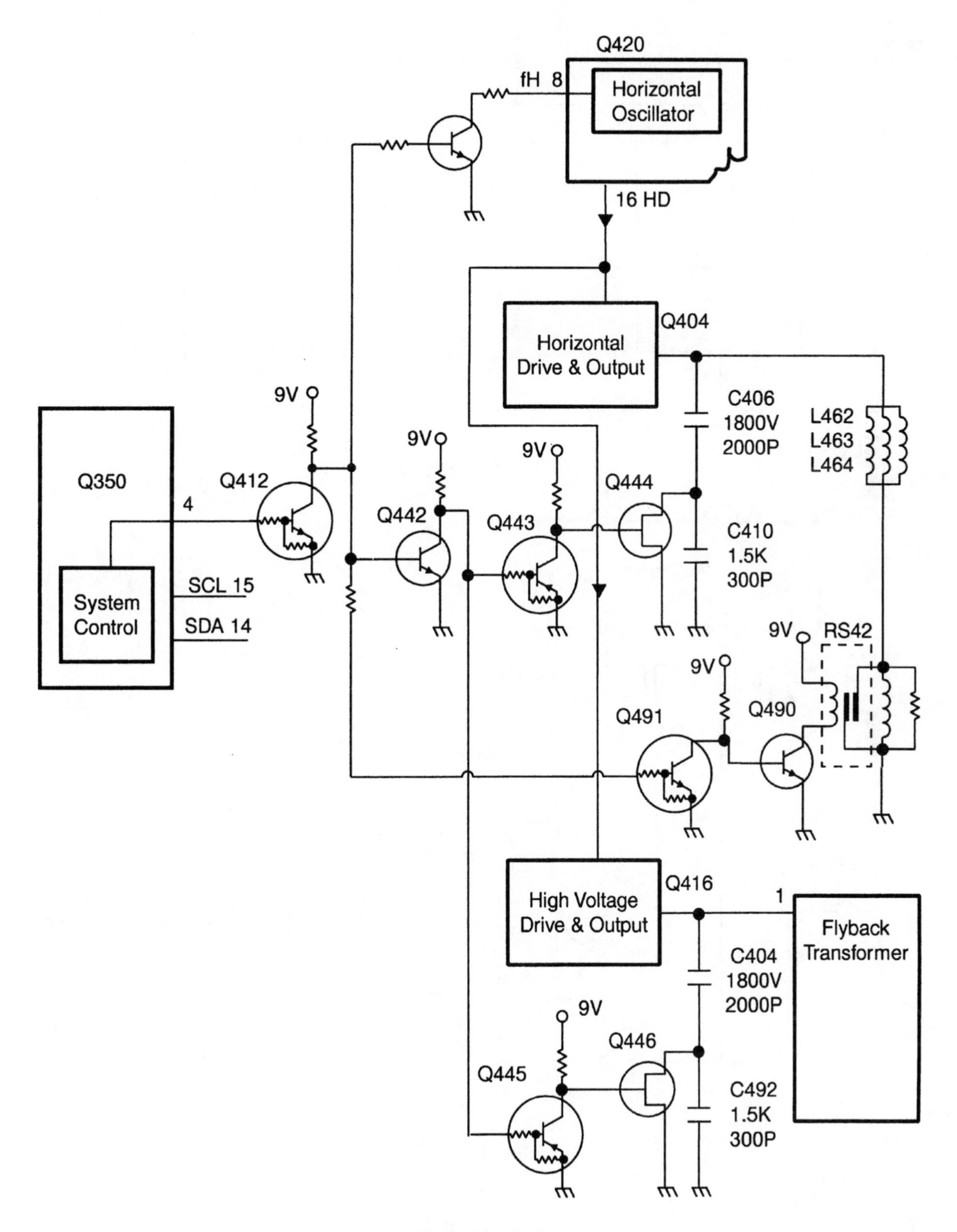

Q420
fH 8
Horizontal Oscillator
16 HD
Q404
Horizontal Drive & Output
C406
1800V
2000P
L462
L463
L464
9V
Q350
Q412
4
System Control
SCL 15
SDA 14
Q442
Q443
Q444
C410
1.5K
300P
RS42
Q491
Q490
Q416
High Voltage Drive & Output
1
Flyback Transformer
C404
1800V
2000P
Q446
Q445
C492
1.5K
300P

FIGURE 7-12

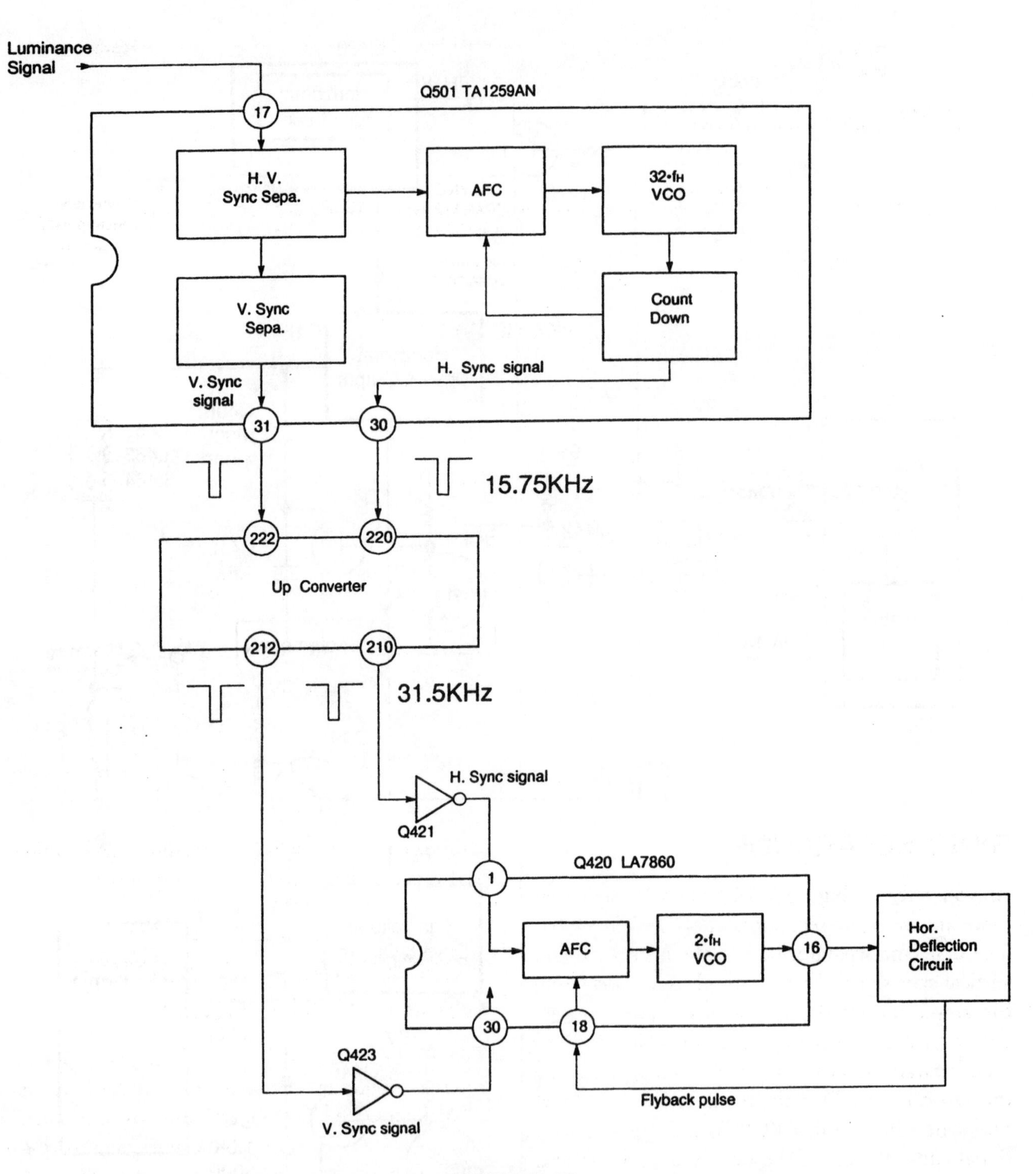

FIGURE 7-13

PC. The horizontal deflection pulses are synchronized by the internal AFC signal and applied to the up-converter via pin 30.

The up-converter uses the vertical pulses for synchronization. The vertical frequency in NOT doubled. The up-converter doubles the horizontal frequency and sends the doubled signal out of pin 210.

Both the horizontal and vertical output pulses from the "up-con" are inverted and applied to Q420. The outputs of Q420 are applied to the horizontal and vertical drive circuits.

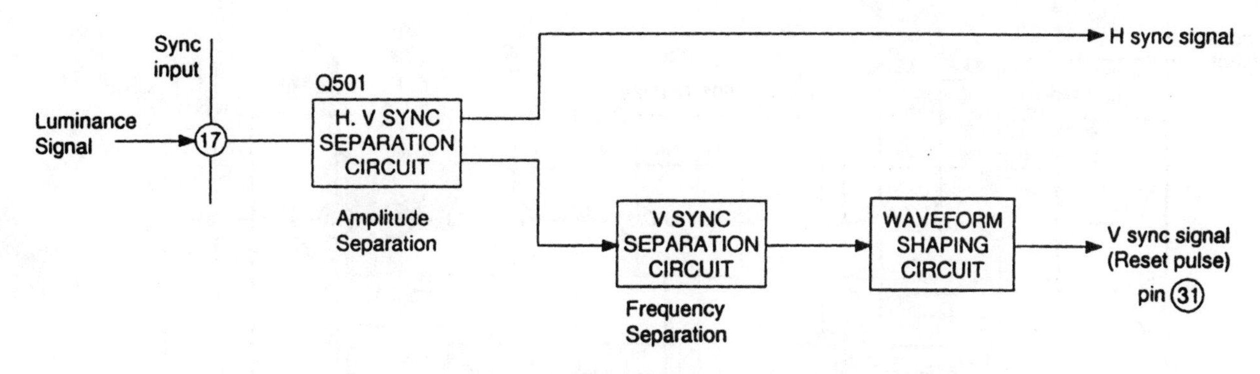

FIGURE 7-14

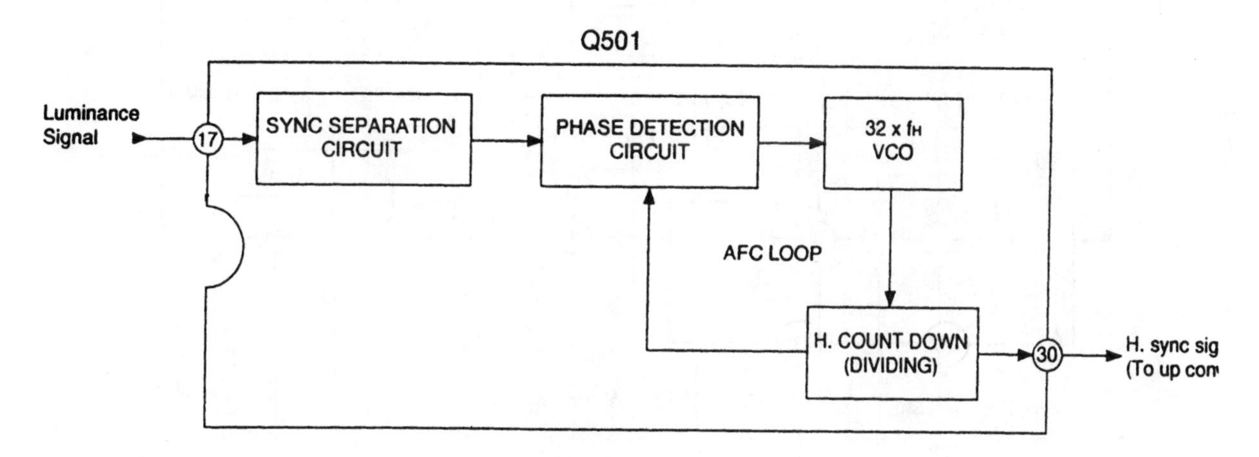

FIGURE 7-15

SYNC SEPARATION

As you refer to **Figure** 7-14, you will note that there are two sync separators contained in Q501. The first sync separator detects the horizontal and vertical sync signals by their amplitudes. Detecting the amplitudes of the sync signals improves sync detection when receiving weak video signals. This sync separator removes the horizontal sync leaving the vertical sync. This circuit maintains good vertical sync when using a VCR with a copy-guarded signal and stabilizes the vertical sync under special field conditions (ghosting, sync depression, and adjacent channel beat).

THE 32 FH OSCILLATOR

The 32 fH Oscillator is shown in the **Figure** 7-15 block diagram. The internal voltage controlled oscillator (VCO) and AFT circuit of Q501 synchronize the vertical and horizontal sync pulses and remove the vertical equalizing pulses.

HORIZONTAL AND VERTICAL OSCILLATION CIRCUIT

Because the horizontal frequency is doubled, Q501's internal oscillator cannot function as the horizontal oscillator. The IC chip, LA7860, has been added and operates as the vertical and horizontal oscillator. The horizontal (H-sync) and vertical (V-sync) deflection pulses (HD and VD) from the up-converter PC are applied to pins 1 and 30 of the LA7860 chip shown in **Figure 7-16**. The pin outs for this IC are also included in Figure 7-16.

Pin 1 of Q420 (LA7860) is the input for the horizontal sync signal. The input receives positive pulses at approximately 2 volts P-P. Q420 triggers from the front (leading) edge of the pulse.

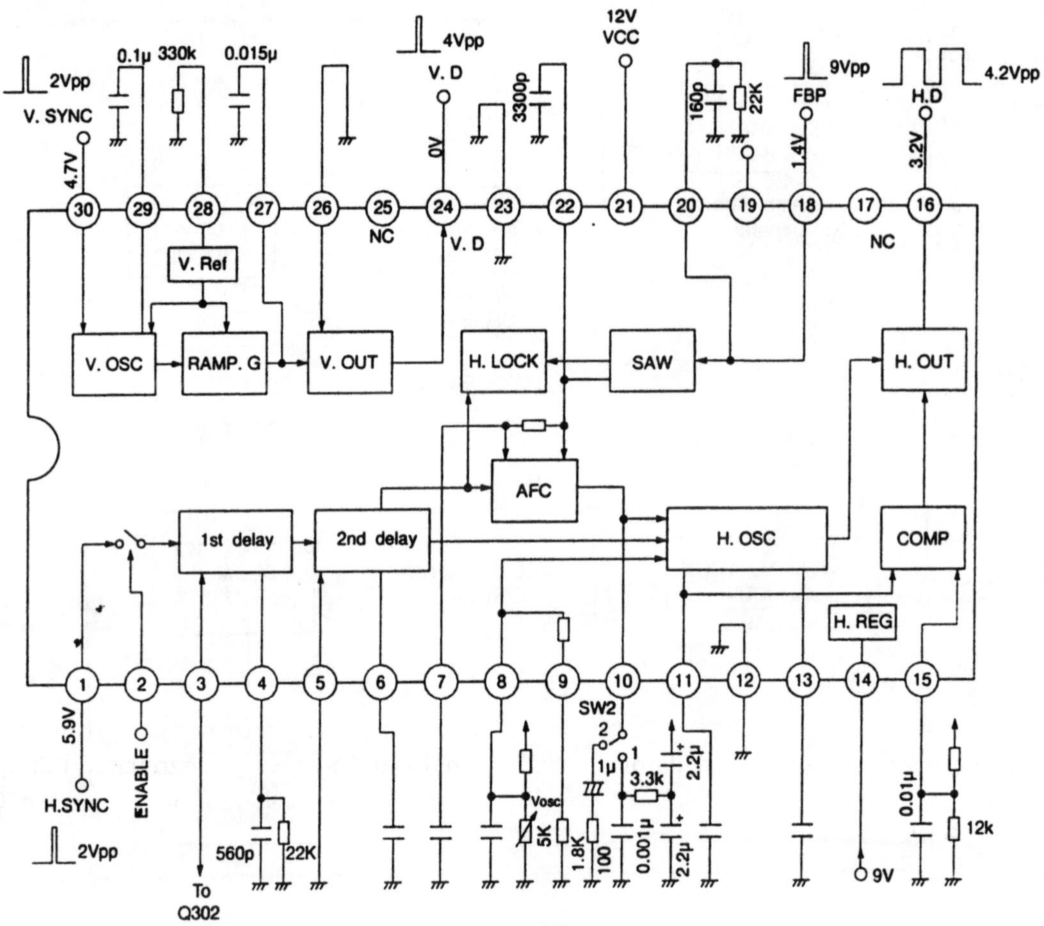

FIGURE 7-16

Pin 2 is an ENABLE terminal. This pin is a switch to terminate the horizontal sync input. Pin 2 is connected to test terminal "H." Applying 3 volts to the "H" test terminal interrupts the sync to the horizontal oscillator. Then, the horizontal oscillator loses sync and becomes a free-running oscillator.

Pin 3 is the controlling input for horizontal phase shift. When adjusting the HPOS (horizontal positioning) in the service mode, the DC voltage on this pin changes. This voltage change controls the phase of the horizontal oscillator. As the block diagram illustrates in **Figure** 7-17, the phase of the oscillator determines the horizontal centering of the picture.

LA7860 IC Pinouts

As you refer back to Figure 7-17 of the LA7860 IC, let's see how these pins function.

- Pin 4 determines the RC time constant for the horizontal phase shift, which is controlled by pin 3. The time constant is a reference point for pin 3.
- A capacitor at pin 7 smooths the AFC comparing the waveform developed at pin 22.
- Pin 8 controls the frequency of the horizontal oscillator. R4019 adjusts the frequency. If either Q420 or C441 are replaced, adjusting the horizontal drive out of pin 16 will be required.

ADJUSTMENT TIP: After adjusting R4019, enter the video mode and check the on-screen display with no input. If the on screen blinks, further adjustment of R4019 is needed.

- Pin 10 connects to an RC filter circuit for the AFC. Two separate RC circuits filter the AFC. One is for start-up and the other is for normal operation.

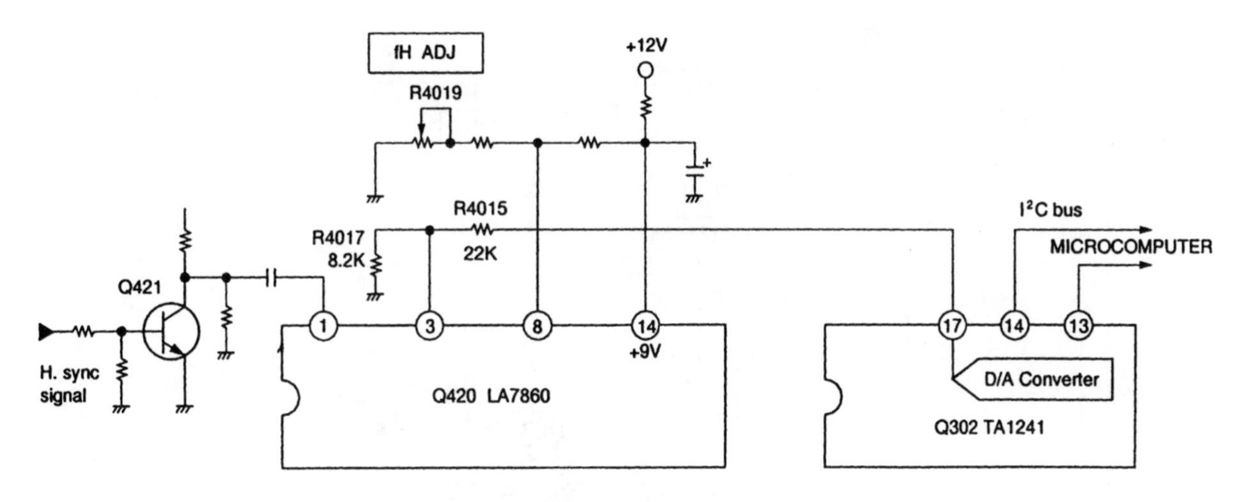

FIGURE 7-17

- The capacitor connected to pin 11 of Q420 determines the horizontal oscillator frequency.
- Pin 13 is the input to a low pass filter that limits the band pass of the oscillator.
- The circuit connected to pin 15 of Q420 determines the duty cycle of the horizontal drive output on pin 16.
- Pin 16 is the horizontal output terminal. The output of this pin should be a 5-volt P-P pulse with a frequency of 31.5 KHz.
- Pin 18 is the input for the feedback pulse (FBP, 9.2 volts P-P at 31.5 KHZ) developed from the flyback transformer for the AFC circuit.
- The RC circuit network on pin 20 is a delay for the feedback pulse.
- Pin 22 develops a sawtooth waveform (4.2 volt P-P at 31.5 KHz) for comparison with the AFT.
- Pin 24 is the vertical deflection output (VD). This signal is a 60 Hz, 5-volt P-P pulse.
- The capacitor connected to pin 27 produces a sawtooth waveform for the vertical ramp generator. The sawtooth is 5-volt P-P and 60 Hz.
- Pin 28 is a reference for the vertical oscillator and ramp generator.
- The capacitor connected at pin 29 determines the vertical oscillator frequency.
- Pin 30 is the vertical sync input. This signal is a positive going pulse at 60 Hz with a P-P 2.2 voltage.

THE DOUBLE-SPEED CIRCUIT

The Toshiba double-speed circuit converts the NTSC interlaced, 15.75 KHz horizontal scanning frequency to the progressively scanned signal of 31.5 KHz. Using frame and line doubling, the double speed circuit also doubles the video information.

Double-Speed Circuit Operation

Refer to the up-con block diagram in **Figure 7-18** as the circuit operation is explained.

As we look at the signal flow, note that the Y, Q, and I signals are applied to the double-speed (up-con) circuit on pins 214, 216, and 218 of terminal plug PX02. Horizontal deflection (HD) and vertical deflection (VD) pulses are applied to an A/D converter (QX100). QX100 converts the Y, Q, and I signals to digital luminance (Y) and digital chrominance (C) signals. The digital Y signal then goes to QX300. QX300 detects whether the video information contains motion or is stationary. After passing QX300, along with the C signal from QX100, the Y signal is applied to projection format converter (jforce) IC (QX400). QX400 performs the line and field doubling of the video information and doubles the horizontal frequency.

QX400 doubles the frequencies of the following outputs: Y out, I out, Q out, HD out, and HBLK. QX400 also outputs. VBLK and (VD) Vertical Deflection Pulse, but the frequencies are unchanged. The digital Y, I, and Q output signals

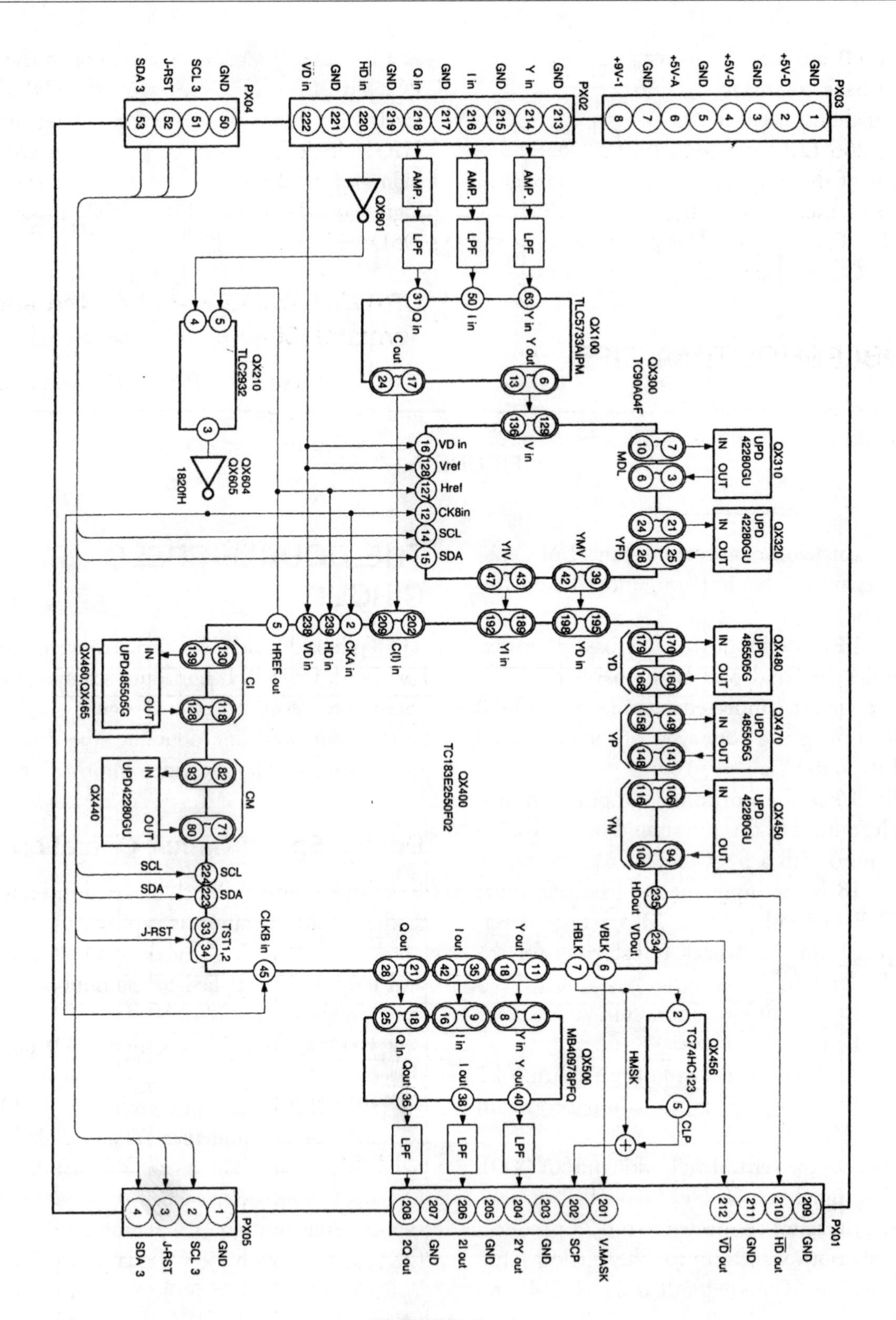

FIGURE 7-18

from QX400 are sent to an A/D converter and converted back to analog signals.

The final outputs of the up-con are at Pins 204 (2Y out), 206 (2I out), and 208 (2Q out). The waveforms of the outputs will appear very similar to the inputs, except for the frequencies, which will be doubled.The up-con PC is controlled by the microprocessor through the I square C bus.

TROUBLESHOOTING TIPS

The up-con is not repairable at the component level. Should the up-con fail, it should be replaced as a module. If a problem develops in the up-con PC, pin 201 of PX01 may go to a HIGH logic level. Pin 201 is the vertical blanking output. A HIGH logic level on this pin causes continuous blanking of the picture. Because of this HIGH logic level, the picture, PIP, closed captioning, and OSD will be blanked out.

UP-CON Troubleshooting Pin and Terminal Chart

Tables 7-1 and 7-2 show the terminal description and pin voltages that are for normal operation of the up-con microprocessor controller.

Connector No.	Pin No.	Terminal name	Signal name	Voltage, etc.	No power supply	No Y output	Abnormal Contrast	No color output	Abnormal color signal	Screen noise	Wavering screen
PX01	201	V. MASK	V mask pulse output	5V, positive polarity							
	202	SCP	Clamp/ mask	4.3V/2.0V							
	203	GND	GND	0V							
	204	2YOUT	Doubld speed Y signal output	1V(p-p)		O	O			O	
	205	GND	GND	0V							
	206	2IOUT	Double speed I signal output	1V(p-p)				O	O	O	
	207	GND	GND	0V							
	208	2QOUT	Double speed Q signal output	1V(p-p)				O	O	O	
PX02	213	GND	GND	0V							
	214	Y IN	Y signal input	1V(p-p)		O	O			O	
	215	GND	GND	0V							
	216	I IN	I signal input	1V(p-p)				O	O	O	
	217	GND	GND	0V							
	218	Q IN	Q signal input	1V(p-p)				O	O	O	
	219	GND	GND	0V							
	220	$\overline{HD}$ IN	HD signal input	5V, negative polarity							O
	221	GND	GND	0V							
	222	$\overline{VD}$ IN	VD signal input	5V, negative polarity							O
PX03	1	GND	GND	0V							
	2	+5V – D	+5V power supply	+5V, ±0.25V	O	O	O	O	O	O	O
	3	GND	GND	0V							
	4	+5V – D	5V power supply	+5V, ±0.25V	O	O	O	O	O	O	O
	5	GND	GND	0V							
	6	+5V – A	5V power supply	+5V, ±0.25V	O	O	O	O	O	O	O
	7	GND	GND	0V							
	8	+9V – 1	9V power supply	+9V, ±0.5V	O	O	O	O	O	O	O
PX04	50	GND	GND	0V							
	51	SCL2	I^2C bus clock	5V		O	O	O	O	O	O
	52	J-RST	J-RESET	0V		O	O	O	O	O	O
	53	SDA2	I^2C bus data	5V		O	O	O	O	O	O

TABLE 7-1

Connector No.	Pin No.	Terminal name	Signal name	Voltage, etc.	No power supply	No Y output	Abnormal Contrast	No color output	Abnormal color signal	Screen noise	Wavering screen
PX01	209	GND	GND	5V, negative polarity							
	210	$\overline{HD}$ OUT	HD signal output	0V							O
	211	GND	GND	0V							
	212	$\overline{VD}$ OUT	VD signal output	5V, negative polarity							O
		N.C.	Not connected.	————							

TABLE 7-2

Some information and diagrams in this chapter are courtesy of the TOSHIBA ELECTRONICS CORP.

CHAPTER 8

Electronic HDTV Tuners

INTRODUCTION

In this chapter we will review operation of electronic tuners now being used and details of Microtune's "tuner-on-a-chip" concept. This one-chip-tuner technology is now being found in HDTV receivers, cable set-top boxes, satellite set-top boxes, cable modems, VCRs, and other digital video systems.

We will view a basic electronic tuner block operational diagram and then discuss various circuit operations within the tuner. These sections of the tuner will include the RF amplifier, RF bandpass network, phase-locked loop, IF bandpass, and oscillator/mixer circuits.

Then we'll review Microtune, Inc. solid-state tuners that feature the high-performance, dual-conversion single-chip tuners for HDTV receivers, plus many more applications for digital video reception.

We'll then detail operation of the MicroTuner 2000 intergrated bandwidth tuner/receiver contained on a single monolithic microcircuit, followed by a review of the MicroTuner technology, product family, and chip architecture.

Next, we'll cover MicroTuner's tuner chip application notes for the series MT2030 and MT2032 Micro Tuner ICs. This will include drawing of chip pin outs and tuner block diagrams.

This chapter concludes with circuit information on Toshiba's progressive scan receiver's electronic tuner and IF systems. We will also discuss troubleshooting information on Toshiba HDTV tuner RF/IF circuitry, and tuners used in the RCA MM01 HDTV chassis.

BASIC ELECTRONIC TUNERS

Tuners used in conventional and HDTV receivers at present are small and compact, have no moving parts and are controlled by a microprocessor. And, of course, they contain solid-state components. Because of cost and special equipment required for repair, most are replaced or exchanged/rebuilt when they are found to be defective. The tuner must not only select TV station signals from an antenna, but also 50 or more cable channels that are only 6 MHz in bandwidth before converting it to an intermediate IF frequency. The tuner accomplishes all of this with transistors, ICs, and a series of tuned circuits. These circuits are tuned to a resonant frequency that is different for each RF TV channel carrier that has been selected by the TV viewer.

Basically, a tuned circuit is made up of an inductor and capacitor that is in a parallel or a series circuit. An inductor is a coil of wire that will oppose a change of current flowing through it. This opposition is called "inductance."

A capacitor is made from closely spaced plates with insulation between them, called "dielectric" material. The capacitor will store an electrical charge. Capacitors are used in tuned circuits as they cause the current to lead the voltage.

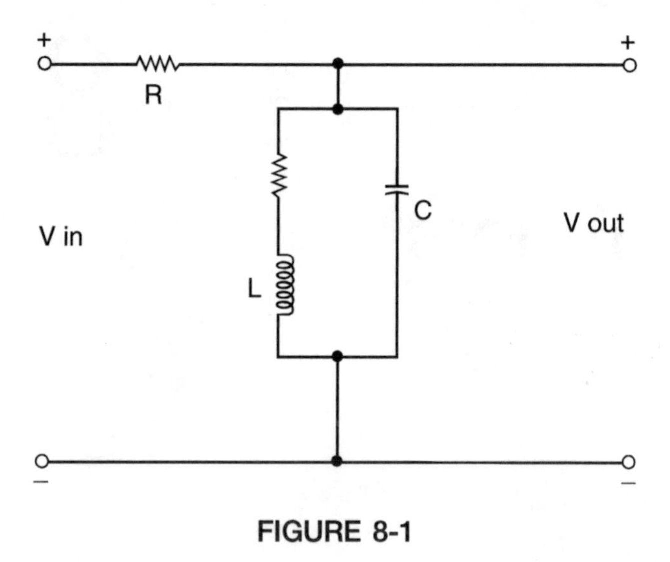

FIGURE 8-1

As you may recall from basic electronics, inductors and capacitors can be connected in either series or parallel configurations. And, you can compute the total inductance of coils in series as you would resistors in series. When coils are in parallel, the inductance is computed as you would resistors in parallel. Conversely, total capacitance of capacitors in series is calculated like resistors in parallel, while total capacitance of capacitors in parallel are computed the same as resistors in series.

The basic parallel resonant circuit shown in **Figure 8-1** can be found in most TV tuners. The term "resonance" is a condition existing in a circuit where the inductive reactance cancels the capacitive reactance that occurs at the tuned frequency. The impedance of this circuit at the "tuned" or resonant frequency will be maximum, and a maximum signal voltage will be developed across it. When the capacitance and/or inductance is decreased, the resonant frequency will increase. When you increase the capacitance and/or the inductance then the resonant frequency will be lowered.

The older twist-and-turn tuners used a switch to change capacitors and inductors in and out of the circuits for different channels. Electronic tuners perform the same function, only electronically. Referring to **Figure 8-2**, note that the diode in this circuit is a varactor. The varactor diode acts like a voltage-variable capacitor. When the reverse bias across it is increased, the capacitance is increased. Conversely, the decreasing the reverse bias causes the capacitance to decrease. In this way the resonant frequency can be raised or lowered electronically. A block diagram of a typical solid-state electronic TV tuner is illustrated in **Figure 8-3**.

INPUT FILTER TRAP

The first circuit the TV signal encounters in a tuner is the input filter. The tuner contains a filter network to "trap-out" unwanted FM and IF frequencies that may come into the tuner. This device is called a single-tuned filter or "trap." It tunes the frequency of the desired channel and passes on the selected frequency to an RF amplifier Q1 shown in the **Figure 8-4** drawing. Most RF amplifiers are dual-gate, depletion mode, metal-oxide semiconductor field-effect transistors (MOSFET). These MOSFETs are voltage-controlled semiconductors that operate somewhat like vacuum tubes as they have very high input impedance. The depletion varieties are normally "ON" without any gate bias while the enhancement types require bias to operate. When a negative voltage is applied to the gate (with respect to the source) of a depletion type MOSFET, the current is reduced by "pinch off" if the negative voltage is large enough. When a positive voltage is applied, the current flow is increased to a value predetermined in the design phase.

The MOSFETs described are dual gate devices. Both gates affect drain current, which is another feature making them good RF amplifiers. When used in these tuners, the RF signal from the tuner input is applied to gate #1, while automatic gain control (AGC) voltage is applied to gate #2.

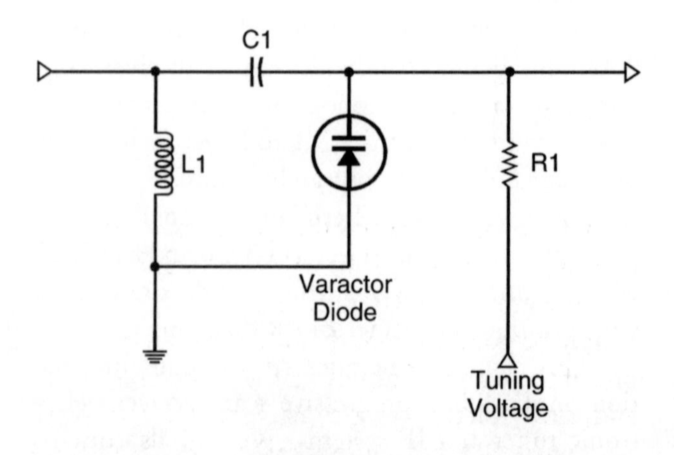

FIGURE 8-2

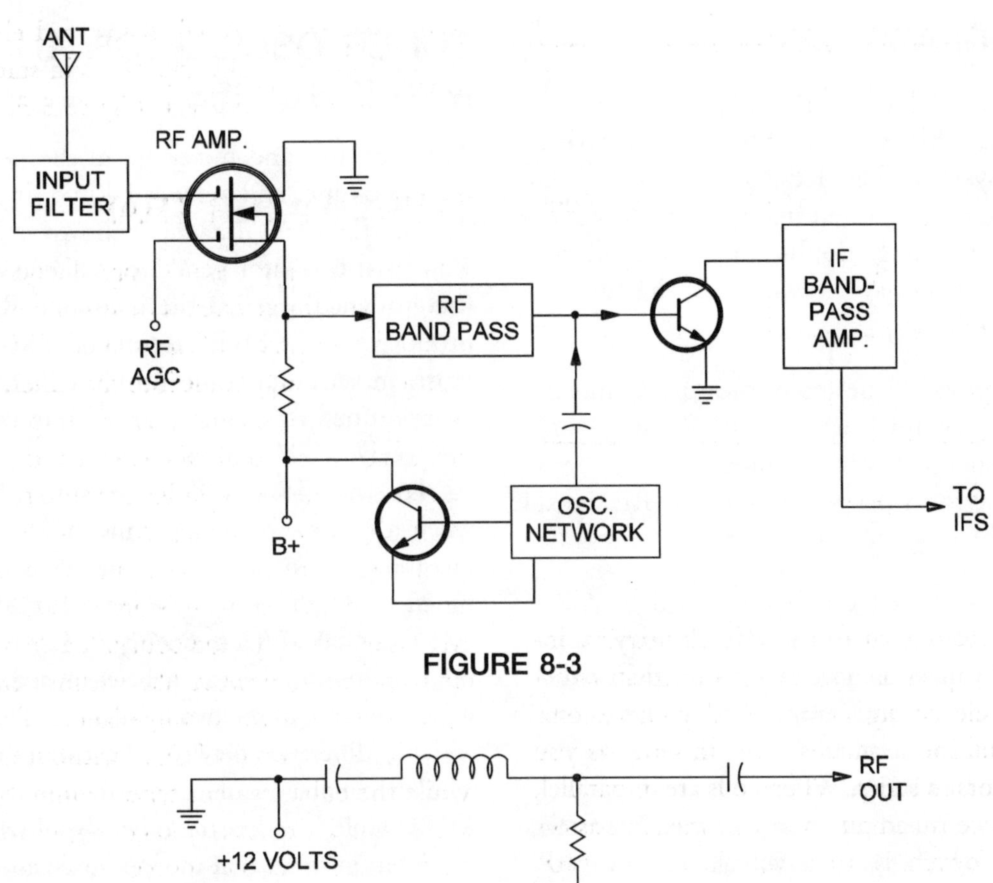

FIGURE 8-3

FIGURE 8-4

If AGC voltage increases, the RF signal at the drain increases in strength. If the AGC voltage decreases, the strength of the RF signal at the drain decreases. It may be good to note this when troubleshooting one of these type of tuners that is producing a weak and snowy picture on a conventional TV receiver.

Figure 8-4 notes the location of Q2 and Q3 and a block labeled "BV/U switch." These are band-switching transistors that are used when VHF channels are selected, and when turned ON provide a ground path for the source of the VHF, RF amplifier. When a UHF channel is selected, the band switch turns OFF, removing the ground path for the VHF amplifier, turning it OFF. The UHF amplifier works the same as the VHF amplifier.

RF BANDPASS NETWORK

Now for a brief look at the RF bandpass network circuit. This is the next block in the basic TV tuner system. The RF bandpass network, shown in **Figure 8-5**, is referred to as the "double-tuned primary" and "double-tuned secondary" circuits. The bandpass network has the following dual functions:

1. To provide a means of measuring impedance matching between the RF stage and the later stages in the tuner.
2. To perform sharp tuning of the RF signal to obtain greater selectivity.

The double-tuned primary is tuned by CR1 via a voltage is applied to the "RF Primary" line. The control voltage is developed from the phase-locked-loop circuit and summed with the digital alignment voltage developed from the parameters stored in the EEPROM.

The double-tuned secondary is tuned in a similar manner, which is by a voltage on the "RF Secondary" line, and applied to the cathodes of CR2 and CR5. Band switching is accomplished by toggling the voltage on the band switch line.

TUNER OSCILLATOR AND MIXER STAGES

The oscillator and mixer are in the next block of the basic tuner and are located in a single IC chip. This partial tuner drawing is shown in **Figure 8-6**. The local oscillator generates a signal that is beat (heterodyned) against the incoming RF signal to produce a 45.75 MHz intermediate frequency (IF) at the mixer output. The local oscillators output is always tuned to a signal 45.75 MHz that is above the received TV station signal.

Let's now look at some frequencies in the tuner when it's set to receive a channel 6 TV station. The incoming video portion of the RF signal will be tuned to 83.25 MHz. The local oscillator will develop a signal with a frequency of 129 MHz. When the two are mixed (beat or heterodyned), the results will be the sum of the two frequencies (212.25 MHz) and the difference between the two (45.75 MHz). The IF bandpass extracts the desired signal, 45.75 MHz, while it rejects the other signal frequencies.

Keep in mind that the local oscillator frequency must change over a wide range in order to tune in all of the channels. Most of today's electronic tuners use a frequency synthesizer to control the local

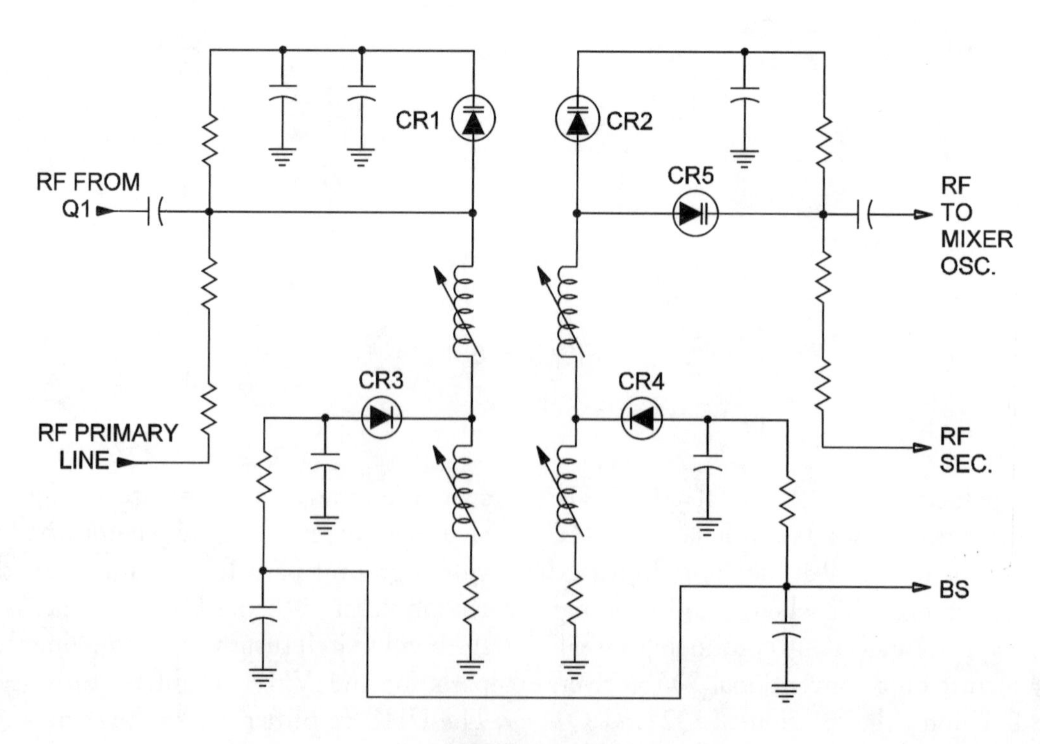

FIGURE 8-5

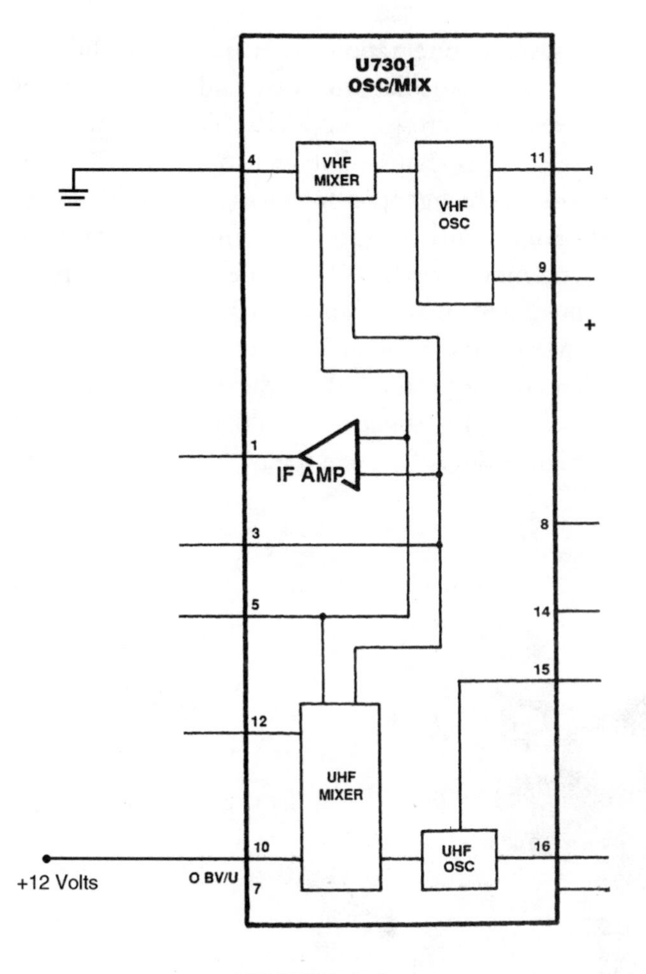

FIGURE 8-6

oscillator. A frequency synthesizer consists of a phase lock loop (PLL) and a programmable divider.

A voltage-controlled oscillator (VCO) sends a frequency sample to a comparator that compares it to a reference frequency, generated by a crystal-controlled oscillator. Should the VCO go off frequency, the comparator sends it an error voltage to correct its output frequency. With this set-up the VCO stays locked to the crystal oscillator frequency.

A programmable divider enables the PLL to lock the oscillator onto a huge range of frequencies. The block diagram of the VCO and PLL is shown in **Figure** 8-7. The microprocessor sends "division logic" to the programmable divider. The division logic sets the divider ratio within the frequency divider, permitting the VCO to change frequency. The divided down frequency is then compared to the reference frequency. If it is off frequency, the comparator generates an error voltage that the VCO uses to correct the new frequency.

MICROTUNERS — TUNER ON A CHIP

The evolution of the TV tuner found in the early model sets has evolved from the mechanical wafer-

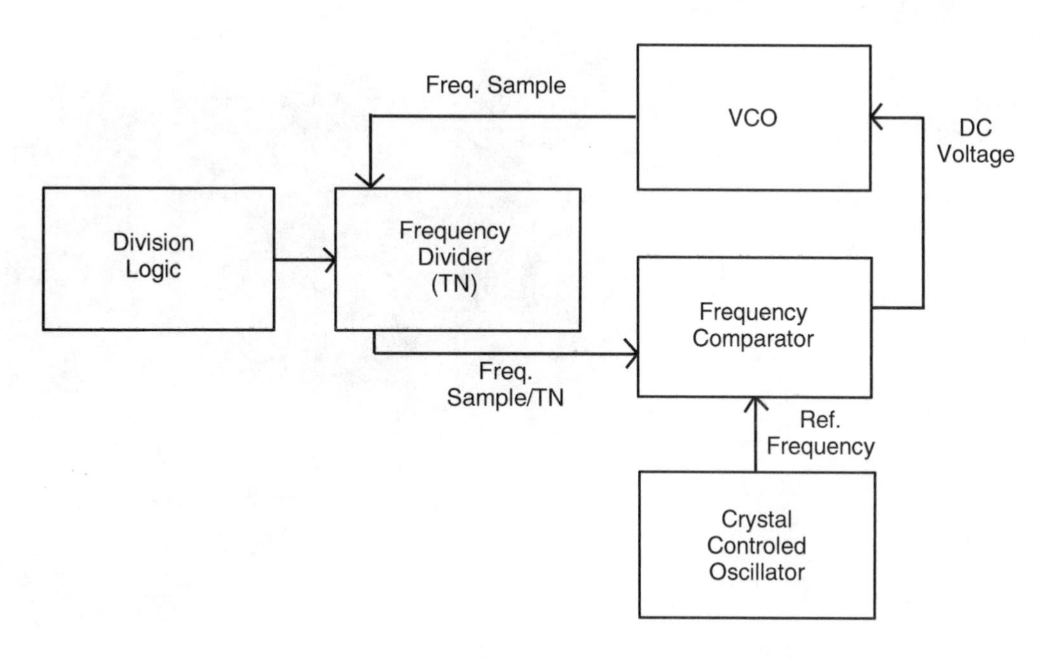

FIGURE 8-7

type-switch unit shown in the **Figure 8-8** photo, to the present day electronic tuner shown in **Figure 8-9**. To emphasize the small size of the Microtune tuner-on-a-chip refer to **Figure 8-11** for the MicroTuner unit.

Microtune, Inc. is a pioneer and leader in manufacturing highly integrated broadband media access solutions for home, business, and mobile applications for the digital consumer electronics markets.

Many leading-edge products and services were founded due to pioneered RF systems-in-silicon and systems integration expertise. Microtune enables digital media broadcast and entertainment applications across a new class of consumer devices: cable modems, digital set-top boxes, PC multimedia, convergence appliances, TV/DTV, IP telephony, and home/ automotive personal entertainment systems. This wide-ranging spectrum is illustrated in the **Figure 8-10** drawing.

Microtune is the inventor of the revolutionary, patented MicroTuner, the industry's first complete broadband TV tuner-on-a-chip that is illustrated in **Figure 8-11**. The company combines its core,

FIGURE 8-8

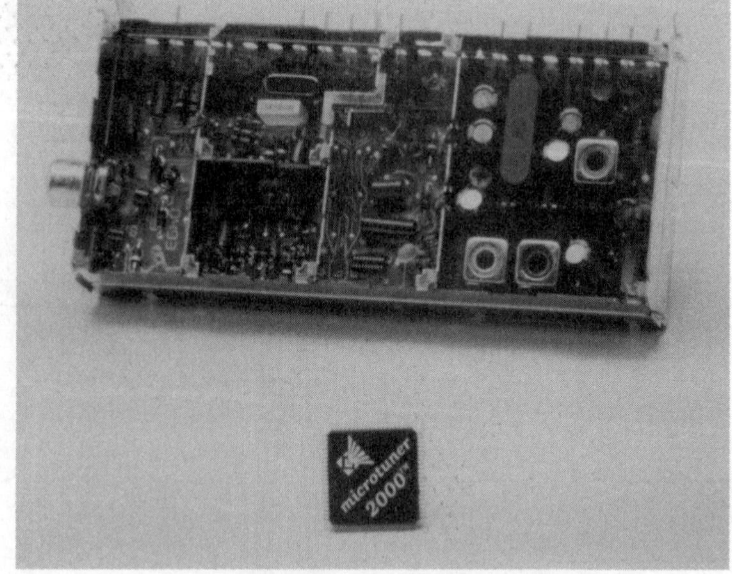

FIGURE 8-9

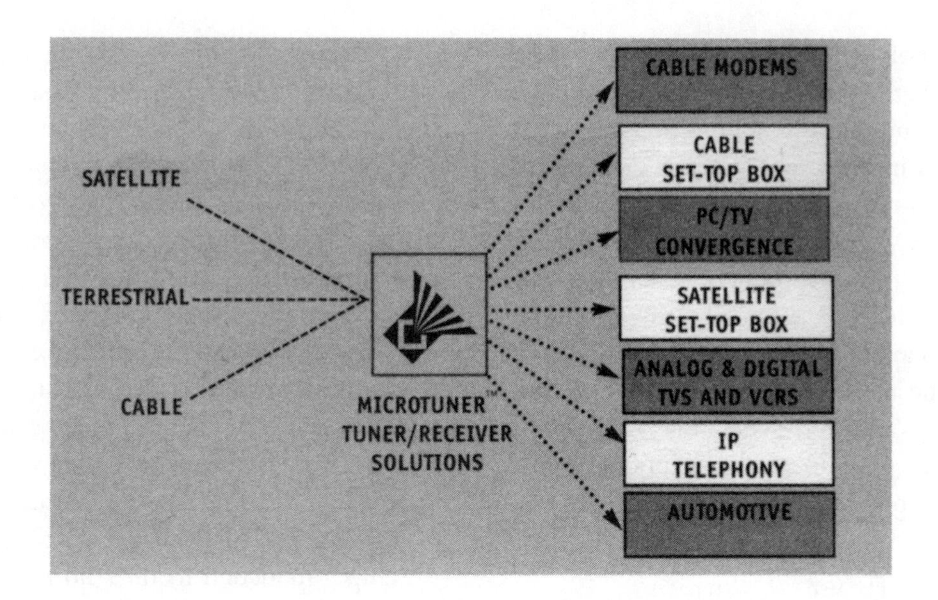

FIGURE 8-10

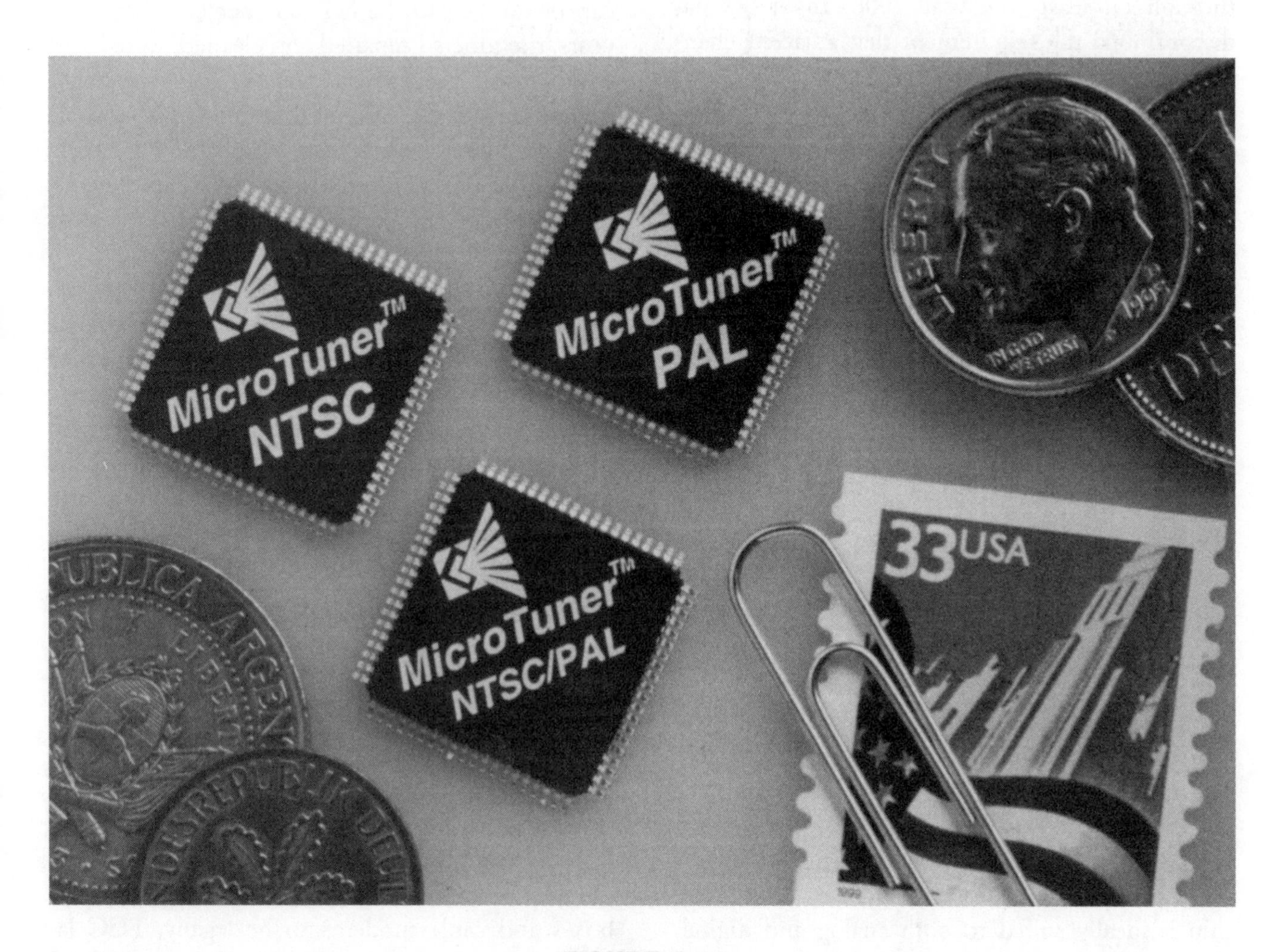

FIGURE 8-11

solid-state technologies with complementary functions to deliver highly-integrated and tightly-coupled broadband communications solutions. These products function as information "gateways," funneling high-speed, high-resolution video and data to consumers on a PC or TV via terrestrial, cable, or satellite networks. With products that range from tuner-on-board ICs, to manufacturing-ready RF MicroModules, to complete Reference Design solutions, Microtune accelerates customers' transition to solid-state tuners and next-generation digital products. This enables them to develop solutions rapidly, providing a competitive advantage.

HDTV Digital Tuner Requirements

Because of the broadcast of Digital Television (DTV) in the existing TV frequency bands, along with the continued broadcast of analog television through (at least) the year 2006, the FCC has discontinued the requirement that adjacent channels be left vacant. It is expected that in major television markets, the entire TV spectrum will become "packed." With the advent of digital television, including multicasting and HDTV, television receivers and VCRs must be capable of accurately receiving and discriminating tightly packed analog and digital signals delivered by broadcasters or 100-plus channel cable television systems. Conventional electronic tuners are not capable of handling closely spaced, complex signals or high quality, interference-free video and sound. Tuners for digital HDTV receivers will require higher close-in phase noise performance than is typical for existing dual-conversion tuners.

MICROTUNER TECHNOLOGY OVERVIEW

With phase-noise performance of -85 dBc (dB relative to the carrier level) at 10 KHz offset, composite triple beats (CTBS) of less that -57 dBc with 100 channels at 15 dBmV input, 8 dB maximum noise figure, and 30 dBmV maximum input for 1 percent cross-modulation, the MicroTuner 2000 chip is ideally suited to both analog and digital, off-air and cable receiver applications.

Microtune's chip meets the performance requirements for the crowded off-air and cable TV spectrums. With multiple systems and circuit patents, including the only patent for an integrated television tuner on a single microchip, Microtune has successfully addressed the complex problems of cable analog/digital tuners, offering them in one cost-effective, monolithic device. The MicroTuner is made feasible and cost-effective with innovative RF-VLSI architecture that eliminates interference and distortion prevalent in many environments. As well, it provides excellent image rejection, impedance matching, and wide dynamic range.

Because the MicroTuner is a single semiconductor chip, production requires no mechanical tuning, calibration, or human interaction, making the device immune to drift and greatly increasing reliability and process repeatability and cost reduction.

The analogy of the quantum leap between vacuum tubes and the invention of the transistor can be applied to the MicroTuner chip. For fifty years, the tuners I repaired and cleaned, were large, cumbersome mechanical devices. TV tuner technology has not changed all that much during these 50 years. They have basically remained a scaled-down collection of discrete high-frequency devices. The highly-integrated MicroTuner chip represents a quantum leap in TV tuner technology.

Microtune has demonstrated that not only can multiple LOs (local oscillators) be placed on a single chip, but an entire high-performance TV tuner RF system can be integrated on one as well. Microtune invented a novel architectural approach and proprietary circuit design techniques that enabled the company to integrate all of these functions to solve the tuner/receiver problem.

SPECIAL DUAL-CONVERSION TECHNIQUE

A substantial portion of the MicroTuner's architecture is comprised of a special dual-conversion tuner design that combines the best features from single and dual-conversion tuners as summarized in the **Figure 8-12** diagram.

The MicroTuner chips will now allow set-top-boxes and cable modems to be legally, FCC labeled as "cable ready" and/or "cable compatible."

MicroTuner Combines the Best Features from Single- and Dual-Conversion Tuners

TUNER FEATURE	OFF-AIR-TV TUNER	CABLE-TV TUNER	MICROTUNER
ARCHITECTURE	SINGLE-CONVERSION	DUAL-CONVERSION	SUPERSET
Cable *Tuner-of-Choice*		✓	✓
Off-Air *Tuner-of-Choice*	✓		✓
Low Distortion		✓	✓
Matched Input Impedance		✓	✓
Good Selectivity		✓	✓
High Image Reject		✓	✓
Internal Signal Containment		✓	✓
FCC Part 15.118 Compliant		✓	✓
Wide Dynamic-range	✓		✓
Low Noise Figure	✓		✓
Low Phase Noise	✓		✓
Low Material Costs	✓		✓
Low Labor Costs			✓
High Reliability			✓

FIGURE 8-12

Single-conversion tuners are common in television sets and VCRs; dual-conversion tuners are used in systems requiring higher performance, such as cable modems or cable set-top boxes. Single-conversion tuners require the use of front-end tracking filters. To have the lowest cost, these tracking filters use discrete components and thus require complex adjustments when manufactured. Additionally, because their band-pass filter characteristic moves across the band to "track" the selected channel, their performance is not ideal, resulting in less-than-optimal image rejection and selectivity, along with poor adjacent channel rejection.

MICROTUNER CHIP DESIGN

The MicroTuner's dual-conversion design eliminates the need for the first filtering operation to be performed by a tracking, band-pass filter. It converts the entire input band to a first intermediate frequency and performs filtering with a fixed band-pass filter. Since the first IF filter is fixed, no manual adjustments are required. A fixed filter provides higher performance at a lower cost. The primary functions of the first filtering operation in a tuner are image rejection and band limiting. Band limiting is desirable because it reduces the aggregate signal level into the receiver's low noise amplifier (LNA) and first mixer. This reduces the power requirements and the cost of linear components that are required for single conversion TV tuners.

With the MicroTuner chip, the LNA and first mixer receive the entire input band and handle the linearity issues through innovative circuit design. The first IF filter is complemented by an on-chip image reject mixer so that a simple, low-cost, two-pole ceramic resonator filter is adequate to achieve an overall 65 dB image rejection across the entire TV band. By way of comparison, single-conversion design typically has excellent image rejection in the low VHF band, acceptable performance in the high VHF band, and poor performance in the UHF band.

MICROTUNE PRODUCT FAMILY

As the first members of its product family, Microtune offers two MicroTuners, each a fully integrated single chip targeted at different market segments. The first MicroTuner, the MT2000, has an RF input and second IF output, and it is optimized for the consumer electronics, set-top box (STB), and cable modem markets. A block diagram of the MT2000 chip is illustrated in **Figure 8-13**.

The second chip, called the MT2500, includes the MT2000 functionality as well as IF demodulation, and it is targeted at the PC/TV convergence

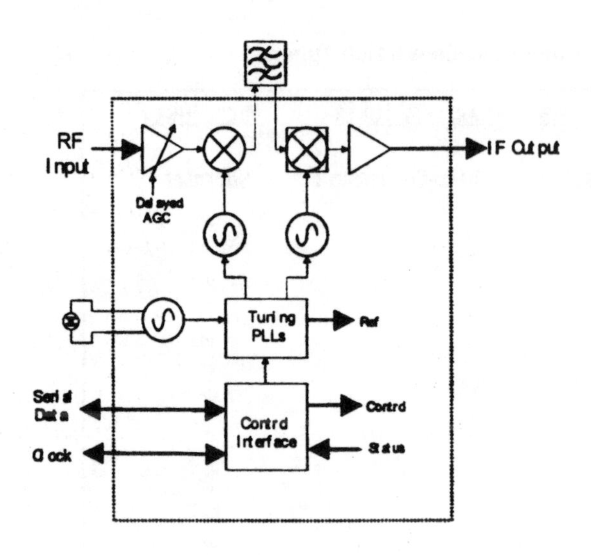

FIGURE 8-13

market. It has full multimedia tuner functions and provides both RF, IF, and baseband processing.

The drawing in **Figure 8-14** features the Microtune broadband TV receiver chips (MT2000 and MT2500) with RF input frequencies of 50 to 860 MHz. Both chips include a variable-gain low-noise amplifier (VLNA), first and second mixers, and frequency synthesis systems. The MT2500 adds NTSC video detection, and FM sound demodulation, fully synchronous demodulation with carrier regeneration, AGC and AFT. It also has separate sound conversion circuits (including a sound IF limiter) and an FM detector function.

On both the MT2000 and the MT2500, the input LNA's variable gain provides excellent dynamic range allowing the single chips to be used in off-air, hybrid fiber coax (HFC), and cable environments without an external tracking filter. Superior image rejection and channel selection of the up-converted IF spectrum is achieved via an external 15 MHz wide, narrowband IF filter, coupled to an internal image reject mixer. The down-converted, industry-standard second IF center frequency of 44 MHz allows for precise signal processing and detection of the video signal. On the MT2500, a gated video AGC maintains a constant peak-to-peak baseband video output signal. FM detection is accomplished using a unique internal delay line, eliminating the need for external FM discriminators.

Solving The Interference Problems

The MicroTuner design includes a number of oscillators and mixers, running large signals on a common substrate, without creating interference problems. Typically, multiple oscillators on a chip create spurious signals that may include harmonics, parasitic emissions, or intermodulation products. Unwanted electronic coupling between circuits causes interfering spurious signals. To resolve this problem, the MicroTuner relies on its patented circuitry to minimize interference. It deploys circuits that minimize the generation of spurious signals while maximizing the rejection of those that

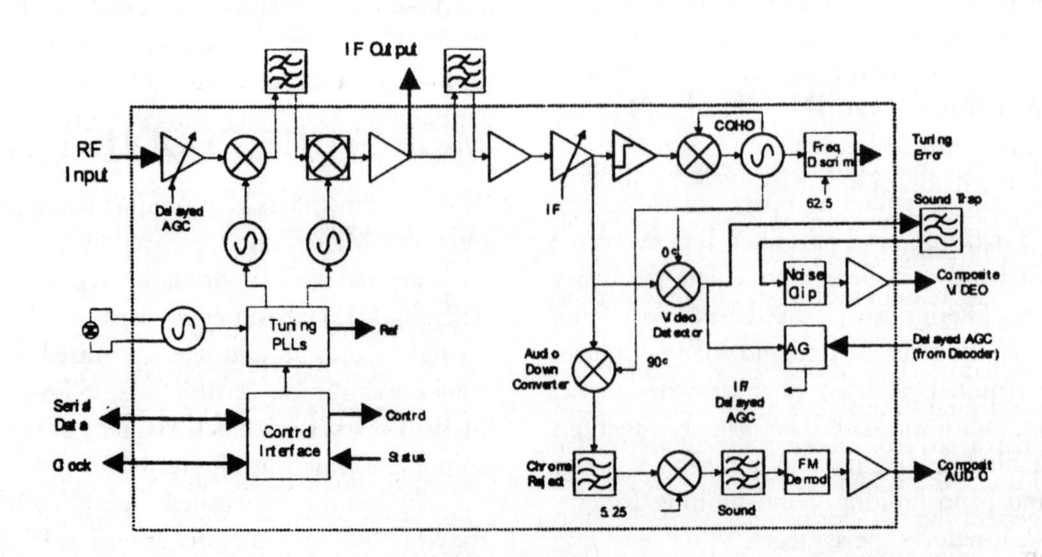

FIGURE 8-14

might exist. It also uses microarchitectures, and frequency plans that generate signals that are out-of-band to down-stream processing.

MicroTuner Circuit Chip Details

The RF input circuit consists of a variable-gain noise amplifier (VLNA), the design of which is critical because it plays a large role in setting the receiver's noise figure. The noise figure for the MicroTuner VLNA is typically 5 dB. It is important to consider that a TV receiver is a broadband system, meaning the input band covers several octaves (50 to 860 MHz), as opposed to a narrowband system, such as a cell phone, which covers much less than one octave. In narrowband applications, reactive components can be used to create LNAs. This is not practical in broadband systems because a circuit cannot be tuned across the entire multi-octave band.

The MicroTuner's first mixer up-converts the entire spectrum to the first IF frequency. The fully-integrated VLNA and first mixer, operating together, achieve a composite triple beat (CTB) distortion in excess of -57 dBc for any channel. In engineering parlance, a CTB refers to a major distortion caused by signals beating against each other.

The external first IF filter operates at a frequency well in excess of 1 Ghz. The second mixer is an image rejection mixer, which down-converts to the second intermediate frequency. A major concern of RF designers is image rejection, and this second IF mixer is of a specialized type with built-in image rejection. Its purpose is to reject the image frequency of the first mixer. In amplitude modulation systems, every signal has an image (which is usually not desired). When down conversion is performed, filtering can be used to attenuate the image signal. However, the combination of a mixer and a filtering function in an image reject mixer provides a greater amount of image rejection (in this instance it would be 65 dB).

Local oscillators (L01 and L02) drive the mixers. In conventional tuners, L0s are frequency synthesizers constructed from discrete components. In the MicroTuner, these L0s are fully integrated on the chip, generating all frequencies required for tuning from a single external crystal with a frequency resolution of 62.5 KHz. The L0s are part of a complex frequency synthesis system in which the VCOs, including the varactors, charge pumps, phase-frequency detectors, and dividers, are fully integrated. In terms of phase noise, the chip performs a robust -85 dBc at 10KHz offset, which is regarded as excellent for both analog and digital TV reception, for both off-air and cable systems.

The second IF filter is a standard 44 MHz TV filter. The IF amplifier is a variable-gain amplifier, and automatic gain control (AGC) is applied between the IF amplifier and VLNA. The AGC gain range of the MicroTuner is 96 dB. The video detector consists of a PLL, which generates in-phase (zero degree) and out-of-phase (90 degree) signals. The PLL locks to the incoming picture carrier and generates in-phase and quadrature components. It uses the in-phase component to demodulate the IF video signal, and uses the quadrature component to demodulate the IF audio.

Continuing on the video path, a sound trap eliminates residual audio. Next, the noise clipper, as the name applies, clips occasional video signal noise spikes. After filtering, the output is a full, 4.2 MHz baseband video. The output of the video detector is sampled to determine its magnitude and is then fed back to the AGC, IF amplifier, and VLNA.

On the audio side, there is an audio down-converter and chroma reject filter. An integrated, continuous-time, self-tuned filter serves as the sound filter. This highly advanced filter automatically adjusts its own characteristics. The FM demodulator provides the composite audio output signal.

The entire MicroTuner chip is controlled through a standard serial-interface, which is compatible with 12C. This allows the read-back of all status registers on the chip, as well as permits device programming such as channel tuning. To tune a channel, registers are loaded into the tuning PLLs.

SINGLE VERSUS DUAL-CONVERSION MICROTUNER CHIPS

Generally, single-conversion tuners have depended on adjacent channel vacancy to achieve acceptable performance. Changes will be necessary in the tracking filter to improve this performance to levels typical of tuners designed for cable television. These performances, including composite triple beats (CTBs), composite second order distortion products (CSOs), and cross-modulation products (XMOD), must be improved to provide acceptable performance for the new, fully-utilized spectrum of the digital era.

Cable set-top boxes have predominantly used dual-conversion tuners for a variety of reasons including in-band local oscillator leakage (the LO of a single conversion is in the TV band typically), impedance matching throughout the input band (the out-of-band impedance of a tracking filter typically provides a poor impedance match to a cable), and superior distortion performance.

MicroTuner 2030 Series Application Notes

- Cable modems
- Cable set-top boxes
- TV receivers
- VCRs

The MicroTuner series 2030 chip's topside is illustrated in **Figure 8-15**.

MicroTuner Features

- DOCSIS capable
- Fully integrated tuner with on-chip low-noise amplifier (LNA), mixers and frequency synthesizers, varactors, and serial interface.
- Single, 5-volt supply
- Minimal external components
- No manually-tunable parts required
- Variable gain LNA with dynamic range
- 48 MHz to 860 MHz input frequency range
- Low noise figure
- Excellent CTB/CSO distortion performance
- Fully programmable frequency synthesizers via serial-control interface
- Intermediate frequency (IF) output fully compatible with NTSC, PAL, and other standards
- 80-pin ePad TQFP chip

The MicroTuner 2030 tuner block diagram is shown in the **Figure 8-16** drawing. The pin diagram for this chip is shown in **Figure 8-17**. The

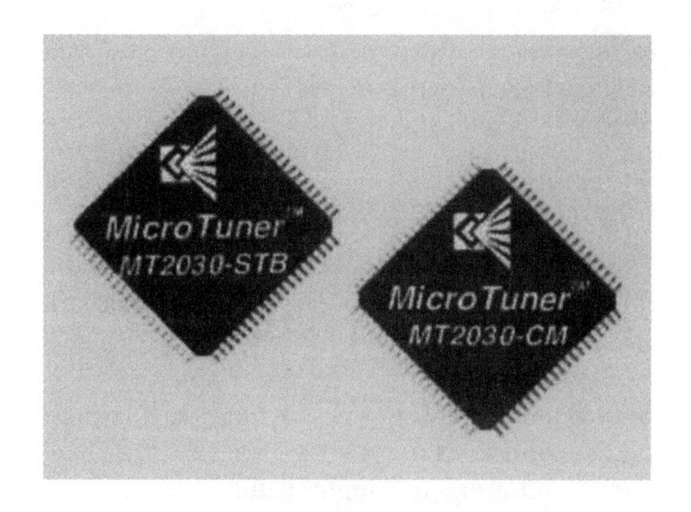

FIGURE 8-15

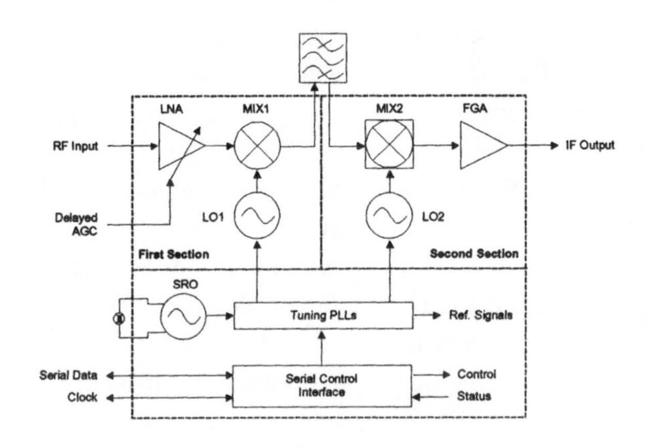

FIGURE 8-16

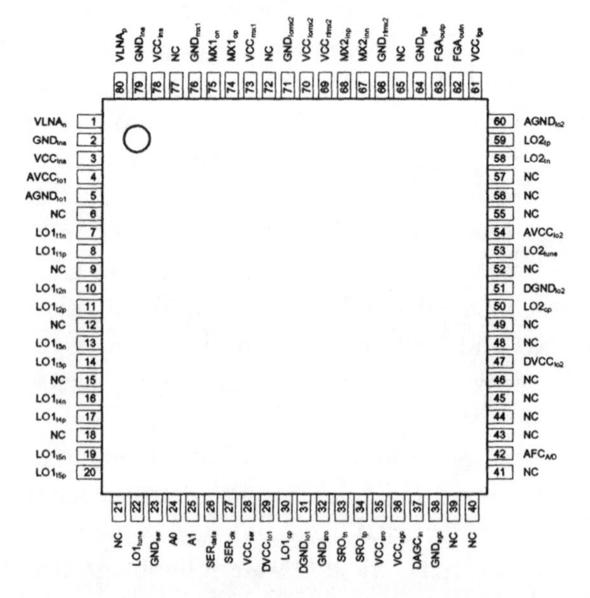

NC = No connect. Solder pins to a pad, but do not connect the pad to any other points, including ground.

FIGURE 8-17

package detail of the 2030 series chip in shown in **Figure 8-18**.

MicroTuner 2032 Series Chip

The MicroTuner series 2032 is a single-chip broadband tuner capable of receiving frequencies in the 48 MHz to 860 MHz range and of converting a selected channel to an intermediate frequency (IF). These include NTSC, PAL, and digital standards including ATSC (DTV), DOCSTS, DAVIC, and DVB-C. The tuner block diagram for the 2032 series MicroTuner is shown in **Figure 8-19**.

The MT2032's low close-in phase noise allows it to be used for both digital and analog video signals. The input LNA's variable gain provides an automatic gain control (AGC) range of 50 dB, making the MT2032 suitable for off-air as well as cable TV applications. Its dual-conversion architecture, with no

Symbol	Min	Nominal (mm)	Max	Notes
A		1.20		
A1	0.05	0.10	0.15	13
A2	0.95	1.00	1.05	
D	16.00 BSC.			4
D1	14.00 BSC.			7, 8
E	16.00 BSC.			4
E1	14.00 BSC.			7, 8

Symbol	Min	Nominal (mm)	Max	Notes
L	0.45	0.60	0.75	
N	80			
e	0.65 BSC.			
b	0.22	0.32	0.38	9
b1	0.22	0.30	0.33	
ccc			0.10	
ddd			0.13	

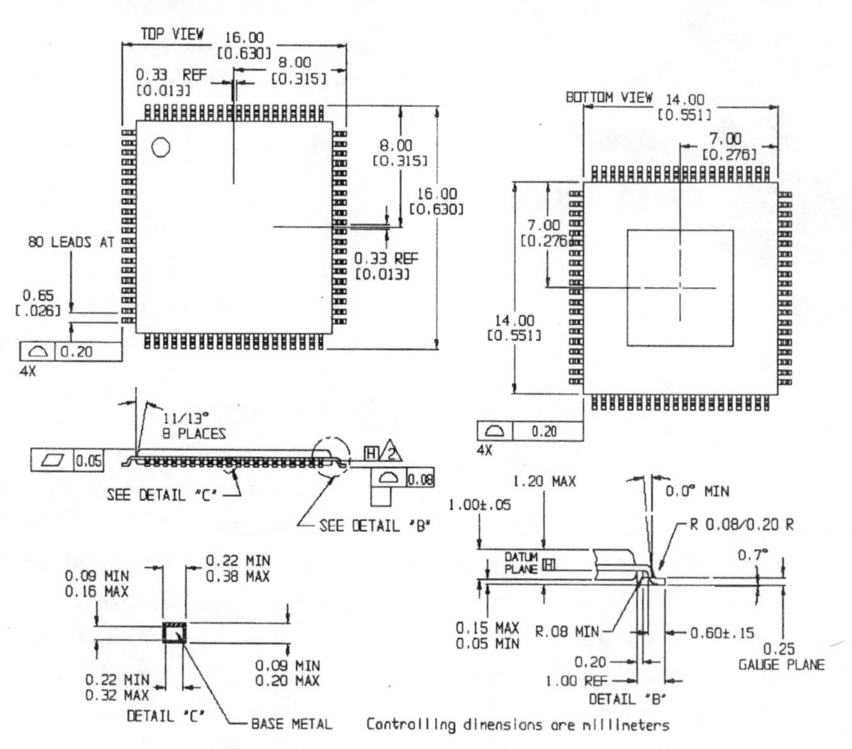

NOTES:

1. All dimensioning and tolerancing conform to ASME Y14.5M-1994.
2. Datum plane H located at mold parting line and coincident with lead, where lead exits plastic body at bottom of parting line.
3. Datums A-B and D to be determined at centerline between leads where leads exit plastic body at datum plane H.
4. To be determined at seating plane C.
5. Dimensions D1 and E1 do not include mold protrusion. Allowable mold protrusion is 0.254 mm on D1 and E1 dimensions.
6. "N" is the total number of terminals.
7. These dimensions to be determined at datum plane H.
8. The bottom of package is smaller than the top of package by 0.15 millimeters.
9. Dimension b does not include dambar protrusion. Allowable dambar protrusion shall be 0.08mm total in excess of the b dimension at maximum material condition. dambar cannot be located on the lower radius or the foot.
10. Controlling dimension: millimeter.
11. Maximum allowable die thickness to be assembled in this package family is 0.50 mm.
12. This outline is not yet JEDEC registered.
13. A1 is defined as the distance from the seating plane to the lowest point of the package body.
14. Exposed die pad shall be coplanar with bottom of package within 0.05 mm.
15. Metal area of exposed die pad shall be within 0.30 mm of the nominal die pad size.

FIGURE 8-18

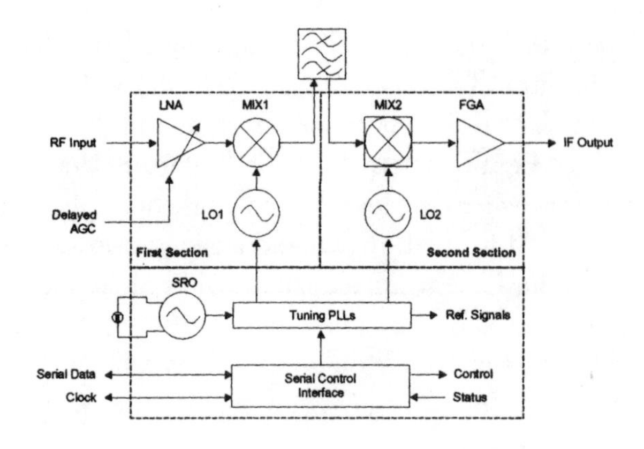

FIGURE 8-19

FIGURE 8-20

requirement for tracking filters, yields the desirable characteristics of traditional cable television tuners: controlled input impedance across the input band, low in-band emissions, and good selectivity.

The MT2032 device is highly integrated, with on-chip components including a variable gain LNA, a first mixer that up-converts the input spectrum to a first IF of 1090 MHz, and an image-reject mixer that down-converts the filtered first IF spectrum to a standard second IF. It also includes a buffer amplifier for the IF output, a complete frequency synthesis system to create the first and second local oscillator (LO) frequencies from an external 5.25 MHz crystal, and a serial interface to program the device as well as read back its status.

The PLLs of the frequency synthesis system are fully integrated, including the varactors. The image-reject mixer provides excellent image rejection and eases the requirements of the first IF filter. The chip requires only a 5-volt supply, thus eliminating the need for the 28 volts to 33 volts supplies typically required by traditional TV tuners. The trademark logo for the Microtune company is shown in **Figure 8-20**.

TOSHIBA HDTV TUNER/RF SYSTEMS

The Toshiba HDTV, progressively scanned televisions have two RF inputs: ANT 1 and ANT 2. Refer to block diagram in **Figure 8-21**. These TV receivers also have two tuners. One tuner is designated for the main picture and the other for the PIP (picture-in-picture) picture mode. The PIP tuner receives its source from ANT 1 ONLY. The main tuner may receive its signal from either ANT 1 or ANT 2.

The RF switch (H003) is a splitter for the ANT 1 input signal and a switch for the main tuner. This RF switch is illustrated in **Figure 8-22**. The ANT 1 input signal is split by a divider and applied to two amplifiers. The amplifiers provide isolation to reduce cross interference from the main and PIP tuners. One output of the divider supplies signal to the PIP tuner. This is the only TV signal source for the PIP tuner.

The other divider output supplies the signal to internal switches. If ANT 1 is chosen for the main tuner source, the RF signal from ANT 1 is applied to the main tuner through the relay switch. If ANT 2 is chosen for the main tuner source, the relay switch toggles, making ANT 2 the RF source for the main tuner, and the diode switch closes, feeding the RF signal from ANT 1 to the OUT terminal.

As you refer back to **Figure 8-21**, you will note that the main tuning circuit consists of H001 main tuner, H002 IF module, QA01 microprocessor, and Q151/Q152 AGC switch. The microprocessor is a D/A converter for the automatic fine tuning (AFT) circuit. It controls the tuner and IF module through the I square C bus. The outputs of the IF module are applied to the A/V circuit.

A nonvolatile memory is built into H001. The memory contains all alignment information.

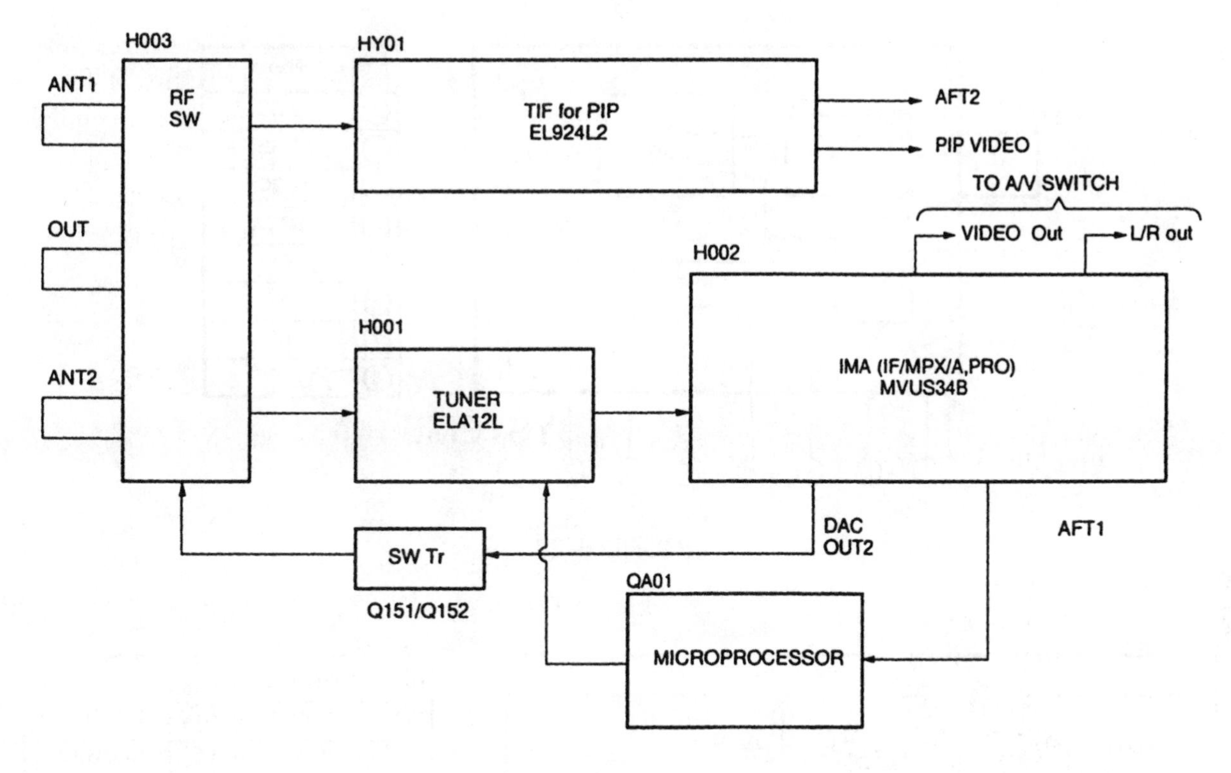

FIGURE 8-21

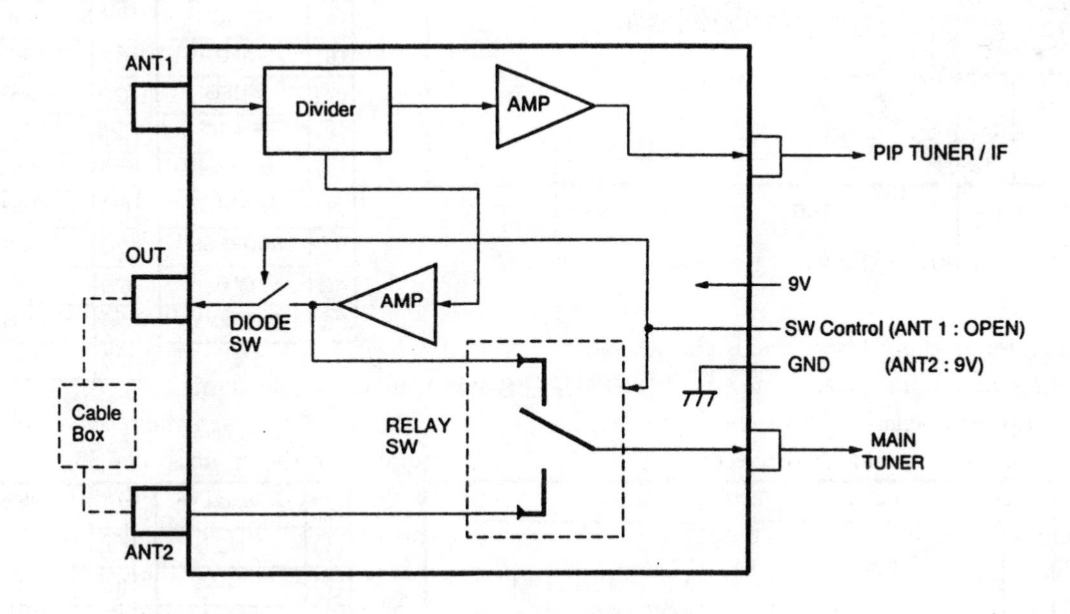

FIGURE 8-22

Therefore, no IF or Sound IF adjustments are necessary after replacing H001. Note the block diagram of this unit in **Figure 8-23**. The IF module block diagram for the model EL924C2 is shown in **Figure 8-24**.

Picture-in-Picture Tuner

Refer to the block diagram in **Figure 8-25** for the HY01 PIP tuner and IF module. The tuner and IF circuits are located in one module. The module is controlled by the microprocessor for channel selec-

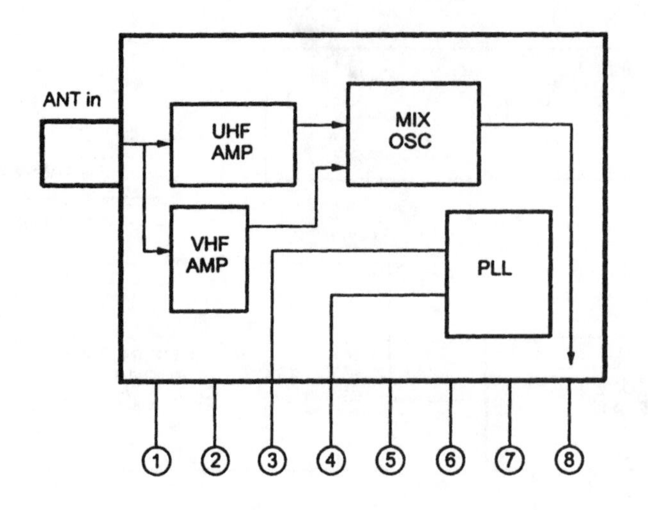

Pin No.	Name
1	AGC
2	ADDRESS
3	CLOCK
4	DATA
5	9V
6	5V
7	32V
8	IF

FIGURE 8-23

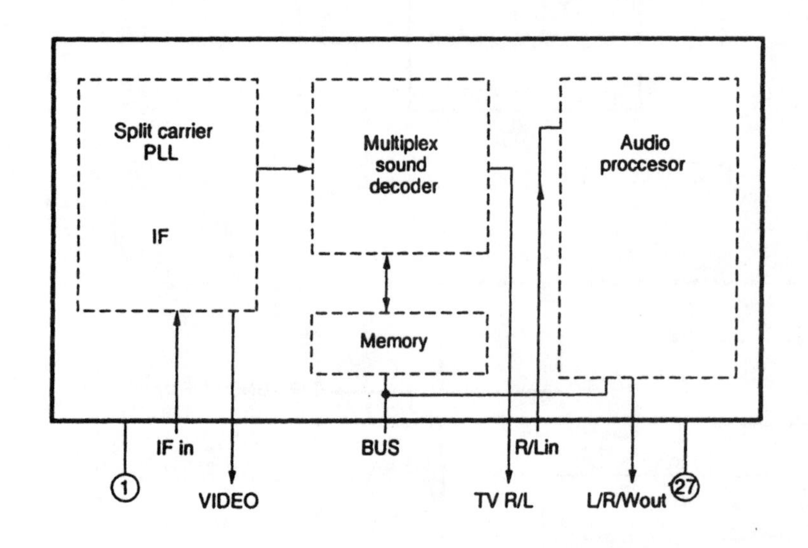

Pin No.	Name
1	GND
2	IF in
3	NC
4	9V
5	RF AGC
6	AFC
7	VIDEO
8	IF AGC
9	MPX out
10	SAP VCO
11	ADDRESS
12	TV R
13	DAC out 1

Pin No.	Name
14	TV L
15	DAC out 2
16	Rin
17	Cin
18	Lin
19	GND
20	CLOCK
21	DATA
22	W out
23	C out
24	L out
25	GND
26	R out
27	9V

FIGURE 8-24

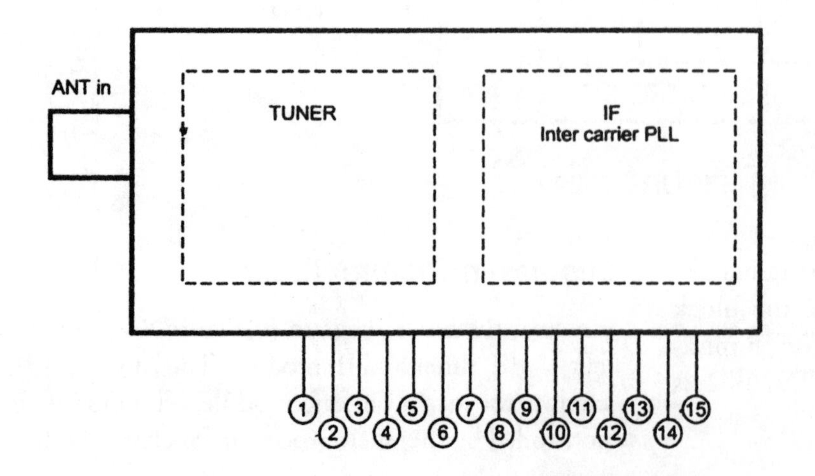

Pin No.	Name
1	NC
2	32V
3	CLOCK
4	DATA
5	NC
6	ADDRESS
7	5V
8	RF AGC

Pin No.	Name
9	9V
10	NC
11	GND
12	AFT
13	NC
14	GND
15	VIDEO

FIGURE 8-25

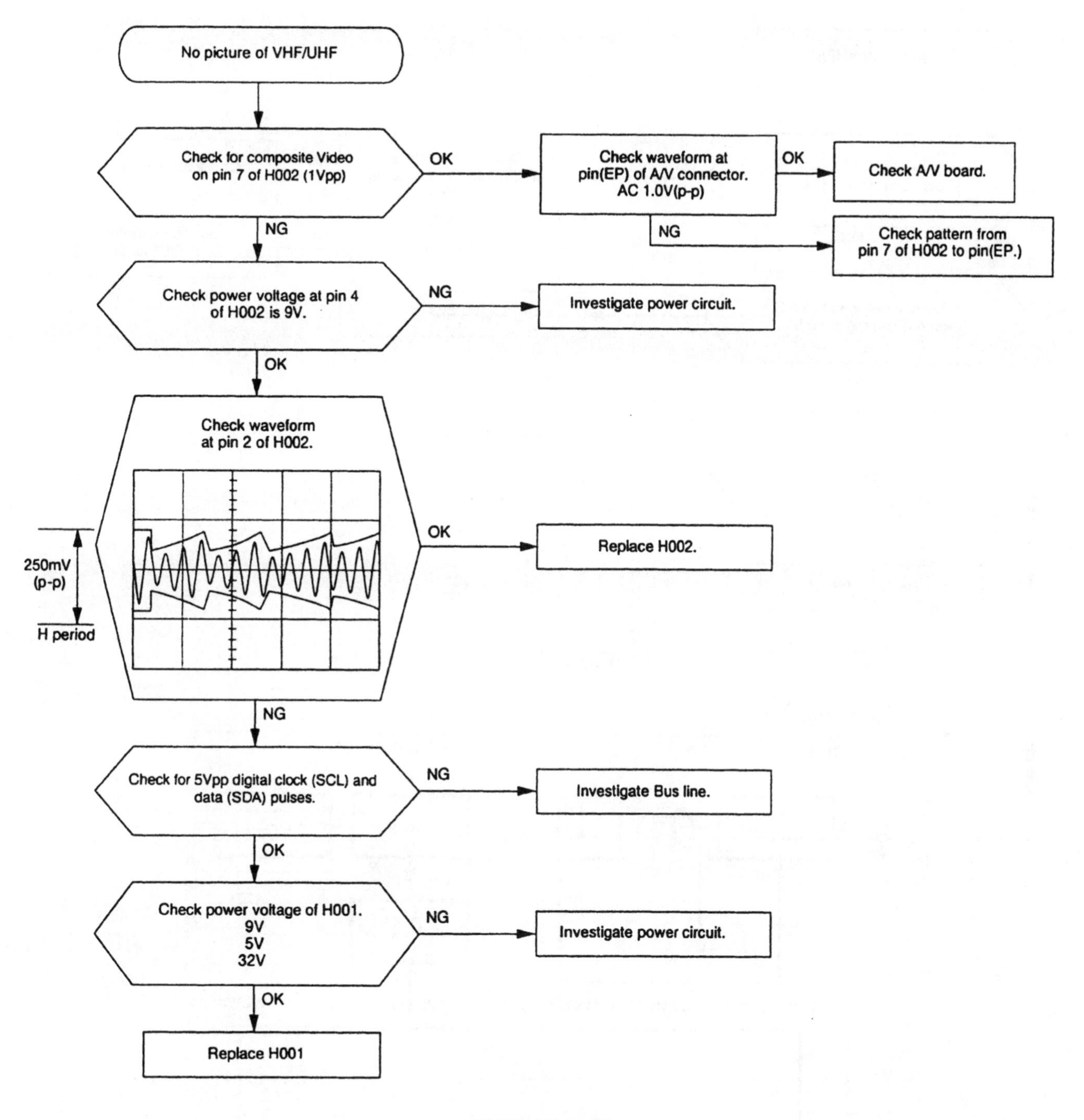

FIGURE 8-26

tion and AFT (auto fine tuning). There are two outputs from the PIP TIF module: AFT and composite video. The video signal is applied to A/V switch circuit, and the AFT signal is sent to the microprocessor. HY01 does not process audio. Thus, no audio is available for the PIP picture.

Toshiba HDTV Troubleshooting Information

For "no picture of VHF/UHF on the main TV screen" problems, refer to the troubleshooting tree information in **Figure 8-26.**

For "no picture of the VHF/UHV with the PIP screen," refer to the troubleshooting tree information in **Figure 8-27.**

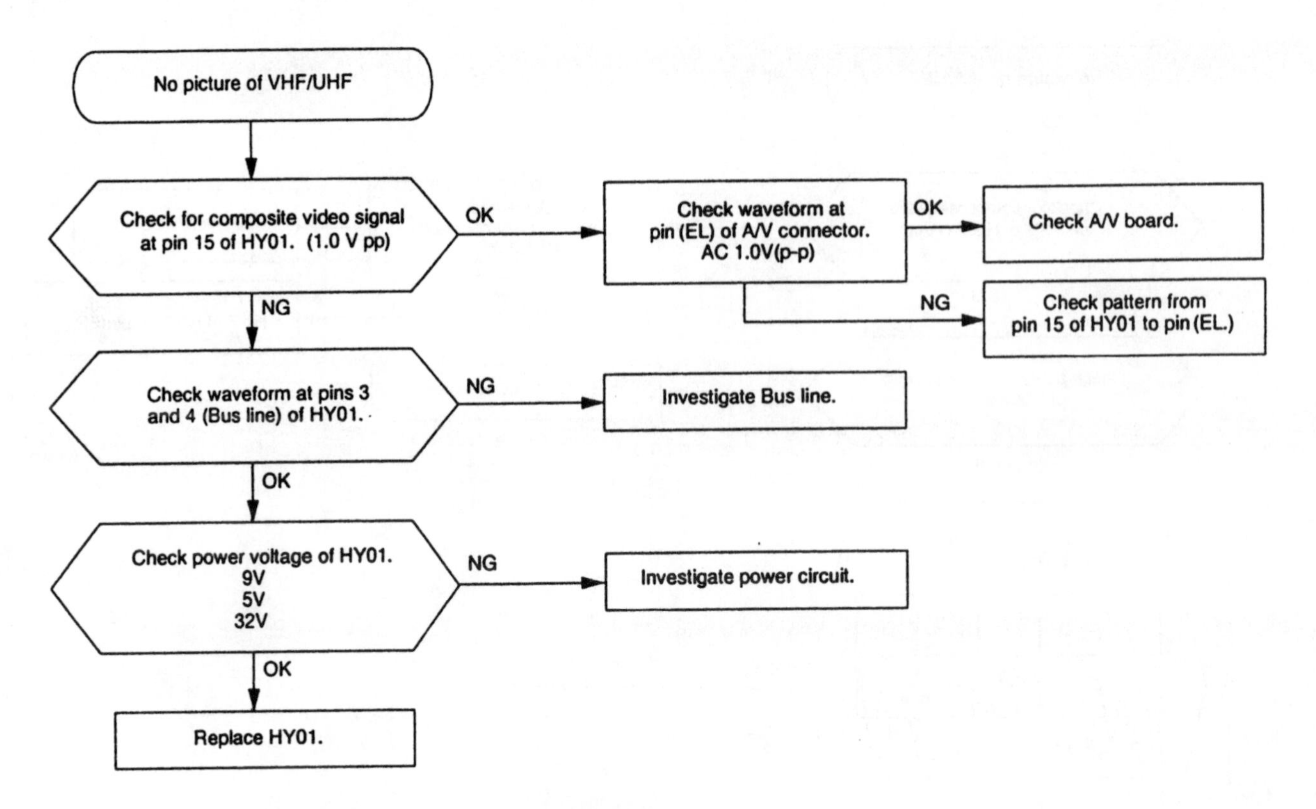

FIGURE 8-27

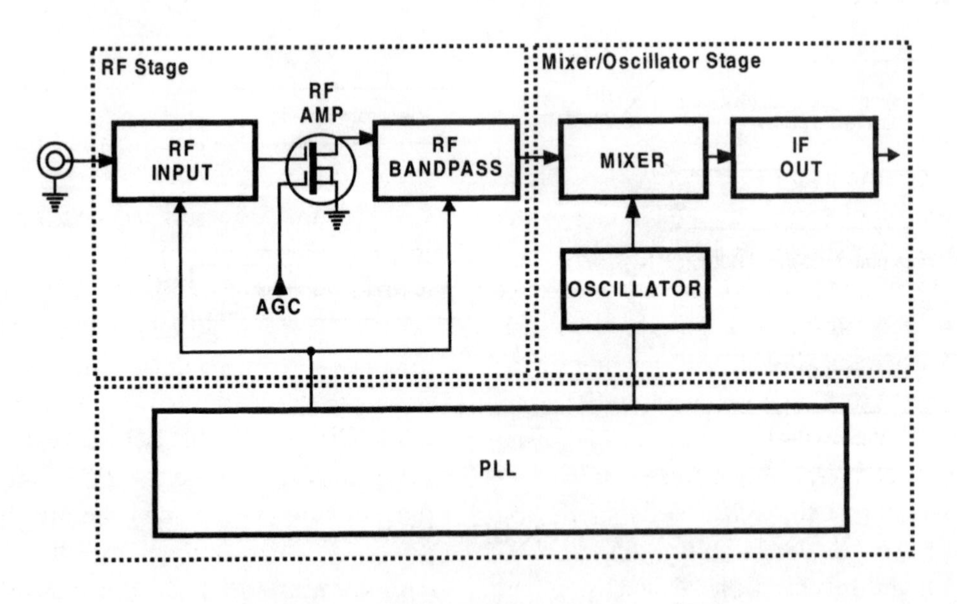

FIGURE 8-28

MAIN TUNER OVERVIEW FOR RCA MM101 HDTV CHASSIS

The MM101 tuner uses TOB (tuner on board) topography with a zinc tuner wrap. It is identical to the RCA CTC195/197 tuner. However, there are two variations:

1. A single input tuner
2. A single input tuner with PIP RF output

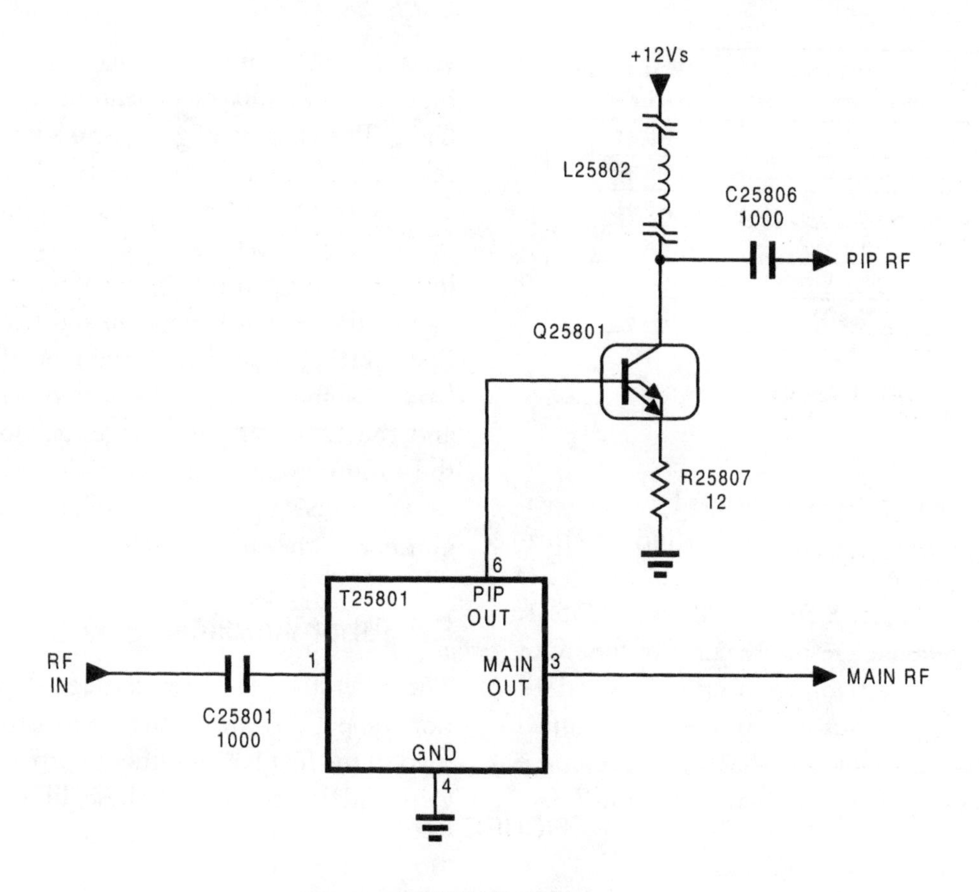

FIGURE 8-29

The second tuner is not based on the CTC197 but is a "cold" version of the CTC185 tuner. There are many similarities between the two. Refer to the tuner block diagram shown in **Figure 8-28**.

This tuner can be separated into three distinct sections for its technical operation. First, the RF stage, which processes the incoming antenna or cable RF signal. This stage captures, filters, and amplifies the RF for further processing. Next, the mixer/oscillator converts the different high-frequency RF carriers to a single IF frequency for use by the remainder of the TV circuitry. The PLL IC controls the RF and mixer/oscillator switching and tuning circuits. The PLL communicates with the main microprocessor for channel selection information, and then converts the digital information to analog voltages needed to tune the RF and mixer/oscillator to the proper frequency.

RF Input Splitter

The input splitter feeds the RF input signal to both the main and PIP tuners as shown in **Figure 8-29**. The splitter consists of asymmetrical power divider transformer T25801 and broadband amplifier Q25801 and its associated circuitry. The arrangement minimizes loss in the main RF path at the expense of higher loss in the PIP tuner path, thus optimizing the noise figure of the more important tuner. Note that the main tuner always supplies the signal for the main picture. During PIP operation, if the PIP and main pictures are "swapped," the main tuner re-tunes to the PIP channel and the PIP tuner re-tunes to the main channel. Broadband amplifier Q25801 amplifies the PIP RF path signal, compensating for the loss in transformer T25801.

Input Filtering

The main RF input from the input splitter is fed to a diplexer that routes the VHF and UHF range signals to the single-tuned tracking filters of the VHF and UHF RF amplifiers, respectively. The single-tuned filters are voltage-tuned tank circuits tuned to the selected channel by voltages generated from DACs on PLL synthesizer IC, U25501. The

Channel	U25501 Pin 6 (Single-tuned Filter)	U25501 Pin 14 (Band Switching)
2	1.2 V	HIGH
6	7.6 V	HIGH
7	4.5 V	HIGH
13	6.9 V	HIGH
14	5.1 V	LOW
69	25.4 V	LOW

FIGURE 8-30

filters reject signals from unwanted frequencies, reducing intermodulation distortion and minimizing image response.

Referring to **Figure 8-31**, the PLL IC, U25501, controls the frequency response curve of the input filter by using output voltage from pins 6 and 14 to change the characteristics of the single-tuned input filter tank circuit and shaping the response of the appropriate RF amplifier. While pin 6 is a variable DAC output voltage, pin 14 is either a high or low voltage, depending on the selected band. The chart in **Figure 8-30** shows the voltage selection of the different RF bands and channels. The chart is a representation of tuning voltages for off-air channel selection. Note that as channel selection goes up through the VHF and the UHF bands, the tuning voltage on pin 6 rises until the first UHF channel 14 selection. At that point the band switching on pin 14 goes from high to low and the DAC output voltages go down to begin the tuning cycle again, increasing as the channel selection goes up the band. This becomes the beginning of the tuning cycle.

RF Tuner Amplifier

The MM101 uses a single-stage dual-gate depletion type FET (field effect transistor). FETs are used in the first RF amplifier to provide the highest gain and lowest noise. These FETs are voltage-

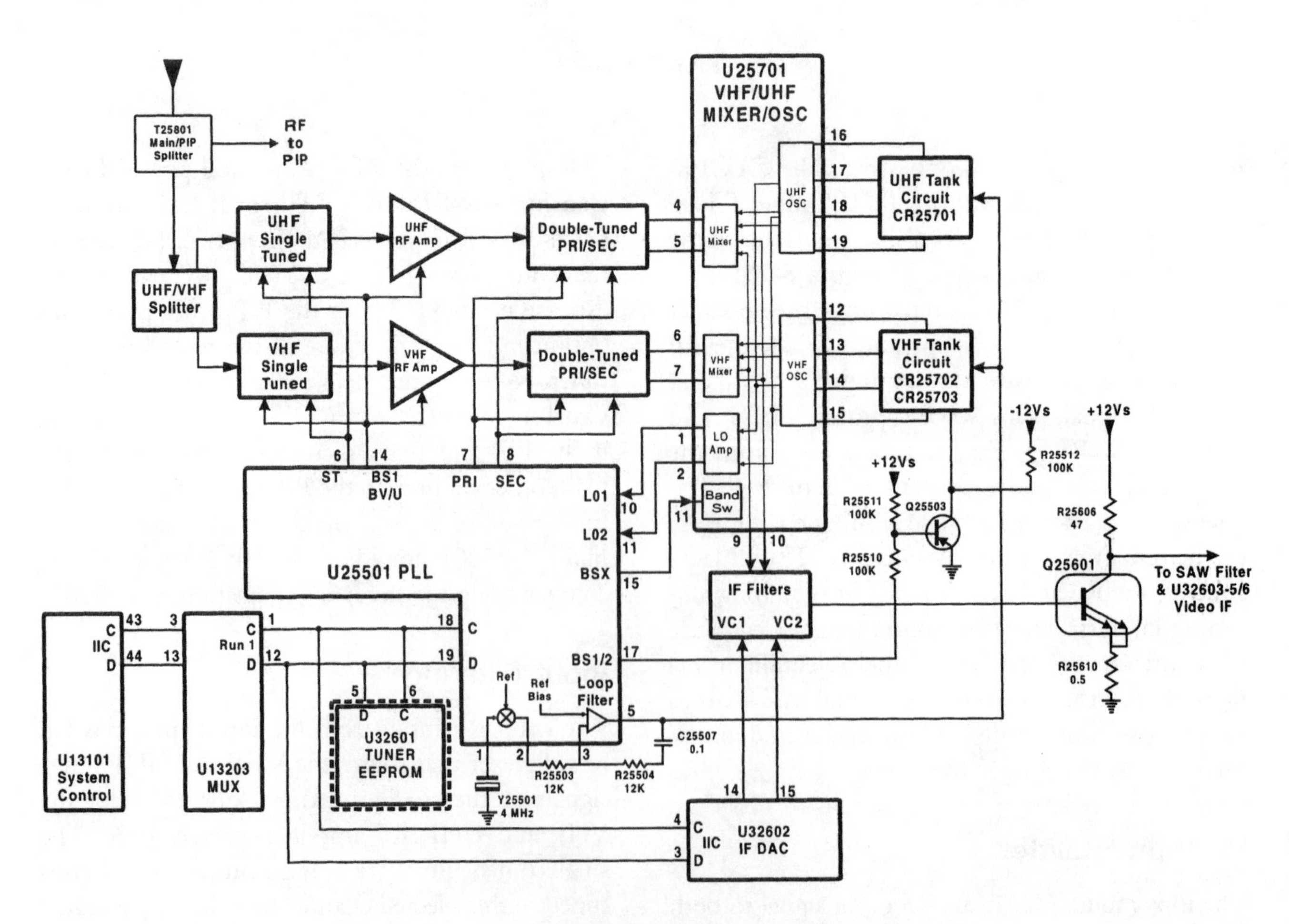

FIGURE 8-31

controlled devices that operate very similar to vacuum tubes. When negative voltage is applied to the gate with respect to the source, drain current is reduced. If the negative voltage is high enough, drain current is pinched off completely. Positive voltage on the gate, with respect to the source, will increase the drain current.

Both gates on the dual-gate MOSFETs affect drain current. In this chassis, the RF input is on gate #1 and the AGC (automatic gain control) is placed on gate #2. As the AGC voltage increases (positive with respect to the source) drain current also increases. When AGC voltage decreases, drain current decreases. The AGC voltage is generated from the video IF IC, U32603 pin 10, which monitors the IF signal level. If the IF signal level increases, AGC voltage is reduced, lowering the gain of the RF amplifier. If the IF signal decreases, AGC voltage increases raising the gain of the RF amplifier.

Double-Tuned Tracking Filter

The double-tuned tracking filter is essentially an air-core, balun transformer with voltage-tuned primary and secondary. The bottom of the primary is returned to ground, whereas the secondary is balanced for driving the balanced mixer inputs of the mixer/oscillator IC, U25701. The double-tuned tracking filter increases selectivity provided by the single-tuned tracking filter. The VHF double-tuned tracking filter is split into two bands and bandswitched with PIN diodes. The UHF double-tuned tracking filter is not bandswitched. Transformers provide impedance matching for the remainder of the RF stage.

Mixer/Oscillator

The U25701 IC comprises a self-contained mixer/oscillator network requiring very few external components. The IC contains two double-balanced mixers and two, balanced voltage-controlled oscillators (VCO), one for VHF, one for UHF, plus bandswitching logic. The VCO tank circuits are external to the IC and contain varactor diodes for tuning, and PIN diodes for bandswitching of the VHF VCO section. As in the case of the tracking filters, the UHF VCO is not bandswitched. The balanced output of the mixer/oscillator feeds the nominal 45 MHz IF signal to the SAW preamp. This IF signal is then sent to the Video IF IC, U32603 pins 5 and 6, for further processing. The video and audio IF signals are separated at that time.

SAW Preamp Circuit

The SAW (surface acoustic wave) preamp receives the balanced mixer IF output from the mixer/oscillator IC, U25701, and bandpass filters and amplifies the IF signal for application to the IF SAW IF filter located just outside the tuner enclosure. At the SAW preamp input is a double-tuned transformer that rejects unwanted mixer products and further augments the selectivity provided by the tracking filters in the tuner front end. The double-tuned transformer primary is balanced, and is tuned by a varactor diode that obtains its tuning voltage from DAC IC, U32602. The secondary is similarly tuned, and has one side grounded to feed the unbalanced input of Q25601, which develops a suitable low impedance drive for the SAW filter. The entire stage is electronically aligned by use of varactor diodes.

PLL FREQUENCY SYNTHESIZER

The PLL frequency synthesizer IC, U25501, controls the frequency of the VCO's in the mixer/oscillator IC, U25701, producing the desired local oscillator frequency for the tuner. The PLL section is of conventional design, consisting of two programmable frequency dividers, a reference oscillator with external crystal, a phase detector, and error amplifier. **Figure 8-32** shows a simplified block diagram of the PLL section in the IC. A reference

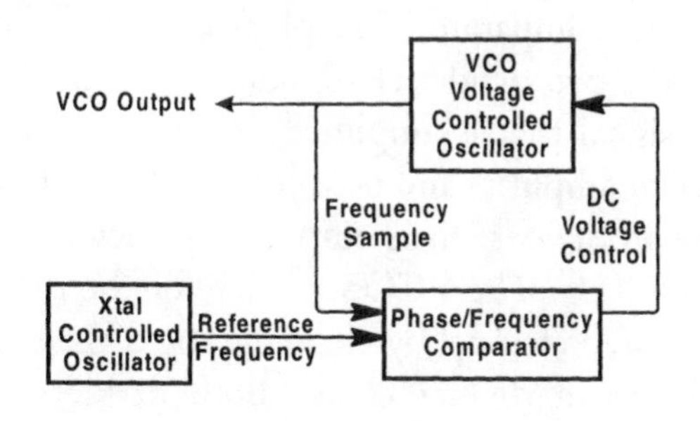

FIGURE 8-32

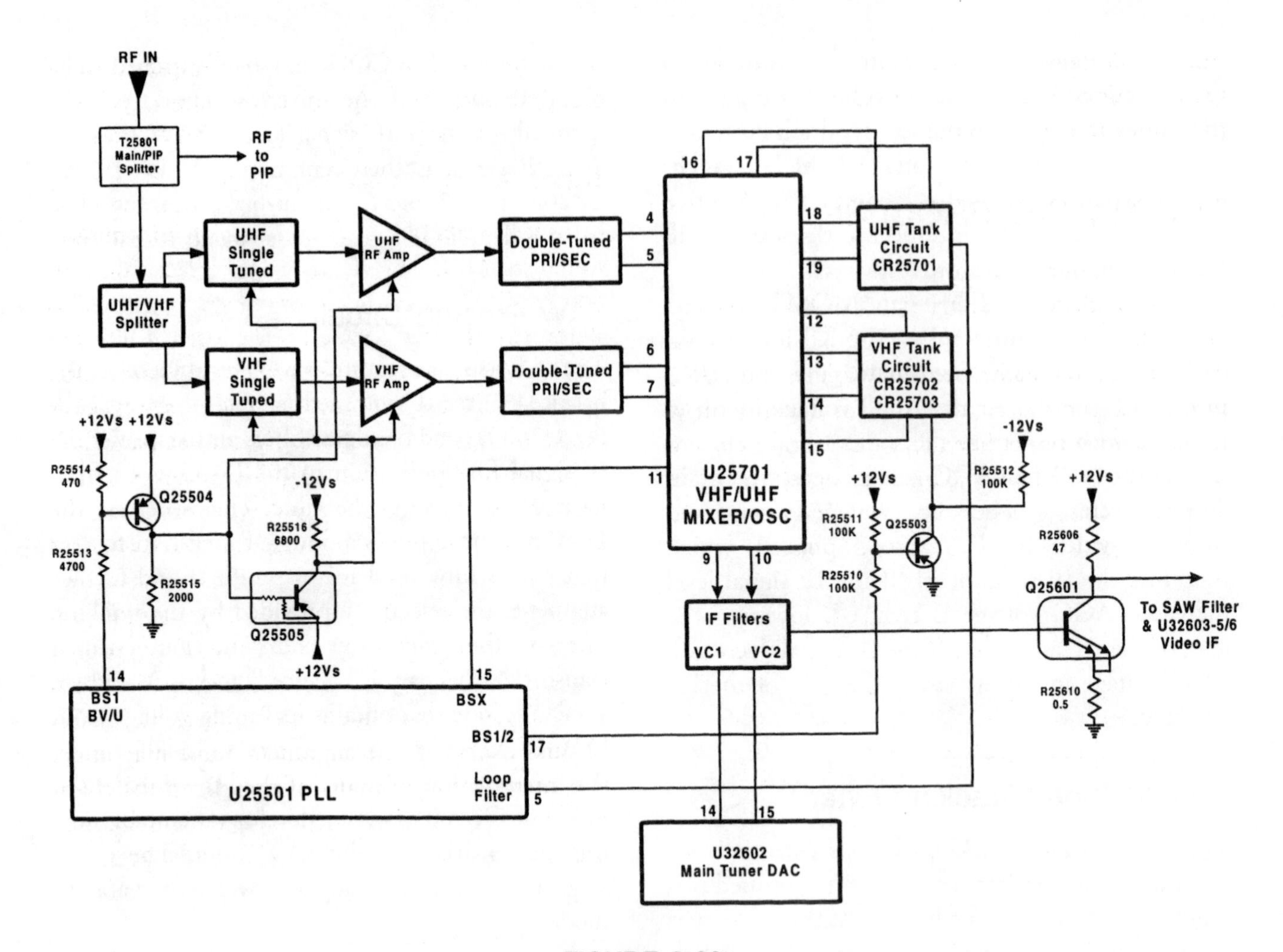

FIGURE 8-33

divider chain divides the output of the 4 MHz crystal reference oscillator to produce a PLL reference frequency of 7.8125 KHz for application to one input of the phase comparator. The main divider chain receives the buffered VCO output from the mixer/oscillator IC and applies it to a fixed divide-by-8 prescaler that feeds a programmable divider section. The main frequency divider thereby divides the VCO frequency down to a value near 7.8125 KHz for application to the other input of the phase comparator. The phase comparator compares the two divider chain outputs, producing an error signal that is amplified, low-pass filtered by the error amplifier and used to tune the VCO until the main divider chain output frequency is exactly 7.8125 KHz. The VCO can therefore be tuned to any desired channel by changing the division ratio of the main divider chain. Both divider chains receive division ratio preset data from the system microprocessor via the IIC bus.

The PLL frequency synthesizer IC also provides the tuning voltages required to tune the tracking filters in the tuner RF section. The tuning voltage is derived from the error amplifier output in the PLL section. The filtered error signal is applied to a series of high-voltage digital-to-analog converters (DACs) that adjust the error voltage as necessary to provide track-tuning voltages that precisely fine-tune the tuner tracking filters to the desired channel frequency. This arrangement provides an exceptionally accurate tuner bandpass response compared to conventional tuners without such electronic alignment. Control data for the DACs is stored in tuner EEPROM U32601 during electronic alignment of the tuner.

The PLL synthesizer IC also contains bandswitching logic, which receives data from the system microprocessor via the IIC bus and sends bandswitch commands to the mixer/ oscillator IC and tuner tracking filters.

THE TUNING PROCESS

Once the bandswitching circuitry has selected the desired band, the PLL section selects the specific channel. It first tunes the selected VCO to the desired local oscillator frequency. Then the DACs in the PLL synthesizer tune the RF tracking filters to the channel center frequency. Referring to **Figure 8-33**, note that the DAC output at pin 6 provides a varactor-tuning voltage for the VHF and UHF single-tuned filters. DAC outputs at pins 7 and 8 control the primary and secondary, respectively, of the VHF and UHF double-tuned filters.

Channel	Band	Midrange (MHz)	Pix Carrier (MHz)	Local Oscillator (MHz)
2	1	57	55.25	101
3	1	63	61.25	107
6	1	85	83.25	129
98	1	111	109.25	155
14	1	123	121.25	167
17	1	141	139.25	185
18	2	147	145.25	191
13	2	213	211.25	257
29	2	255	253.25	299
35	2	291	289.25	335
41	2	327	325.25	371
45	2	351	349.25	395
48	2	369	367.25	413
50	2	381	379.25	425
51	3	387	385.25	431
57	3	423	421.25	467
60	3	441	439.25	485
64	3	465	463.25	509
68	3	489	487.25	533
76	3	537	535.25	581
83	3	579	577.25	623
88	3	609	601.25	653
93	3	639	637.25	683
105	3	681	679.25	125
110	3	711	709.25	755
115	3	741	739.25	785
120	3	771	769.25	815
123	3	789	787.25	833
125	3	801	799.25	845

FIGURE 8-34

Tuner Alignment Channels

In order to compress the amount of information stored in the EEPROM, only the exact information required for tuning a few channels, known as alignment channels, have been chosen. Only the exact values needed to tune these channels are stored by the EEPROM. When a channel selection is made, the microprocessor decides what band it is in, and then what two alignment values it lies between. It must then interpolate, or calculate, the DAC values required to tune the exact channel frequency. This information is then sent via the IIC bus to the PLL IC, which changes the frequency response of the tuner and the LO for proper channel reception. If any of the alignment values are changed, every value must be checked. This is because changing one alignment value may affect the interpolation of many of the alignment channels. Information for the tuner alignment channels is shown in **Figure 8-34**.

Channel Selection

The microprocessor goes through a fixed routine to affect channel selection. Although the instruction routine is lengthy, it is accomplished in less than 150 milliseconds. First, all information necessary to select the channel is retrieved from the tuner EEPROM. This includes the local oscillator (LO) data for the channel, the band switch information and upper and lower alignment channel DAC values for the frequency range that the channel lies within. Now the actual electrical tuning of the RF receiver section can now start functioning.

The LO and band switch information is delivered to the PLL IC and it sets the RF bandpass filters and the LO frequency to the desired values. Next, the interpolation process begins. The correct DAC values for the specific channel selection are calculated by the microprocessor and sent to the PLL IC, which then sets the proper voltages on the RF tuning filters to correctly center the tuning frequency response for the selected channel. Refer to the main tuner block diagram in **Figure 8-35**.

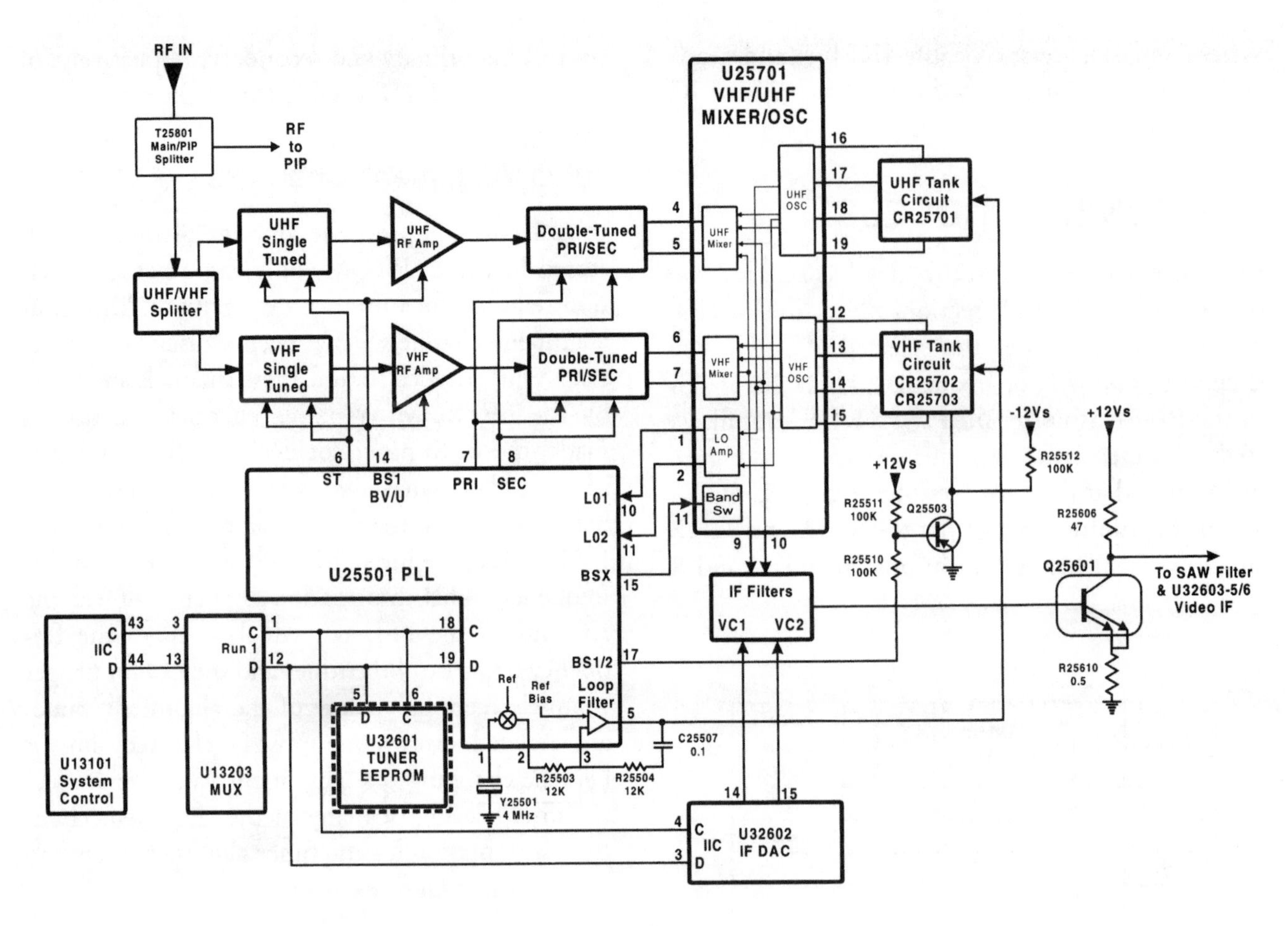

FIGURE 8-35

For example, the microprocessor has a request from the IR remote control or sets keyboard, to tune in channel 53. First, the local oscillator frequency is retrieved and sent to the PLL (U25501) for output from the loop filter on pin 5. A feedback loop from U25701 pins 10 and 11 to the PLL ensures the local oscillator remains on frequency. Then the bandswitch values are retrieved and sent to the PLL and the outputs on pins 14, 15, and 17 are set. In this case, channel 53 is within tuning band 3, so all three pins would be set high. Channel 53 lies between alignment channels 51 and 57. The microprocessor must now calculate the exact DAC values to send the PLL in order to place the proper voltages on the single-tuned, and primary/secondary double-tuned filters. These voltages come from the DACs on pins 6, 7, and 8. They tune the RF stage to the exact channel requested. The microprocessor may use any combination of adjacent frequencies to interpolate the required channel values as long as the frequencies remain in the same band.

Software Control

The PLL/DAC IC, U25501, is controlled from the main microprocessor over the IIC bus. Data is sent, according to IIC bus specifications, in packets of two to five bytes with the first byte being the address byte. There is a "start" condition at the beginning of an address byte and a "stop" condition at the end of the data with an "acknowledge" condition at the end of each byte.

EEPROM Requirements

Because the tuner's RF filters are electronically aligned, those alignment values need to be stored in nonvolatile memory to be used when tuning channels. The main tuner now has its own EEPROM to store these alignment values. Note

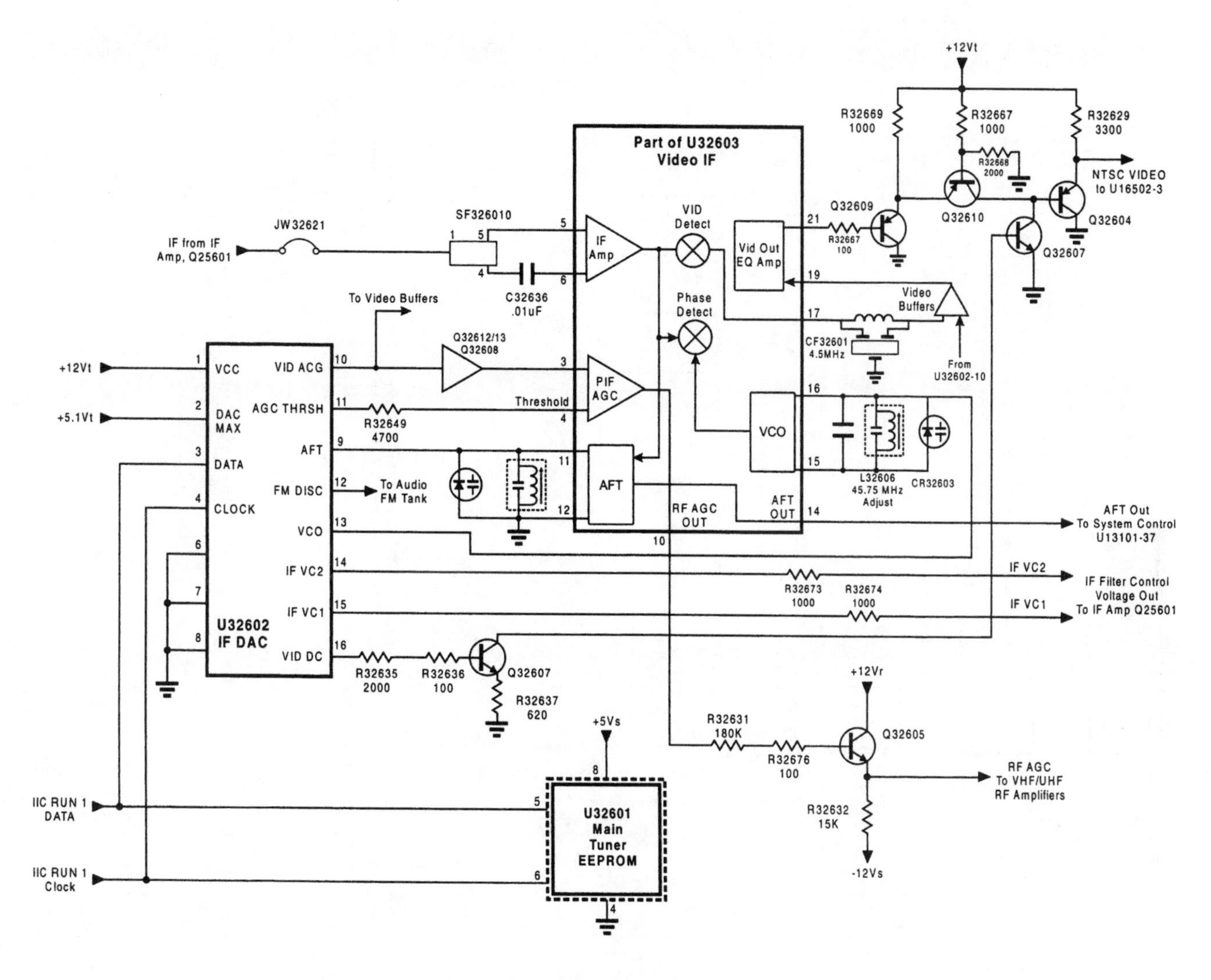

FIGURE 8-36

the diagram in **Figure 8-36**. The EEPROM contains the data format and amount of memory required for alignment data stored in the tuners EEPROM. Three bytes are required for each alignment channel. There are 29 alignment channels, which total 87 bytes of memory. The segment of memory for alignment data is stored in the order of frequency of the alignment channels. The lowest three bytes contain the lowest frequency alignment channel data; the highest three bytes contain the highest frequency alignment channel data. The alignment values for a second PIP tuner are different and therefore require a second set of storage locations. The DAC values stored in the EEPROM, that show up on the Chipper Check tuner alignment screen, return values from 0 to 63. The actual alignment values are -31 to +31. The alignment values are translated before storing them in the EEPROM by adding 31.

Refer to chapter two (2) test equipment information, for more details on the "Chipper Check" testing technique.

Technical information and drawings in this chapter are courtesy of THOMSON multimedia, TOSHIBA ELECTRONICS COMPANY OF AMERICA, and MICROTUNE, Inc. of Plano, Texas. My thanks to Rick Blumberg, Alise Perkins, and Kathleen Padula of MICROTUNE for locating and sending me the information on the Microtuner System.

CHAPTER 9

Audio Systems Digital Signal Processing and Troubleshooting Techniques

INTRODUCTION

The HDTV audio system has built-in provisions to transmit six channels of high fidelity audio for a full home theater stereo surround sound experience. The audio subsystem consists of the complete audio path, from the point where it enters the audio encoders at the HDTV transmitter, to the audio decoder output in the HDTV receiver.

This chapter begins with a brief overview of the basic digital audio signal processing. Then we'll look at how an actual progressive scan HDTV receiver's audio circuit operates.

In the next section, you will find techniques on how to perform stereo frequency response checks in a HDTV receiver. This will include techniques for aligning the stereo decoder and sound IF and detector stages.

The chapter concludes with actual "real world" HDTV and conventional TV receiver audio troubleshooting information.

BASIC DIGITAL AUDIO SIGNAL PROCESSING

The audio portion of a HDTV system is transmitted digitally. However, by nature the original audio signal is analog, and the human ear also receives sound in an analog form. In order to make the total HDTV system work, the audio has to have some type of conversion process. This process is called analog-to-digital, or A/D conversion. The complimentary process in the receiver is called digital-to-analog, or D/A conversion. Let's now take a few moments for a simple explanation of audio digital signal processing (DSP).

The digital signal processing is the conversion of analog signals to digital form for computer processing. Regardless of the type of analog signal, the basic blocks of the system are the same. For a simple understanding of this refer to the basic blocks shown in the **Figure 9-1** drawing.

Digital System Blocks

The degree of digital signal processing will vary from simple EQ operations to the complex audio system used in HDTV operation. To start the process, the analog signal is first A/D converted. A/D conversion has three processes:

- Sampling
- Quantization
- Binary notation

Sampling Process

The audio analog signal is sampled with a frequency that is above two times the maximum frequency found in the signal. The sampling process is illustrated in the **Figure 9-2** drawing. This produces a more accurate conversion of the audio signal.

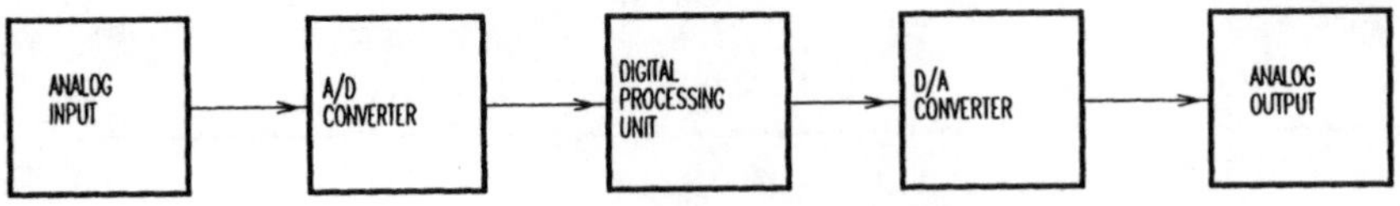

ANALOG INPUT → A/D CONVERTER → DIGITAL PROCESSING UNIT → D/A CONVERTER → ANALOG OUTPUT

FIGURE 9-1

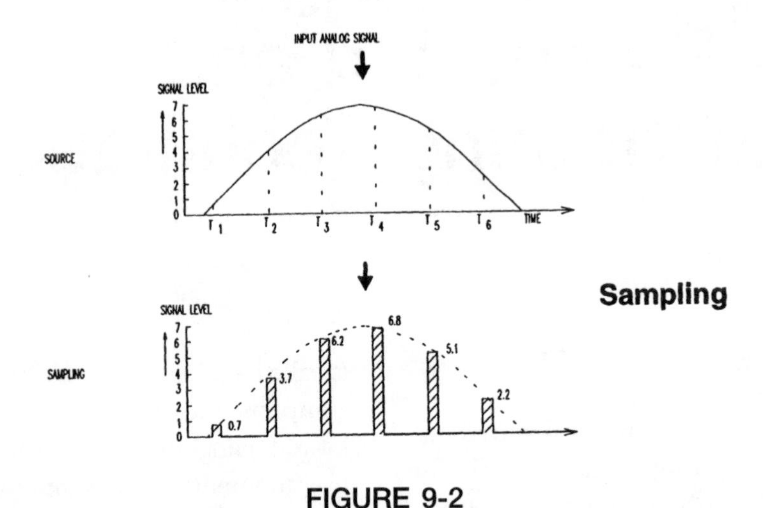

FIGURE 9-2

Quantization Coding

In this process, the sampled signal is divided into certain number levels, and each level of binary is coded. Refer to the 3-bit coding drawing in **Figure 9-3**. In this drawing the amplitude of the signal is quantized into eight levels, each corresponding to a 3-bit binary code between 000 and 111. If the maximum analog amplitude level is 7 volts, then eight voltage levels can be expressed in binary code (0 to 7 volts). However, errors occur in quantization because number values between the whole numbers are considered as the numbers above or below them. As an example, 6.2 volts and 5.7 volts would be rounded off to 6 volts. Quantization

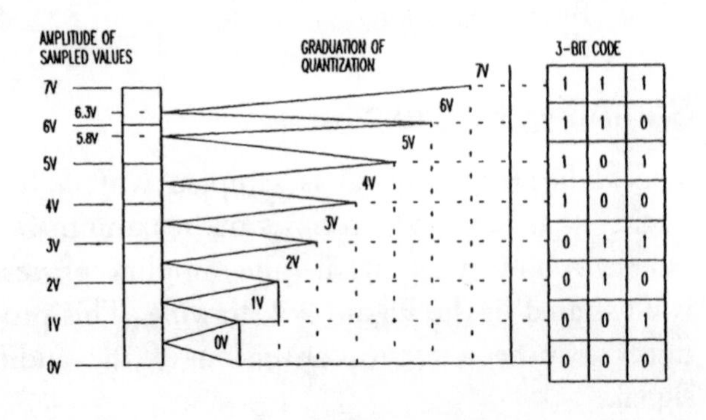

FIGURE 9-3

errors are reduced by increasing the number of bits for quantization. In practice, a sampling frequency of 44.1 kHz is used. This frequency is just above 2 X 20KHz, as 20KHz is considered to be being the theoretical highest frequency of the human hearing range. In addition, this frequency solves the problem of aliasing frequencies. These frequencies are lower than the sampled frequency and are created when the sampling frequency is less than 2 times the highest frequency that is sampled. In audio sampling, the alias frequencies are audible.

BINARY CODING FOR AUDIO

With each audio analog voltage level sampled and given a binary code, a serial data stream is formed. The example shown in **Figure 9-4** uses 3-bit coding. As many as 20 bits are used in digital audio HDTV systems to produce over 1 million levels. Parity bits are added to this data for error checking and correction. A "parity bit" is a binary digit that is added to an array of bits to make the sum of the bits always odd, or always even, to ensure accuracy. This process is called ECC encoding. The ECC encoding and modulation waveforms are shown in **Figure 9-5**.

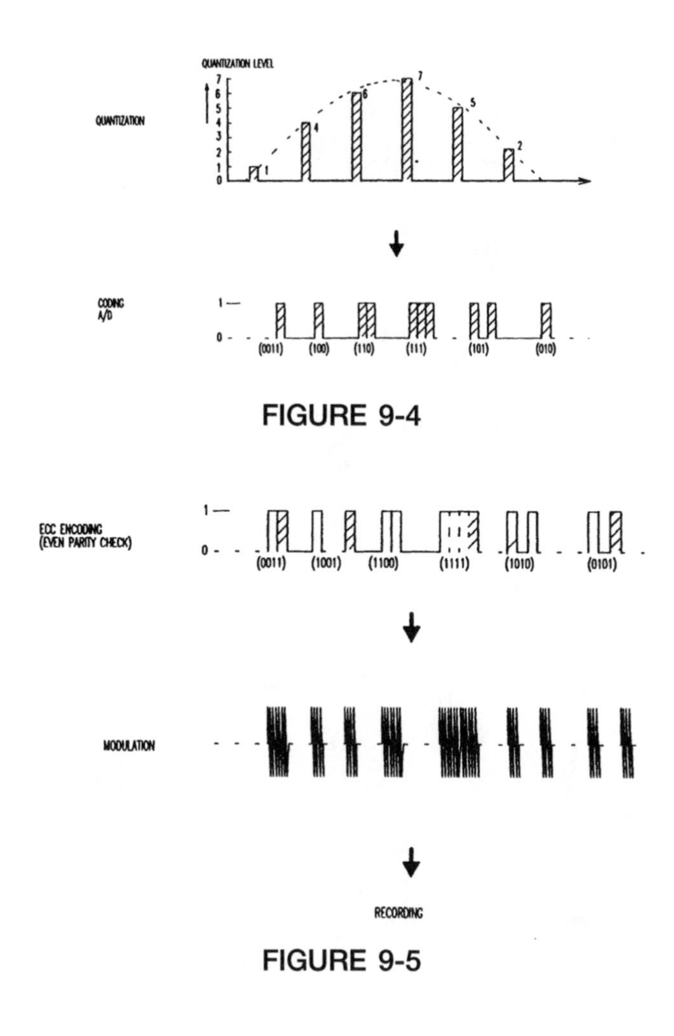

FIGURE 9-4

FIGURE 9-5

The audio signal is finally modulated in a format called EFM (8 to 14-bit modulation). EFM is a system in which 8-bit data is converted to 14-bit data for the purpose of avoiding continuous ones and zeros in the data stream during transmission.

The preceding information may not necessarily represent the way an actual HDTV audio systems operates. It is used for a simplified digital audio system concept. Actual HDTV audio systems will no doubt differ from this explanation. Some audio data systems' processing will be divided into two blocks. One would be the PES packetization, and another one the transport packetization. Also, some of the functions of the transport subsystem may be included in the audio coder or the transmission subsystem.

Some of the HDTV audio systems may contain six audio channels dedicated to audio programming. These six channels, also referred to as 5.1 channels, are as follows:

1. Left channel
2. Center channel
3. Right channel
4. Left surround channel
5. Right surround channel
6. Low frequency enhancement (LFE)

When required, the transport can actually transmit more than one of these elementary audio bit streams. Note that the bandwidth of the LFE channel is usually limited to the range of 3 kHz to 120 kHz.

Audio Compression Technique

In order to keep the digital HDTV video signal within the 6 MHz TV channel bandwidth, the video signal must be compressed. And the same technique is also used for the audio system. The compression of the audio portion of the HDTV system is desired for two reasons. The first reason is to make the channel bandwidth more efficient. And the second reason is to reduce the bit memory and bandwidth space needed to store the program material. The purpose of audio compression is to reproduce an audio signal faithfully, with as few bits as possible, while maintaining a high level of stereo sound quality.

Audio Recovery Process

In the audio recovery process, the audio signal is demodulated to produce the ECC coded signal. Following a stage where the parity bits are removed, the digital signal representing the original audio signal is produced. This digital signal is then D/A converted and filtered to reproduce the original analog audio signal. The waveform drawings in **Figure 9-6** illustrate the analog audio recovery process of the digital signal.

Dolby Pro-Logic Signals

In some HDTV and conventional TV stereo receivers, digital signal processing is used to produce a number of sound effects and to decode Dolby Pro Logic signals. In **Figure 9-7**, a generic block diagram of such a system outlines the basic digital A/D conversion, digital signal processing, and D/A signal conversion.

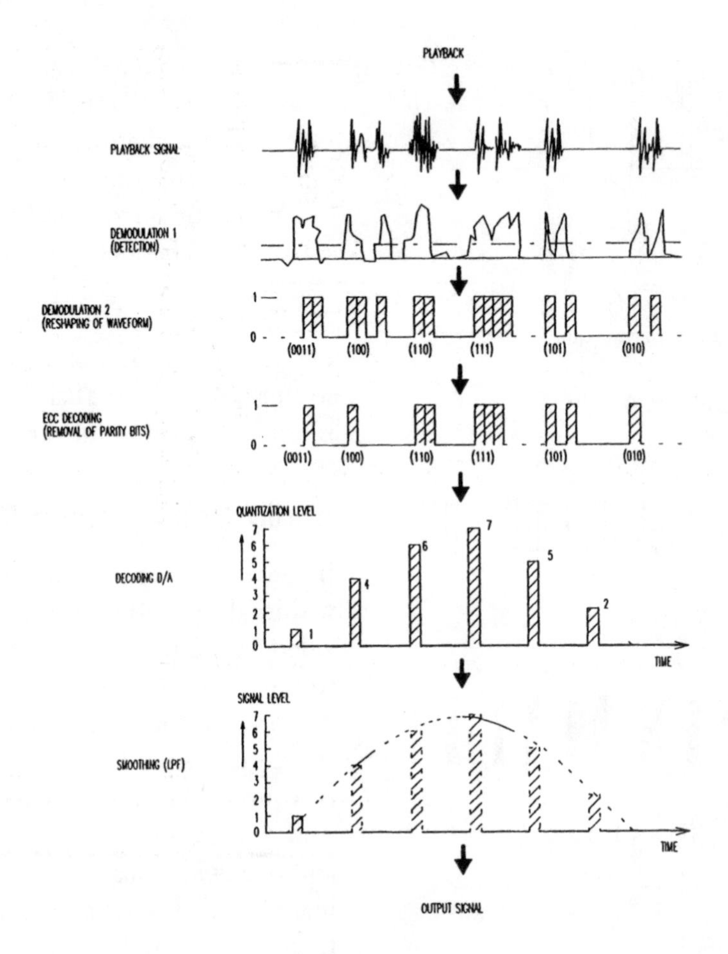

FIGURE 9-6

A/D Conversion

Left and right analog audio signals are A/D converted in IC301 and leave as serial data at pin 16 of the chip. The speed that the A/D converter samples at is determined by the clock signal at pin 20. The bit clock (BCK) and the left/right clock (LRCK) at pins 15 and 14, respectively, are needed for audio data timing and to identify left and right audio data for processing in digital filter IC405, and D/A converter IC307.

DIGITAL SIGNAL PROCESSING AND D/A CONVERSION

The digital signal from IC301 enters DPS IC404/ pin 11. The processing that takes place within this IC depends on three factors:

1. The sound field (effect) that the user selected
2. The type of signal receiver
3. The signal's frequency

Within DPS IC404 the signal is processed digitally to produce some audio sound modifications. These can be used to adjust the receiver for different room acoustics that may be encountered.

After processing, the signal data for the left and right front channels leaves IC404 at pin 62. This data is then A/D converted and filtered in digital filter IC405 and leaves as front left audio at pins 25 and 27, and front right audio at pins 16 and 18. After a stage of mixing and low-pass filtering, the reconstructed audio is sent onto the audio amplifiers.

Signal data for the center channel and Dolby surround channel leave DSP IC404 at pin 64 and are D/A converted in D/A converter IC307. The reconstructed left and right audio signals leave the converter at pins 1 and 28. They are sent onto the audio amplifiers in the TV receiver following a stage of low-pass filtering.

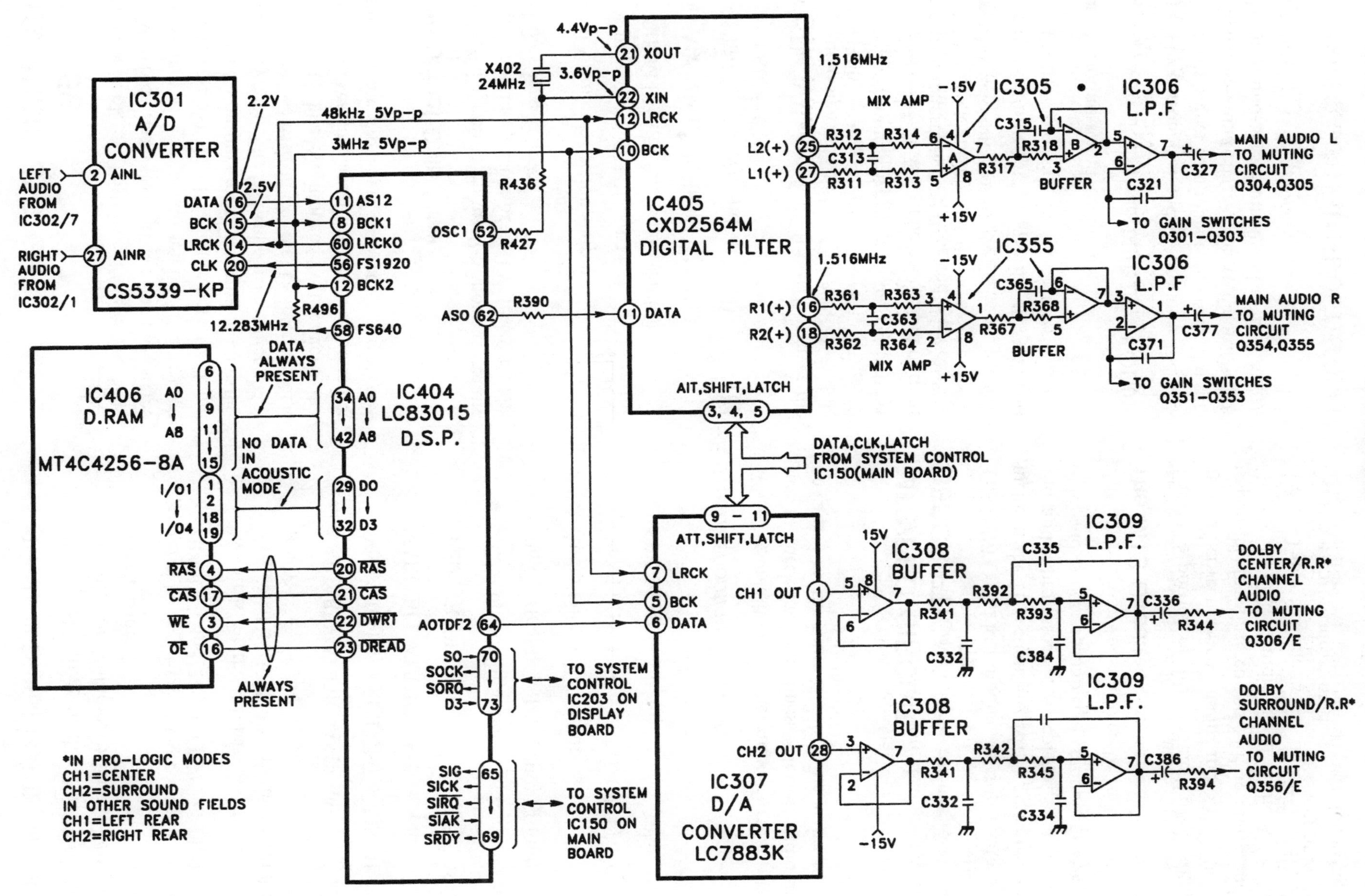
IC301
A/D
CONVERTER
CS5339-KP
LEFT AUDIO FROM IC302/7
RIGHT AUDIO FROM IC302/1
2.2V
2.5V
48kHz 5Vp-p
3MHz 5Vp-p
12.283MHz
X402
24MHz
4.4Vp-p
3.6Vp-p
IC405
CXD2564M
DIGITAL FILTER
AIT,SHIFT,LATCH
1.516MHz
MIX AMP
IC305
IC306
L.P.F
BUFFER
MAIN AUDIO L TO MUTING CIRCUIT Q304,Q305
TO GAIN SWITCHES Q301-Q303
IC355
MAIN AUDIO R TO MUTING CIRCUIT Q354,Q355
TO GAIN SWITCHES Q351-Q353
IC406
D.RAM
MT4C4256-8A
DATA ALWAYS PRESENT
NO DATA IN ACOUSTIC MODE
ALWAYS PRESENT
IC404
LC83015
D.S.P.
DATA,CLK,LATCH FROM SYSTEM CONTROL IC150(MAIN BOARD)
TO SYSTEM CONTROL IC203 ON DISPLAY BOARD
TO SYSTEM CONTROL IC150 ON MAIN BOARD
ATT,SHIFT,LATCH
IC307
D/A
CONVERTER
LC7883K
IC308
BUFFER
IC309
L.P.F.
DOLBY CENTER/R.R* CHANNEL AUDIO TO MUTING CIRCUIT Q306/E
DOLBY SURROUND/R.R* CHANNEL AUDIO TO MUTING CIRCUIT Q356/E
*IN PRO-LOGIC MODES
CH1=CENTER
CH2=SURROUND
IN OTHER SOUND FIELDS
CH1=LEFT REAR
CH2=RIGHT REAR

FIGURE 9-7

HDTV AUDIO SYSTEM OVERVIEW

Let's now look at Toshiba's Progressive Scan HDTV, which has a 14-watt per channel, stereo sound system with Front surround, SAP, and a sub bass system (SBS). The PSTs have both fixed and variable audio outputs for connecting to an external sound system. The PSTs also have center channel input for use with an external Dolby Pro Logic amplifier. This line level input allows the television's internal speakers and amplifier to act as the center channel for the Pro Logic receiver.

Audio Signal Chip Flow

The block diagram in **Figure 9-8** shows the audio circuit for a Progressive Scan TV receiver. The signal flow is as follows:

- TV tuner detection output

  Pins 12 (R-OUT) and 14 (L-OUT) of the IF/MTS/A.PRO.module (H002)

- A/V selector IC input (QV01)

  Pins 6 (R-IN) and 4(L-IN) of QV01 (TA8851CN)

  The A/V selector IC (QV01) receives the audio signals from IF/MTS/A.PRO.module (H002), Video 1 input (El), Video 2 input(E2), and Video 3 input (E3). QV01 selects the chosen audio source and outputs it on pins 39 and 41 for audio processing.

- A/V selector IV output

  Pins 39 (R-OUT) and 41 (L-OUT) of QV01

- Surround-sound IC input

  Pins 16 (R-IN) and 14 (L-IN) of QD01

  The surround-sound IC produces the necessary signal components to create the surround-sound effect. Refer to Figure 9-8 for Surround-Sound pinouts as follows:

- Surround-sound IC output

  Pins 9 (R-OUT) and 11 (L-OUT) of QD01

- IF/MTS/A. PRO. module input

  Pins 16 (R-IN) and 18 (L-IN) of H002

  The signal returns to the IF/MTS/A.PRO module for Volume, Balance, Bass, and Treble control.

- IF/MTS/A. PRO. module output Pins 26 (R-OUT) and 22 (W-OUT) of H002

  W-out is for the sub-bass system. The sub bass signal is sent directly to the Audio Amp (Q601).

- VARIABLE-AUDIO-AMP input

  Pins 3 (R-IN) and (L-IN) of QS101

- VARIABLE-AUDIO-AMP output

  Pins 1 (R-OUT) and 7 (L-OUT) of QS101

- Audio output IC input

  Pins 2 (R-IN) and 5 (L-IN) of Q601

- Audio output IC output

  Pins 11 (R-OUT) and 7 (L-OUT) of Q601

- SPK (speaker output)

Troubleshooting Tip Note:

When troubleshooting the audio circuits, make sure to check the mute lines. A loss of audio may be caused by the mute circuit being activated.

MTS/PRO CIRCUIT DESCRIPTION

The IF/MTS/A.PRO module, refer to **Figure 9-9**, includes the IF section, the MPX stereo decoder, and audio processor. It is replaced as a module only, and is not repairable to component level.

The MTS Circuit

The MTS portion of this module (H002), distinguishes between stereo, mono, and SAP. H002 is responsible for all stereo and SAP decoding, and it outputs the desired audio signal to the AV switch circuit. **Table 9-1** shows the different audio options in each broadcast mode.

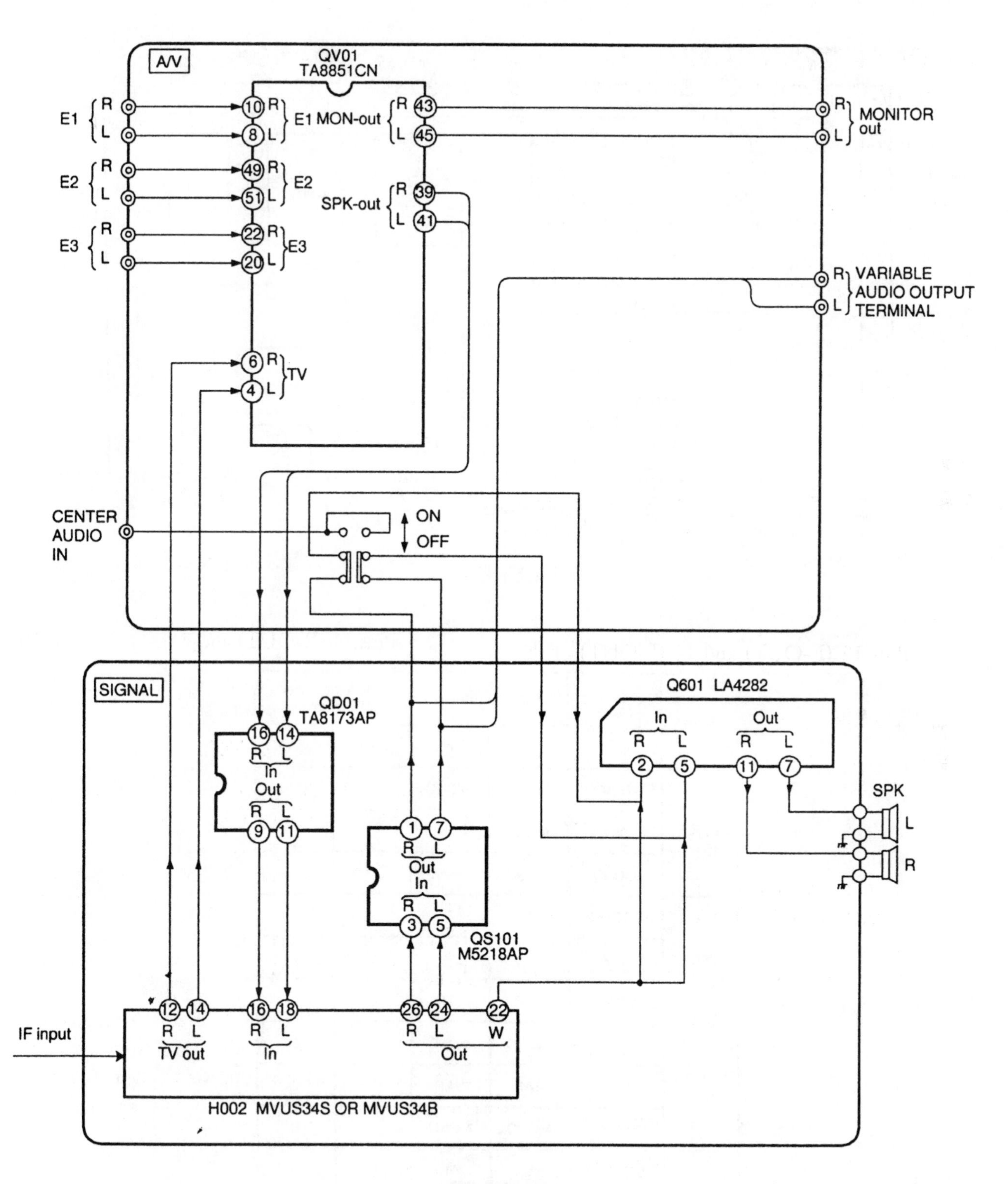

FIGURE 9-8

A.PRO Circuit

The audio processor (A.PRO) section of H002 includes the following controls:

- volume control
- tone control (bass, treble, balance)
- SBS level control
- SBS on/off

The microprocessor communicates the level of each setting to the TA1217AN in H002

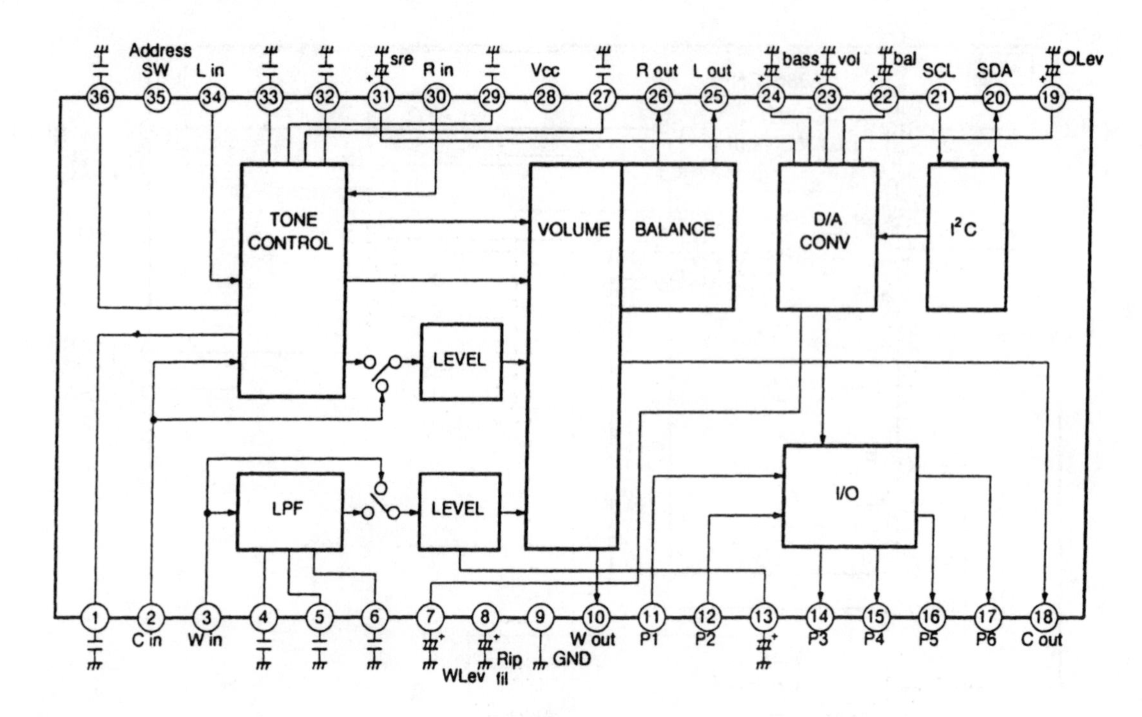

FIGURE 9-9

Signal	Mode Selection	Speaker Output		Multi-Channel Display	
		L	R	Stereo	SAP
MONO	STEREO	MONO	MONO		
	SAP	MONO	MONO		
	MONO	MONO	MONO		
STEREO	STEREO	L	R	O	
	SAP	L	R	O	
	MONO	L+R	L+R	O	
MONO + SAP	STEREO	MONO	MONO		O
	SAP	SAP	SAP		O
	MONO	MONO	MONO		O
STEREO + SAP	STEREO	L	R	O	O
	SAP	SAP	SAP	O	O
	MONO	L+R	L+R	O	O

TABLE 9-1

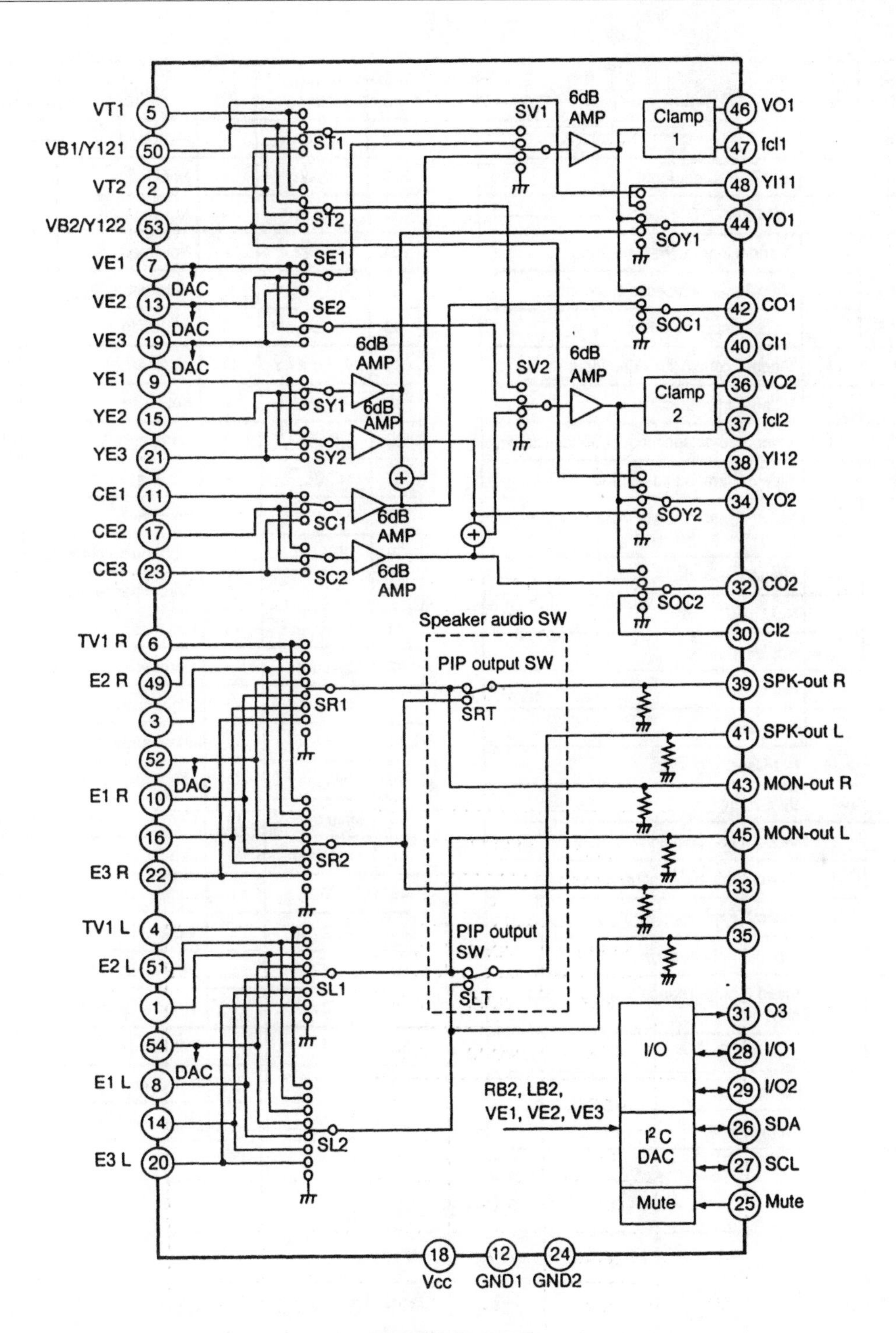

FIGURE 9-10

through the IC bus. Refer to **Figure 9-9** for the block diagram of the A.PRO IC.

Table 9-1 shows the different audio features in each broadcast mode. When tuning stations, or pushing the recall buttons on the remote, the on-screen display indicates stereo or SAP. No display is shown when the signal is in the monaural mode.

A/V Selector Circuit

The main component of the audio/video switch is TA8855CN (QV01). The microprocessor controls QV01 through the bus line. QV01 selects the desired signals and sends them to the video and audio circuits.

Pin no.	Terminal Name	I/O	Function
1	TV2 / TV L	-	Not Used
2	TV2 / TV V	I	PIP/POP Tuner Input
3	TV2 / TV R	-	Not Used
4	TV1 / TV L	I	R audio input from Main Tuner
5	TV1 / TV V	I	Main Tuner composite video input
6	TV1 / TV R	I	L audio input from Main Tuner
7	E1 / V	I	Video 1 composite video input
8	E1 / L	I	Video 1 left audio input.
9	E1 / Y	I	Video 1 luminance input (S-Video only)
10	E1 / R	I	Video 1 right audio input
11	E1 / C	I	Video 1 chrominance input (S-Video only)
12	GND	-	Ground
13	–	-	Not Used
14	–	-	Not Used
15	–	-	Not Used
16	–	-	Not Used
17	–	-	Not Used
18	VCC	I	9V Supply
19	E3 / V	I	Video 3 composite video input
20	E3 / L	I	Video 3 left audio input.
21	E3 / Y	I	Video 3 luminance input (S-Video only)
22	E3 / R	I	Video 3 right audio input
23	E3 / C	I	Video 3 chrominance input (S-Video only)
24	GND	-	Ground
26	SDA1	I/O	I²C bus data
27	SCL	I	I²C bus clock
28	I/O	-	Not Used
29	SW1	-	Not Used
30	Cin PIP	-	Not Used
31	SW2	-	Not Used
32	PIP out / C	O	Chrominance out for PIP/POP
33	PIP out / R	-	Not Used
34	PIP out / Y	O	Luminance out for POP/PIP
35	PIP out / L	-	Not Used
36	PIP out / V	O	Composite Video out for PIP/POP
37	CL	-	Ground
38	Y in PIP	-	Not Used
39	SPK out R	O	Right audio out to QD01
40	C in	I	Chrominance in from comb filter
41	SPK out L	O	Left audio out to QD01
42	Main out / C	O	Chrominance out to Q501
43	Main out / R	O	Right audio out to monitor out
44	Main out / Y	O	Luminance out to Q501
45	Main out / L	O	Left audio out to monitor out
46	Main out / V	O	Composite out to comb filter
47	Clamp	-	Ground
48	Y in	I	Luminance in from comb filter
49	E2 / R	I	Video 2 right audio input
50	E2 / V	I	Video 2 composite video input
51	E2 / L	I	Video 2 left audio input.
52	I2	-	Not Used
53	I3	-	Not Used
54	I4	I	Indecates Colorstream input

TABLE 9-2

Troubleshooting Tip:

If the audio is lost at QV01, make sure the mute is turned off. Pin 25 is 5 volts when muted and 0 volts during normal operation. The pin outs for the QV01 chip are shown in **Table 9-2.**

Surround-Sound Circuit (QD01)

TA8173AP (QD01) processes the phase shift and delay effect for the surround-sound TV receiver feature. The drawing in **Figure 9-11** shows the internal block diagram of the TA8173AP chip.

Audio Output Circuit

LA4282 (Q601) is the final audio output stage for audio amplification. This stage provides 14 watts of power to drive the speakers. In **Figure 9-12** you will find the block diagram of the internal workings of this audio output chip.

Q601 Audio Output Chip Troubleshooting Tip

If the TV set in question has repeated audio output chip failure, there may be a problem with the

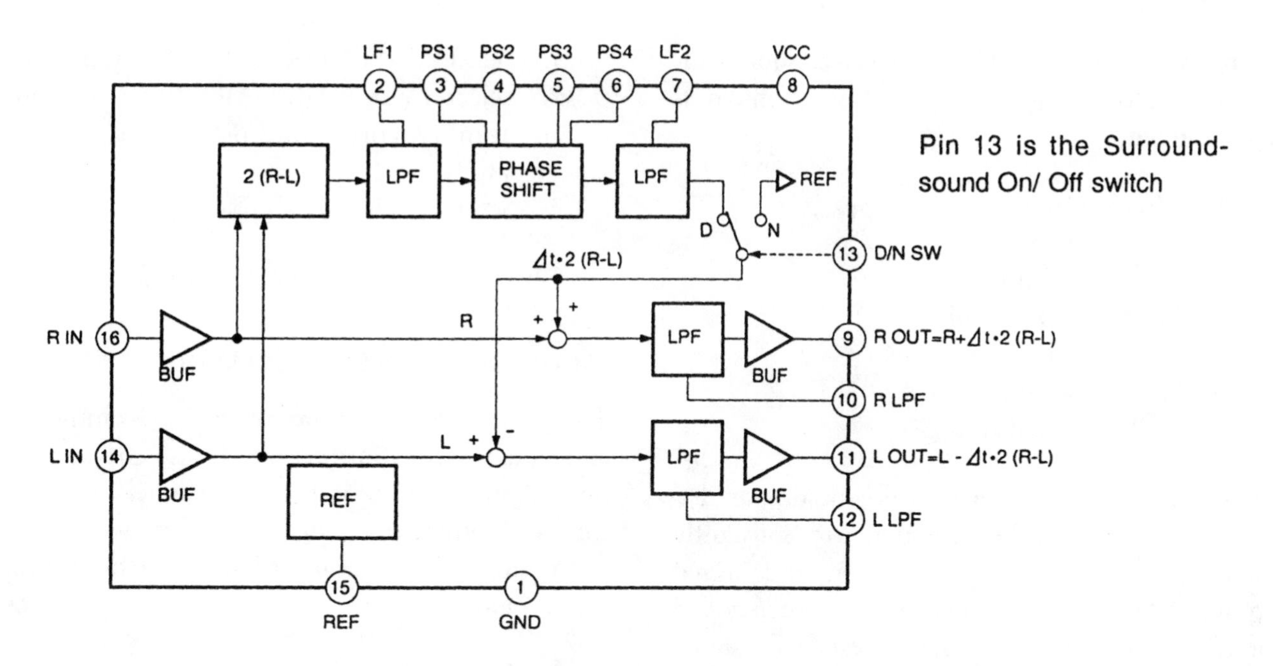

Pin 13 is the Surround-sound On/ Off switch

FIGURE 9-11

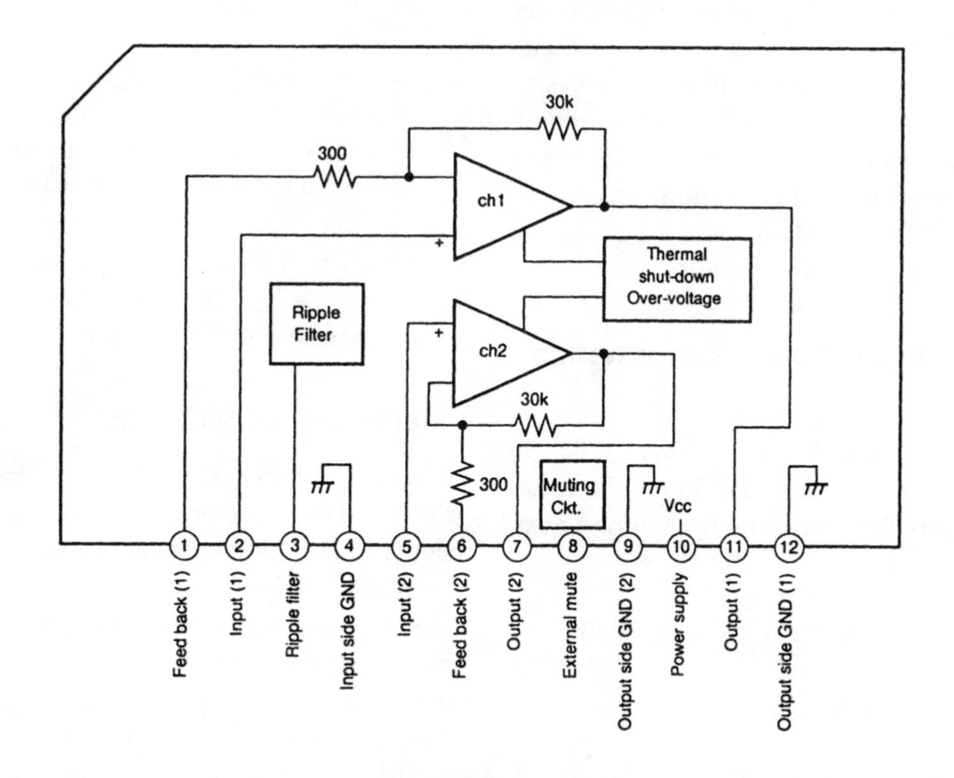

FIGURE 9-12

connections of external equipment, such as the speakers. If the set is connected for center channel input operation, make sure it's connected for a line level signal. If an amplified signal is supplied to the center channel input, Q601 may be damaged. Connecting external speakers to the TV improperly may also damage transistor Q601.

AUDIO PERFORMANCE TESTS

For the following performance test procedures we will be using the Sencore VG91 Universal Video Generator. Problems in the TV tuner, IF stages, sound detector, or MTS decoder can affect the quality of the recovered audio signal. A full performance test of a stereo HDTV receiver should include the following when an audio problem is encountered:

1. Mono operation
2. Stereo operation
3. Stereo separation
4. Stereo signal-to-noise test
5. SAP operation
6. SAP signal-to-noise test
7. SAP threshold test

Over-the-air TV signals will not do for TV audio receiver tests because they are constantly varying in reflections, separation, and signal strength. The Sencore VG91 provides a controlled signal to accurately and reliably test MTS stereo HDTV receivers. These performance tests should be performed by connecting the test signal to the tuner input. This will test the entire receiver's audio circuits under actual operating conditions. Any RF channel may be used to check the operation of the audio stages. As with a video performance test, a quick operational test of mono and stereo operation should be a standard procedure before and after a circuit repair or adjustment.

Setting Up for an Audio Performance Test

Connect the RF cable from the VG91's RF/IF output to the TV sets tuner RF input. Refer to drawing in **Figure 9-13** for audio test setup. Adjust the VG91 as follows:

1. RF-IF Signal to STD TV or CABLE.
2. Select a TV channel.
3. Set RF-IF range to HI.
4. RF-IF level to 1.
5. Video patter to RASTER, and turn off all colors for a black raster.
6. SAP and STEREO PILOT to 100 percent, unless otherwise specified.
7. MTS STEREO MODE to specified setting.
8. AUDIO FREQUENCY to specified setting.

Monitor the AUDIO OUT jacks of the receiver with either an oscilloscope or an audio voltmeter. Set up the receiver as follows:

1. Select ANTENNA or TUNER input.
2. Select TV or CABLE MODE and channel number corresponding to that on the VG91.
3. Set the receiver to MONO audio operation.

Check the audio output for proper results.

Mono and Stereo Operation

This test checks the ability of the stereo indicator and stereo circuits for turn-on. It also checks for a normal output from both audio channels. Perform the basic setup for a performance test. Set the TV for mono audio operation. Monitor the output for proper audio.

1. Check for an equal output from both the left and right audio channels of a stereo system.
2. If the levels are different, adjust the balance control, or the internal gain controls, for each channel to make the levels the same.
3. Switch the MTS STEREO MODE to L+R.
4. Set receiver for stereo audio operation. Check for an equal output from both the left and right audio channels, and see if the stereo indicator is activated.

Stereo Separation Test

This test checks how well the receiver can reproduce a signal in one channel while rejecting the signal in the other channel. The modulation is set to produce a signal in just one channel. Any signal that appears in the other channel is crosstalk and indicates poor separation. The best separation occurs when the difference between the two audio output channels is greatest.

Many factors can cause poor stereo separation: a noisy signal, improper IF alignment, a defective or misaligned stereo decoder, a defective matrix, a defective or misaligned dBx decoder, and even a faulty power supply problem may cause poor stereo separation.

Stereo separation should be tested at two different audio frequencies because the dBx noise reduction circuits process low frequencies differently than high frequencies. Thus, problems can develop

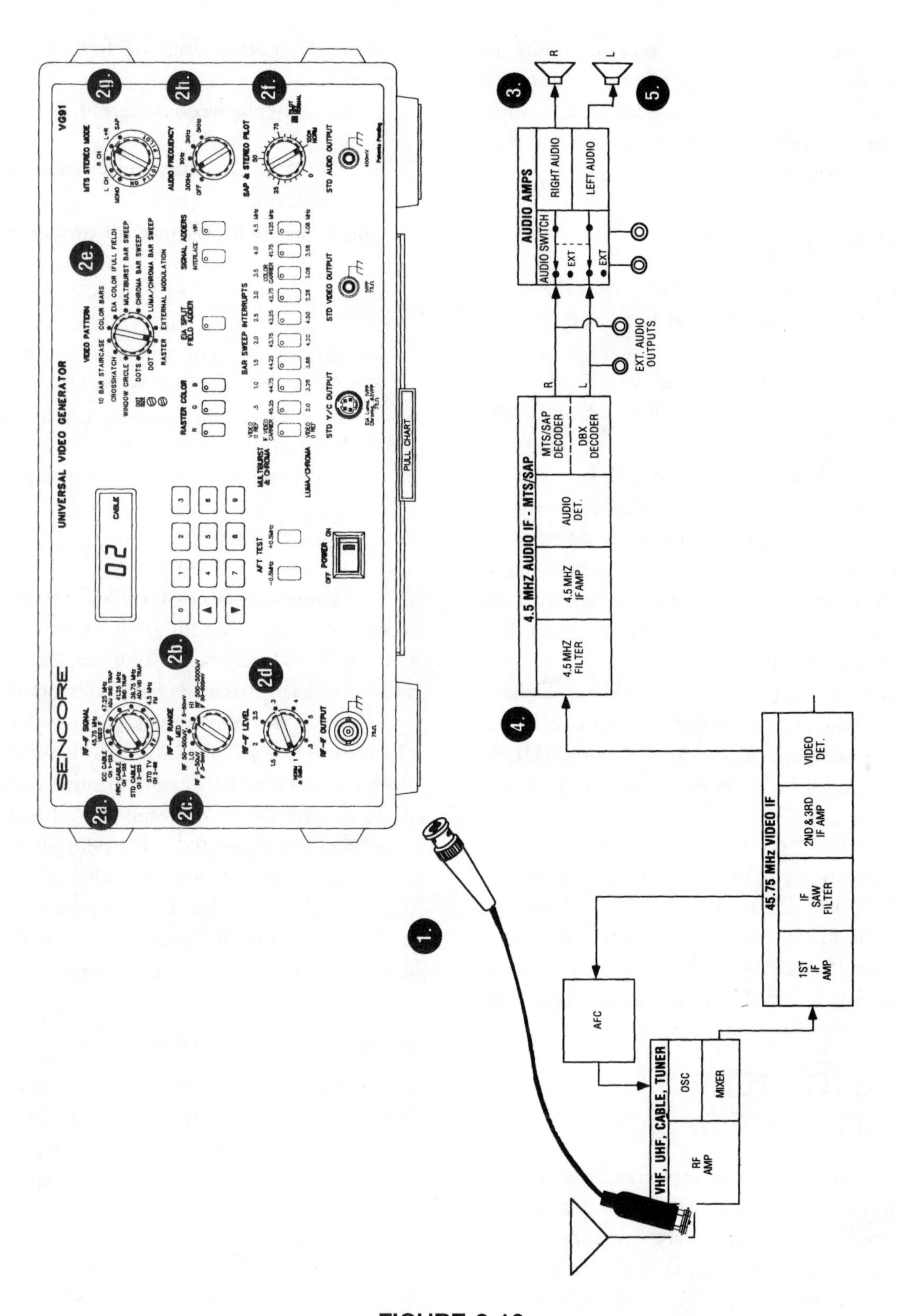
SENCORE
UNIVERSAL VIDEO GENERATOR
VG91
02
CABLE
PULL CHART
POWER
2a.
2b.
2c.
2d.
2e.
2f.
2g.
2h.
1.
3.
4.
5.
AUDIO AMPS
AUDIO SWITCH
RIGHT AUDIO
LEFT AUDIO
EXT
EXT. AUDIO OUTPUTS
R
L
4.5 MHZ AUDIO IF - MTS/SAP
4.5 MHZ FILTER
4.5 MHZ IF AMP
AUDIO DET.
MTS/SAP DECODER
DBX DECODER
45.75 MHz VIDEO IF
1ST IF AMP
IF SAW FILTER
2ND & 3RD IF AMP
VIDEO DET.
AFC
VHF, UHF, CABLE, TUNER
RF AMP
OSC
MIXER

FIGURE 9-13

that affect the high frequency separation but do not affect the low frequency separation and vice versa. The Sencore VG91 provides several audio frequencies to check both low frequency and high frequency separation.

Testing Stereo Separation

1. Confirm that the receiver passes the stereo operation test and that both output levels are the same.
2. Set-up the VG91 as follows:
   a. AUDIO FREQUENCY set to 11300 Hz.
   b. MTS stereo mode to LEFT CHANNEL.
3. Select STEREO audio mode on the TV receiver.
4. Measure the output voltage at both the left and right audio outputs. Calculate separation.
5. Repeat step 4 with the MTS STEREO MODE set to the RIGHT CHANNEL.
6. Repeat steps 4 and 6 with the AUDIO FREQUENCY set to "13 kHz and 5 kHz."

Most MTS receivers should produce 20 dB or more of stereo separation at 300 Hz. This may typically drop to 15 dB at 5 kHz. Check out the specs for the HDTV receiver you are servicing. If the receiver's performance is lower than than its specs, the stereo decoder may require alignment.

CHECKING STEREO FREQUENCY RESPONSE

Most TV receivers use a deemphasis network to reduce the output of the high audio frequencies. This deemphasis network reverses the preemphasis (higher audio frequency boost) that was added at the TV transmitter. In mono receivers, the deemphasis network is after the FM detector output and in stereo receivers it is in the L+R signal path.

Problems in the RC deemphasis network, or in other critical audio response components, will degrade the audio fidelity. The VG91 generates four audio test frequencies to verify that the audio circuits are properly reproducing the full range of audio frequencies.

Perform stereo frequency response as follows:

1. Confirm that the receiver passes the stereo separation test.
2. Set up the VG91 as follows:
   a. MTS STEREO MODE to "L+R."
   b. AUDIO FREQUENCY to "300 Hz."
3. Measure the audio voltage level at either the left or right output connections.
4. Adjust the VA91 AUDIO FREQUENCY to 3 kHz, and measure the audio output voltage.
5. Calculate the frequency response.
6. Repeat steps 4 and 5 with the AUDIO FREQUENCY switch set to 5 kHz.

The audio output of an MTS TV receiver should decrease at higher audio frequencies when tested with the VG91 generator. This decrease occurs because the deemphasis network rolls off the output since it expects audio signals to have preemphasis. The output test signal of the VG91, however, does not have preemphasis. Therefore the audio output of the receiver should correspond to the deemphasis curve shown in **Figure 9-14**. For example, you should see an approximate reduction of 5 dB between 1 kHz to 3 kHz of audio. If the output levels do not conform to the deemphasis curve, a problem exists in the sets audio system's circuitry.

Stereo Signal-to-Noise Test

This test is to determine how much internal noise is being generated by the receiver's stereo circuits, from the antenna to the speakers. Excessive noise reduces the quality of the audio signal.

How to Test the S/N Ratio

1. Confirm that the TV receiver has passed the stereo frequency response test.
2. Set the Sencore VG91 instrument as follows:
   a. Set MTS STEREO MODE to L + R.
   b. Set AUDIO FREQUENCY to 300 Hz.

Frequency	RF out (mv)	dB reduction in gain
300	99.0	−.1 dB
1,000	90.4	−.9 dB
2,120	70.7	−3.0 dB
3,000	57.6	−4.8 dB
5,000	38.9	−8.2 dB
8,000	25.6	−11.8 dB
10,000	20.7	−13.7 dB
15,000	13.9	−17.0 dB

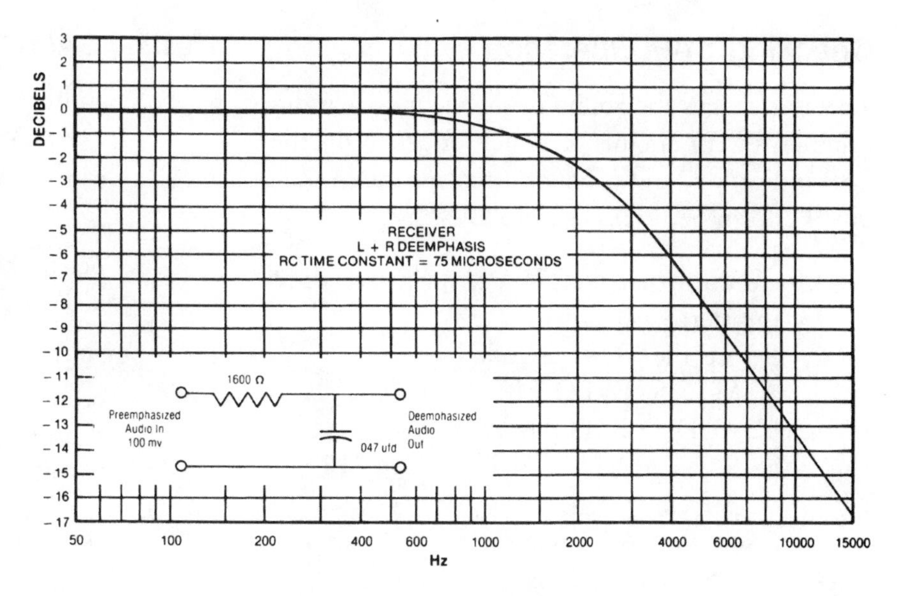

FIGURE 9-14

3. Measure the audio voltage level at the output of either audio channel.
4. Set the VG91 AUDIO FREQUENCY to "OFF." Measure the output noise voltage level.
5. Calculate the signal-to-noise level.

The signal-to-noise ratio should be greater than 33 dB. Defects in the bandpass filter, a misaligned IF, or a dBx decoder can cause poor signal-to-noise ratio.

Stereo Pilot Switching Test

This test checks that the stereo circuits switch on only when a stereo signal is being received, and then switches off on a mono signal. If the stereo circuits are incorrectly switched ON when no pilot is present, the receiver will have very noisy audio. If the pilot detect circuit fails, the TV receiver will remain in the mono mode.

The Sencore VG91 has an adjustable stereo pilot signal that you can use to check the level at which the stereo pilot detect circuit will turn on. During this procedure you should check the turn-on and turn-off pilot detect levels.

Checking The Stereo Pilot Switching Threshold

1. Confirm that the receiver passes the stereo signal-to-noise ratio test.
2. Set up the Sencore VG91 as follows:
   a. MTS STEREO MODE to "L + R."
   b. AUDIO FREQUENCY to "OFF."
   c. SAP and STEREO PILOT to "100 percent normal."
3. Confirm that the receiver's stereo indication is on.
4. Reduce the setting of the VG91's SAP and STEREO PILOT level control until the receiver's stereo indication goes off. Note the setting of the level control. This is the pilot turn-off threshold.
5. Increase the setting of the VG91's SAP and STEREO PILOT level control until the receiver's stereo indication comes back on. Note the setting of the level control. This is the pilot turn-on threshold. The pilot level turn-off threshold should be 50 percent or a little less. The pilot level turn-on threshold should be greater than 50 percent for reliable turn-on.

Checking SAP Operation

This test checks the ability of the secondary audio program (SAP) circuits to switch on, turn on the SAP indication, and produce a normal audio output. Perform SAP checks as follows:

1. Confirm that the TV receiver passes all of the stereo performance tests.
2. Set up the VG91 as follows:
   a. MTS STEREO MODE to "SAP."
   b. AUDIO FREQUENCY to "1 kHz."
   c. SAP and STEREO PILOT to "100 percent."
3. Select the SAP mode on the TV receiver.
4. Check for the proper audio output:
   a. The SAP indicator on the receiver should come on.
   b. A 1 kHz tone should be present at the output.

SAP Signal-to-Noise Ratio

If excessive noise enters the SAP channel the quality of the SAP audio signal is reduced. The SAP signal-to-noise test determines how much noise is generated at the receiver's SAP circuits from the antenna to the audio output.

Testing SAP S/N Ratio

1. Confirm SAP operation.
2. Measure the voltage at the output of either audio channel.
3. Set the Sencore VG91 AUDIO FREQUENCY to "OFF." Measure the output noise voltage.
4. Calculate the signal-to-noise ratio. It should be greater than 33 dB.

SAP Switching Threshold Test

This SAP switching test checks the ability of the SAP circuits to properly turn on with a SAP signal present, and to turn off when no SAP signal is received. Both the SAP pilot detect turn-on and turn-off levels should be checked for proper operation.

Testing SAP Carrier Threshold

1. Confirm that the receiver passes the SAP S/N ratio test.
2. Set the VG91's AUDIO FREQUENCY to "OFF."
3. Confirm that the receiver's SAP indication is "ON."
4. Reduce the setting of the VG91's SAP and STEREO PILOT level control until the receiver's SAP indication goes off. Note the setting of the level control. This will be the pilot turn-off threshold.
5. Increase the setting of the VG91's SAP and STEREO PILOT level control until the receiver's SAP indication comes back on. Note the setting of the level control. This is the pilot turn-on threshold.

The SAP carrier turn-off threshold level should be 50 percent, or a little less for reliable operation. The turn-on threshold should be greater than 50 percent.

ALIGNING THE STEREO DECODER

Proper alignment of the MTS decoder is necessary for optimum performance. Misalignment can cause poor stereo separation, poor signal-to-noise, and poor frequency response. The VG91 provides a reference MTS signal so that you can properly adjust the stereo decoder for optimum performance.

Steps For Aligning The Stereo Decoder

1. Set the VG91 test unit and tuner for a basic setup and audio performance test.
2. Set up the VG91 as follows:
   a. MTS STEREO MODE to "L CH or R CH."
   b. AUDIO FREQUENCY to "300 Hz".
   c. SAP and STEREO PILOT to "100 percent."
3. Adjust the low frequency separation adjustments to obtain maximum separation.

4. Set the VG91's AUDIO FREQUENCY to "3 kHz" or "5 kHz" as noted for receiver under test, and set the high frequency separation adjustments for maximum separation. Note that: terminology of separation controls varies among set manufacturers. Refer to the service manual to decide which control is adjusted for separation at the low and high audio test frequencies.

You may notice some interaction between the low and high frequency separation adjustments. Adjust for the best separation at the high and low test frequencies.

SOUND IF AND DETECTOR CHECKS

Problems in the 4.5 MHz sound IF or in the sound detector can cause poor audio reproduction or degraded MTS performance. The VG91 provides a 4.5 MHz FM that can be fully modulated with any of the audio MTS signals and audio frequencies. Use this signal to isolate defects in an MTS stereo TV receiver, or to align the sound IF or detector stages for optimum performance.

Testing or Aligning the 4.5 MHz IF/Detector

1. Set the VG91 and TV tuner as for a "basic set-up and audio performance test." Monitor the audio output at the speakers.
2. Set up the VG91 as follows:
   a. RF-IF Signal set to "4.5 MHz FM."
   b. RF-IF Range to "LO" position.
   c. RF-IF Level to 1-5 mV as required.
   d. MTS STEREO MODE to "L + R."
   e. AUDIO FREQUENCY set to "300 Hz."
3. Confirm that you have a good audio output signal.
4. Decrease the setting of the RF-IF Level control until noise or distortion is heard.
5. Adjust the sound IF and detector for maximum output with the least distortion.

TROUBLESHOOTING AUDIO PROBLEMS

In conventional TV, or HDTV receivers, the audio circuits detect the audio modulation from the 4.5 MHz sound IF carrier, amplify it, and drive the speakers. MTS receivers contain a decoder stage that recovers the left, right, and SAP audio infor-

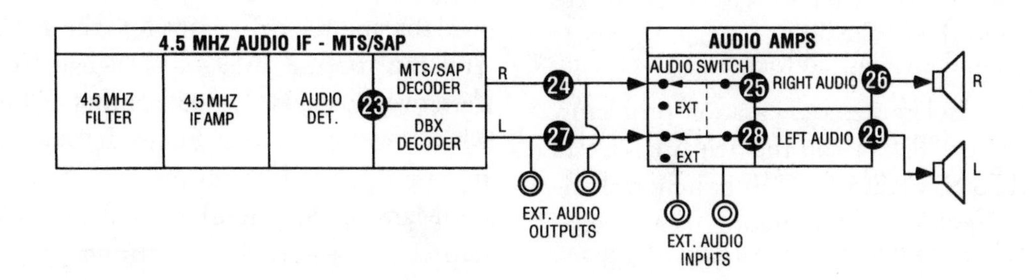

Test Point	TVA92 Signal
23	MTS Composite (Stereo Receiver)
23	Audio (Mono Receiver)
24, 25, 26	Audio
27, 28, 29	

FIGURE 9-15

mation from the composite audio signal at the detector output. With HDTV sets, these tests are performed after the audio D/A converter stage.

Audio circuit faults may result in a complete loss of audio, very weak sound, or distorted audio. The Sencore TVA92 Analyzer provides you with test signals that may be substituted in stages following the audio detector. Use the "MTS composite" drive to substitute for composite MTS signal input to the MTS/SAP decoder with MTS receivers. Use the "audio" drive to inject into any audio stage after the 4.5 MHz audio detector (monaural only receivers). The audio troubleshooting test points are shown in the **Figure 9-15** drawing.

A good place to start isolating an audio problem is to inject the "audio" drive signal into the audio detector output. Monitor the output by listening to the audio at the speakers. You can also connect an oscilloscope to the speaker or AUDIO OUTPUT jacks, or use an audio voltmeter or power meter to monitor the condition of the audio when it returns.

The audio detector output in MTS receivers is a combination of audio signals (composite audio) that must be further decoded by the MTS and SAP decoders. To test MTS and SAP decoders for proper operation, inject the "MTS composite" drive signal at the audio detector output. Monitor both audio outputs with a dual-trace scope or dual-channel audio power meter. MTS decoders are very sensitive to the level of input signal, therefore adjust the "MTS composite" drive level for the best sound. To isolate stereo separation problems, select "L CH" or "R CH" on the Sencore VG91s generator MTS STEREO MODE switch and adjust the drive level for best separation. If any separation is noted, the MTS decoder circuits are functioning.

Injecting Test Signal to Output of Audio Detector

1. Connect the TVA92 to the TV receiver for injecting into the audio circuits after the audio detector as shown in the **Figure 9-16** test set-up.
2. Set up the VG91 generator as follows:
   a. MTS STEREO MODE to "L + R."
   b. Set AUDIO FREQUENCY to "300 Hz."
   c. SAP and STEREO PILOT to 100 percent.
3. Set up the TVA92 as follows:
   a. AUDIO AND VIDEO DRIVE SIGNAL to "audio." Use "MTS composite" for MTS receivers.
   b. AUDIO AND VIDEO DRIVE RANGE to "3 volt."
   c. AUDIO AND VIDEO DRIVE LEVEL to "0."
   d. OUTPUT LEVEL/DVM to "audio and video drive."
4. Connect the direct test lead to the AUDIO AND VIDEO DRIVE OUTPUT jack.
5. Connect the direct test lead to the injection point.
6. Monitor the output sound level by listening to the speakers or an audio power meter connected to the audio output jacks.
7. Adjust the AUDIO AND VIDEO DRIVE LEVEL control while monitoring the audio output.

Audio Troubleshooting Tips

When using these injection techniques, a good audio tone at the output (speakers) indicates that the stages from your injection point to the output are working and the defect is before the audio detector. You can use the VG91's IF signals to isolate the defective IF stages. If good output is not returned when you inject at the audio detector, move your injection point closer to the output point.

Many audio amplifier circuits mute under certain conditions, such as when changing channels. This prevents loud, undesired noise. However, defective stages may also cause the audio to be muted. If you isolate and repair other trouble in the TV receiver before troubleshooting audio problems, you may find that the audio problem has also been eliminated.

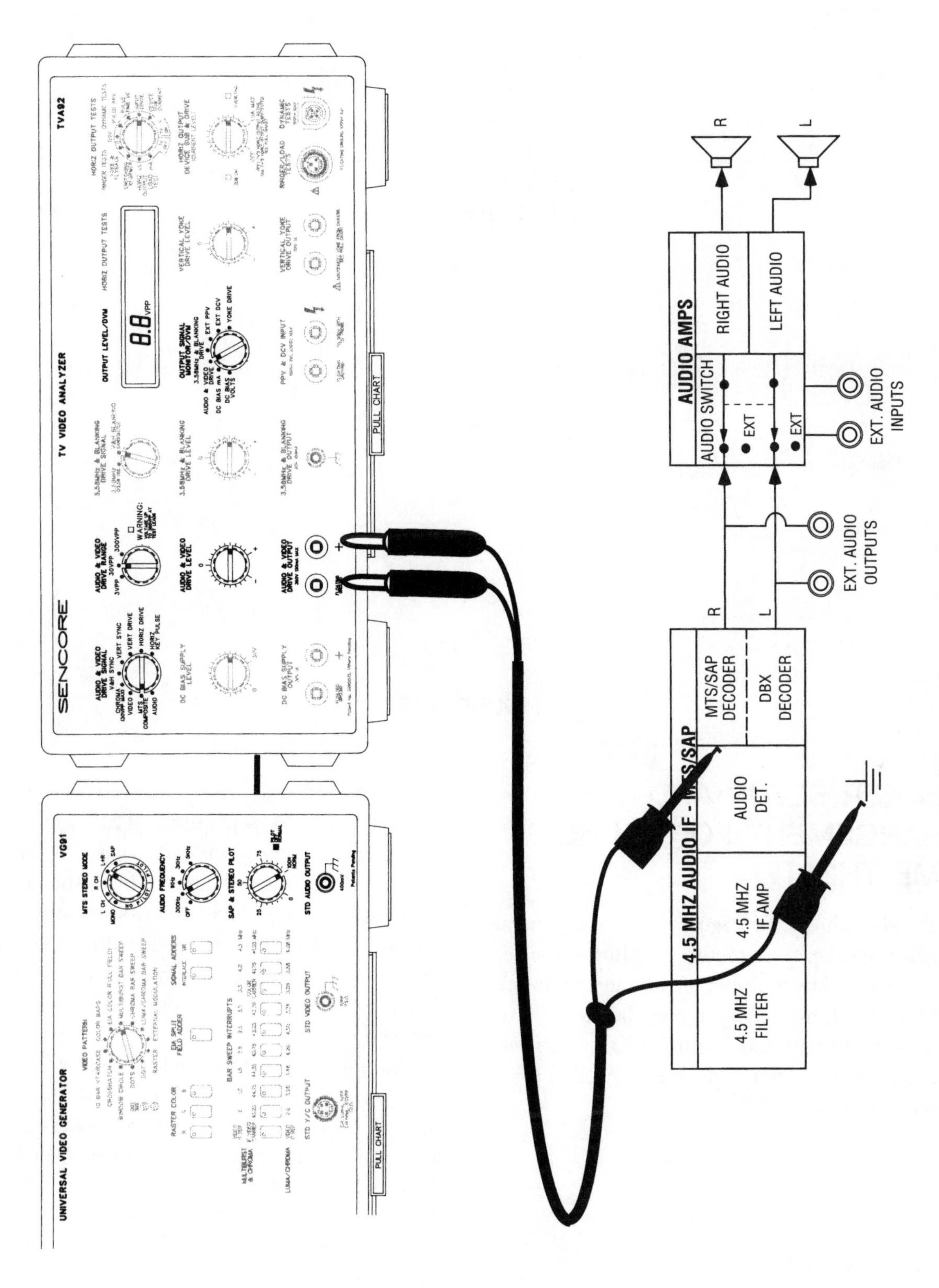
SENCORE
TV VIDEO ANALYZER
TVA92
PULL CHART
UNIVERSAL VIDEO GENERATOR
VG91
PULL CHART
AUDIO AMPS
AUDIO SWITCH
RIGHT AUDIO
LEFT AUDIO
EXT
EXT
R
L
EXT. AUDIO INPUTS
EXT. AUDIO OUTPUTS
R
L
4.5 MHZ AUDIO IF - MTS/SAP
4.5 MHZ FILTER
4.5 MHZ IF AMP
AUDIO DET.
MTS/SAP DECODER
DBX DECODER

FIGURE 9-16

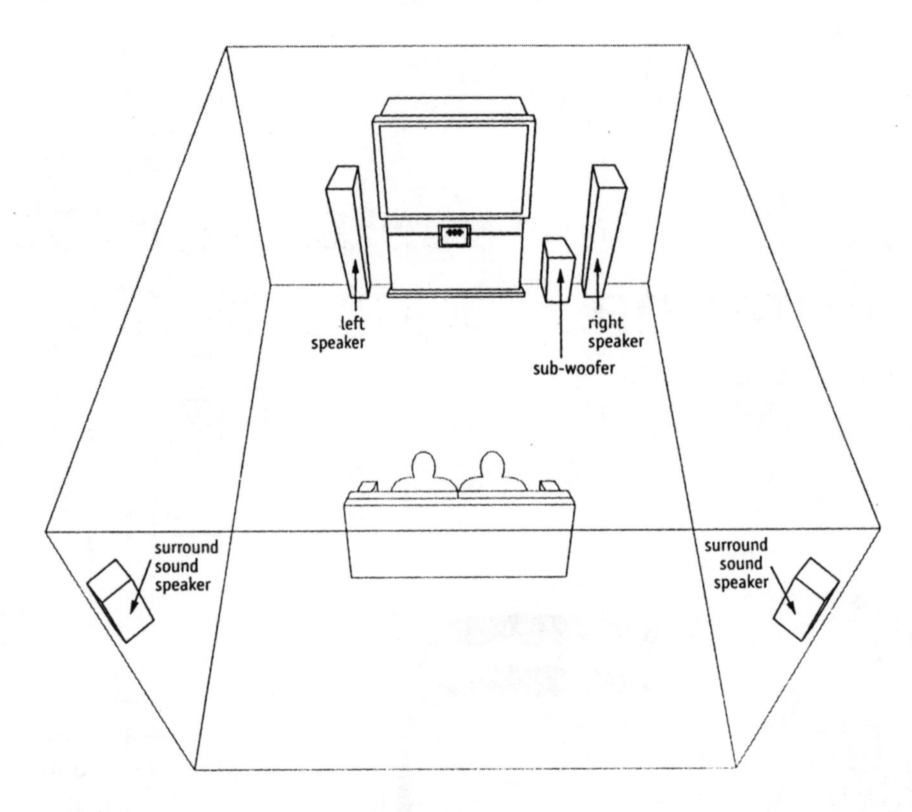

FIGURE 9-17

SPEAKER SETUP AND ARRANGEMENT FOR AN HDTV HOME THEATER

The drawing shown in **Figure 9-17** is a general design for speaker placements of a home theater system set-up. Any number of arrangements are possible, and some changes may be needed to maximize your particular sound quality desires. However, a Dolby Digital receiver is needed for a 5.1 channel audio.

A left and right speaker on either side of the TV set enhances separation. The entertainment equipment "center mode" makes the dialog sound as though it's coming directly from the entertainment equipment. The rear surround-sound speakers provide the majority of the other sounds, like those from special effects in movies. Your subwoofer generates ultra-low frequency sound, for rumbling, low-end audio.

Sound is affected by speaker placement, so make sure nothing is in front of the speakers, and that they are aimed in the right direction. You usually have the option of turning the receiver's internal speakers on or off.

Some information in this chapter courtesy of SENCORE Inc. and TOSHIBA Ainerica Consumer products, Inc.

CHAPTER 10

Digital HDTV Reception Problems and Antenna Requirements

INTRODUCTION

This chapter will cover HDTV signal reception, HDTV antennas, antenna installation, problems that you may encounter, as well as other considerations. We will also give tips concerning the comparison of HDTV and conventional analog TV reception and antennas.

A historical review of the development of the TV antenna should be of interest. This will include the single channel, wideband, multiple channel, all channel VHF/UHF, and circularly polarized antennas.

A brief discussion of digital HDTV reception problems and the pitfalls of using an indoor (rabbit ear) antenna for the DTV 8-VSB signal will follow. We will also review the new Motorola MCT2100 and NXTWAVE VSB-QAM chips that are used within the HDTV receiver to solve delay problems with full equalizer design.

Let's now move onto selecting and installing the HDTV antenna system. We will discuss transmission lines, antenna impedance, bandwidth, TV antenna construction, installation tips, and a signal survey using the CEMA color-coded charts and maps designed for your local installation area.

The chapter continues with a review of the various high-tech HDTV-designed antennas that are presently available. This will include performance specifications, construction details, design features, preamplifiers, and photos of various models.

The conclusion for this chapter will feature a review of combination digital satellite dishes and off-the-air antennas now on the market. The portable/remote digital satellite system will be discussed, as well as the home satellite dish antenna kit. Winegard's mobile Sensar antennas will be covered. These antennas are used on RVs and camper trailers, and receive TV signals from all directions without moving the antenna.

At the conclusion of this chapter, you will find a complete TV frequency chart and a conversion chart used to convert dBM to microvolts.

TV ANTENNA REVIEW

In the early days of TV, the antennas were built into the cabinet interior or onto the TV set's back cover. This method consisted of aluminum foil or hookup wire designed as a folded dipole. A few years later set manufacturers provided built-in and set-top rabbit ears.

The early model TV antennas were low-band and either conical, folded dipoles utilizing a simple reflector, or the, and I am not making this up, "bedspring" colinear arrays that looked like, guess what, bedsprings.

Signal gain was low on these early TV antennas. Snow was common and fringe areas were only 40 to 50 miles out. Of course, TV stations were not high-powered and their antenna towers were short by today's standards. Many of the early TV antenna designs were of the 300 ohm, folded dipole variety with a reflector element.

The original Yagi-Uda array design, as developed by H. Yagi and S. Uda of Japan, used additional director elements and provided excellent performance. It was relatively simple to construct, was low in cost, and offered low wind resistance. The Yagi-Uda array is basically a narrow bandwidth antenna. The length of the elements determines the frequencies to which it is tuned, or responds to. Its bandwidth, determined by its geometry, extends over a full TV channel (6 MHz) and into adjacent channels, but falls off sharply beyond that. It can be made to respond to channels 4, 5, and 6 or 7, 8, and 9, for example, to serve TV market areas where those channels are available. Some compromise is required when the antenna is tuned for channel 8 and is used on channels 7 and 9. However, in strong signal areas, such as large metropolitan areas, this is rarely a problem.

In fringe areas where optimum gain is needed, "cut-to-channel" Yagis were the best way to receive those available channels and blacked-out sports events that would only be on TV stations 90 or 100 miles away.

These narrow-band Yagis worked well for several years until more TV stations came on the air and also began transmitting color programs. As a result, the older antennas could not cover all of the VHF and UHF stations.

One of the first answers to this problem was the high gain VHF all-channel television antenna, developed by Winegard. This new antenna, which was called the "Intercepter," retained the same physical size and design of the single channel Yagi-Uda array, yet it performed equally well on all VHF channels whether at 54 MHz for channel 2, or 216 MHz for channel 13.

The key to this important development was a patented design by Winegard called the Electro-Lens. The Electro-Lens is considered to be the first major improvement to the Yagi design.

Yagi Design Basics

The Yagi antenna actually has two types of dipole elements. One type is the driven or collector elements from which the signal voltage is derived. The others are the director parasitic elements that are connected to nothing. Parasitics influence the signal voltage by re-radiating energy into adjacent or nearby elements. They can be used to reinforce or cancel signals at the collector elements and thereby influence the gain and directivity of the antenna array.

Electro-Lens Focusing

The Electro-Lens director system has solved this problem by allowing a director element to perform efficiently at both the low-band and high-band frequencies in spite of their wide frequency separation. This made it possible for an all-channel VHF-TV antenna having a performance comparable to the famous cut-to-channel Yagi-Uda array, yet retaining the small physical size and rugged mechanical design of the single-channel Yagi antenna.

Antennas Designed for Color Reception

The next Winegard company project was a high-gain all-VHF antenna, expressly designed for color TV reception. This design was called the "ColorCeptor" antenna. This antenna contained an add-on, 5-element unit called a "Power Pack," which was an 18-element, all VHF Yagi with more power for fringe-area reception. This particular antenna incidentally averaged 47 percent more gain on high band and 30 percent more low-band frequencies than any other brand of antennas.

Significant to the development of signal reception technology, of course, was the preamplifier. The Winegard Company introduced this amplified antenna development with its "Powertron." Considered to be the first electronic antenna, the Powertron incorporated a built-in preamplifier.

The amplifier was powered by 24 volts sent up the lead-in wire from a separate power supply. In addition to the clear, sharp pictures it produced on previously unusable channels, the Powertron permitted up to six TV sets to be operated from a single antenna or a remote located antenna.

THE QUARTER-WAVE STUB TO THE RESCUE

At UHF frequencies, the quarter-wave stubs act like a short circuit to allow UHF antenna currents to bridge the gap causing the elements to act like a normal folded dipole element. At VHF frequencies, the stubs are simply open circuits. And what was a folded dipole at UHF is now simply conductors of VHF signals to the UHF transmission line terminals in the center of the lower half of the folded dipole.

If the 300-ohm transmission line from the existing VHF antenna is connected across the quarter-wave stubs, then the VHF signals at that point will be conducted to the UHF folded-dipole terminals and then down the lead-in wire.

The Tetrapole, then, not only receives UHF signals, but serves as a no-loss, VHF-to-UHF antenna coupler at very little added cost, simplifying the antenna installation job.

Improving UHF Reception

Despite continuing refinement, UHF reception remained a problem area for antenna manufacturers. Some of the early model TV sets had very inadequate built-in UHF antennas, consisting of triangular-shaped metallic foil fastened to the inside of the set's cabinet back. Other outdoor UHF antennas were mostly low-sensitivity "bow tie" devices in front of a plane reflector.

Another Winegard innovation was the collapsible parabolic for use on channels 14 through 83. The reflector elements approximated the performance of a true, dish-type parabolic reflector, and sensitivity was greatly improved for the first time.

Another antenna advancement was the "Planar Grid," another variation of the Yagi, and consisted of a rear-fed, multi-element driven array.

CIRCULARLY POLARIZED TV SIGNALS

More than two decades ago the FCC authorized the circularly polarized transmission of television signals. The main advantage of circularly polarized (CP) TV signals over the horizontally polarized (HP) signal was a significant reduction in ghosting. Also, in some areas, coverage and improved picture quality increased.

When a TV station goes to CP, the transmitter power has to be doubled and a new CP antenna installed. The TV set owner can use the same antenna. CP usually performs better, especially with a set-top antenna.

The Circular-Polarization Signal

Let's now see what the CP signal is and how it works. Radio and TV signals are electromagnetic (EM) waves comprised of a magnetic and electric field. The electric field is what we will look at now. It has either a vertical or horizontal orientation. This is similar to signal polarization. Commercial and police two-way radios, AM broadcast, and CB radios all use vertical polarization (VP) for transmitting their electric wave signal. This is easy to tell as the transmitting and receiving antennas are mounted vertically with a rooftop whip antenna. These are very short stub antennas for police cars as most now use the 80OMHz to 90OMHz trunking systems. However, television and FM stations use the horizontally polarized antennas as rooftop antennas that are horizontally positioned in a horizontal plane.

Horizontally polarized electric-field signals also lie in a horizontal plane, and the amplitude of the electric field varies sinusoidally as it propagates away from the transmitter. This is illustrated in the (**Fig. 10-1**) drawing. Rotating a receiving antenna 90 degrees on its longitudinal axis (boom) will virtually null a received HP signal. A CP signal will actually rotate with a constant amplitude as it is propagated. The electric-field vector creates a pattern, if it could be seen, that resembles an auger-type wood bit as shown in the (**Fig. 10-2**) drawing.

Circular polarization causes the electric field vector to rotate either with a right-hand or left-hand rotation as it propagates away from the transmitting antenna.

As we look at a circularly polarized wave at any particular instant, the field vector may be anywhere in between a horizontal and vertical orientation as it rotates. Thus, during one cycle of rotation, it excites both the vertically and horizontally polarized antennas as well as those oriented between, such as indoor "rabbit ear" antennas or bent rooftop antennas.

This CF mode of signal transmission is not new as FM radio broadcasters have switched to circular polarization to accommodate these new markets.

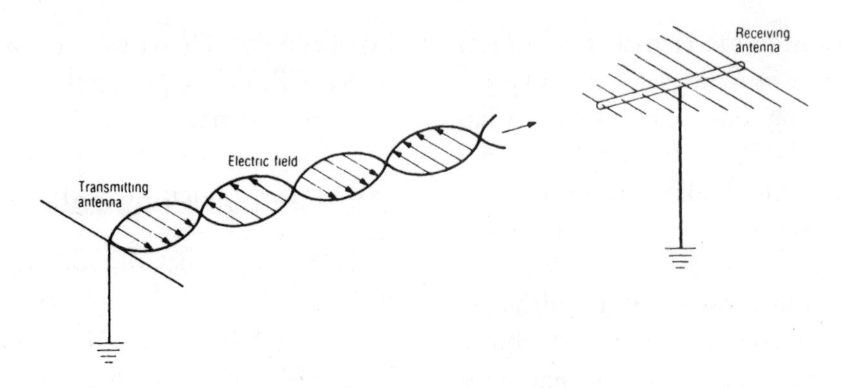

FIGURE 10-1

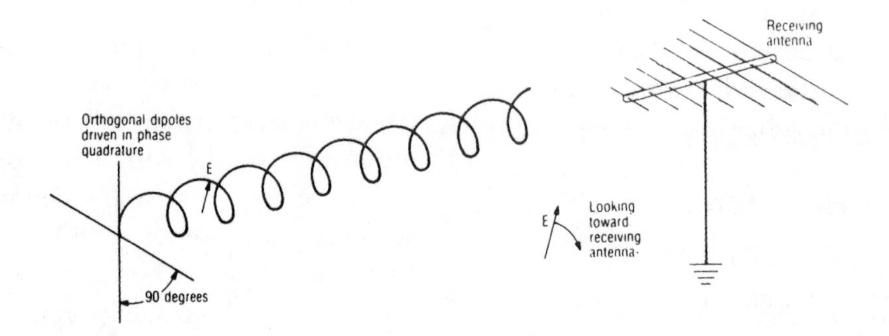

FIGURE 10-2

CP Benefits

The improved reception from circular polarization of FM broadcasts are essentially the same as those from TV except that in the case of the former, the improvement is aural and the latter is visual.

The principal benefits to TV viewers is reduced ghosting of the picture. To FM listeners, it is reduced multipath reception resulting in less distortion of the FM station audio.

Those who stand to benefit the most from CP are those who have indoor antennas. These are the apartment dwellers, and residents with restrictive covenants who are prohibited from using outdoor TV antennas.

CP Outdoor Antenna Reception

If the TV viewer has a quality, properly installed outdoor TV antenna in good working condition, the picture received should be just as good on CP broadcasts as they are for HP broadcasts. The viewer will not have to buy a specially designed antenna to receive CP except in cases where you experience difficult reception problems involving such things as severe ghosting, and co-channel and adjacent channel interference. However, in most cases, ghosting problems can be solved with a special purpose anti-ghosting antenna, or a large, highly directional one.

RECEIVING THE DIGITAL TV SIGNAL

For the best Digital Television Signal (DTV) a high quality, directional outside antenna will be required for terrestrial reception. The digital (DTV) terrestrial television broadcasting depends on properly delivering a modulating RF energy for a discrete data base rather than an analog variable signal.

The data TV signal must be delivered at a constant and reliable rate. The employment of MPEG coding reduces the required bit-rate. Also, the MPEG data is more sensitive to bit errors and needs better than a one bit-per-hour error to allow acceptable picture viewing. A real-world TV channel will not deliver that kind of performance without help—error correction will be used. Error correction will eventually be offset by an increase in reliability, along with an increase in bandwidth.

In regards to an RF signal, reflecting objects produces a delayed signal at the receiving antenna, in addition to the direct signal. With analog transmissions this will cause ghosting. But with digital transmissions, the bit-rate is so high that the reflected signal may be several bits behind the direct signal, causing a type of inter-symbol interference. This type of interference is not like noise because it is statistical, due to reflections. It is also continuous and results in a high bit-rate error.

Increasing the transmitter power will not help, as the power of the reflection will also go higher proportionally. As with analog UHF TV transmissions, a directional antenna is also required for good reception and transmission of digital HDTV, so as to reject the reflection. Generally, best results will be obtained by aligning the antenna so that the reflection is in a null of the polar diagram, rather than for a maximum signal strength. However, directional antennas tend to be large and require a good permanent installation for the best results.

The ATSC (Advanced Television Standards Committe) standard adopted in the U.S. uses a simple 8-VSB (vestigial sideband) modulation, where a single carrier is modulated with a high bit-rate. A couple of packets in each frame are replaced with a "training" sequence, which the TV receiver can identify. Comparison of the spectrum of the received signal training sequence with the ideal sequence will reveal the amount of equalization required. It is also possible to detect the presence of a delayed signal and cancel it out.

More Digital Signal Reflection Comments

As you may have ascertained, such a simple system has little resistance to any strong reflections. Thus, a properly installed directional antenna will be needed. Some actual tests that I have performed confirm that the standard indoor "rabbit ear" antenna does not give the viewer a good enough picture with the 8-VSB HDTV signal format. If the HDTV picture is not good enough, then the viewing public may just turn to the digital satellite dish, cable, or the Internet when NTSC reception is shut down.

The reason that the DVB-Tls OFDM (orthogonal frequency division multiplex) system ticks is that a carrier sending a very low bit rate is less troubled by a reflection because it arrives during the same bit (see drawing illustrated in **Fig. 10-3**). Instead of sending one carrier with a high bit rate, OFDM sends many carriers, each having a low bit rate.

CHIP MAY SOLVE HDTV MULTIPATH PROBLEM

Two "chip" makers may have solved the HDTV digital debate over the ATSC reception standard. In late 1999, Motorola and NxtWave Communications have separately introduced TV receiver chips said to produce better 8-VSB reception with an indoor antenna than those that have been available to date. The chips contain enhanced adaptive equalizers that attack static and dynamic multipath received signals.

Data released by Motorola, Sarnoff laboratory and field tests have demonstrated that this new chip, the MCT2100, enables indoor reception as with COFDM in large urban areas. NxtWave has reported similar results.

DTV reception fails when a TV receiver is unable to recover the datastream being transmitted. This can happen for a few different reasons. But for indoor reception in a strong signal area, it generally occurs when the DTV demodulator is unable to lock to the incoming data stream. To correct this, Motorola and NxtWave had to look at why the receiver could not recover the datastream clock and synchronize to it.

No Training Signal Required

Motorola's chip has a full equalizer designed to cover a range of delays significantly longer than those for early model receivers.

In the technical notes, Motorola claims the MCT2100 has the most advanced adaptive equalizer ever fielded in a commercial product. The

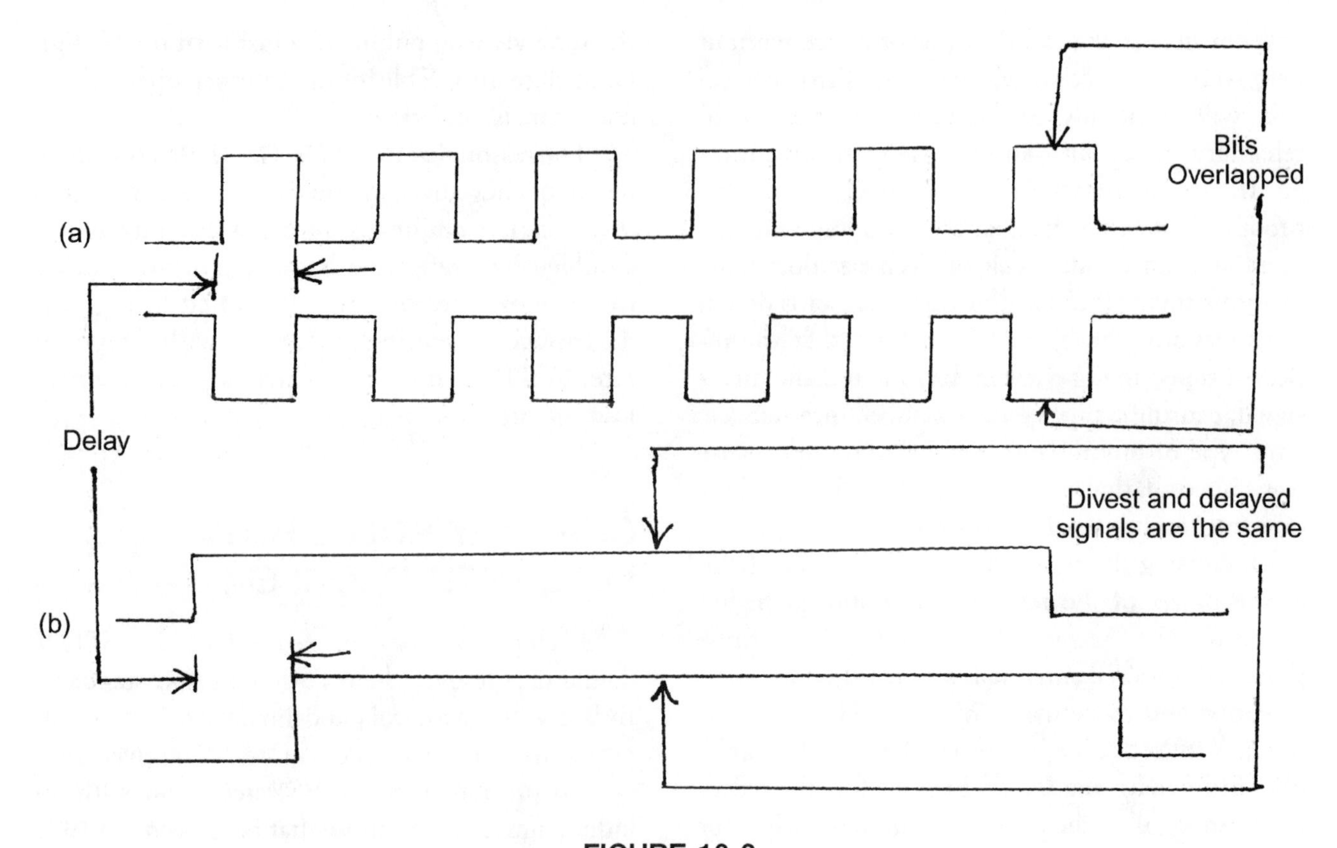

FIGURE 10-3

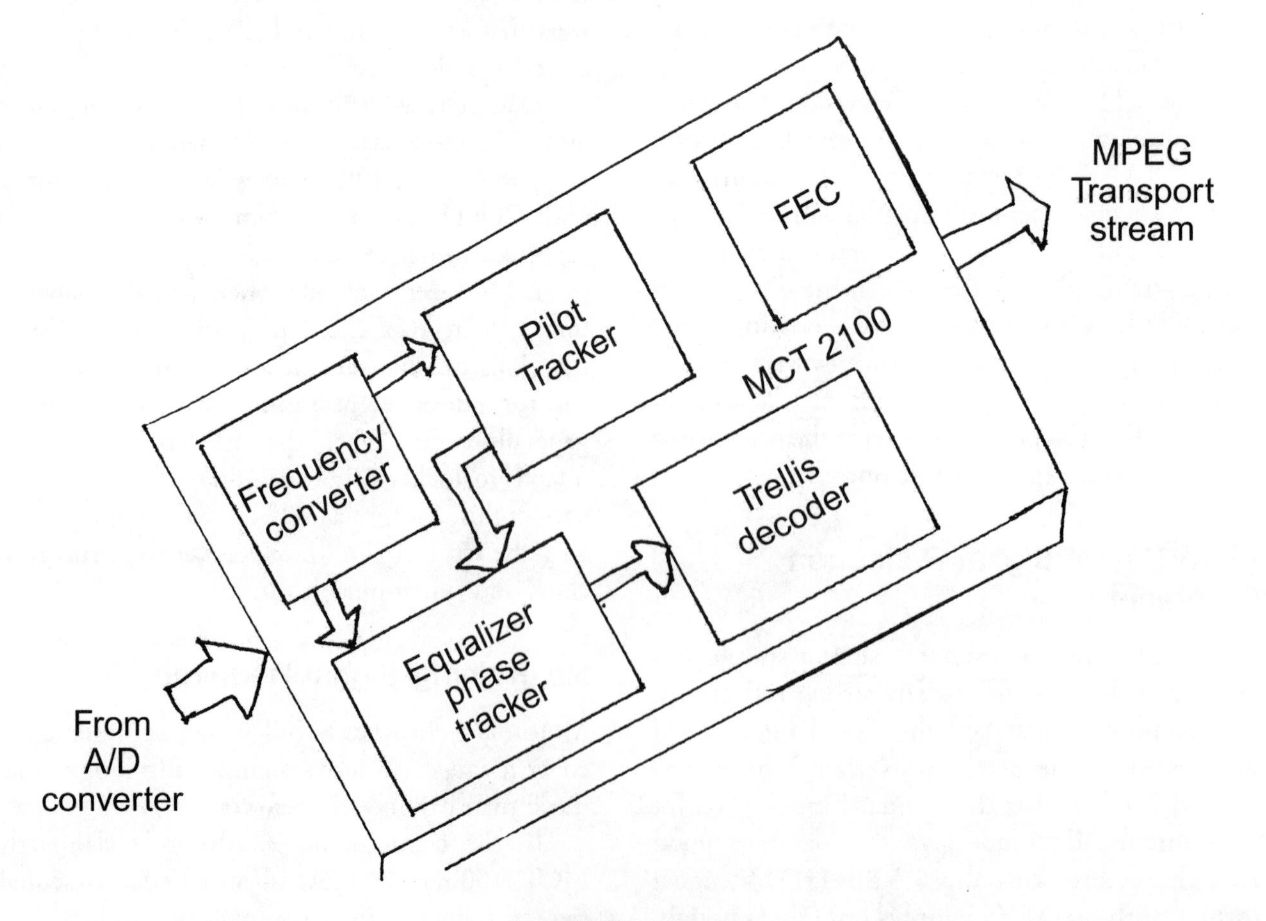

FIGURE 10-4

notes include a layout block drawing (see Fig. 10-4) of the Motorola MCT2100 chip, designed to eliminate multipath problems for 8-VSB DTV receiver signals. The Motorola chip does not rely on a training signal for the MPEG stream for equalization as used in early technology. It appears as if an entire block of information comes in and extracts from the characteristics of the data block required for decoding. In the MCT2100 chip it is not looking for the training signals, it is looking for global parameters for the chip to develop the information required for decoding.

Although the NxtWave communications solution — the NXT2000 — also uses improved equalization, there are key differences from the Motorola chip. One difference is the NxtWavels chip is a multimode VSB/QAM receiver capable of performing in 8-VSB, 64 QAM, 256 QAM, or 16-VSB. The NXT2000 also uses sparse equalization. Refer to the block diagram drawing of the NxtWave NXT2000 chip shown in (**Fig. 10-5**). According to information released by NxtWave, the sparse equalization employed by the NXT2000 chip covers the same range of delays as full equalization, but with less equalizer taps. Because of NxtWavel's technology, they get the benefits of a sparse equalization, which is a cost-reducing element with no compromise in performance. NxtWavel's chip also comes with an A/D converter, which is a feature that the Motorola chip does not have.

It should be noted that other companies are using similar technologies and obtaining breakthrough improvements that will bring even better performance in the future. The good folks at Zenith Electronics Corp., the original developer of the VSB standard, agree that multiple venders will be good for the HDTV industry.

HDTV Transmission Standards Comments

Here's a comment from Richard Lewis, the senior vice-president of Technology and Research of the Zenith Electronics Corp: "We are looking at other companies' products and they are using some of the same techniques that Zenith is using in our third-generation HDTV receivers. It is making us push faster and harder on these projects. We, at

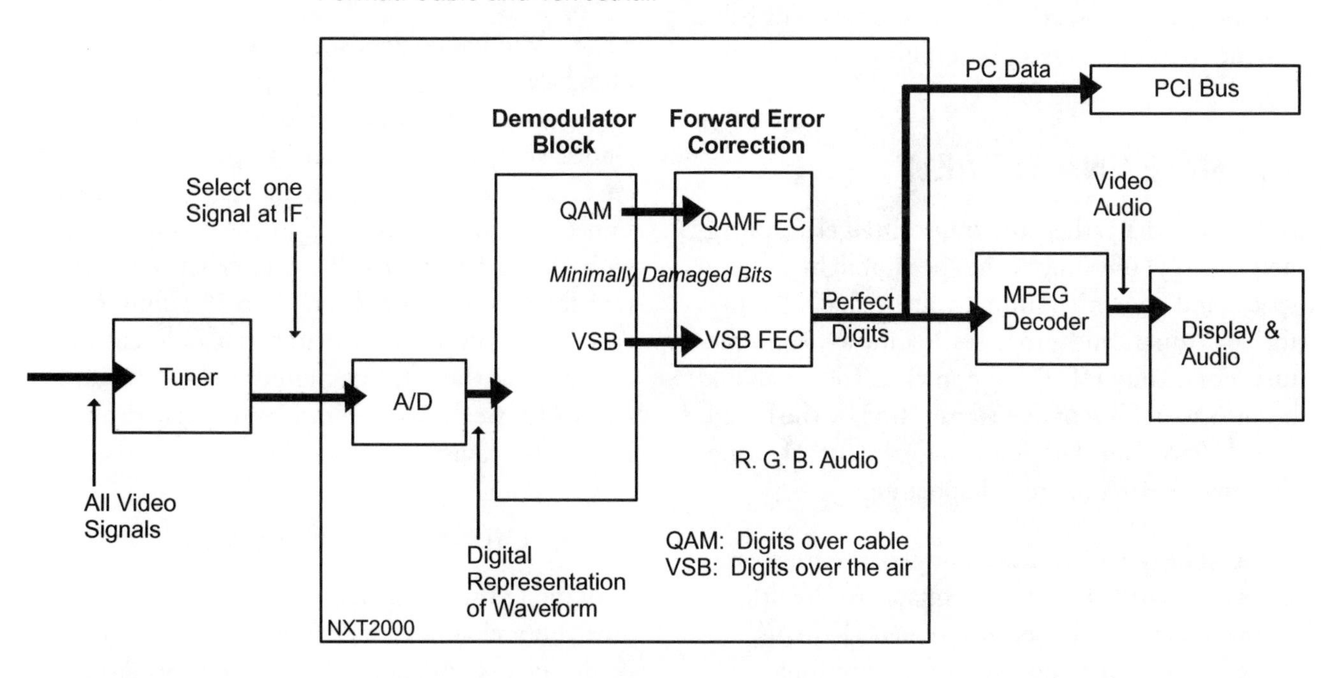

FIGURE 10-5

Zenith, are concerned with obtaining the best performance out of our receivers because that is one place where you are going to see a difference in the marketplace."

"There is absolutely nothing wrong with the transmission standard — what we have is the early adapter learning curve that anyone has with any new technology."

Field Testing These Chips

NxtWave expects to start field testing their chip at the end of 1999 with production, by STMicroelectronics in Europe, to follow. The NXT2000 will sell for approximately $22 each in quantities of 10,000 units. HDTV set manufacturers will decide when they will start using them in the HDTV receivers. For a few months the NXT2000 chips will be available to set makers for evaluation.

Motorola will also have their MCT2100 in the hands of customers for field testing in several cities. Motorola expects HDTV receivers to have their chip and be on the dealers' sales floors by the second quarter of 2000. Motorola has also indicated that it will have receiver circuit boards with the MCT2100 mounted chip by late 1999. The cost of the Motorola chip is pegged at $20 each in quantities of 10,000 units or more. Time and field testing will tell if these chips will solve the indoor antenna reception multipath problems.

TV ANTENNA REVIEW

A TV antenna gathers the transmitted electromagnetic energy, traveling at the speed of light through space, and converts it into a voltage that's fed into the TV tuner, and processes it into a visible picture. For quality HDTV, the antenna voltage should be an exact replica of that transmitted by the broadcast TV station. The following is a list of things that may deter this from happening:

- The downlead has an impedance mismatch.
- Antenna has an inadequate bandwidth.
- Damaged, loose, or crooked elements.
- Corrosion at joints and connections.
- Improper phasing.
- Too close to metal objects.
- Reflected signals.
- Antenna not pointed in correct direction.

Any of the above can be avoided by selecting the proper antenna and correctly installing it. Also, making the proper repairs of an older or damaged antenna can eliminate the problems. Let's now look at how these above problems can be avoided.

Antenna Lead-In Comments

For most TV antenna systems the delivery system (transmission lines) consists of a length of 75-ohm coaxial cable or 300-ohm twinlead that is connected to the antenna and is run to the TV set. In some installations where more than one TV receiver is connected to one antenna, there may be splatters, set couplers, or line-drop taps distributing the signals throughout the viewer's home. This would be considered a simple master-antenna television (MATV) system. Except where amplifiers are used or where connections are bad, there is little chance for any trouble. Where 300-ohm unshielded twinlead is used, keep it at least 6 inches from metal objects and DO NOT COIL it. Now, let's take a more detailed look at the TV antennas.

Antenna Impedance

Most TV antennas are usually either 75 ohm or 300 ohm balanced. It's generally impossible to look at an antenna and determine its impedance. The data included in the box will usually specify the impedance. You can then match the impedance to the impedance of the downlead or, when not included with the antenna, install an impedance-matching transformer to convert from 75 to 300 ohms or vice versa. Mismatched impedances not only cut down on signal to the set, but cause standing waves that result in a weak, washed-out, and smeared picture. It can also cause loss of a color picture.

Notes On Antenna Bandwidth

Looking at the construction of an antenna will not give you a clue as to its bandwidth. It can only be determined with sophisticated engineering instruments. This information should be available from the manufacturer who has the measurement equipment to specify the antenna bandwidth.

An antenna is like a tuned circuit. It will respond to frequencies within the bandwidth to which it is tuned and is virtually unresponsive to frequencies outside that bandwidth.

Ideally, the antenna's response curve should be flat, with steep skirts at the cutoff frequencies. If it cuts off the edge of a channel too soon or too sharply it will attenuate either the picture, sound, or color carrier. This will degrade picture quality on that channel. On the other hand, an antenna that does not cut off sharply outside the passband will make the antenna responsive to frequencies other than the TV signals, such as land-mobile radio, aircraft, and FM broadcasts.

Any of these can cause interference patterns in the picture. They represent extraneous signals not part of the original TV signal. It is always best to install a good, quality antenna that is well constructed.

ANTENNA CONSTRUCTION TIPS

The looks of an antenna may not always show that it is well constructed. The test comes when they perform in the real world. Severe weather affects them, causing their weaknesses to be discovered. Elements bend or break under stresses caused by strong winds or heavy snow and ice loads. Correct element spacing and alignment are critical to phasing and will affect the frequency response, impedance matching, gain uniformity, and directivity.

Look for heavy-duty construction and reinforcing supports at points of maximum stress. Look for strength in mountings and fabricated parts—particularly for 3/8-inch diameter aluminum tubing used for long elements. Look for bracing and heavy-gauge materials where strong forces will be applied as a result of high wind or ice buildup.

Where elements rotate into position and plastic or aluminum parts hold them captive for proper alignment, test the captive positioning device for rigidity and strength by subjecting the elements to bending and twisting forces.

Downlead connections, impedance-matching components, and preamplifiers should be enclosed and protected from the weather. Typically they are enclosed in plastic housings to keep dirt, rain, and snow out.

Well-designed antennas are not only mechanically strong and electrically sound, but can be assembled quickly and easily. Fold-out elements that are self-locking and secure, involving a minimum of assembly time, have a desirable feature. The Winegard Chromstar 2000 digital ready TV antenna shown in (**Fig. 10-6**) is an example of a good, quality antenna with all of the above features.

Because of an installer's labor costs, the ease of antenna assembly and speed of installation is as important as the antenna price. It is a good idea to have the installer do a signal survey before the antenna is installed. A quality antenna, installed correctly, may prove less costly over the years of service it will provide, than a cheap, poorly installed one.

The mast clamp, since it supports the entire antenna array, should be a major concern. A simple clamp with a heavy bolt is not good enough. The antenna load with all its twisting and bending motions must be transferred from the boom to the mast through that mast clamp. The boom and the mast must be strong enough to withstand these forces while the clamp locks them together rigidly.

INSTALLING THE ANTENNA

A quality TV antenna that is installed poorly is bad news. The antenna installation can make or break the TV reception quality. Good picture reception depends on a good antenna installed properly to transform the received signal into a faithfully reproduced voltage fed to the TV receiver.

For a good HDTV picture, the set needs 1000 microvolts (O dBmV) or more at the antenna terminals. The AGC circuits in most sets can handle up to a few hundred thousand microvolts. To measure these levels, a good signal-level meter (SLM) is necessary.

When purchasing an instrument like an SLM, it pays to obtain a high quality instrument. Like a high-quality antenna, a good SLM will pay for itself many times over. And should you plan to some day do MATV work (master antenna televi-

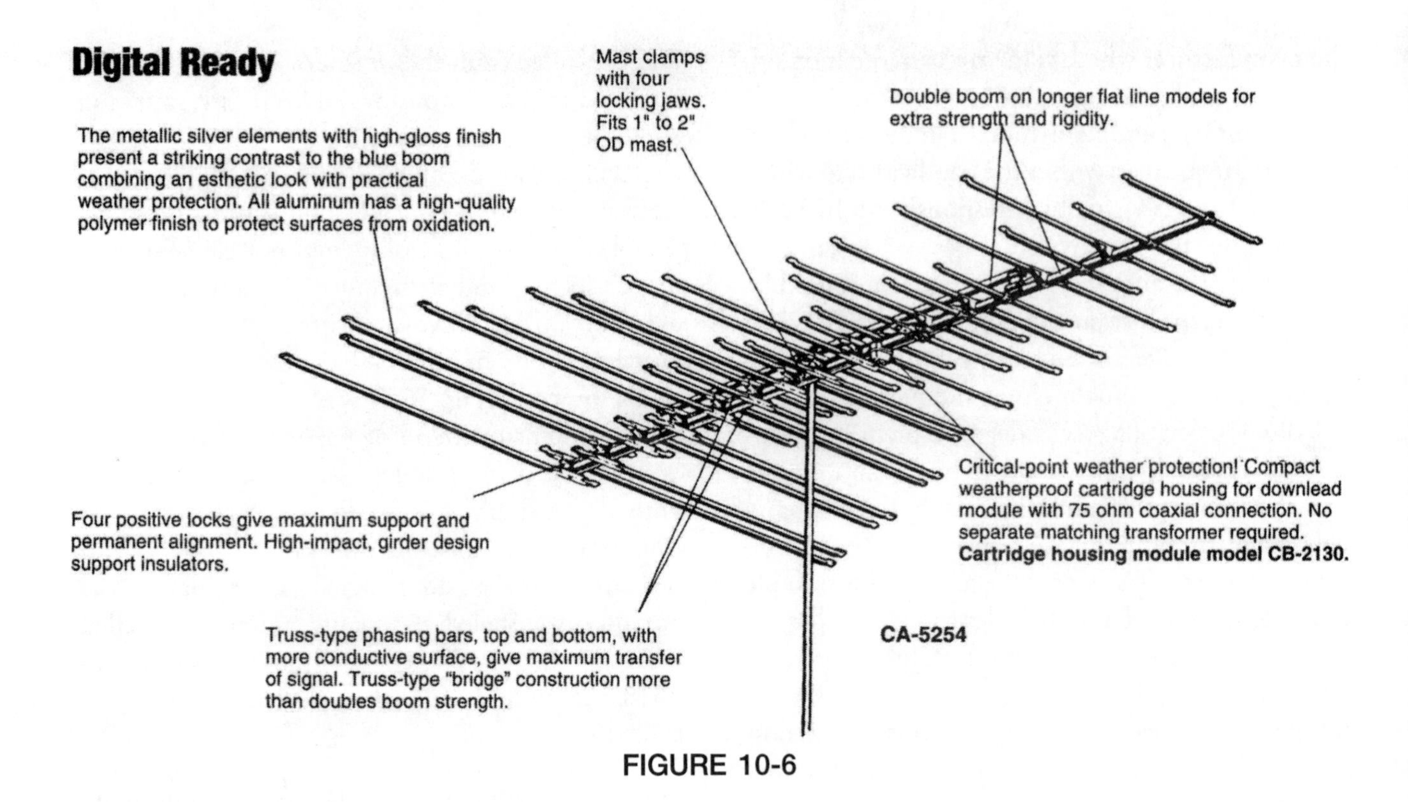

Digital Ready
The metallic silver elements with high-gloss finish present a striking contrast to the blue boom combining an esthetic look with practical weather protection. All aluminum has a high-quality polymer finish to protect surfaces from oxidation.
Mast clamps with four locking jaws. Fits 1" to 2" OD mast.
Double boom on longer flat line models for extra strength and rigidity.
Four positive locks give maximum support and permanent alignment. High-impact, girder design support insulators.
Critical-point weather protection! Compact weatherproof cartridge housing for downlead module with 75 ohm coaxial connection. No separate matching transformer required.
Cartridge housing module model CB-2130.
Truss-type phasing bars, top and bottom, with more conductive surface, give maximum transfer of signal. Truss-type "bridge" construction more than doubles boom strength.
CA-5254

FIGURE 10-6

sion systems), then purchase the more expensive and accurate SLM now, not later.

SENCORE FS74 CHANNELIZER-SIGNAL LEVEL METER (SLM)

One good, quality SLM is the Sencore FS74 Channelizer, Sr. or the FS73 Channelizer, Jr. which does not have the built-in video monitor. The FS74 Channelizer, Sr. is shown in the (**Fig. 10-7**) photo. Some of the FS74 SLM features are as follows:

This SLM unit has microprocessor control that makes all performance tests fast and simple. The microprocessor technology allows all tests to be performed on any in-use channel without removing or decreasing modulation, or adding special carriers. A signal-to-noise test automatically compares the signal level to the actual in-channel noise level. Making audio-to-video level tests are very simple. The FS74 automatically tunes both carriers and automatically reads out the separation in dB. Hum tests are made directly also, which is an FS74 exclusive.

The built-in, wide-band monitor allows picture quality checks anytime or anywhere. The FS74's integral wide-band monitor lets you see tough system problems like ghosting and interference, and pin-point the trouble source quickly. With the monitor turned on you can view any channel in full detail. The 4 MHz bandwidth means you can isolate problems that would go unnoticed on a portable TV. The FS74 also has a handy, built-in autoranging AC/DC voltmeter and ohmmeter.

If you cannot afford a high-quality SLM instrument, then you should consider a good portable color TV receiver so that you can actually see what grade of picture you are receiving. Keep in mind that a good picture is what you are shooting for. Because of AGC control in the TV set, it cannot provide you with an indication of the signal's relative strength.

TV SIGNAL SURVEY

Now that a good antenna has been selected, it's time for the installation. However, you should first take a signal survey. With the antenna assembled you can mount it on a short mast, and different locations can be tried until the best picture, free of ghosts, is found. Just connect the lead-in from the antenna to a portable TV set or the SLM instrument to find the best picture and signal strength.

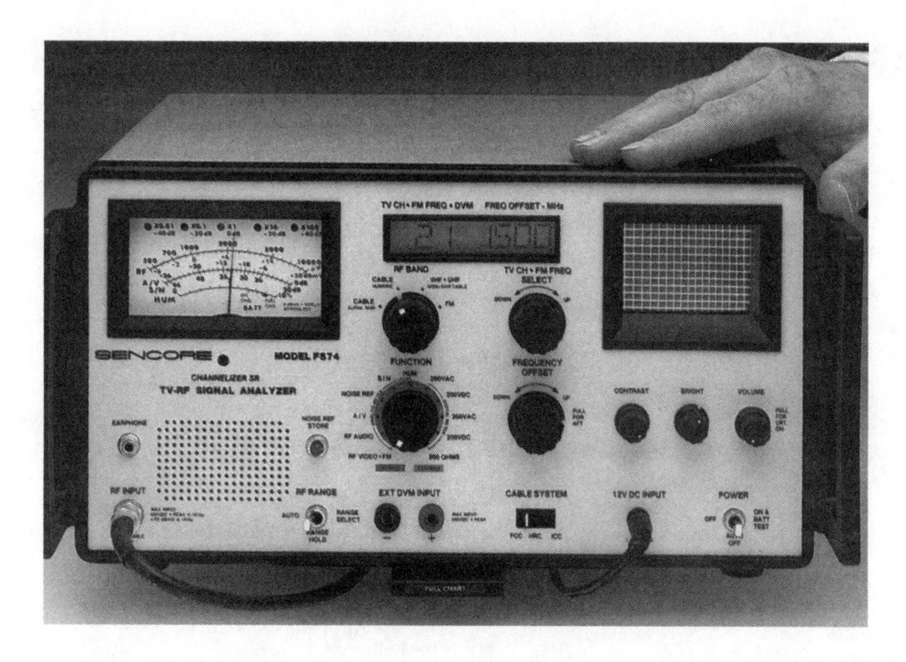

FIGURE 10-7

Always check the picture quality as the SLM may indicate adequate signal strength but the picture may be full of interference patterns or ghosts. And be sure to check all channels, paying close attention to the weakest ones.

When you have located the best point, rotate the mast to obtain the best picture, one free of ghosts and interference. If you cannot eliminate the ghosts, you may try stacking antennas for greater directivity. Now with the antenna oriented properly, make sure the mast is secure and will not turn. A large, well- designed antenna is highly directional, and being off just a few degrees can degrade the picture considerably.

SELECTOR MAP PROGRAM

There is a CEMA selector map program available for choosing the correct antenna for your location, allowing you to receive the best digital HDTV reception. The Winegard antenna company has a complete chart and map guide for their various antenna models. This guide is shown in (**Fig. 10-8**). These CEMA color-coded maps can be obtained by calling (703) 907-7600. It is important to note that large directional antennas can be used in all map areas yellow through blue, but require an optional antenna mounted preamp and roof-top mounting when needed in the violet or pink map areas.

NOTE: For certain markets, area special antennas are available from the Winegard factory.

Digital HDTV Reception Note:

With digital HDTV reception the picture does not go weak or snowy. It will just go blank or have the appearance of a "brick wall falling apart," and then go blank. A good, well-installed antenna that has sufficient signal strength is a must for HDTV reception.

WINEGARD CHROMSTAR 2000 ANTENNA SERIES

These Chromstar TV antennas have been designed for easy, quick installation and long life. The 75-ohm downlead coupler housing simply snaps into place for fast, easy assembly. No fasteners are needed. The steel hardware is zinc plated for maximum corrosion resistance, and flexible polyenthylene boot covers keep connections clean, dry-and protected from the weather.

These antennas deliver powerful VHF performance, and offer an additional ldB to 2dB higher gain on VHF and UHF for greatly im-

WINEGARD® ANTENNAS FOR CEMA SELECTOR MAP PROGRAM

Antenna Selector
This antenna works in the following zone(s).
YELLOW
©Consumer Electronics Manufacturers Association.
See the TV Antenna Selector Map to find the zone in which you live.

MULTI-DIRECTIONAL
Non-amplified
MS-1000
GS-1000
If more signal strength is desired, larger directional antennas can be used for these areas.

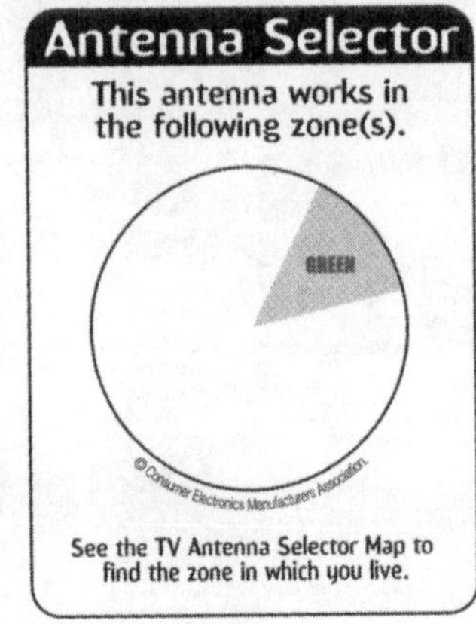

MULTI-DIRECTIONAL
Amplified
MS-2000
GS-2000
If more signal strength is desired, larger directional antennas can be used for these areas. Caution: Using an amplified antenna too close to signal source may cause overload.

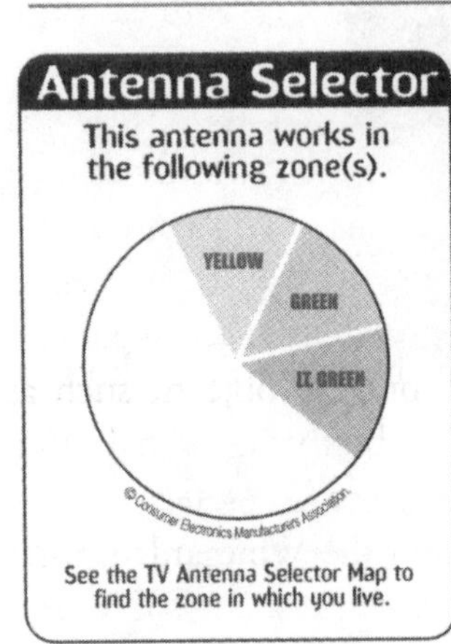

SMALL DIRECTIONAL
VHF/UHF: CA-7210, CA-7074, DS-7033, DS-7055, DS-7067, PR-7000, PR-7005, PR-7010, PR-5646
VHF: PR-5000
UHF: AC-3050, PR-4400, PR-9012, PR-9016
For more signal, use any medium or large directional model listed without a preamp.

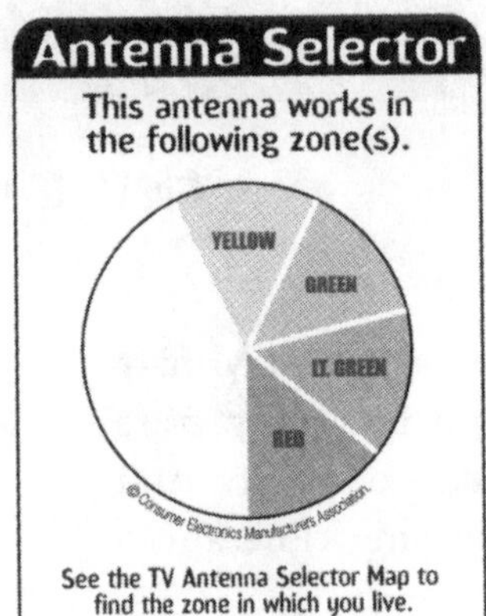

MEDIUM DIRECTIONAL
VHF/UHF: CA-7210, CA-7078, CA-7080, DS-7088, PR-7015, PR-7032
VHF: CA-4053, PR-5030
UHF: CA-9065, DS-8040, DS-8050, PR-9014, PR-9018, PR-8800
For more signal, use any of the large directional models listed without a preamp. With use of an optional antenna mounted preamp and rooftop mounting these antennas can be used in the blue map area as well.

Antenna Selector
This antenna works in the following zone(s).
YELLOW
GREEN
LT. GREEN
BLUE
RED
©Consumer Electronics Manufacturers Association.
See the TV Antenna Selector Map to find the zone in which you live.

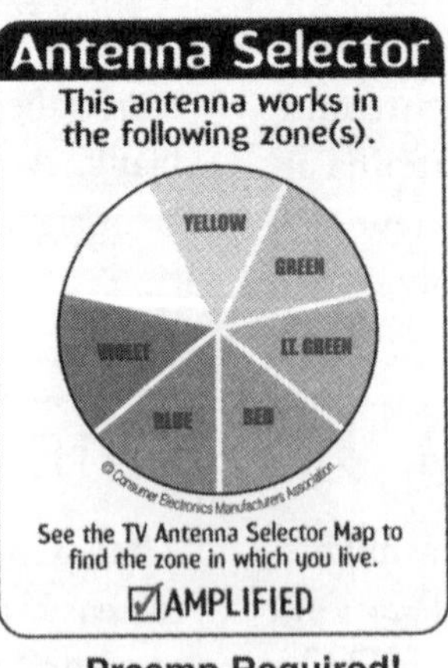

Preamp Required!

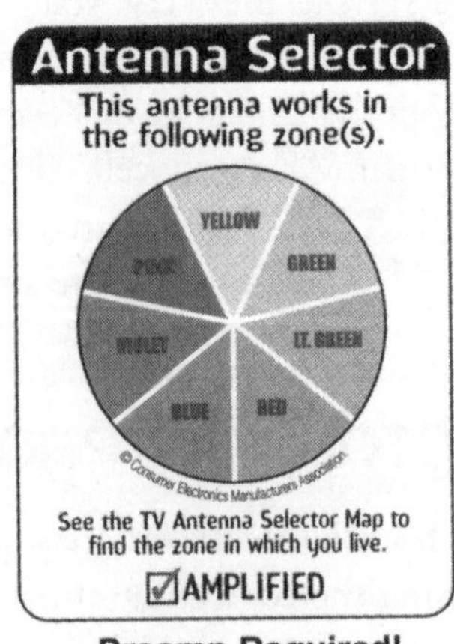

Preamp Required!

LARGE DIRECTIONAL

FIGURE 10-8

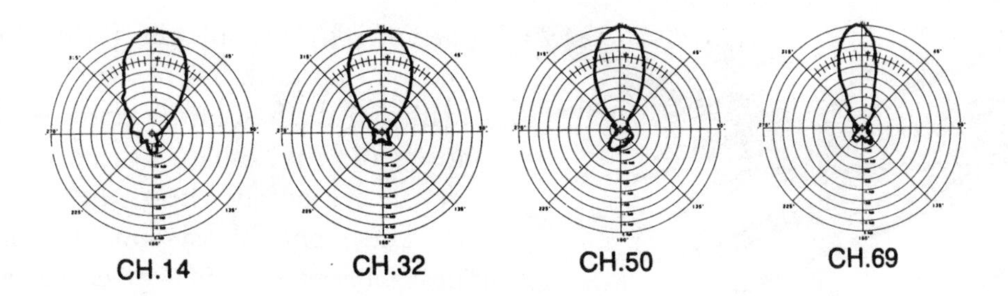

FIGURE 10-9

proved picture quality, especially important for weak signal areas. And the Electro-Lens director system combines with cross-phased end-fire driven elements, along with low Q dual directors, for increased capture area for peaking the antenna on each channel.

Outstanding UHF reception is achieved by precise director spacing and highly efficient, pre-assembled snap-out corner reflectors. They capture and reflect weak UHF signals with minimal loss to the driven element for increased uniform gain on all frequencies.

One of the many antenna models offered in the Chromstar line can resolve any difficult reception situation you may encounter.

Chromstar Antenna Features

- ABS weather coupler housing snaps into place with no fasteners.
- Boot collars weatherproof downlead connections.
- 75-ohm downlead connection from antenna to preamp.
- Double boom braces.
- Truss-type phasing line.
- Mast clamp with four locking jaws.
- Dual band Electro-Lens director system.
- Preassembled snap-out corner reflectors for high gain on UHF channels.
- Aluminum protected with a high-quality polymeric finish.
- Scissors-type struts for wedges.
- High-impact ABS support insulators.
- Zinc-plated steel hardware for extra protection.
- Wrap-around insulators.
- Unique phasing lines.

Chromstar Antenna Specifications

The Winegard model CA-9065 chromstar antenna is shown in the (**Fig. 10-9**) drawing with its engineering specifications and polar patterns. The drawing in **Figure 10-10** is the Winegard CA-7210 "ghost killer" antenna that is suitable for cities and suburbs. This antenna offers an effective solution for eliminating or drastically reducing TV ghost problems.

Most ghost problems occur where there is enough signal for good reception, but tall objects, such as buildings or trees, produce bouncing signals that result in multiple images on the TV screen.

The best way to stop these extraneous signals from reaching the TV receiver is to install a relatively small antenna with medium gain, but with the same high front-to-back ratio as a large, deep fringe area antenna. The drawing in (**Fig. 10-11**) shows the models CA-7084, CA-7082, and CA-7080 Winegard Chromstar 2000 antennas along with their specifications.

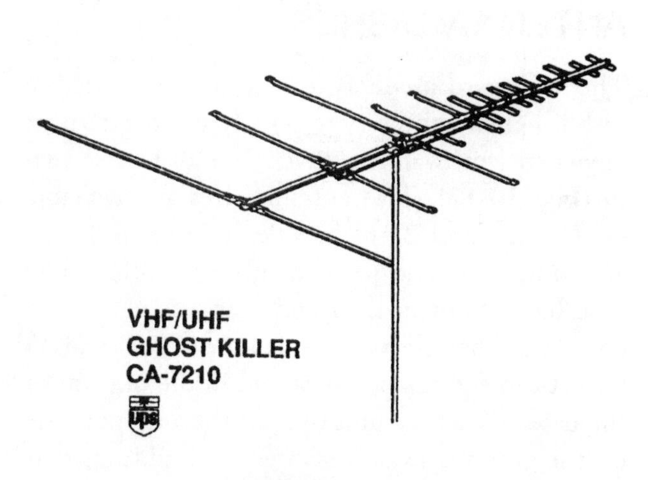

FIGURE 10-10

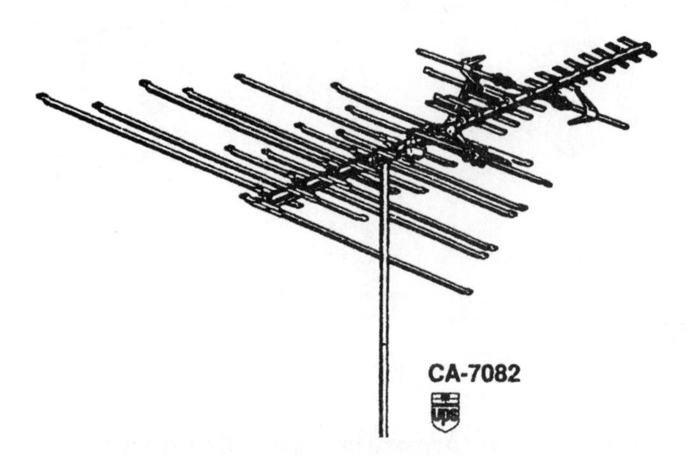

FIGURE 10-11

FIGURE 10-12

PORTABLE/REMOTE SATELLITE ANTENNA DISHES

The Winegard model RD-9046 portable satellite dish is lightweight, easy to set up, and ready to go anywhere you want to travel. The dish is shown in (**Fig. 10-12**). Two antenna sizes are available — 18-inch and 24-inch. The antenna and base are completely and permanently assembled. The base has a built-in level and compass. There's a convenient handle on the side for carrying and 25 feet of cable permanently stored in the bottom of the base. To set the unit up, put the antenna/base on the ground, away from trees or buildings that might block the satellite signal. Turn the base to the correct position, grab the top of the antenna, and pull upward. There are elevation positioning marks on the base. After positioning the antenna, attach the coax cable to the receiver and you can view television anywhere when you are traveling.

These portable systems are also suitable for areas where local codes or covenants prevent permanent installation of an antenna.

Satellite Antenna Dish Kit

The Winegard DS-4184 is designed for RV owners' digital satellite system with one-receiver viewing provision. This kit features the same 18-inch antenna and single-feed LNBF as the Winegard mobile satellite system, plus a wall/roof/pole mount that adapts to almost any kind of installation. This antenna unit can be installed on a house or any other appropriate place. The digital receiver from your RV can then be used for satellite TV viewing. Enjoy watching digital satellite TV at home and get double use from your digital receiver and still have only one programming fee. The DS-4148 includes antenna, feed support, single output LNBF, combination wall/roof/pole mount, and installation hardware. The model DS-4248 is a dual-dish kit that is used for multiple receivers that are located at the same place. With this setup you can watch two different satellite dish channels (programs) at the same time.

Combination Satellite/Off-Air Antenna

The Winegard RD-4610 combination mount lets you install the 18-inch digital satellite antenna and the omnidirectional RS-1000 RoadStar antenna together. With the RD-4610, you have one mount and only one installation. This allows both antennas to be raised and lowered together with the hand crank located inside your RV, thus giving you the choice of satellite or local TV broadcasts. This unit is shown in (**Fig. 10-13**).

Crank-Up Satellite Antenna Dish

The Winegard model RD-4600 crank-up satellite system for RVs is easy to install and simple to operate. The LNBF is attached to a folding feed

FIGURE 10-13

FIGURE 10-14

support and the field-proven lift assembly has an operating radius of only 17 inches. The antenna is raised, rotated, and lowered to a travel position using the hand crank located inside the RV. This system is so compact it will fit on almost any recreational vehicle — an area 30 inches by 20 inches is all that's required for installation. Control parts are made of sturdy, molded plastic. The system is 27.5 inches high when raised, 12 inches high in the travel position, and weighs a mere 13 pounds.

Automatic Satellite Antenna

The Winegard model RD-9946 satellite system is totally automatic. This unit includes the 18-inch antenna dish, the lift system, LNBF, and positioner. This unit is illustrated in (**Fig. 10-14**). The white lift assembly is injection molded to provide more strength with much less weight, and the strong dish support bracket is cast aluminum with an aluminum feed arm. This automatic system has two motors: one for lift, the other for rotation. The antenna is easily positioned for the best reception wherever your RV may be parked. Use the positioner (or optional wallplate control panel) and the antenna automatically raises, rotates, and lowers to the travel position.

Winegard's Elevation Sensor for Satellite Dish

This elevation dish sensor takes the guesswork out of finding the satellite dish when on RV trips at different locations around the country. With the Digital Magic elevation sensor, you can find the satellite faster than with most high-priced automatic systems. This device will mount on Winegard dish models RD-4600, RD-4646, RD-4604, and the RD-4610.

The model DE-4600 digital elevation sensor gives the exact elevation of the antenna on an LCD readout inside the vehicle. It's accurate to within 1 degree, even when the vehicle is not level. The sensor is attached to the elevating tube of the antenna and wired to a wallplate with LCD readout located inside the your vehicle. The unit includes 20 feet of 4-conductor cable and requires +12 volts of power or a 9-volt battery.

Roadstar and Sensar RV Antennas

Unlike low-frequency signals, which tend to follow the curvature of the earth, high-frequency TV signals in the VHF and UHF band travel in a tangent to the earth. If you are located behind a building, hill, mountain, or down in a valley, the TV signal will not reach your antenna. This is why reception is affected by the height of your installation in certain areas. As you can see in the (**Fig. 10-15**) illustration, a higher antenna on your RV would provide better signal reception.

The Winegard Company offers two different RV television antennas for TV viewing when traveling down the highway or stopped for the night at an RV center.

The Winegard Sensar antenna provides the optimum in gain, resulting in a sharp, clear picture, but it must be pointed toward the TV station you want to receive. The omnidirectional Roadstar antenna will receive signals from all directions without moving the antenna. A photo of these two antennas is shown in (**Fig. 10-16**).

Winegard's Roadstar antenna is unique because, once properly mounted, it doesn't need adjusting. It is omnidirectional, designed to pick up VHF/UHF/FM signals from any point on the compass without being rotated. This means your RV passengers can watch TV while traveling down the road without worrying that a change in direction will degrade the reception and cause the picture to fade, become snowy, or have ghosts.

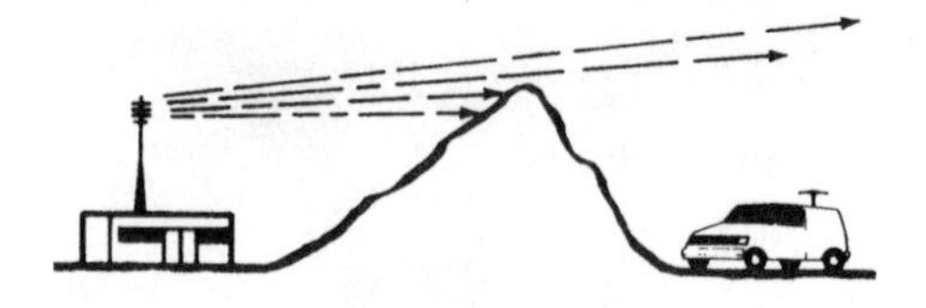

FIGURE 10-15

The Roadstar antenna is contained in a stylish, high-impact weatherproof housing that is coated with Korad to protect against the sun's damaging ultraviolet rays.

The Roadstar antenna has no moving parts to break or corrode. It measures 21 inches in diameter, 4 inches deep, and mounts in a fixed position just 9 inches above the roof. And, it only weighs 11 pounds.

If additional height is needed to optimize performance, a lift is available for the RoadStar antenna to provide better reception in fringe areas.

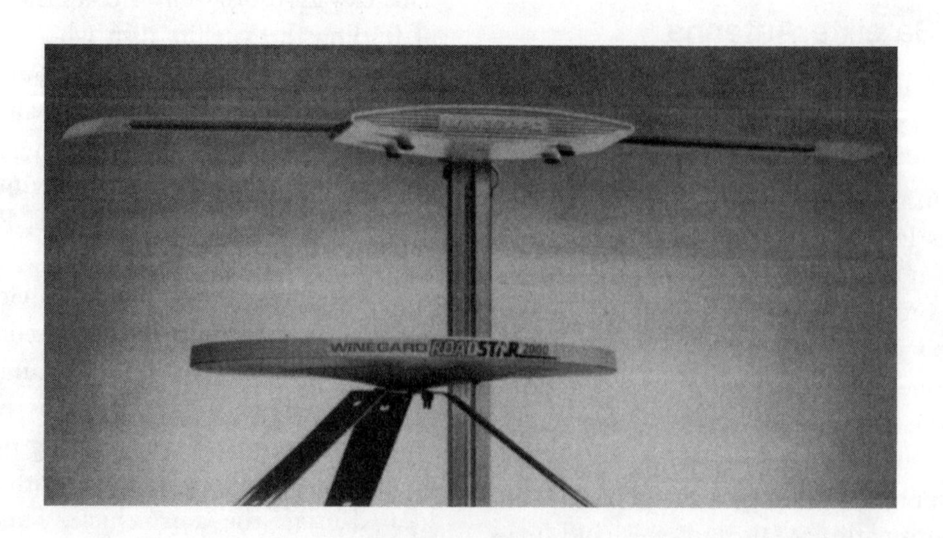

FIGURE 10-16

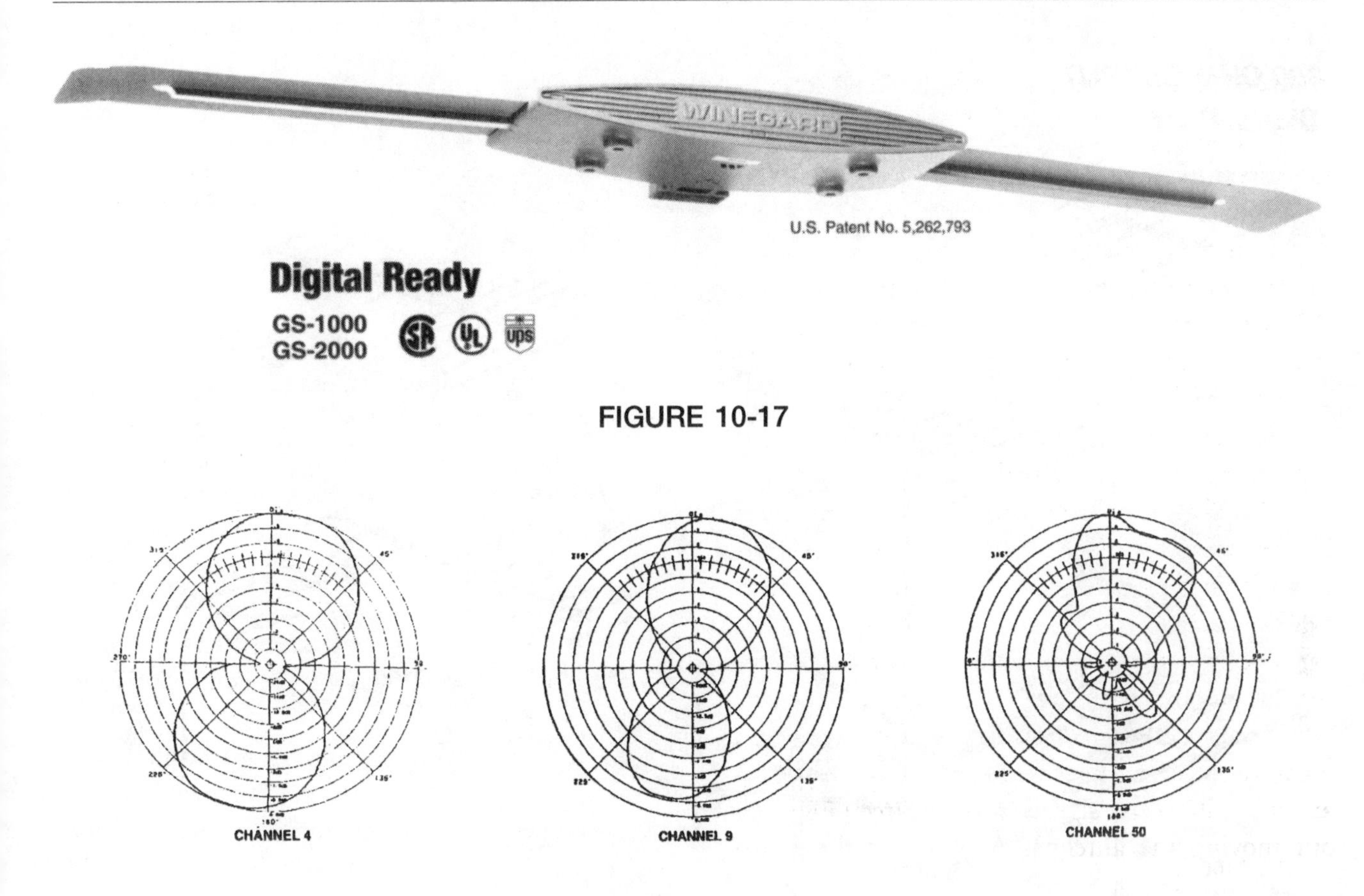

FIGURE 10-17

FIGURE 10-18

The lift, which is a model RV-0003, can increase the reception of good, clear signals by lifting the antenna 21 inches above the rooftop while parked.

The Winegard's reliable Sensar RV TV antenna line continues to provide excellent reception for motor homes, travel trailers, fifth wheels, and other RVs. The Sensar antenna rests just 4 inches above the roof in the lowered travel position. But when raised to its full height of 44 inches, and directed toward the TV station, this antenna will provide the ultimate in reception. For optimum reception and safety, the Sensar antenna *MUST ONLY* be used while the RV is parked. The combination of the antenna's built-in amplifier, directivity, and height advantage provide excellent reception and clear VHF/UHF TV pictures.

All Sensar models raise, lower, and rotate from inside the parked RV, and are easy to install with simple hand tools. The housing is made from ABS polymer and the elements are aircraft grade aluminum with polyester powder coating finish. And, all hardware is corrosion resistant.

Both antennas feature a high-gain, low-noise amplifier for clearer TV reception. Additonally, both antennas can utilize the same control panel as the RV-6000 and RV-700 series. The same TV outlets are applicable to RV-type installations.

The Winegard Sensar II antennas for the home are shown in the (**Fig. 10-17**) photo. The model GS-2000 is an amplified unit recommended for a distance 10 to 55 miles from the TV transmitter, while the GS-1000 is a non-amplified Sensar antenna ideal for metropolitan reception. The polar patterns for these Sensar II antennas are shown in the (**Fig. 10-18**) illustrations.

The drawings in (**Fig. 10-19**) is of the various Winegard PROSTAR 1000 VHF/UHF Yagi digital-ready TV antennas.

300 OHM OUTPUT
Digital Ready

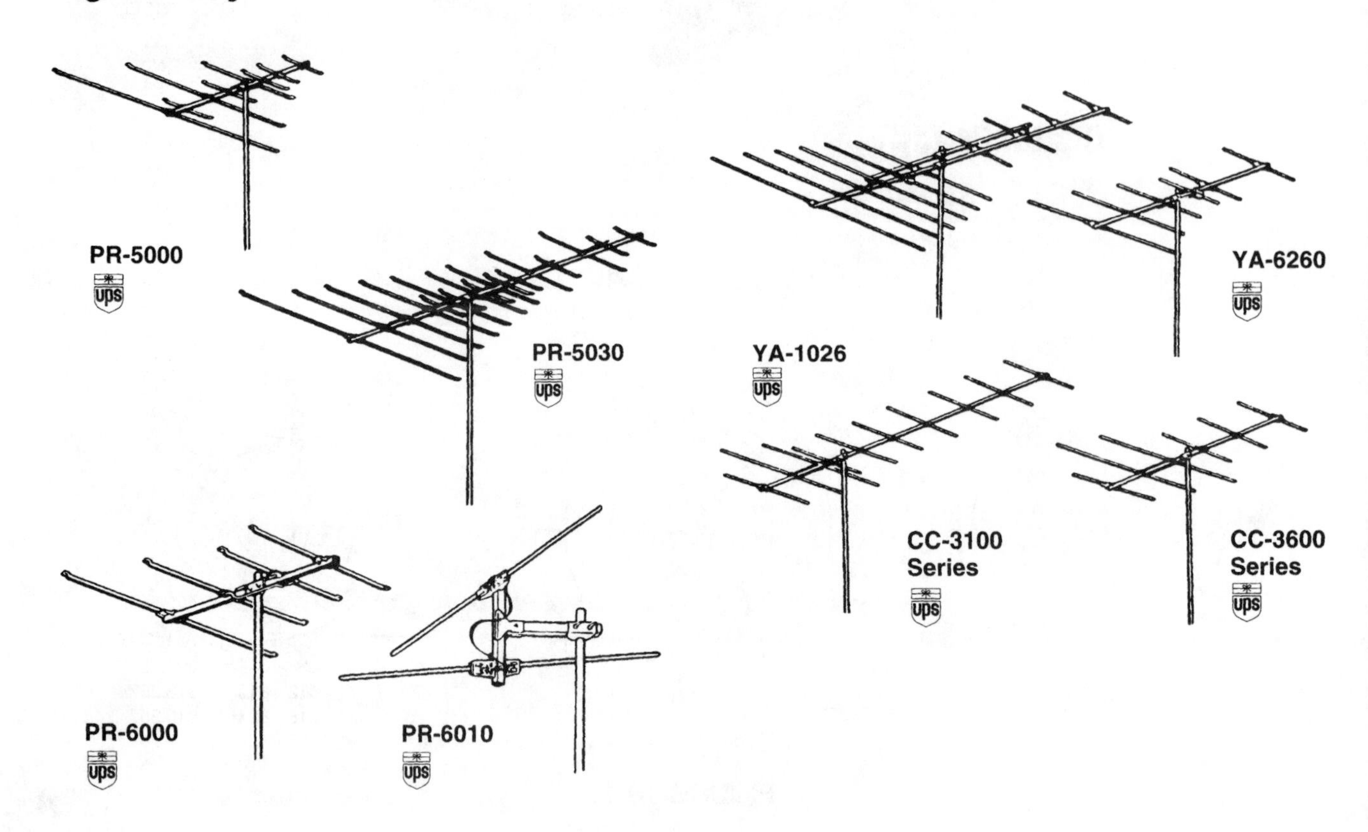

FIGURE 10-19

Complete List of All TV Frequencies

The frequency chart shown in **Figure 10-20** covers the entire TV/FM/translator and satellite LNB frequencies. In **Figure 10-21**, you will find a handy conversion chart to calculate dBM to microvolts. Note that ODBMV = 1,000 microvolts, which will usually give you a good, clear color TV picture.

Author's Antenna Setup

The photo in **Figure 10-22** is of the author's antenna, which consists of a VHF/UHF antenna on top of the mast and an 18-inch satellite digital dish mounted below. This small-gain, VHF/UHF antenna pulls in sharp VHF and UHF TV stations from more than 60 miles away. Its secret is that it sits on top of a 1,880-foot mountain.

Portions of the information and illustrations used in this chapter is courtesy of the WINEGARD COMPANY, Burlington, Iowa

TV FREQUENCIES

CH.	BAND	WIDTH	VIDEO	AUDIO
		SUB-BAND		
T-7	5.75	11.75	7	11.5
T-8	11.75	17.75	13	17.5
T-9	17.75	23.75	19	23.5
T-10	23.75	29.75	25	29.3
T-11	29.75	35.75	31	35.5
T-12	35.75	41.75	37	41.5
T-13	41.75	47.75	43	47.5
TV-1F	40	46	45.75	41.25
		LOW-BAND		
2	54	60	55.25	59.75
3	60	66	61.25	65.75
4	66	72	67.25	71.25
54 IRC	72	78	73.25	77.25
5	76	82	77.25	81.75
55 IRC	78	84	79.25	83.75
6	82	88	83.25	87.75
56 IRC	84	90	85.25	89.75
		FM RADIO BAND 88-108 Mz		
57 IA-5	90	96	91.25	95.75
58 IA-4	96	102	97.25	101.75
59 11A-3	102	108	103.25	107.25
98 A-2	108	114	109.275	113.775
99 A-1	114	120	115.275	119.775
		MID-BAND		
57 IA-5	90	96	91.25	95.75
15 B	126	132	127.2625	131.7625
16 C	132	138	133.2625	137.7625
17 D	138	144	139.25	143.75
18 E	144	150	145.25	149.75
19 F	150	156	151.25	155.75
20 G	156	162	157.25	161.75
21 H	162	168	163.25	167.75
22 I	168	174	169.25	173.75
		HI-BAND		
7	174	180	175.25	179.75
8	180	186	181.25	185.75
9	186	192	187.25	191.75
10	192	198	193.25	197.75
11	198	204	199.25	203.75
12	204	210	205.25	209.75
13	210	216	211.25	215.75
		SUPER-BAND		
23 J	216	222	217.25	221.75
24 K	222	228	223.25	227.75
25 L	228	234	229.2625	223.7625
26M	234	240	235.2625	239.7625
27 N	240	246	241.2625	245.7625
28 O	246	252	247.2625	251.7625
29 P	252	258	253.2625	257.7625
30 Q	258	264	259.2625	263.7625
31 R	264	270	265.2625	263.7625
32 S	270	276	271.2625	275.7625
33 T	276	282	277.2625	281.7625
34 U	282	288	283.2625	287.7625
35 V	288	294	289.2625	293.7625
36 W	294	300	295.2625	299.7625
		HYPER-BAND		
37 AA	300	306	301.2625	305.7625
38 BB	306	312	307.2625	311.7625
39 CC	312	318	313.2625	317.7625
40 DD	318	324	319.2625	323.7625
41 EE	324	330	325.2625	329.7625
42 FF	330	336	331.275	335.775
43 GG	336	342	337.2625	341.7625
44 HH	342	348	343.2625	347.7625
45 II	348	354	349.2625	353.7625
46 JJ	354	360	355.2625	359.7625
47 KK	360	366	361.2625	365.7625
48 LL	366	372	367.2625	371.7625
49 MM	372	378	373.2625	377.7625
50 NN	378	384	379.2625	383.7625
51 OO	384	390	385.2625	389.7625
52 PP	390	396	391.2625	401.7625
53 QQ	396	402	397.2625	401.7625
54 RR	402	408	403.25	407.75
55 SS	408	414	409.25	413.75
56 TT	414	420	415.25	419.75
57 UU	420	426	421.25	425.75
58 W	426	432	427.25	431.75
59 WW	432	438	433.25	437.75
60 XX	438	444	439.25	443.75
61 Y	444	450	445.25	449.75
62 ZZ	450	456	451.25	455.75
		UHF-BAND		
14	470	476	471.25	475.75
15	476	482	477.25	481.75
16	482	48	483.25	487.75
17	488	494	489.25	493.75
18	494	500	495.25	499.75
19	500	506	501.25	505.75
20	506	512	507.25	511.75
21	512	518	513.25	517.75
22	518	524	519.25	523.75
23	524	530	525.25	529.75
24	530	536	531.25	535.75
25	536	542	537.25	541.75
26	542	548	543.25	547.75
27	548	554	549.25	553.75
28	554	560	555.25	559.75
29	560	566	561.25	565.75
30	566	572	567.25	571.75
31	572	578	573.25	577.75
32	578	584	579.25	583.25
33	584	590	585.25	589.75
34	590	596	591.25	595.75
35	596	602	597.25	601.75
36	602	608	603.25	607.75
37	608	614	609.25	613.75
38	614	620	615.25	619.75
39	620	626	621.25	625.75
40	626	632	627.25	631.75
41	632	638	633.25	637.75
42	638	644	639.25	643.75
43	644	650	645.25	649.75
44	650	656	651.25	655.75
45	656	662	657.25	661.75
46	662	668	663.25	667.75
47	668	674	669.25	673.75
48	674	680	675.25	679.75
49	680	686	681.25	685.75
50	686	692	687.25	691.75
51	692	698	693.25	697.75
52	698	704	699.25	703.75
53	704	710	705.25	709.75
54	710	716	711.25	715.75
55	716	722	717.25	721.75
56	722	728	723.25	727.75
57	728	734	729.25	733.75
58	734	740	735.25	739.75
59	740	746	741.25	745.75
60	746	752	747.25	751.75
61	752	758	753.25	757.75
62	758	764	759.25	763.75
63	764	770	765.25	769.75
64	770	776	771.25	775.75
65	776	782	777.25	781.75
66	782	788	783.25	787.75
67	788	794	789.25	793.75
68	794	800	795.25	799.75
69	800	806	801.25	805.75
		TRANSLATOR BAND		
70	806	812	807.25	811.75
71	812	818	813.25	817.75
72	818	824	819.25	813.75
73	824	830	825.25	829.75
74	830	836	831.25	835.75
75	836	842	837.25	841.75
76	842	848	843.25	847.75
77	848	854	849.25	853.75
78	854	860	855.25	859.75
79	860	866	861.25	865.75
80	866	872	867.25	871.75
81	872	878	873.25	877.75
82	878	884	879.25	883.75
83	884	890	885.25	889.75

SATELLITE LNB FREQUENCIES

950 - 2250 MHz

FIGURE 10-20

FROM dBM to MICROVOLTS

0dBmV = 1,000 MICROVOLTS = GOOD PICTURE

dB	uV	dB	uV	dB	uV	dB	uV	dB	uV	dB	uV
-40	10.00	-19	112.2	0dBmV	1000	21	11220	42	125900	62	1259000
-39	11.22	-18	125.9	1	1122	22	12590	43	141300	63	1413000
-38	12.59	-17	141.3	2	1259	23	14130	44	158500	64	1585000
-37	14.13	-16	158.5	3	1413	24	15850	45	177800	65	1778000
-36	15.85	-15	177.8	4	1585	25	17780	46	199500	66	1995000
-35	17.78	-14	199.5	5	1778	26	19950	47	223900	67	2239000
-34	19.95	-13	223.9	6	1995	27	22390	48	251200	68	2512000
-33	22.39	-12	251.2	7	2239	28	25120	49	281200	69	2818000
-32	25.12	-11	281.8	8	2512	29	28180	50	316200	70	3162000
-31	28.18	-10	316.2	9	2818	30	31620	51	354800	71	3548000
-30	31.62	- 9	354.8	10	3162	31	35480	52	398100	72	3981000
-29	35.48	- 8	398.1	11	3548	32	39180	53	446700	73	4467000
-28	39.81	- 7	446.7	12	3981	33	44670	54	501200	74	5012000
-27	44.67	- 6	501.2	13	4467	34	50120	55	562300	75	5623000
-26	50.12	- 5	562.3	14	5012	35	56230	56	631000	76	6310000
-25	56.23	- 4	631.0	15	5623	36	63100	57	707900	77	7079000
-24	63.10	- 3	707.9	16	6310	37	70790	58	794300	78	7943000
-23	70.79	- 2	794.3	17	7079	38	79430	59	891300	79	8931000
-22	79.43	- 1	891.3	18	7943	39	89130	60	1000000	80	100 000 000
-21	89.13	0dBmV	1000.0	19	8913	40	100000	61	1122000		
-20	100.00			20	10000	41	112200				

FIGURE 10-21

FIGURE 10-22

CHAPTER 11

HDTV Power Supplies

INTRODUCTION

This chapter briefly explains conventional TV set power supply operations and goes into more detail for the newer scan-derived, switch-mode (chopper) systems found in the HDTV receivers.

The power supply is the heart of all electronic devices, so we will briefly explain the simpler power supplies. These basic power supplies are called half-wave, full-wave, bridge, and voltage doubler power supplies. You will find some versions of these supplies in most HDTV receivers.

The next portion of this chapter is devoted to operation and troubleshooting of the switch-mode power supply system. This may also be referred to as a scan-derived, sweep system power supply. Also, we'll discuss the voltage regulation in these power supplies. Next, we'll look at ways to troubleshoot various problems in these switch-mode devices. Problems with trickle mode, kick-start mode, and circuit shutdown will also be investigated.

The chapter continues with circuit operation of some actual production model HDTV receivers' main power supplies, as well as subsystems of those supplies. This section will also include operational and troubleshooting flow charts.

The oscilloscope will be referred to in various portions of this chapter to illustrate troubleshooting techniques for locating power supply faults in HDTV sets. The scope is used for checking waveforms in the switching modes of regulated power supplies, finding the breakdown voltage of switching transistors and detecting ripple in filtered DC voltages.

The chapter concludes with detailed information you need to know when replacing failed "high efficiency special rectifier diodes" now used in modern, scan-derived HDTV power supply systems.

SOME BASIC POWER SUPPLY CIRCUITS

In order to operate any electronic device, including an HDTV set, a power supply or some source of voltage is required. Usually the power is taken from an AC power line before diodes and filters are used to produce a DC voltage. In most instances a DC, or direct current voltage, is needed to operate these electronic circuits. Follow along now as we look at a few of these basic power supply circuits.

The Half-Wave Power Supply

A half-wave power supply circuit is referred to in the (**Fig. 11-1**) circuit drawing. Directing your attention to the top right, note that the negative-going part of the sine wave is missing and only the positive-going part is being used. The bottom part of the waveform has been removed by diode filtering, which only allows current to pass in one direction. This DC voltage pulses at 60 times per second and can now be smoothed out with filter capacitors.

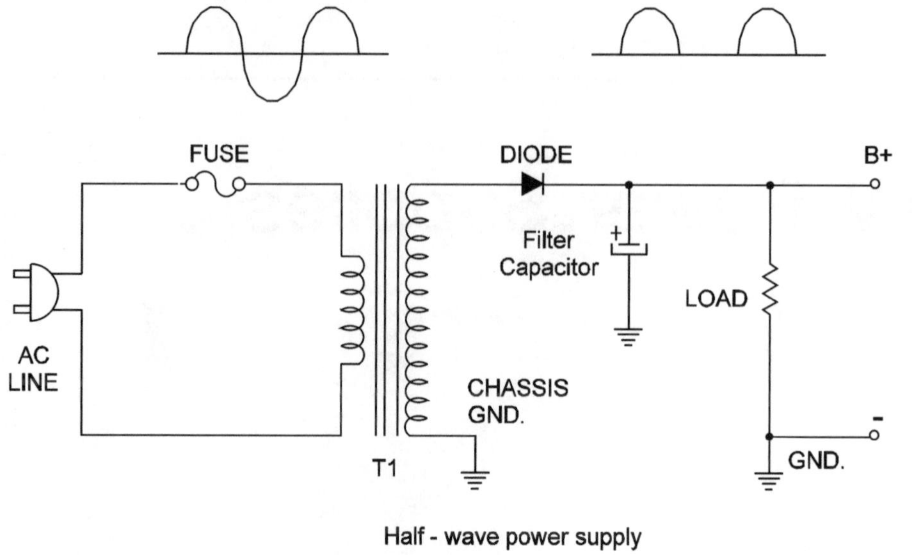

Half - wave power supply

FIGURE 11-1

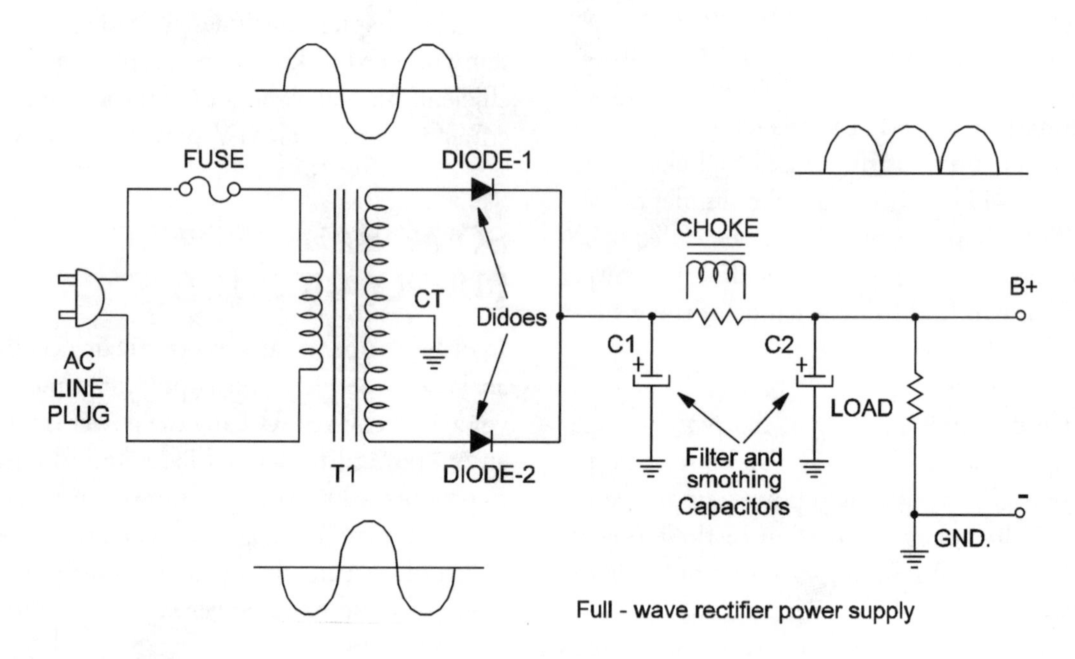

Full - wave rectifier power supply

FIGURE 11-2

The Full-Wave Power Supply

Illustrated in the (**Fig. 11-2**) drawing, this full-wave power supply circuit allows both halves of the AC sine wave to be used. Thus, you have an output ripple of 120 hertz, and not 60 cycles (hertz), as with the half-wave power supply. These two diodes are connected so that one diode conducts on each half cycle. This 120-cycle ripple now must be smoothed out with a resistor or choke coil and two filter capacitors. The choke helps prevent sudden changes of current through it, and a second electrolytic capacitor (C2) provides even more filtering.

The Full-Wave Bridge Power Supply

The diode bridge-configured power supply circuits are used in many types of electronic equipment. The bridge circuit power supply is unique because it can produce a full-wave voltage output without using a center-tapped transformer. The diamond,

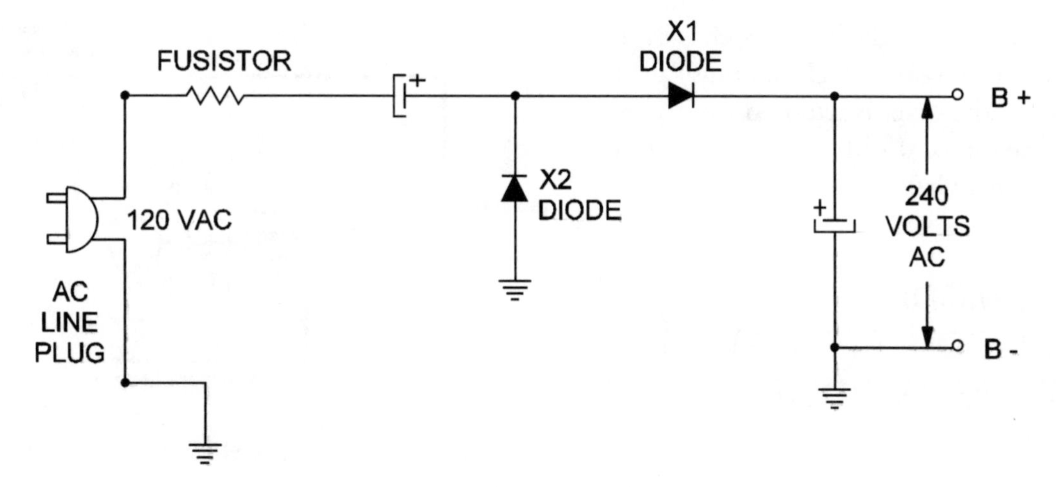

Voltage - Doubler Power Supply.

FIGURE 11-3

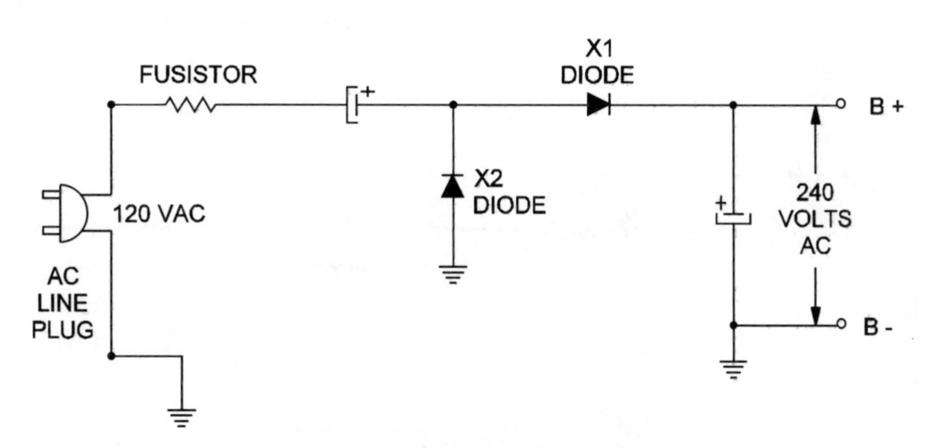

Voltage - Doubler Power Supply.
NOTE: No power transformer is used.
CAUTION: The chassis will be "HOT".
Use an isolation transformer to service TV set.

FIGURE 11-4

shaped diode layout of this power supply in shown in (**Fig. 11-3**). The bridge circuit is actually an electronic switching device. Consider the diode rectifiers as switching all of the positive AC pulses to the B+ line and all of the negative AC pulses to the B- line or to chassis ground.

The Voltage Doubler Power Supply

The voltage doubler power supply circuit may use a power transformer or it can be connected directly to the AC line. A power supply is used without a transformer when equipment requires a higher DC voltage than the AC line can supply and to reduce the cost and weight of the device.

A typical voltage doubler circuit, without a transformer, is shown in (**Fig. 11-4**). To observe its operation, assume that the half-wave diode (Xl) is connected to produce a positive voltage on the B+ line of 120 volts. Diode (X2) is then added to the circuit but is connected with the opposite polarity. This will cause diode (X2) to be -120 volts, with respect to ground. Note that these voltages will "add," and then develop a voltage of approximately 240 volts between the B+ and B- supply output terminals. A drawback of this type of supply is that the B- is connected to the TV set's chassis. This makes it a "hot" chassis, which can cause a shock hazard. When the cover is removed from a "hot" chassis device, proceed with caution when servicing and use an isolation transformer.

SWITCH-MODE/PULSE WIDTH MODULATED (PWM) SWITCHING TV POWER SUPPLIES

The sophisticated protection devices, and feedback loops found in HDTV switching power supplies, can be a challenge to troubleshoot. Follow along now as the operational theory of each block within a pulse-width modulated (PWM) switching power supply is explained.

Operation of a Basic Switch-Mode Circuit

A switching power supply includes a DC voltage source, a high-frequency transformer, a power transistor, and a pulse generator. This basic switching circuit is shown in (**Fig. 11-5**). Now, let's analyze the circuit. The transistor is switched on and off by the drive signal closing and opening the path for DC current flow. This sets up a changing magnetic field in the transformer's primary winding. This changing magnetic field induces voltage in the secondary winding, where it is rectified and filtered into an AC voltage.

The power transistor operates as a switch; it is either on (saturated) or off (cut off). The amount of energy delivered to the secondary is determined by the "on time" of the transistor. The output voltage can be varied by changing the frequency or duty cycle of the transistor drive signal. These two techniques of regulating the output are called pulse-width modulation (varying duty cycle) or pulse rate modulated regulation (varying frequency of pulses). The waveform drawings in (**Figures 11-6 and 11-7**) illustrate how each form of regulation changes the total "on time" of the switcher's transistor. The big advantage of using switchmode power supplies (SMPSs) is the ability to closely and efficiently regulate the power delivered to the load. Regulation in an SMPS is obtained by comparing a DC voltage from the output with a reference voltage. This comparison provides feedback to the pulse generator, which alters the drive signal to compensate for the output voltage change. This

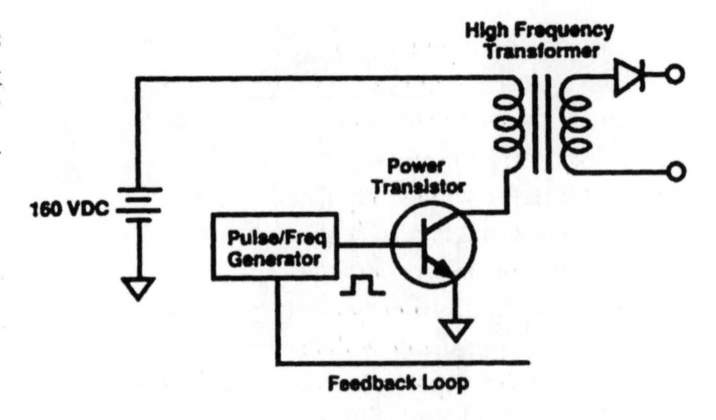

FIGURE 11-5

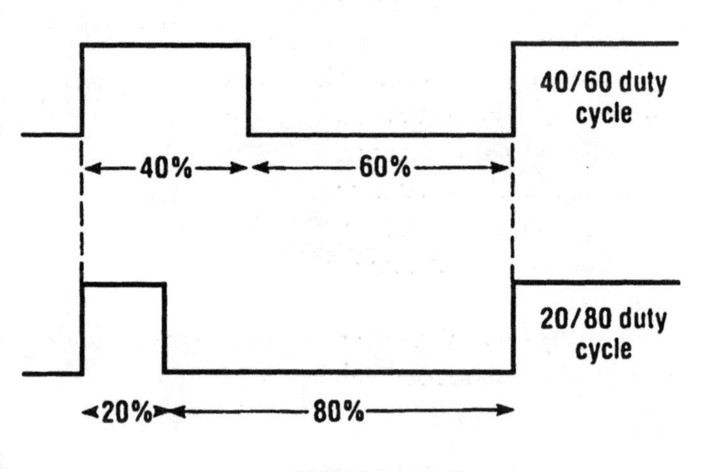

FIGURE 11-6

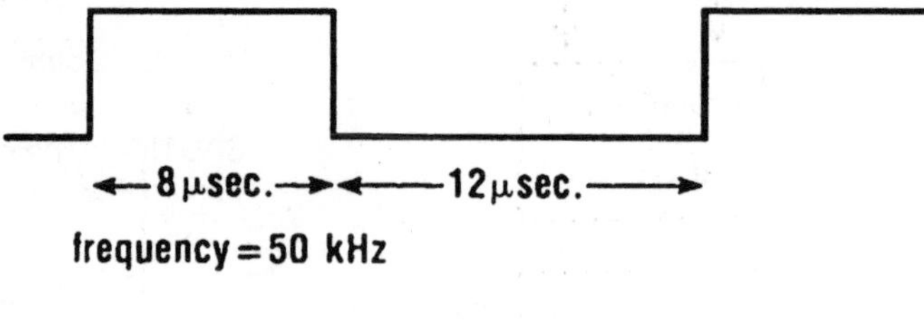

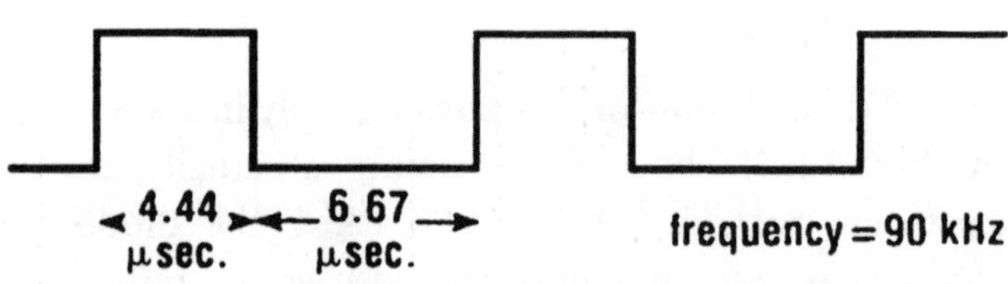

FIGURE 11-7

comparison/compensation occurs continually and provides a closely regulated output. Note that only enough of the necessary voltage is provided. Thus, no excess power has to be dissipated as in conventional shunts or voltage regulated systems.

The block diagram in (**Fig. 11-8**) shows a typical switchmode power supply. The basic blocks include the transformer, power transistor, pulse generator, and feedback voltage regulator blocks. The transformer input is an unregulated B+ supplied by rectified and filtered AC line voltage. Optoisolators may be used to isolate the transformer secondary circuits from the AC line.

Switch mode circuits will usually have overcurrent protection. These circuits usually sample the voltage drop across a resistor in series with the switching transistor. Should current become high and exceed a predetermined level, the pulse generator will be shut down. This circuit provides component protection should a malfunction occur.

Usually HDTV receivers have a standby power supply that is operational when the TV set is plugged in. This small amount of continuous power is needed for the microprocessor, memory chips, remote control receiver, and the SMPS's startup circuit operation.

BASIC SWITCHING POWER SUPPLY

Switching power supplies found in HDTV sets offer the advantage of increased efficiency, better output regulation, smaller size, and less weight compared to conventional power supplies. A pulse-width power supply includes the blocks shown in the (**Fig. 11-9**) drawing. These blocks are as follows: (1) an input AC filter/raw DC voltage supply. (2) switched primary current path. (3) pulse-width drive generator. (4) transformer/secondary outputs. (5) a regulation/feedback control.

The 120-volt line voltage is input to the power supply through an input AC filter network to the raw DC power supply. The raw DC power supply consists of a four-diode arrangement, or

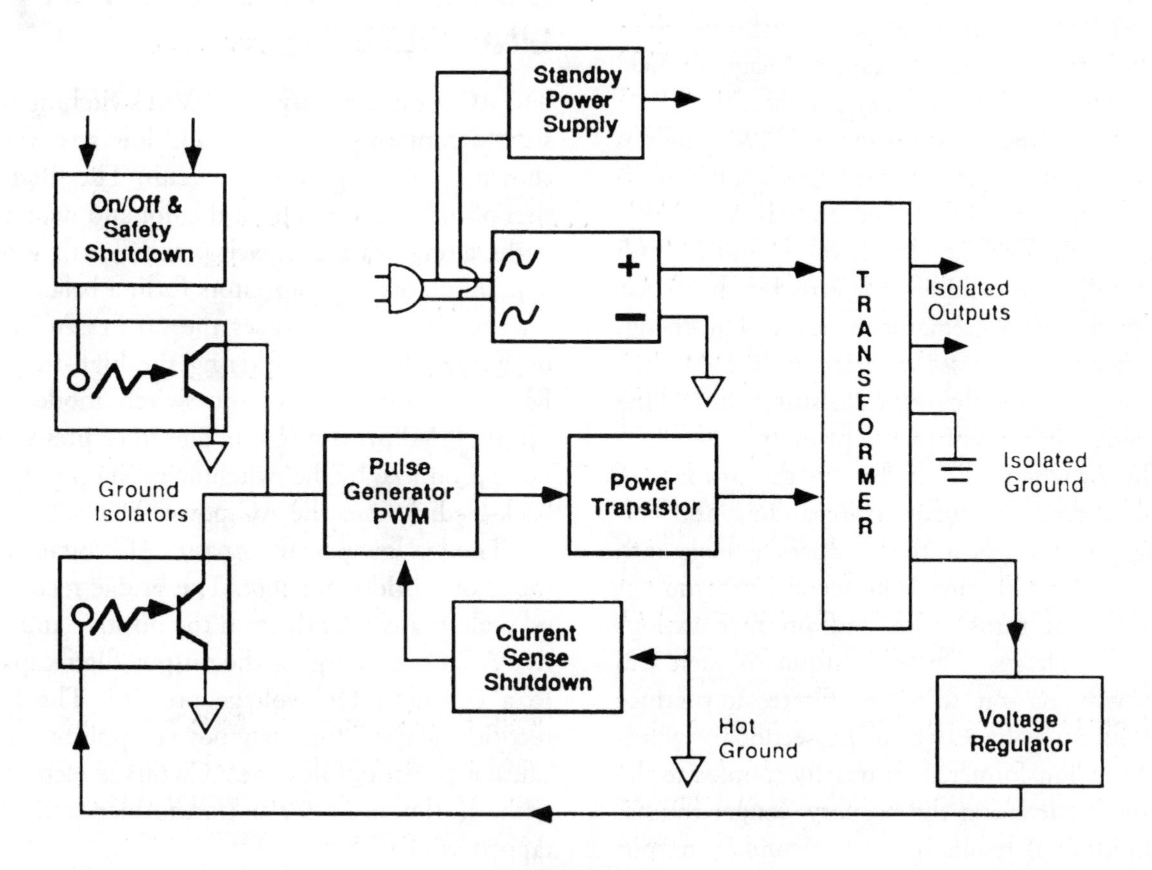

FIGURE 11-8

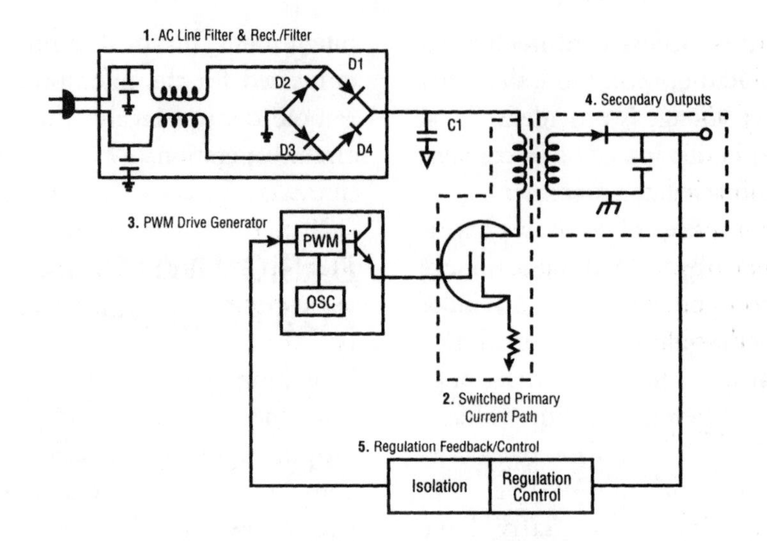

FIGURE 11-9

bridge rectifier, and a filter capacitor. The raw DC power supply provides a DC voltage to the primary windings of the pulse-width modulated switching transformer.

A switching transformer acts as an on/off switch to provide a current path through the source resistor and primary winding of the transformer. The switched on/off current in the transformer produces an expanding and collapsing magnetic field that energizes the transformer primary.

The switching transistor in the PWM system is turned off and on with a drive signal applied to its base or gate. The drive is generated by a pulse-width modulator IC. The drive is "pulse-width modulated," or is continuously varied in duty cycle while the frequency remains the same. The changing duty cycle increases or decreases the conduction time of the switching transistor, controlling the buildup of current in the primary.

The switched current flow in the primary of the transformer produces an expanding and collapsing magnetic field that induces voltage into each secondary winding. The secondary windings have different turns ratios and produce various amplitude voltages. The AC output of each secondary winding is rectified and filtered to produce a DC output voltage. Since the secondary windings of the transformer are mutually coupled to the core, more current in the primary (longer transistor conduction) results in more secondary output voltage. Less primary current (shorter transistor conduction) results in less output voltage. By actively changing the duty cycle and the conduction time of the switching transistor, the PWM IC regulates the secondary output DC voltages.

DC SUPPLY AND AC INPUT LINE FILTER

The AC input circuitry to a PWM switching power supply contains a balanced LC low pass filter as shown in the (**Fig. 11-10**) circuit. The filter consists of inductor(s) in both the hot and neutral AC paths along with a capacitor(s) to earth ground. The inductors and capacitors form a balanced LC low pass filter that passes the 60 Hz AC to the bridge rectifier while filtering the high-frequency RF noise produced by the switch mode power supply (SMPS) to ground. The filter prevents RF noise produced by the switching power supply from back-feeding into the AC power line.

The LC low pass filter passes AC voltage to the input of a bridge rectifier. The bridge rectifier diodes alternately conduct on the positive and negative AC cycle charging the output filter capacitor to a common DC voltage polarity. The bridge rectifier is used for high power applications because it provides full wave AC voltage rectification without the need for a transformer and center tapped winding.

The bridge rectifier is typically connected directly through the inductors to an AC outlet for "off-line" rectification. Off-line rectification eliminates a heavy, expensive, and inefficient power transformer. However, eliminating the transformer results in a "hot" common reference point or ground at the negative terminal of the filter capacitor. A "hot" ground has a direct path to one or both sides of the electrical outlet that allows a personal and equipment safety hazard.

> CAUTION: To ensure your safety and equipment safety, always plug the devices into an isolation transformer when servicing electronic equipment with a "hot" ground. The normal voltages and waveforms in the raw DC power supply can best be understood by considering the operation of the bridge rectifier and output filter capacitor. Refer back to circuit in (Fig. 11-9). Approximately 120 volts of AC is applied to the input of the bridge at the anode and cathode diode junctions.

When troubleshooting these circuits, you need to verify that this input AC voltage is present. To check this input AC voltage, connect your scope probe to the bridge input as shown in (Fig. 11-10). This voltage should measure between 115 and 125 VAC. Now set your scope's controls to view the 60 Hz AC waveform. Little or no AC voltage indicates a faulty AC outlet, line cord, fuse, switch, or input filter component.

On the positive alternation of the AC voltage to the bridge rectifier diodes, Dl and D3 conduct, charging capacitor Cl to the positive voltage peak (120 x 1.414 = 169 volts). During the negative alternation, diodes D2 and D4 conduct, recharging capacitor Cl to the negative peak voltage of 169 volts.

> NOTE: In some supply circuits, voltage doublers, using two capacitors, produce nearly two times the output voltage.

As the filter capacitor (Cl) supplies current to the switching supply, and is recharged by the action of the bridge diodes, a 120 Hz sawtooth waveform is formed across the output capacitor (Cl). The peak-to-peak amplitude of the sawtooth waveform is determined by how much the capacitor discharges. As the load and discharge current increases, the ripple waveform amplitude increases.

> NOTE: This increase results in a decrease of the DC voltage as measured by a DC voltmeter.

To analyze the output of the raw DC supply, connect your scope's test probe across Cl as illustrated in the (Fig. 11-10) drawing. You should have a reading of 150 to 165 volts DC depending on the power supply load.

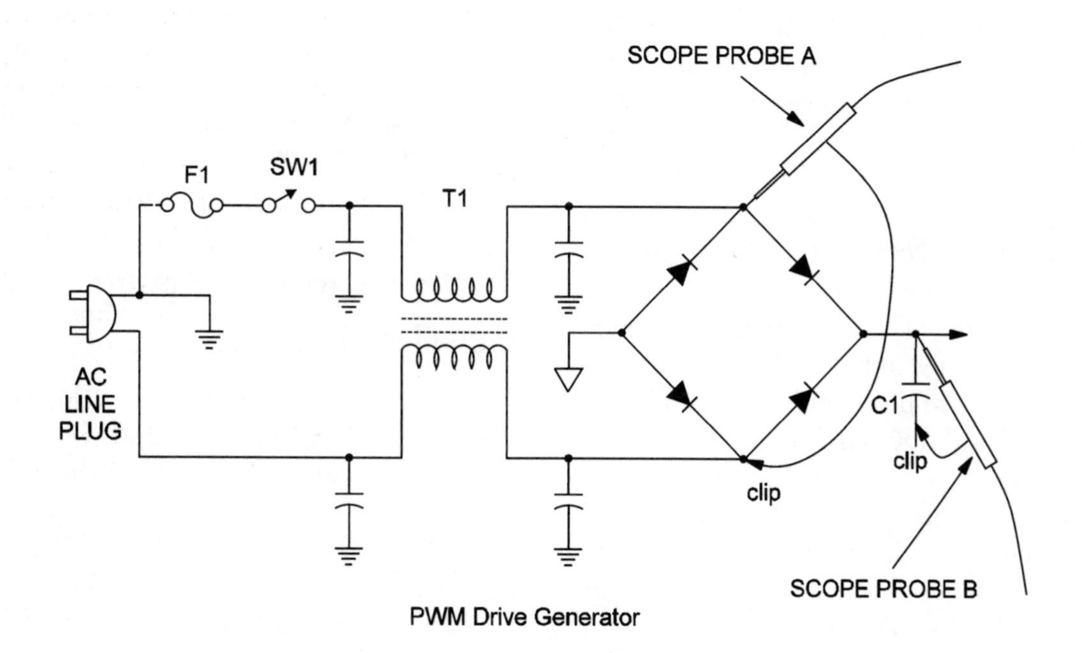

FIGURE 11-10

NOTE: If a voltage doubler is used, the voltage should be twice this value.

Adjust the scope controls to view the 120 Hz sawtooth ripple waveform. Now measure the peak-to-peak amplitude of the ripple. Typical ripple peak-to-peak voltages should measure zero to 10 volts P-P depending on the current demand to the switching supply and its secondary loads. A DC voltage reading nearing 169 volts, with little or no ripple, indicates a small current flow to the switching supply. A doubler circuit would read about 340 volts.

SWITCHING TRANSFORMER OPERATION

The raw DC power circuit supplies voltage to the primary of the switching transformer. The primary of the transistor consists of the winding between the raw DC voltage and drain/collector of the switching transistor. The switching transistor acts as an on/off switch providing a path for primary current flow. When the transistor is biased on, current flows through the source resistor of the transistor, the source and drain terminals and primary winding. Refer again to the **(Fig. 11-9)** circuit. When the transistor is biased off, the current path is opened and no current flows through the switching transistor or transformer primary. The switching transistor in a PWM switching supply is commonly a power FET transistor because of its low power gate drive requirements. However, bipolar transistors may also be used in this stage.

The switching transistor opens and closes at a fast rate ranging from 20,000 to 200,000 times per second (20KHz 200KHz). This is a result of the drive signal applied to the gate or base of the switching transistor. The rapid switching action causes an increasing and decreasing current to flow in the inductive primary of the transformer. The "chopped up" DC current in the primary acts just like an AC current in the primary winding of the transformer, producing an expanding and collapsing magnetic field.

For normal system operation, the DC voltage from the raw DC supply must be present at the drain of the switching transistor. To measure this voltage, connect a scope probe to the drain of the switching transistor and the ground lead to the "hot" ground (negative terminal of Cl). Now measure the DC voltage.

You should now observe a DC voltage that is close to the raw DC voltage. This typically ranges from 150 to 165 volts, or twice this amount for a doubler circuit. If the voltage is very low, or missing, but exists on the output of the raw DC supply, the switching transformer primary, or components in series with the primary (such as fuses or fusible resistors), could be open. A voltage reading much lower than 160 volts DC may indicate a leaky or shorted switching transistor.

Proper switching action and switched current flow in the transformer primary produce a waveform at the drain of the switching transistor. The waveform nears that of a square wave as the switching transistor is constantly switching on and off. The amplitude of the waveform is typically at least two times larger — or more — than the raw DC voltage due to the nature of the inductive voltage induced back into the primary. Amplitudes of the drain waveform typically range from 200 to 400 volts peak-to-peak for switching supplies.

You can confirm proper switching action and primary current by analyzing the waveform at the drain of the switching transistor with a scope.

NOTE: Make sure your scope input can withstand up to 1000 volts peak-to-peak. Connect the scope to the drain and "hot" ground and adjust the scope's controls to view the drain waveform. An example of the drain waveform in a switching supply is shown in **(Fig. 11-11)**.

TYPICAL (PWM) DRIVE CONTROLLER

A drive signal to the gate of the switching transistor switches the transistor on and off to energize the transformer primary. In a PWM switching supply, the gate drive is generated with an integrated circuit, called a "current mode controller IC" or a pulse-width modulator IC. The PWM IC has an oscillator, duty cycle control, and amplifier to develop the gate drive waveform.

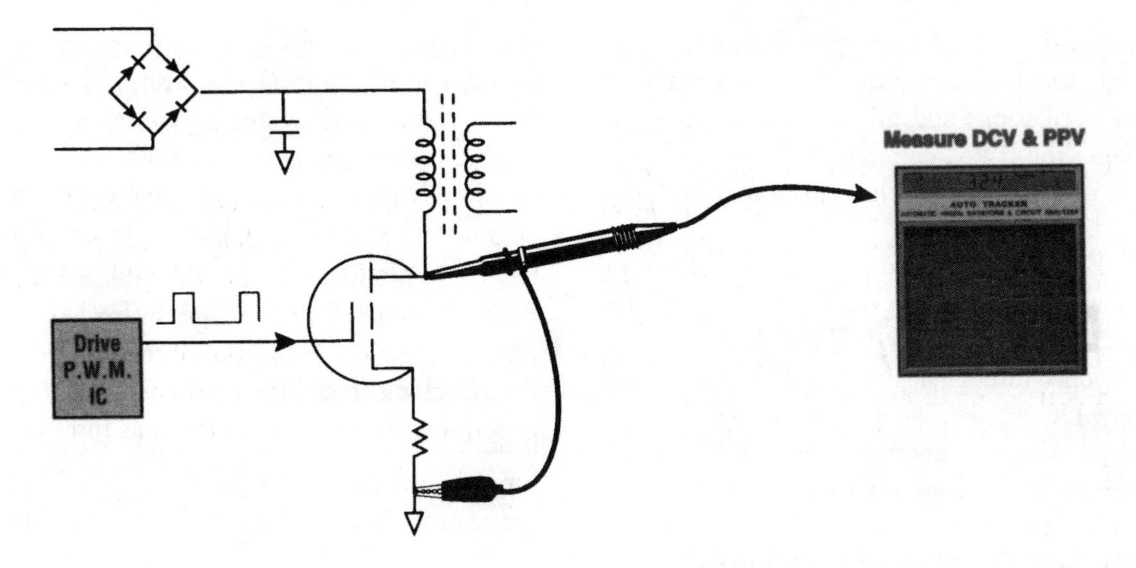

FIGURE 11-11

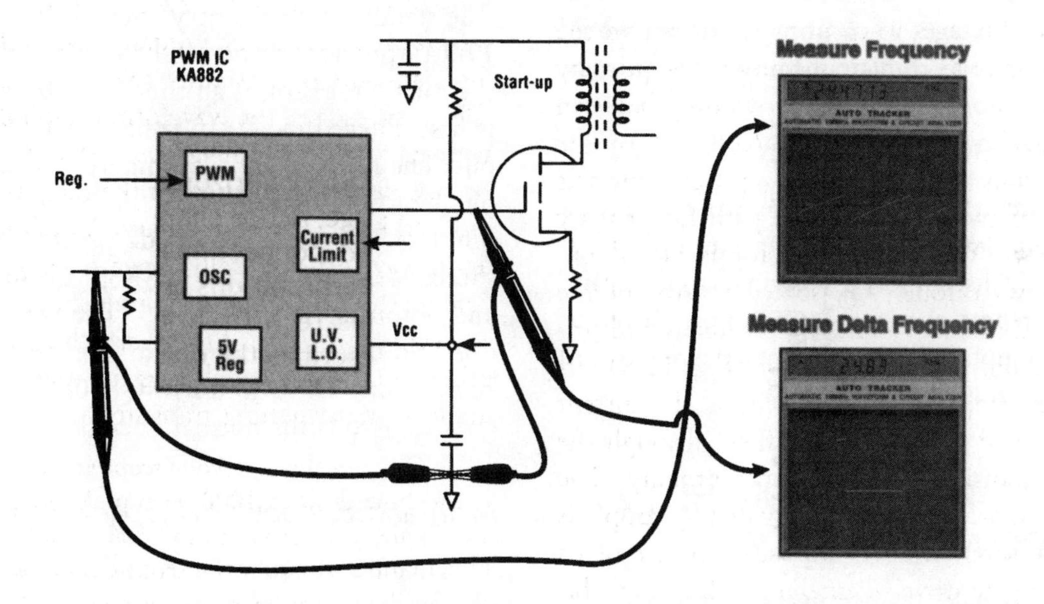

FIGURE 11-12

The IC's oscillator block diagram, along with an external capacitor and resistor, generate the drive signal as shown in (**Fig. 11-12**). Oscillations occur as the external capacitor charges through the resistor to the 5-volt supply and is alternately discharged by the oscillator circuitry inside the PWM IC. Output from the oscillator is fed to the IC's PWM control stages to establish the shape and duty cycle of the output drive waveform. This drive signal is amplified and buffered to the output of the IC, and then fed to the gate of the switching transistor.

The IC's PWM control stages include comparators, logic gates, and RS flip-flops to shape the oscillator signal into a variable pulse-width drive signal output. The drive waveform becomes a rectangular waveform or square waveform, constantly changing in pulse-width while the frequency remains the same. Modulating (varying) the pulse-width as required regulates the DC output voltage.

The voltage needed to power the PWM is commonly derived from the secondary of the switching transformer and is called the "run" voltage. To produce run voltage, the switching transistor must be driven (switched on) to energize the transformer. However, the PWM requires the run voltage to produce oscillations and drive. Thus, a technique for starting the PWM chip, or producing momentary gate drive to energize the transformer primary, is needed.

The PWM IC chip requires a starting voltage, or momentary DC voltage, to the PWM's "VCC" input. The start voltage to the PWM's "VCC" is typically developed through high-value resistors connected to the raw DC supply. The resistors provide charging current to charge a capacitor. Once the capacitor charge exceeds a voltage threshold internal to the PWM IC, referred to as the under-voltage lockout (UVLO), the PWM IC circuits are enabled.

This produces drive to start the supply. UVLO prevents the PWM IC from producing drive at reduced AC voltages when primary current would be too excessive to regulate the secondary outputs.

Switching Supply Problem

Should a circuit defect prevent the switching power supply from starting up, the capacitor, which is now partially discharged, will begin charging again. This will repeat the start-up attempt. Repeated start-up attempts produce momentary oscillator and drive waveforms.

TOSHIBA PROJECTION HDTV MAIN POWER SUPPLY

Let's now analyze the TOSHIBA HDTV-ready projection receiver model TP61G90's main power supply by following the overall block diagram.

Overall Block Diagram

In **Fig. 11-13**, you will find the overall block diagram for the standby, main, and sub power supplies used in the Toshiba progressive scan TV receivers. The standby supply is always active whenever the TV set is plugged into an AC outlet. It produces 5 volts for VDD and 5 volts to reset the microprocessor, keeping it operational at all times -- even when the set is turned off. Transformer T840 isolates the standby supply from the "hot" ground. D840 is a full-wave bridge rectifier that supplies 15 volts DC to the voltage regulator Q840, and relays SR81 and SR83. When the microprocessor receives an on command from the remote control, or power/on button on front TV panel, voltage is sent to the relay drivers to close relays SR81 and SR83. Closing the relays supplies the AC line input to the remaining two power supplies to operate the TV receiver.

Troubleshooting Tip

If both relays never close, check the standby power supply. Both the 5-volt VDD and the 5-volt reset are required for the microprocessor to operate.

Standby Power Supply Circuit

The standby power supply circuit, shown in (**Fig. 11-14**), provides 5-volt VDD and a 5-volt reset to the microprocessor at all times. When the TV set is first plugged in, D840 rectifies the line AC input. The rectified voltage is filtered and applied to pin 1 of Q840. The normal operating voltage on pin 1 is 15 Vdc. The 15Vdc provides a current source for relays SR81 and SR83. Q840 regulates and outputs 5 volts on pin 5. The plug-in of pin 4 is held low for a moment to reset the microprocessor. Once C843 (pin 2 of Q840) charges, pin 4 goes to 5 volts for normal operation.

Main and Subpower Supply Notes

The main and subpower supplies work independently from each other, so one supply can be disabled to check the other one. If the main power supply is disabled, the TV set would not have picture or sound, but the microprocessor would still control the relays. Hence, the sub supply could be turned on and off and its voltages would appear normal. If the sub power supply is disabled, everything would work except that the picture would be out of convergence. Either power supply can easily be disabled by removing its fuse — F860 for the sub supply and F811 for the main power supply.

Main Power Supply Operation

The main power supply is a ringing choke converter and is illustrated in the (**Fig. 11-15**) block

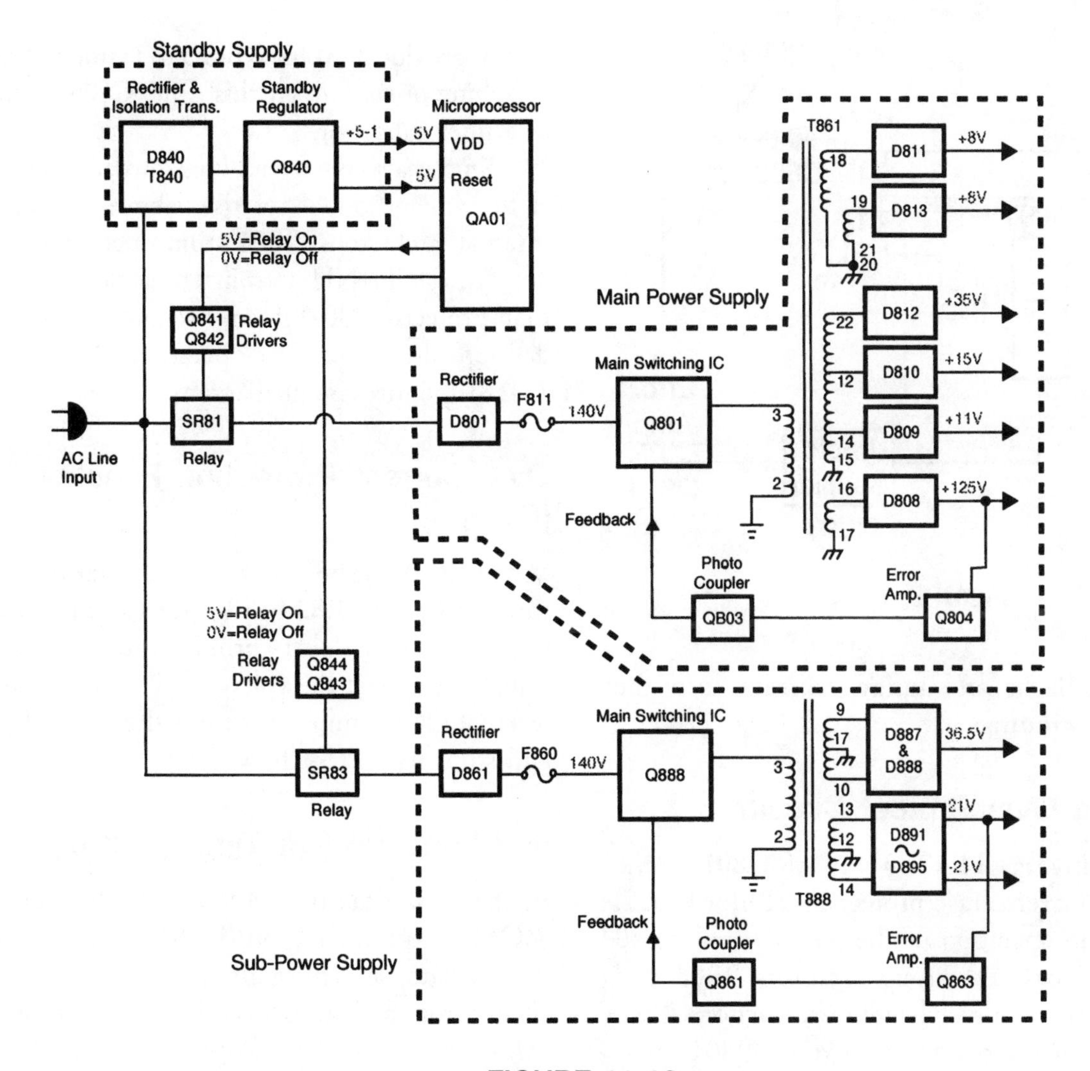

FIGURE 11-13

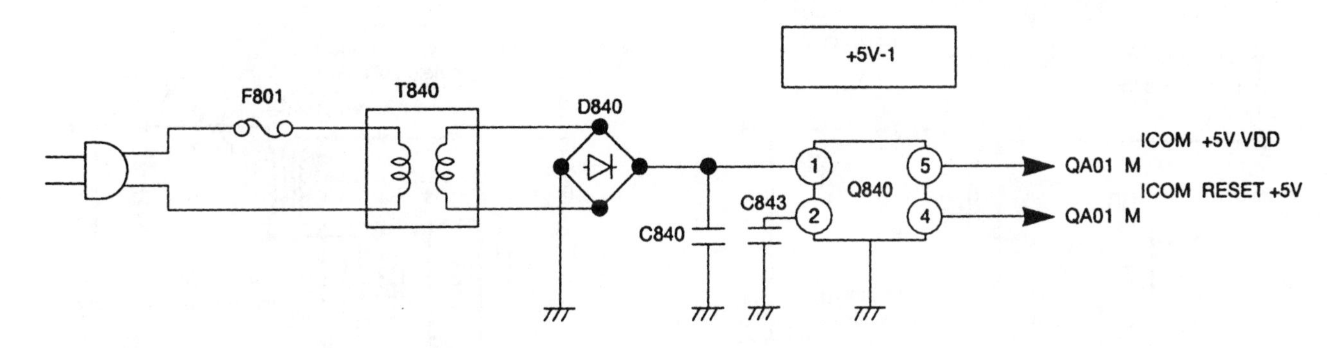

FIGURE 11-14

diagram. An oscillator (OCS) and a switching MOSFET are internal to the switching IC, Q801. During normal operation, D801 rectifies the AC line input to approximately 140 volts. This voltage is applied to the MOSFET through transformer T862's primary winding. As shown in (**Fig. 11-16a**), when the MOSFET conducts, current flows through T862's primary windings and builds an electromagnetic field. Figure (11-16b) demonstrates that, after the field builds, the MOSFET opens to stop the current flow through T862. This causes the electromagnetic

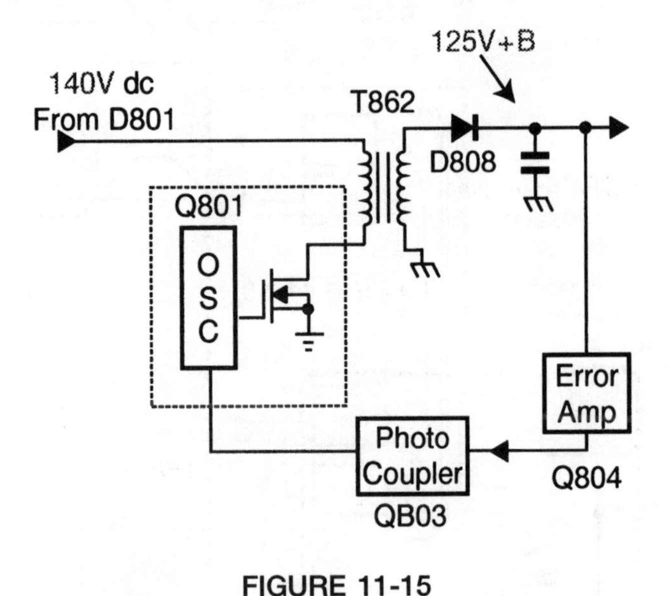

FIGURE 11-15

field to collapse and induce current into the secondary windings.

Start and Over Protect Circuits

As noted in **Fig. 11-17**, V IN of Q801 is the startup and overvoltage protect (OVP) for Q801. To start the operation of the power supply, 140 volts draws current through resistors R802 and R803, which builds a charge on capacitor C825. When the charge reaches 22.5 volts, Q801 operates normally. The ND winding, and rectifier diode D805, provide 25 volts to pin 5 to maintain Q801's switching operation. At this time, C825 functions as a filter capacitor.

Pin 5 is also the overvoltage protection (OVP). If the voltage on V IN rises above 28 volts, an internal latch stops Q801 from operating.

The OCP/INH terminal is a safety terminal that protects Q801 if an overcurrent condition develops during receiver operation. Refer to (**Fig. 11-18**) for this circuit diagram.

Over-Current Protection Function (OCP)

When the MOSFET is turned on, current flows through resistors R827 and R828 causing a voltage to develop at the over-current protect (OCP) terminal at pin 4. If the voltage reaches 0.5 volts, the internal OCP comparator turns the MOSFET off, and stops its current flow.

INH Function (Off Time Control)

At the same time the OCP comparator turns the MOSFET off, the inhibit (IHN) comparator stops the oscillator and prevents the MOSFET and oscillator from operating until the OCP voltage drops below 0.5 volts. The scope waveforms for this operation are shown in **Fig. 11-19**.

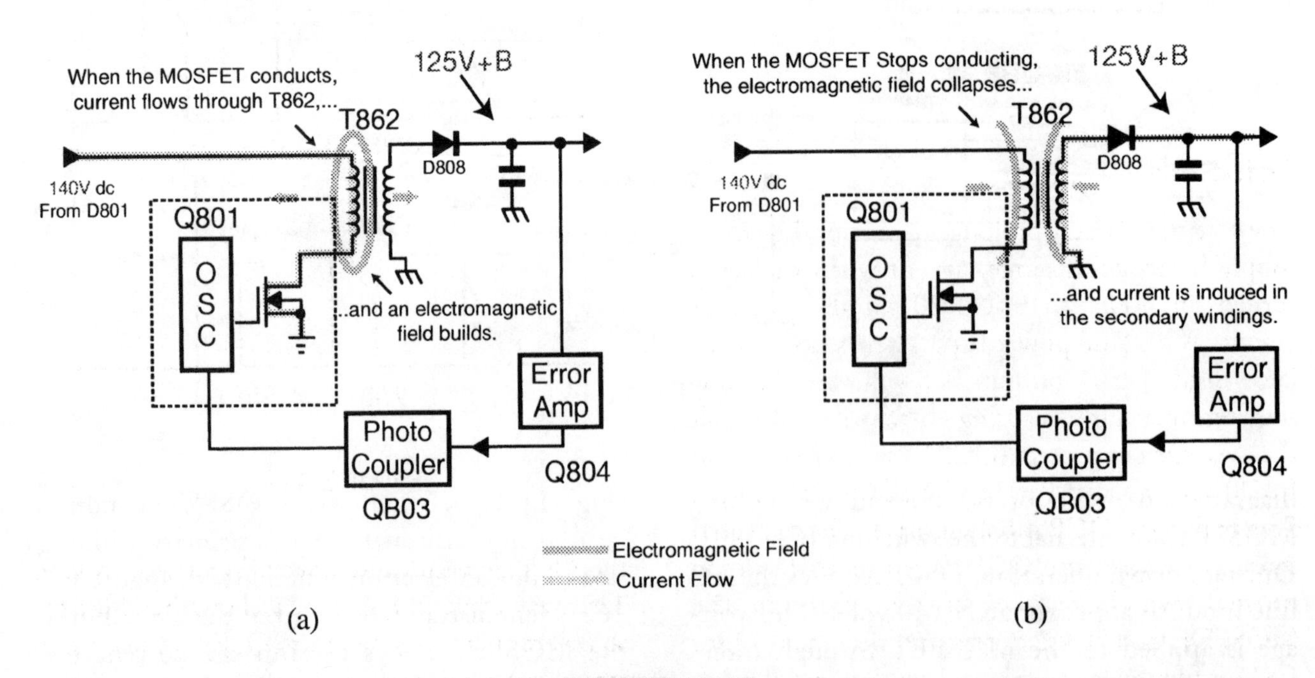

FIGURE 11-16

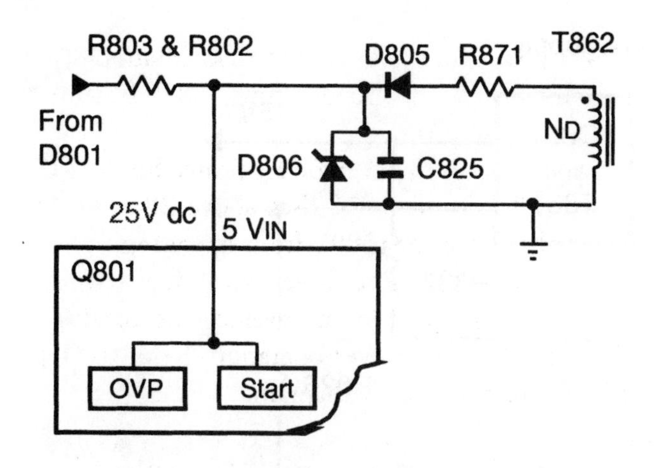

FIGURE 11-17

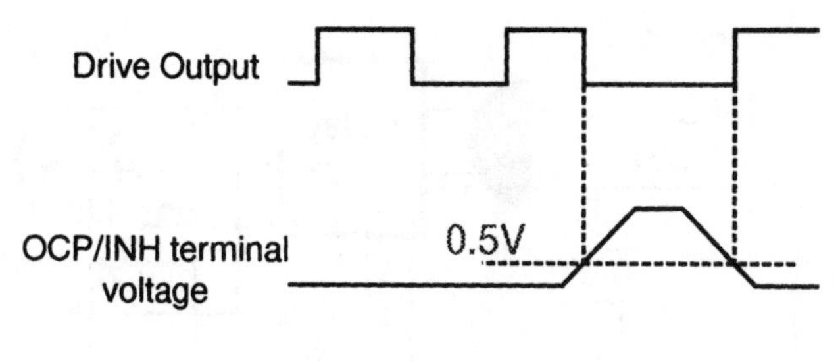

FIGURE 11-19

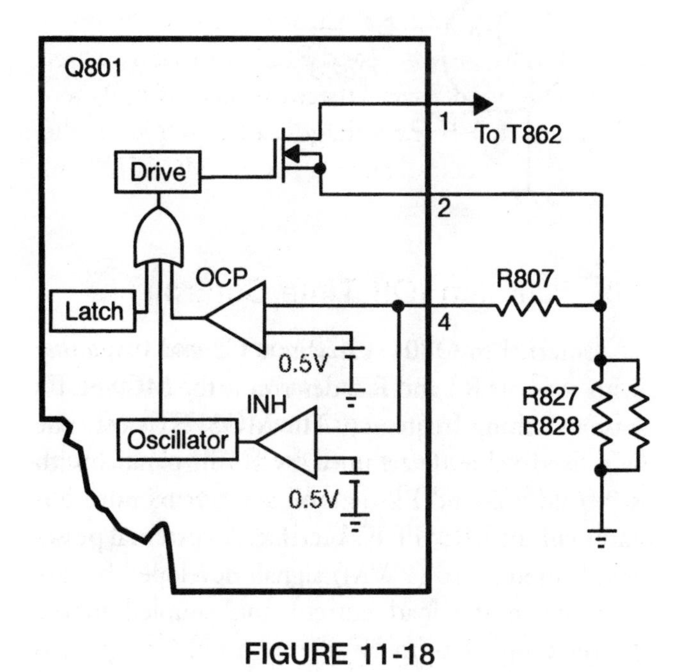

FIGURE 11-18

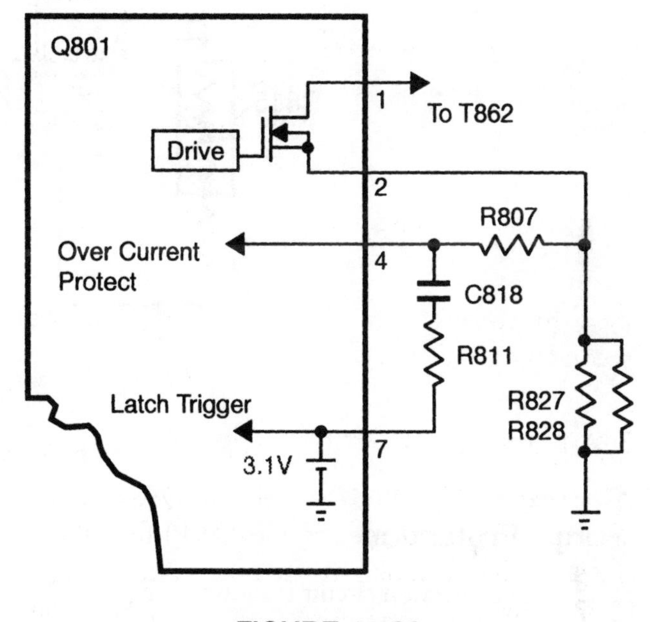

FIGURE 11-20

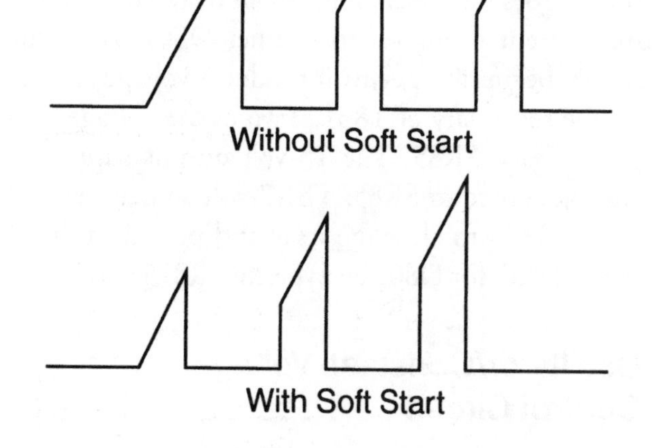

FIGURE 11-21

Soft-Start Circuitry

The soft-start circuit prolongs the life of the power supply by reducing surge current when the set is turned on. Refer to Fig.11-20 for the soft-start circuit. When the power supply starts up, a regulator inside Q801 outputs 3.1 volts on the soft-start terminal pin 7 causing current flow through the over-current protect (OCP) resistors (R827 and R828) and resistor R811. The additional current flow makes the OCP more sensitive. The increased sensitivity causes the OCP to trigger earlier than normal, which reduces current through the MOSFET and T862's primary windings. Once C818 is fully charged, current stops flowing through the OCP resistors and normal operation begins. The waveforms in **Fig. 11-21** show the effect the soft-start circuit has on the MOSFET's drain current during start-up.

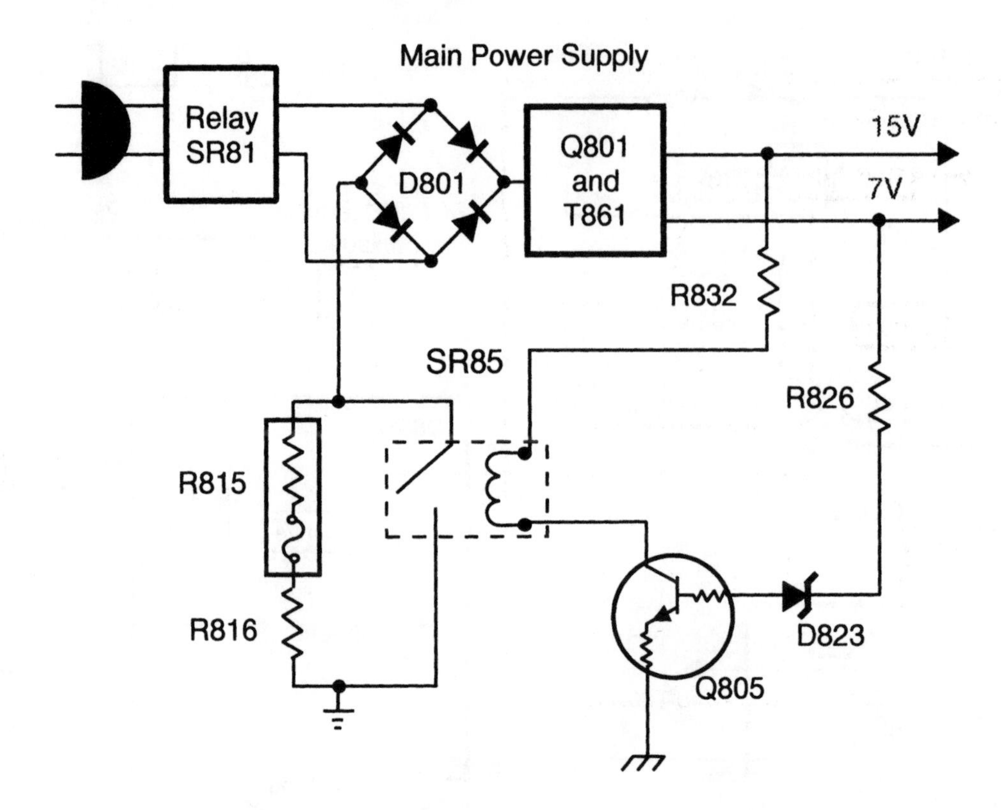

FIGURE 11-22

Surge Protection

The surge protection circuit is shown in **Fig. 11-22.** R815 and R816 reduce surge current through the main power supply at startup. They provide additional resistance in the ground path of rectifier D801. Relays SR85 and SR81 are disengaged before the supply goes into operation. When relay SR81 closes, AC current is applied to rectifier D801. After the supply begins to operate, it produces voltage sources on the secondary of T861. Two of the voltage outputs engage SR85. The 15-volt output supplies a current source to SR85. The 7-volt source forward biases Q805 to close the relay and provide a direct ground path for D801 by bypassing R815 and R816.

Oscillator/Constant Voltage Control Circuit

The voltage control circuit diagram is shown in **Fig. 11-23.** Internal to Q801 is an oscillator and oscillator control circuit. The oscillator controls the switching MOSFET in Q801. To control the power supply's regulation, a feedback signal adjusts the oscillator frequency to maintain a consistent current flow through the load.

Internal to Q801, capacitors C2 and C3, along with resistors R3 and R2, determine the MOSFET's base switching frequency. The MOSFET's off time is a fixed value determined by R3 in parallel with C3. And, C2 and R2 determine the maximum duration of the MOSFET's on time. An external pulse-width modulated (PWM) signal, developed by any variance in the load current and coupled to the power supply's primary side by a photocoupler, is applied to the feedback (F/B) of Q801, pin 6 to adjust the charging time of C2 as required by the load. If the load current decreases, the MOSFET's on time increases to compensate and increase the output of the power supply. The longer the on time, the larger the electromagnetic field that builds around T862's primary windings. The larger electromagnetic field induces more current into the secondary windings when it collapses. On the other hand, if the load current increases, the on time decreases to reduce the overall output of the power supply.

Latch Block Function

A latch internal to Q801 (see **Fig. 11-24**) stops Q801's operation to protect the IC from damage if a fault occurs. Three conditions trigger the latch

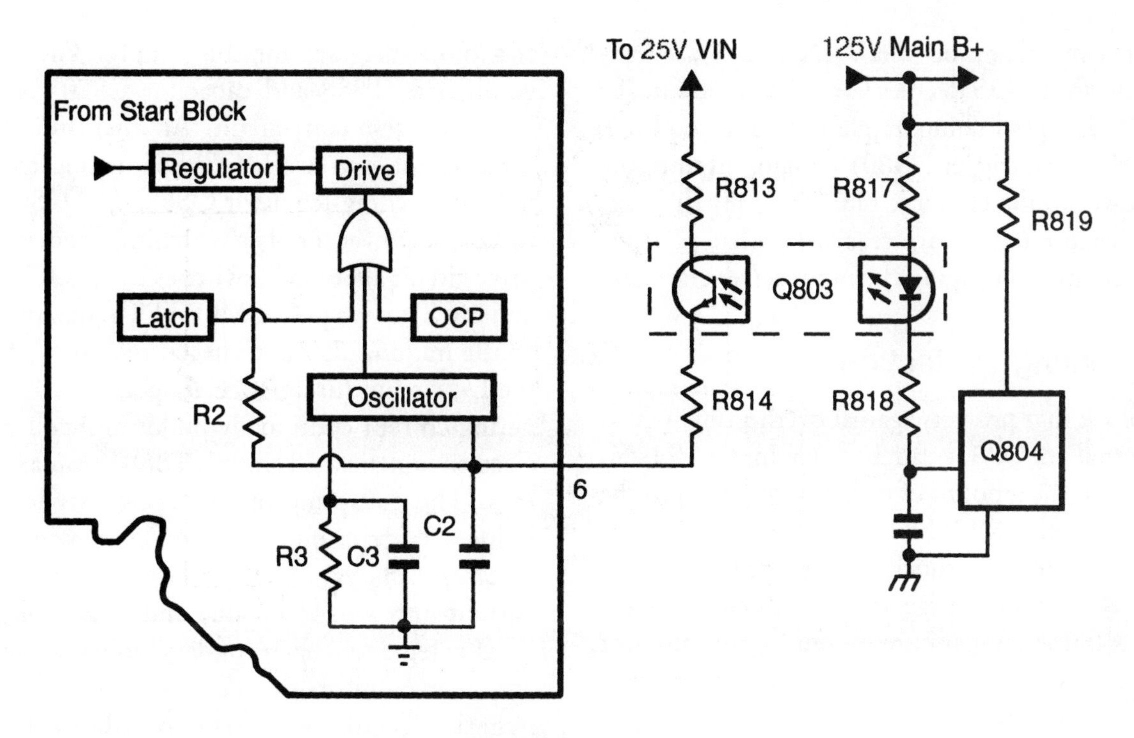

FIGURE 11-23

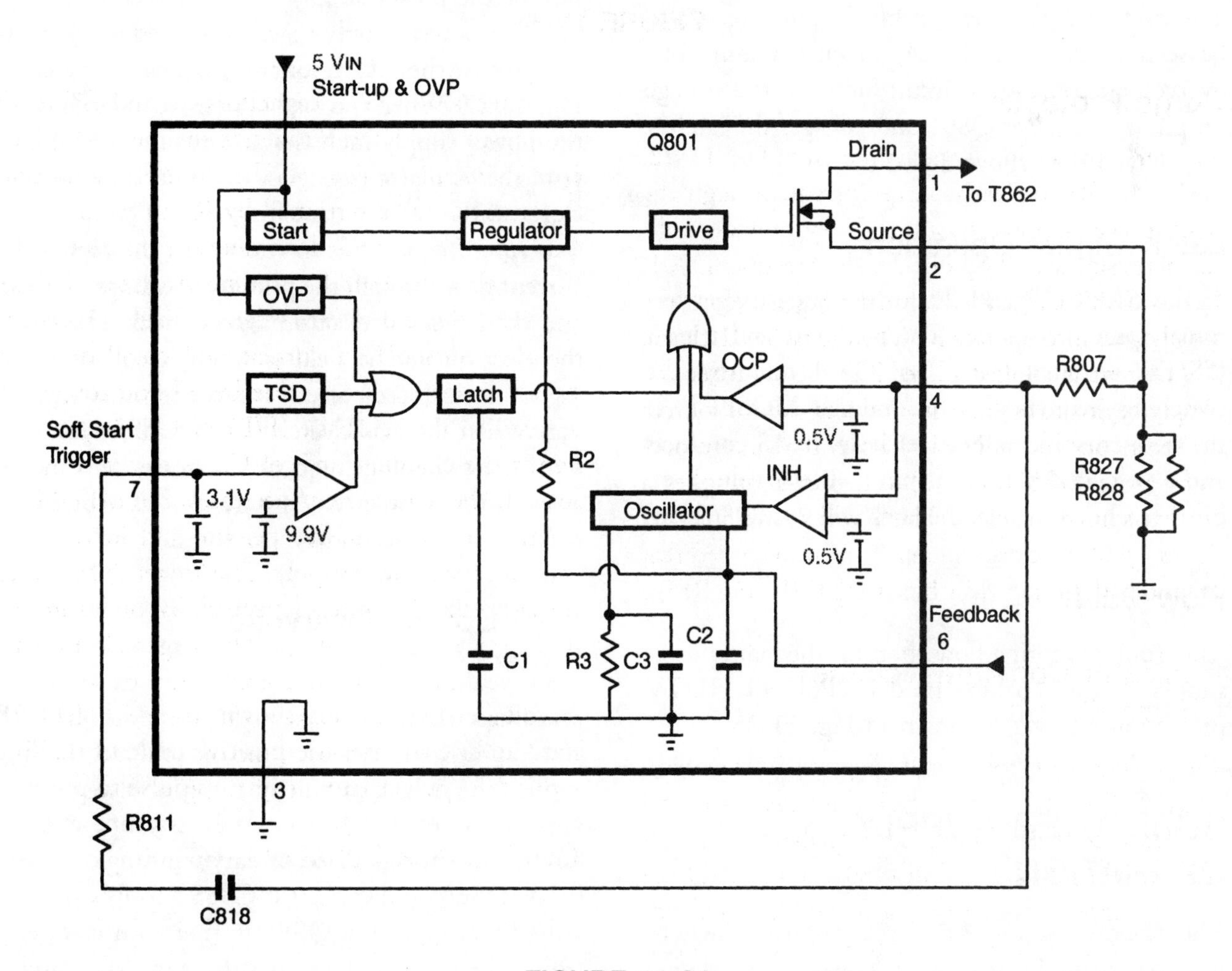

FIGURE 11-24

circuit: overvoltage on the V IN (pin 5), a temperature above 125 degrees C on Q801's frame, or a MOSFET's switching frequency that is too high. Once the latch triggers, Q801 remains off until the AC power to the circuit is removed. Q801's internal capacitor (Cl) provides a delay that prevents the latch from engaging during receiver start-up.

Overheating Protection

The block that protects against overheating, which is internal to Q801, can be seen in **Fig. 11-24.** Because of the amount of current flow through the MOSFET, the MOSFET generates considerable heat. The thermal shock detect (TSD), located internally to Q801, triggers the latch circuit when Q801's frame temperature exceeds a minimum of 125 degrees C.

Overvoltage Protection

The overvoltage protection block (pin 5) can also be seen in the (**Fig. 11-24**) block diagram. The overvoltage protection circuit monitors the voltage on pin 5 (V IN) of Q801 and engages the latch if the voltage rises above 28 volts.

Latch Trigger Operation

Again, refer to **Fig. 11-24** for the latch trigger terminal (pin 7) location. The soft start and trigger (SS/Tri) terminal (pin 7) of Q801 monitors the switching frequency of the internal MOSFET. If the frequency increases excessively, C818 conducts and a voltage develops on pin 7. If the voltage on pin 7 reaches 9.9 volts, the latch will shut Q801 off.

Flow Chart

The troubleshooting flow chart for the main power supply in the Toshiba model TP6lG90 HDTV projection receiver is shown in **Fig. 11-25.**

SUBPOWER SUPPLY OPERATION

The subpower supply is a current resonance switching power supply. It supplies power to the digital convergence and convergence boards. **Figure 11-26** is a block diagram for this supply. The primary winding of T888 and capacitor C870 creates an LC series resonant circuit. An oscillator (OSC), drive circuit, and two MOSFETs are located internal to switching regulator Q888. The OSC determines the power supply's switching frequency. The drive circuit alternately switches the MOSFETs on and off. The two power MOSFETS, in a push-pull configuration, alternate the current flow through the LC circuit during normal operation. The alternating current continually builds and collapses an electromagnetic field around T888's primary windings. The collapsing of the electromagnetic field induces current into the secondary windings of T888. A full-wave rectifier converts the induced current into a +21 volt line and a -21 volt line.

To regulate the secondary voltages, an error amplifier monitors the +21 volt line and supplies negative feedback to the oscillator through photocoupler Q861. Q861 isolates the primary side of the power supply from the secondary.

The power supply's switching frequency operates above the LC resonant frequency. Refer to **Fig. 11-27.** When the load on the secondary side of the power supply increases and requires more current, the oscillator frequency decreases and operates closer to the LC resonant frequency. The closer the switching frequency is to resonance, the higher the current flow through the primary windings of T888 and the larger the electromagnetic field. The larger the electromagnetic field is when it collapses, the higher the induced current is in the secondary winding. When the load decreases and requires less current, the switching frequency increases and moves away from resonance. As a result, less current is induced in the secondary windings.

Start-Up and Overvoltage Protect Circuit

A voltage divider, not shown in **the Fig.11-28** start-up circuit, uses the positive cycle of the line input to supply a 16-volt startup pulse to pin 9 of Q888 via resistor R861. After startup, a drive circuit consisting of a secondary winding of T888, diode D864, and capacitor C868 supply 16 to 20 volts DC to pin 9 of Q888 to maintain its operation. The voltage developed by the drive circuit fluctuates with the switching frequency of the

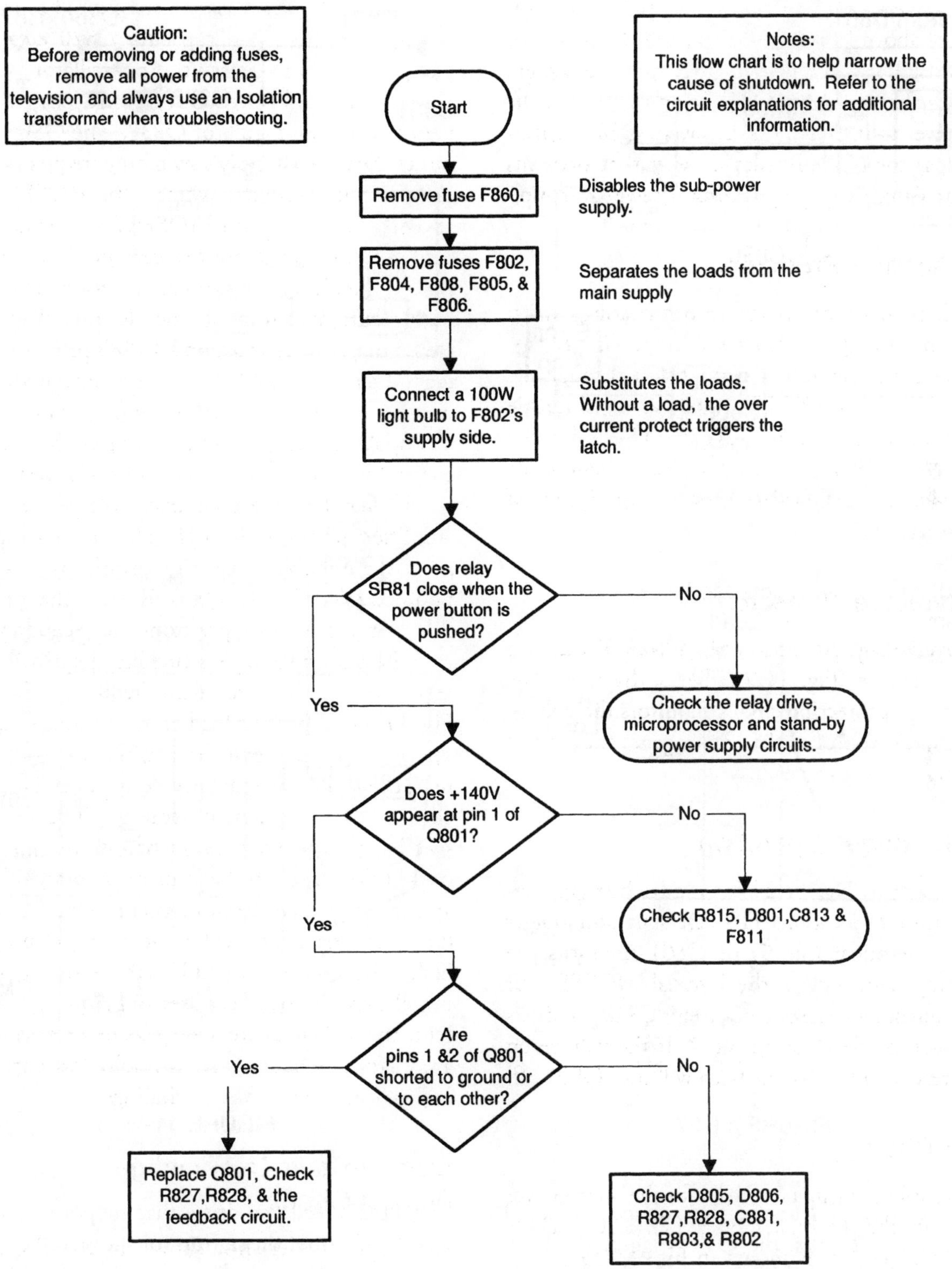

Caution:
Before removing or adding fuses, remove all power from the television and always use an Isolation transformer when troubleshooting.

Notes:
This flow chart is to help narrow the cause of shutdown. Refer to the circuit explanations for additional information.

Start

Remove fuse F860. — Disables the sub-power supply.

Remove fuses F802, F804, F808, F805, & F806. — Separates the loads from the main supply

Connect a 100W light bulb to F802's supply side. — Substitutes the loads. Without a load, the over current protect triggers the latch.

Does relay SR81 close when the power button is pushed?
- No → Check the relay drive, microprocessor and stand-by power supply circuits.
- Yes ↓

Does +140V appear at pin 1 of Q801?
- No → Check R815, D801,C813 & F811
- Yes ↓

Are pins 1 &2 of Q801 shorted to ground or to each other?
- Yes → Replace Q801, Check R827,R828, & the feedback circuit.
- No → Check D805, D806, R827,R828, C881, R803,& R802

FIGURE 11-25

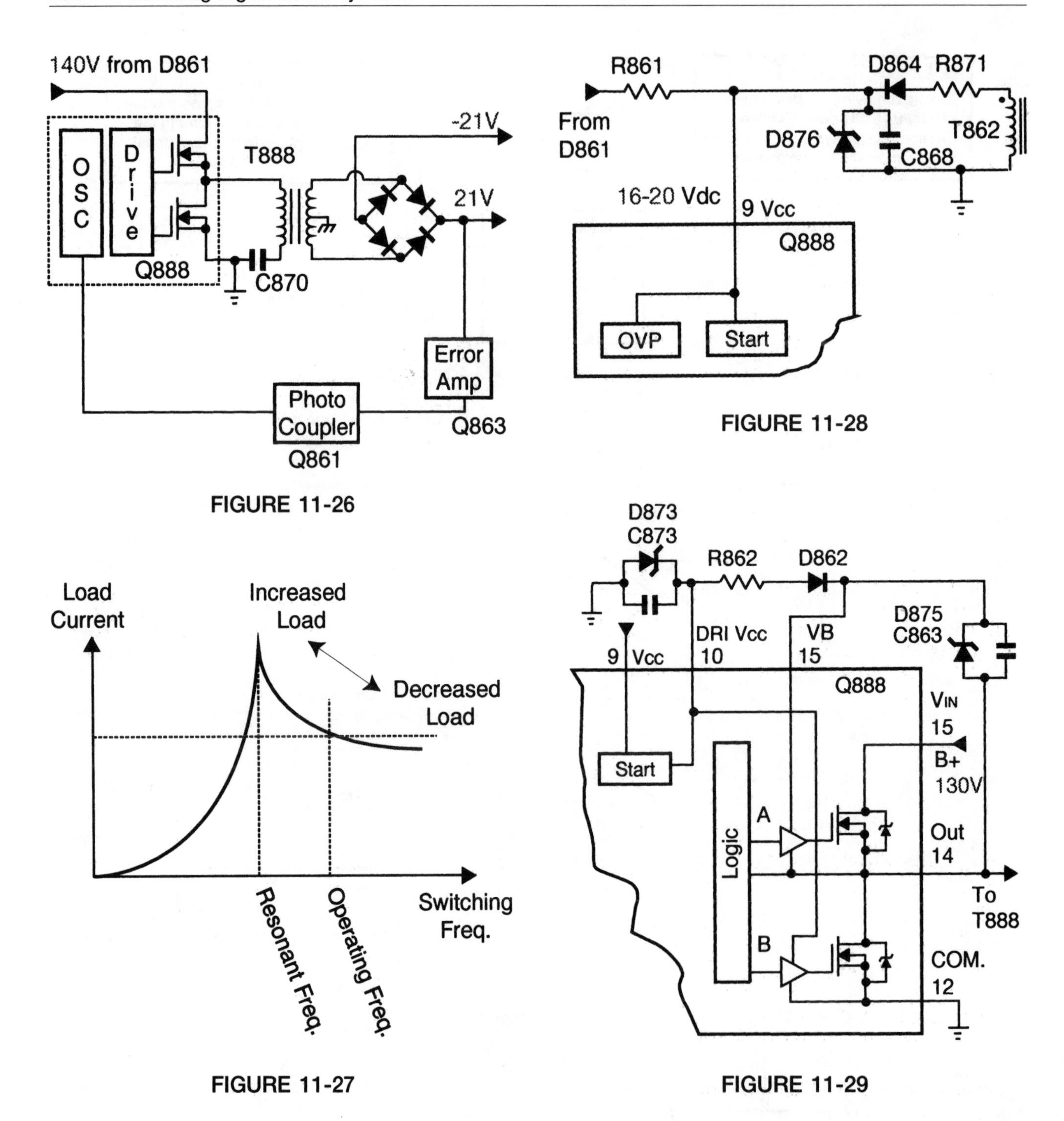

FIGURE 11-26

FIGURE 11-28

FIGURE 11-27

FIGURE 11-29

power supply. Therefore, the voltage on pin 9 increases to 22 volts, the OVP triggers the latch section and the switching stops. Diode D876 is a 27-volt zener diode that protects Q888 by preventing excess voltage on pin 9.

Logic and Driver Circuits

The logic block controls the MOSFETs' switching frequency. The outputs of the logic block, refer to Fig. 11-29, feed two drives that are powered by the start block. After the startup voltage is applied to pin 9, the start block supplies a drive Vcc (DRI Vcc) of approximately 8 volts to pin 10. Delaying the driver supplies at startup prevents damaging the MOSFETs. The 8 volts on pin 10 powers driver B internally. To power driver A, resistor R862, and diode D862 add the voltage from pin 10 to the voltage on pin 15. D875, C863, D873, and C873 are voltage regulators and filters for these supplies.

OSCILLATOR OPERATION

Q888's internal oscillator, shown in (**Fig. 11-30**), develops the power supply's switching frequency by generating a ramp waveform at capacitor terminal (CT) at pin 4. Capacitor C862, connected to pin 4, determines the lowest oscillation frequency. Both MOSFETs are off for a short time when they are alternately switching. This off time is called dead time and is determined by resistor R867 on the dead time (DT) terminal pin 3. Zener diode D872 is used as a clamp.

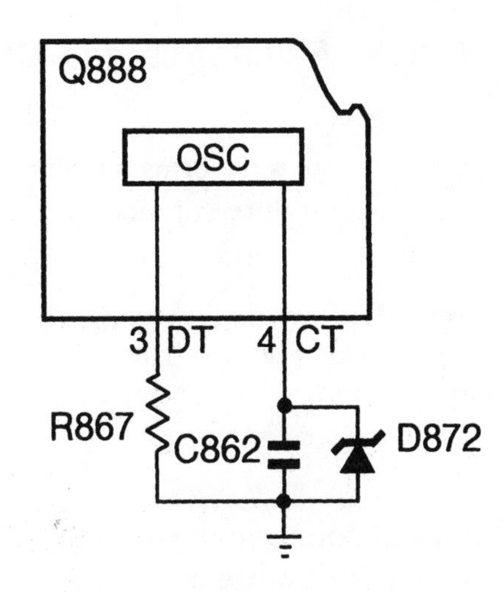

FIGURE 11-30

Oscillator Control

If the load current from the 21 volt dc line increases, the 21 volt dc voltage begins to drop, decreasing the current through Q861's LED side. Refer to **Fig. 11-31**. The current drop causes the LED to couple less light to the photo transistor side and reduces the current flow into pin 5 of Q888. This reduction in current flow varies the OSC frequency, moving it closer to resonance. This increase the supply of current to maintain the 21 volt DC level. Conversely, if the load current decreases, the 21 volt DC rises and increases the light through Q861 and the current into pin 5 of Q888. The increased current causes the OSC operating frequency to move away from resonance to decrease the current supplied to the load and adjust to the 21 volts DC that's required.

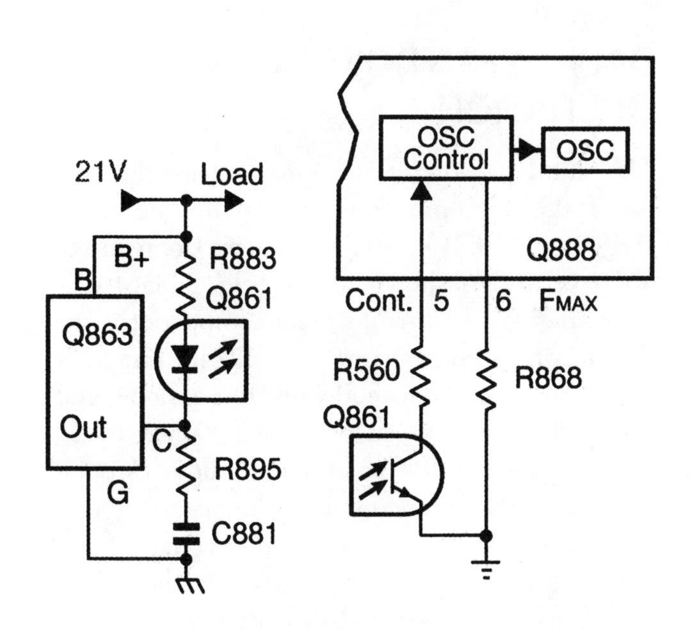

FIGURE 11-31

Latch Block Operation

Refer to the latch block diagram in (**Fig. 11-32**). The latch block stops the operation of Q888 until the voltage on pin 1 of Q888 is removed by turning the television set off. Any of the following detection blocks can trigger the latch operation:

- overvoltage protection (OVP) block
- thermal shock detection (TSD) block
- overcurrent protection (OCP) block

The charging time of capacitor C869, connected to the capacitor delay (CD) terminal pin 8, delays the operation of the latch circuit during start-up.

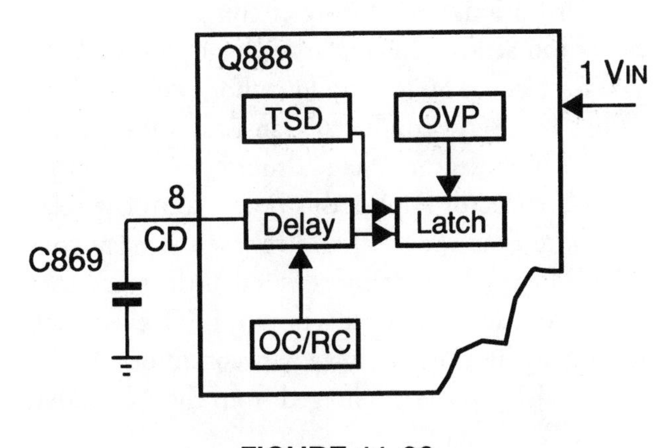

FIGURE 11-32

Power Supply Waveforms and Other Information

The information and waveforms shown in (**Fig. 11-33**) can be used for different models of Toshiba HDTV projection TV receivers.

> CAUTION: Different input signals may cause a variance in voltage readings. The voltages and waveforms shown in (Fig. 11-33) were recorded while displaying a color bar signal on the TV receiver.

The troubleshooting flowcharts shown in (**Figures 11-34 and 11-35** may also be used for several different models of Toshiba HDTV projection receivers.

BASIC SHUTDOWN CIRCUIT OPERATION

The shutdown circuit is a safety device that bypasses the microprocessor to turn off the relays (SR81 and SR83) if certain problems occur in the TV receiver. As shown in the (**Fig. 11-36**) diagram, the protect circuit's main component is the silicon controlled rectifier (SCR), D846. The SCR has an anode and a cathode, just like a diode, and a gate that acts like an on switch. When 0.825 volts appears on D846's gate, current flows through its anode/cathode junction in the same direction as a standard diode. Removing the voltage from D846's gate does not stop the anode/cathode current Once the anode and cathode conduct, they continue to conduct even after the gate voltage is removed. Removing the current flow between the anode/cathode resets the SCR.

Ten various monitoring circuits in the TV receiver can send the necessary voltage to the SCR's gate to start the SCR's anode/cathode current flow. When the SCR conducts, transistor Q845 turns on and its collector voltage drops close to ground. Q845's collector applies this potential to the relay drivers to turn them off and release the relay contacts. The microprocessor senses that the relay drivers are off and blinks the power LED every half second to indicate a shutdown condition. When the television set is unplugged from the AC outlet, the SCR is reset.

Shutdown Circuit Troubleshooting Tip

Because of the speed of the shutdown circuit, technicians may have difficulty getting proper voltage readings when this circuit activates. A peak response or min/max meter is necessary for troubleshooting a shutdown problem. These meters can read a voltage in a split second and store the reading into memory for easy recovery. If a peak response meter is not available, try using an oscilloscope on the DC setting. The scope will react quicker then the digital voltmeter, and the change in DC level can be seen on the scope's CRT. However, most scopes do not have a DC voltage readout or the ability to record the value. This makes obtaining an accurate DC voltage very difficult. Therefore, the peak response meter is the preferred method for measurement.

> CAUTION: Always use an isolation transformer when troubleshooting TV receivers.

Monitoring Circuits

To help find the cause of a shutdown condition, it is necessary to know the operation of each monitoring circuit and the conditions that will trigger shutdown. The following circuit explanations describe the operation of each monitoring circuit, gives a test point for each circuit and provides troubleshooting tips to help you locate the fault. A flowchart is included to help locate which monitoring circuit is causing shutdown.

> A WORD OF CAUTION: Toshiba does not recommend disconnecting the shutdown circuit for troubleshooting. Disconnecting the shutdown circuit increases the possibility of a failure that may damage the TV receiver.

X-Ray Protection Circuit

Illustrated in **Fig. 11-37** is the X-ray monitoring protection circuit that triggers shutdown if it detects excessive radiation, which is produced by an increase in high-voltage. T461's secondary winding across pin 9, diode D471, and capacitor C471 produce a DC voltage directly proportional to the high-voltage (HV). A resistor divider consisting of resistors R451, R452, and R453 reduces the volt-

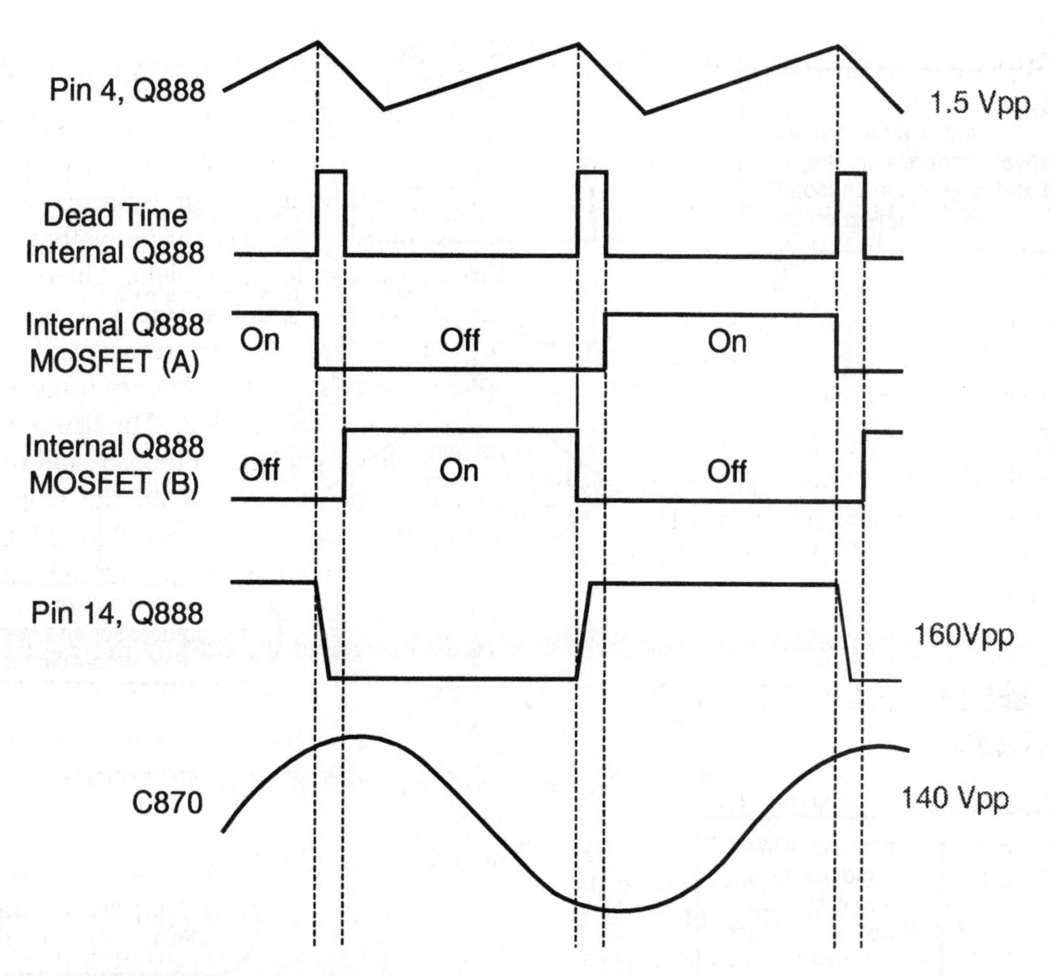

Pin	Name	Vdc	Description
1	VIN	140	Half bridge power input
2	GND	0	Control unit ground
3	DT	6	Dead time resistor terminal
4	CT	2.4	Oscillator capacitor terminal
5	CONT	5.9	Oscillator control terminal
6	FMAX	6.2	Maximun frequency resistor terminal
7	Css	3.7	Soft start capacitor terminal
8	CD	0.4	Delay latch capacitor terminal
9	VCC	18	Control unit power terminal
10	DRI	8	Gate drive power output
11	OC	0.7	Out of resonance / over current detection
12	COM	0	Half bridge GND
14	OUT	71	Half bridge output
15	VB	78	High side gate drive power input

FIGURE 11-33

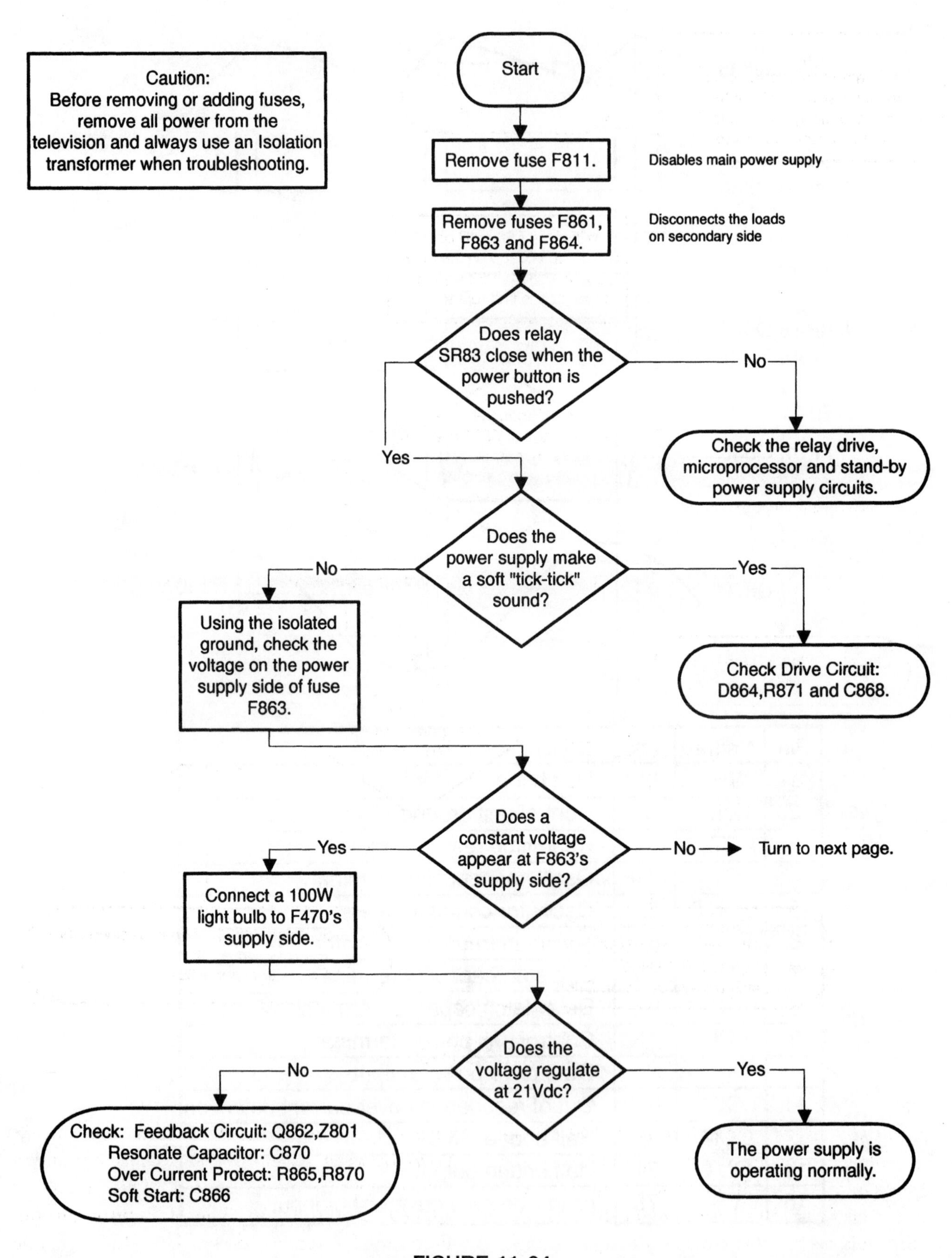
Caution:
Before removing or adding fuses, remove all power from the television and always use an Isolation transformer when troubleshooting.
Start
Remove fuse F811.
Disables main power supply
Remove fuses F861, F863 and F864.
Disconnects the loads on secondary side
Does relay SR83 close when the power button is pushed?
No
Check the relay drive, microprocessor and stand-by power supply circuits.
Yes
Does the power supply make a soft "tick-tick" sound?
No
Using the isolated ground, check the voltage on the power supply side of fuse F863.
Yes
Check Drive Circuit: D864,R871 and C868.
Does a constant voltage appear at F863's supply side?
Yes
Connect a 100W light bulb to F470's supply side.
No
Turn to next page.
Does the voltage regulate at 21Vdc?
No
Check: Feedback Circuit: Q862,Z801
Resonate Capacitor: C870
Over Current Protect: R865,R870
Soft Start: C866
Yes
The power supply is operating normally.

FIGURE 11-34

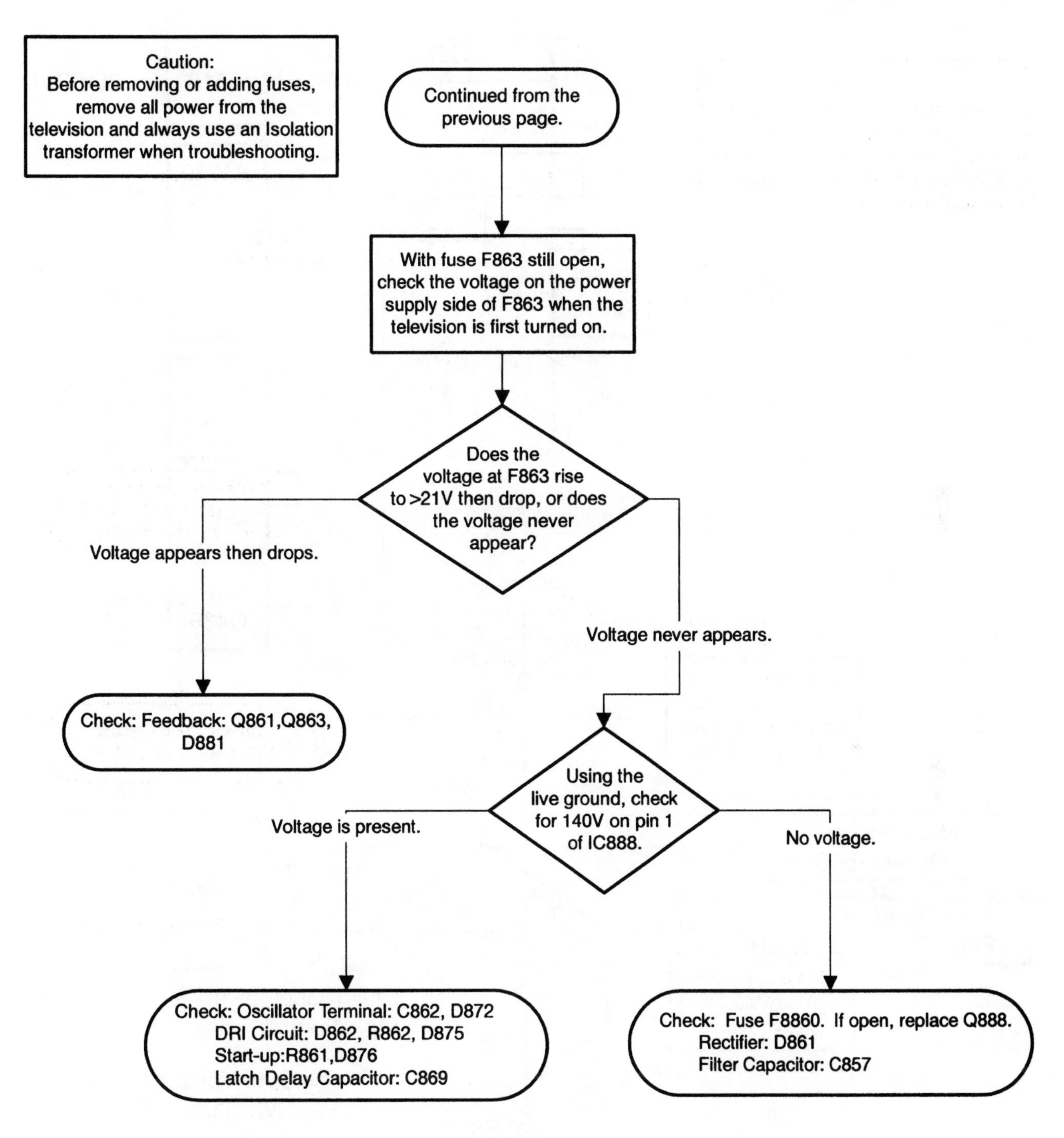

FIGURE 11-35

age and applies it to the emitter of Q430. As the high-voltage increases, the voltages at C471 and on the emitter of Q430 increase proportionately. Connected to Q430's base is zener diode D472. If the voltage on the emitter is large enough, D472 conducts and turns Q430 on. Q430's conduction increases the voltage on its collector to turn Q429 on. When Q429 turns on, current flows between its emitter and collector, and a voltage appears on its emitter. This voltage is applied to SCR D846. The SCR turns on and shuts the TV receiver down. Use D473's anode for the troubleshooting test point.

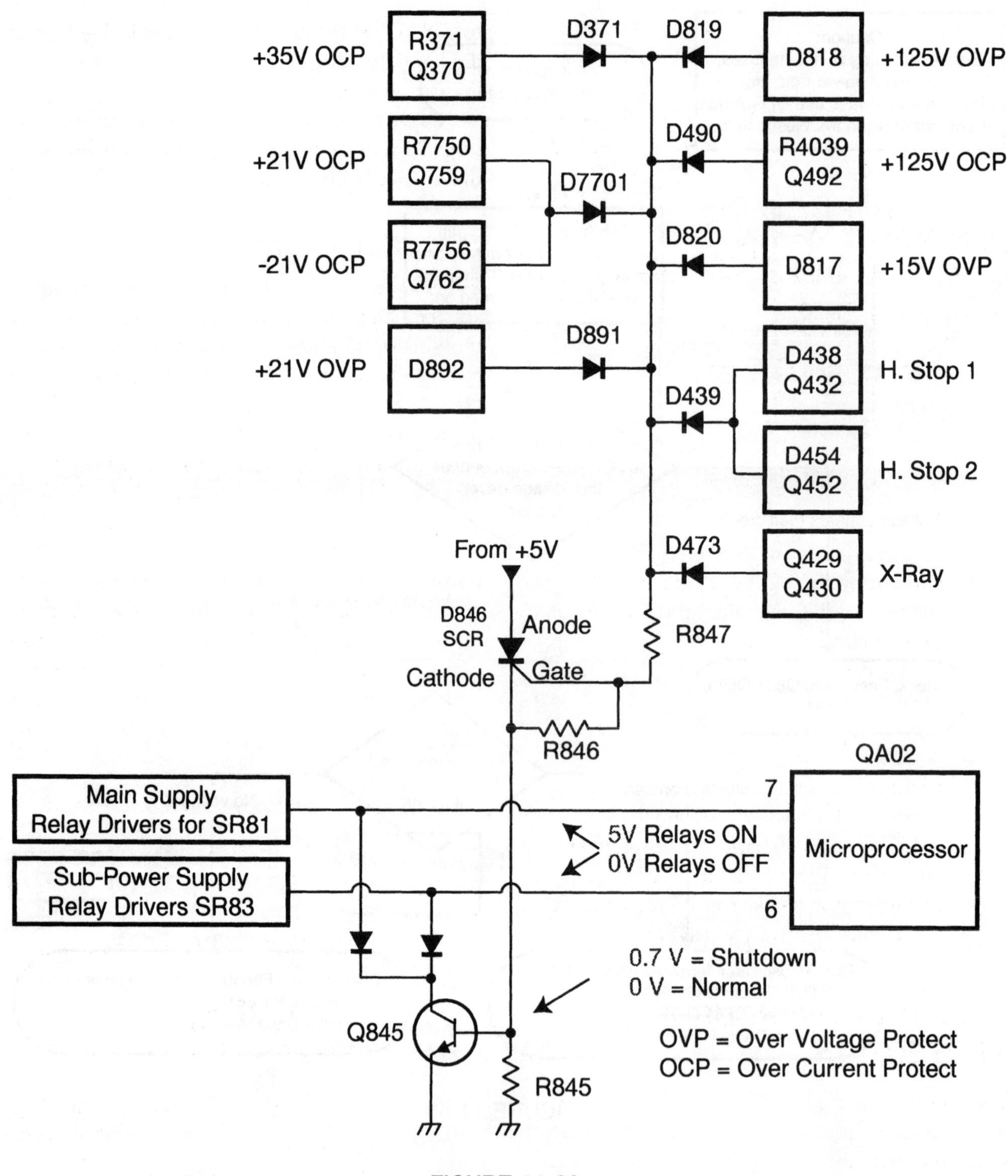

FIGURE 11-36

Troubleshooting Tips

Problems with the horizontal outputs, resonance capacitors, flyback transformer, deflection yokes, or a shorted CRT may trigger shutdown.

- The CRT's are the most likely culprit of an X-ray protection shutdown. Each CRT can be disconnected separately by disconnecting the drive PC board. The TV set can

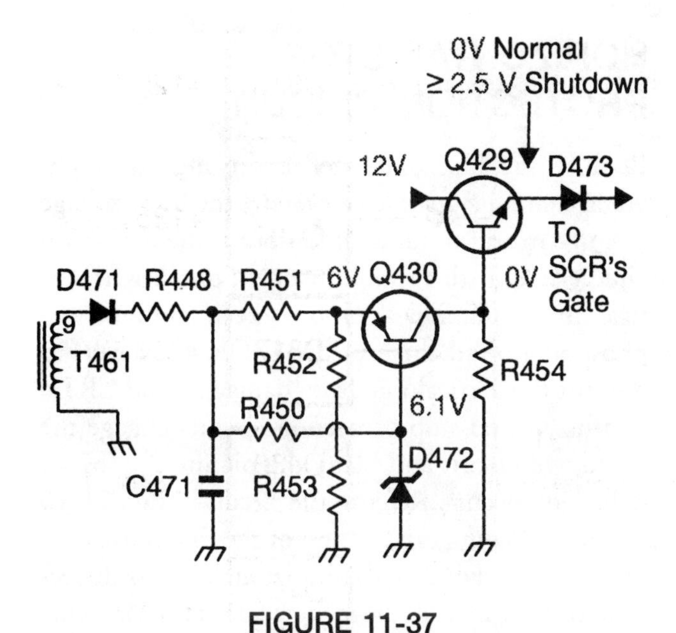

FIGURE 11-37

operate with one of the CRT's disconnected without damaging the remaining CRT's or the set. A CRT may intermittently arc and cause intermittent shutdown. With caution, lightly tap on the neck of the CRT to induce shutdown. Be careful not to tap too hard as the CRT could be damaged.

- A shorted secondary winding of the flyback transformer, or distributor block, can increase the high-voltage. A ringing test may indicate a bad flyback transformer. However, replacement of the flyback transformer or distributor block may be required for a successful defective part analysis.

+125 VOLT OVERCURRENT PROTECTION

You will find in **Fig. 11-38** that resistor R4039 is the overcurrent protection (OCP) sensing resistor which monitors the current flow through the +125 volt line. During normal operation, Q492 is turned off and its collector voltage is zero. An increase in current through the load increases the voltage drop across R4039. If the current increases enough, the voltage across R4039 forward biases Q492 and turns it on. When Q492 turns on, its collector voltage increases towards the supply voltage. To trigger shutdown, resistor R4043 supplies the collector voltage to D846's gate through Zener diode D491 and diode D490. Use D491's anode as the test point for troubleshooting.

Capacitor C498 provides a delay that prevents surge current from triggering the shutdown when the TV set is first turned on.

Troubleshooting Tips

Either a shorted horizontal output transistor, high-voltage output section, flyback transformer, or the horizontal output's resonance capacitors can pull excessive current through R4039 and cause the +125 volt OCP to trigger shutdown. Other possibilities include improper power supply regulation or an increase in the value of resistor R4039. To check the main power supply, refer back to the troubleshooting flow chart in Fig. 11-25. A shorted horizontal output or high-voltage output transistor (Q404 and Q416) is the most likely culprit for this. However, a shorted flyback transformer, arcing in a CRT, or a shorted yoke may have caused one of the outputs to become shorted. Use a ringing test to determine if you have a defective yoke or flyback winding. Replacing the

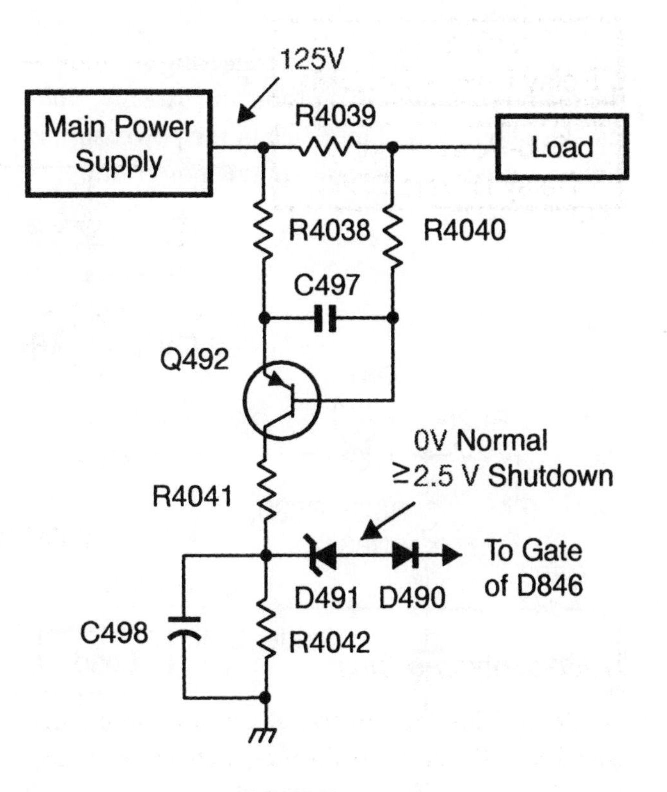

FIGURE 11-38

yoke or flyback may ensure that these components are defective.

Check out the overcurrent sensing resistor R4039. It can slightly increase in value and cause a false shutdown intermittently, or at turn-on, when the high-voltage first develops.

Overvoltage Protection (+125 volts)

The +125 overvoltage protect (OVP), shown in (**Fig. 11-39**), monitors the +125 volt supply and triggers shutdown if the voltage increases excessively. If the supply voltage rises above D818's zener voltage, the diode conducts and delivers a logic high (approximately 2.5 volts or higher) to the anode of diode D819 which applies the voltage to D846's gate to trigger shutdown. Normal voltage at the anode of D819 is about 0 volts. A voltage of 2.5 volts or greater at this point results in shutdown. Use the peak hold meter at this point for troubleshooting.

Troubleshooting Tip

One way a supply voltage increases is by a loss of its load. However, with this supply, loss of load will not increase the supply voltage enough to trigger the OVP. If the OVP is triggering shutdown, the main power supply is producing excessive voltage. Usually, a loss of feedback in the power supply circuit causes an increase in the power supply's output voltage.

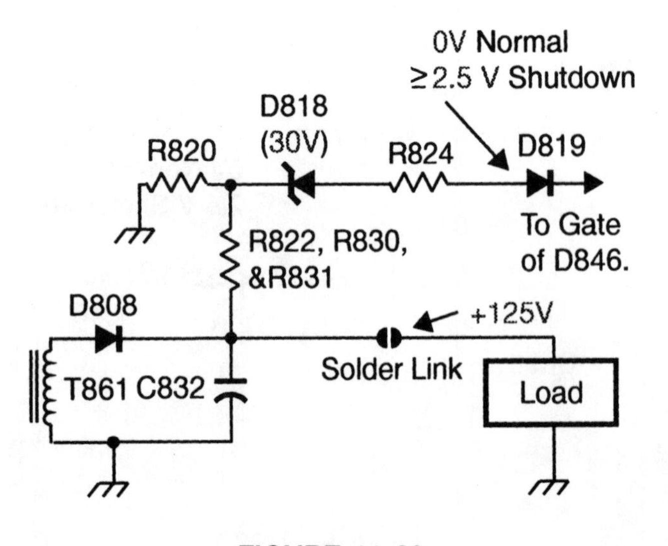

FIGURE 11-39

HORIZONTAL STOP PROTECTION CIRCUIT

Because the horizontal deflection and the high-voltage circuits operate separately, the high-voltage circuit can still produce an output if the deflection circuit fails. If this should happen, one bright vertical line would appear on the screen and burn the phosphors of all three of the CRTs. However, to prevent the vertical line from damaging the CRTs, two horizontal stop protection circuits engage the shutdown circuit and blank the picture if a loss of deflection occurs. Refer to the circuit in **Fig. 11-40** for the following explanation of the first horizontal stop circuit. T462 is the horizontal sweep deflection transformer. During normal operation, current is induced into the secondary windings between pins 3 and 1. Diode D451 rectifies the current, and capacitor C466 filters it to produce a DC voltage that resistor R490 applies to the base of transistor Q451. Q451's emitter connects to the base of transistor Q452. During normal operation, both of these transistors are on, making Q452's collector voltage approximately 3.6 volts. If horizontal deflection is lost, the voltage applied to the base of Q451 drops and both transistors turn off. The voltage on the collector of Q452 increases to 10.8 volts, and diodes D454 and D439 apply the voltage to the gate of SCR D846 to shut down the TV receiver. Q452 also applies the collector voltage to the blanking circuit to black out the picture and protect the CRTs.

The other horizontal stop circuit, shown in **Fig. 11-41**, works in the same way as the first horizontal stop circuit. Transistor Q441, capacitor C450 and diode D440 prevent the shutdown circuit from engaging when the TV set is first turned on. At turn on, the 12 volts appears before the horizontal deflection is fully operational. During this time, Q452 is off and 10.8 volts appears on its collector to engage the blanking circuit. Normally this voltage would also engage the shutdown. But when the power is first applied, capacitor C450 charges and allows current to flow through Q441's base-emitter junction, turning it on. While Q441 is on, its collector voltage is at ground which prevents the 10.8 volts being fed to the SCR. Once C450 reaches its full charge, Q441 turns off. By

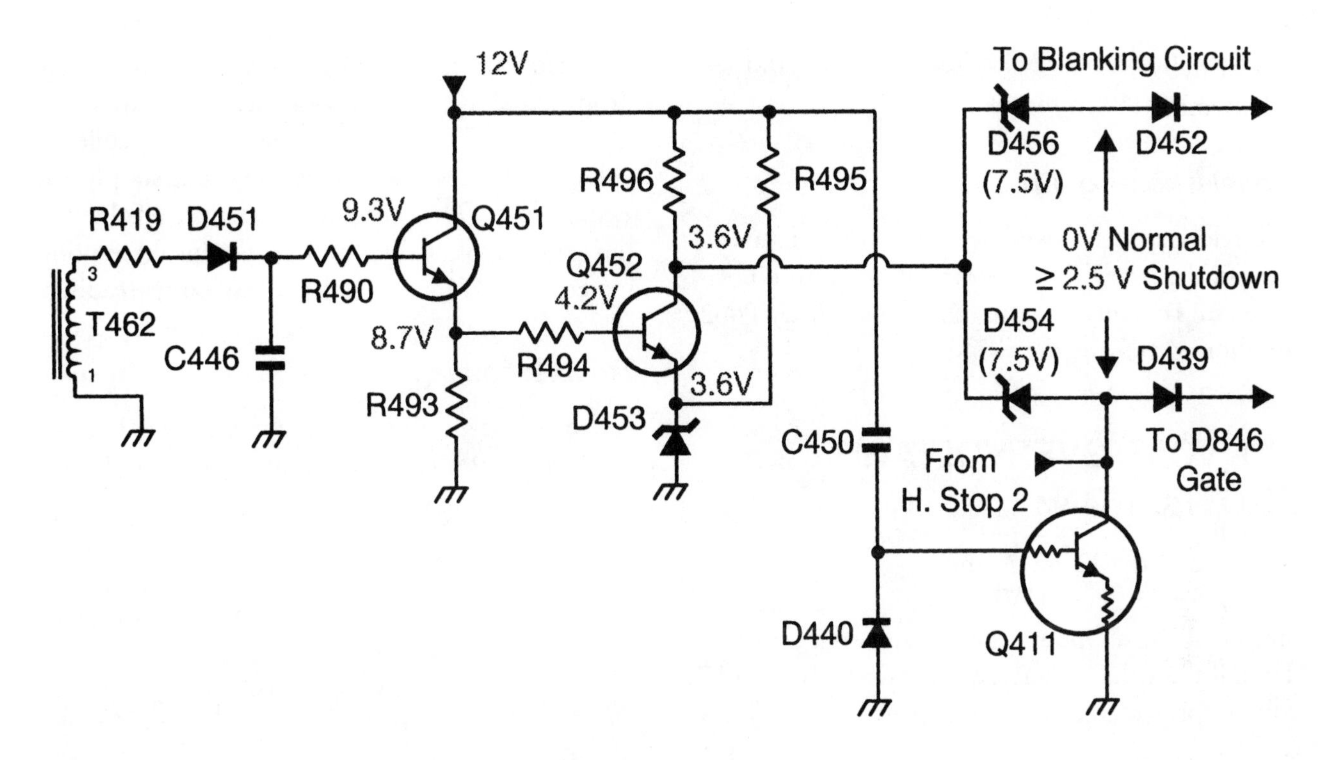

FIGURE 11-40

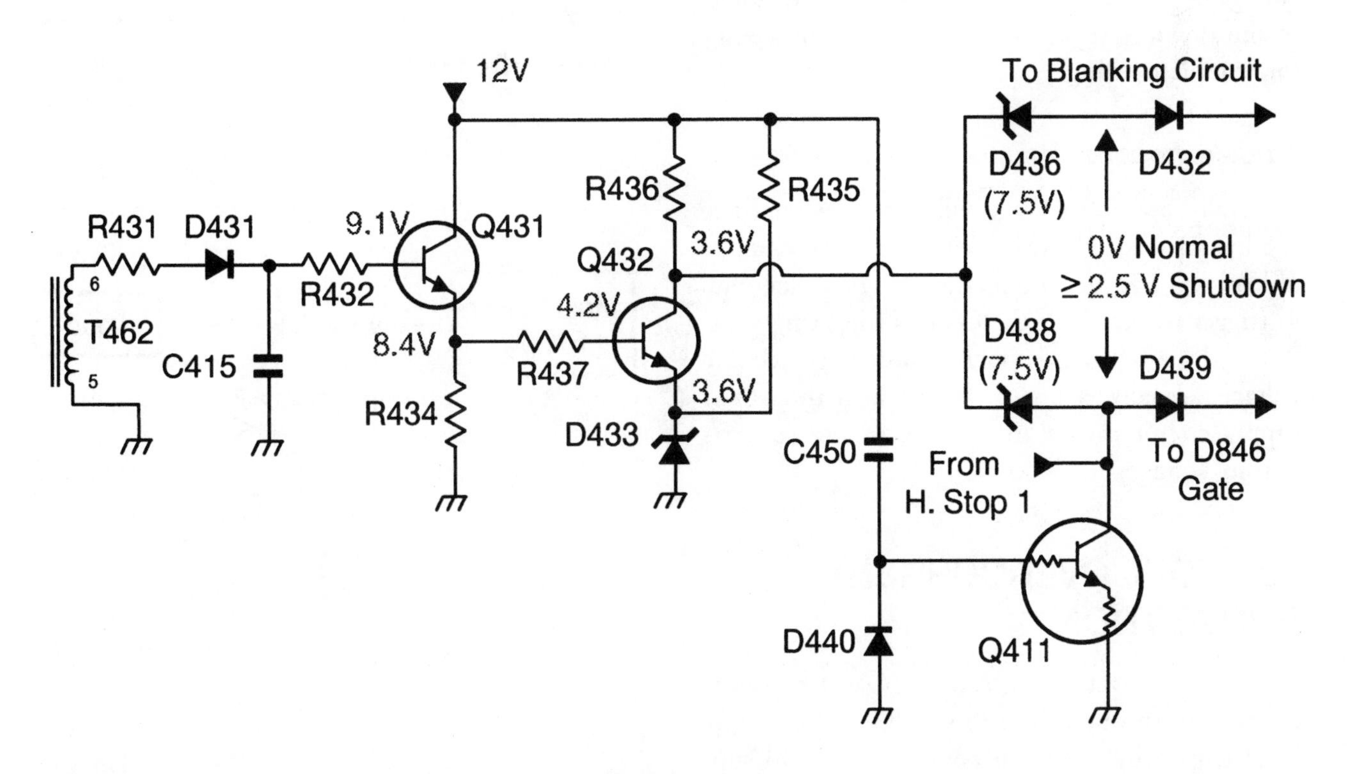

FIGURE 11-41

this time, horizontal deflection has started and the circuit is operating normally.

Troubleshooting Tip

The television uses two horizontal stop circuits to prevent damage if T462 shorts. If one of the horizontal stop circuits is causing shutdown, check T462 for shorted transformer windings.

+21 VOLT OVERVOLTAGE PROTECTION

Shown in **Fig. 11-42** is the overvoltage protect (OVP) circuit. The (OVP) monitors the +21 volt output of the subpower supply and triggers shutdown if the voltage increases excessively. If the +21 volt supply voltage rises enough to break D892's zener voltage, the diode conducts and delivers a logic high (approximately 2.5 volts or higher) to the anode of diode D891. D891 applies the voltage to D846's gate to trigger shutdown. Normal voltage at the anode of D891 is about 0 volts. A voltage of 2.5 volts or greater, at this point, results in shutdown. Use a peak hold meter at this point for troubleshooting.

Troubleshooting Tip

Generally, a supply voltage will increase when there is a loss of load. However, with this supply, a loss of load will not increase the supply voltage enough to trigger the OVP. If the OVP is triggering shutdown, the sub power supply is producing excessive voltage. Usually, a loss of feedback in the power supply circuit causes an increase in the supply's output voltage.

+35 VOLT OVERCURRENT PROTECTION

The +35 volt overcurrent protect (OVP) monitors the current through the +35 volt line. This supply is developed by the main power supply and supplies the vertical output Q301 and other transistor switching circuits. As shown in **Fig. 11-43**, current flows through the current sensing resistor R370. If the load current becomes excessive, the voltage drop across R370 increases and turns transistor Q370 on. When Q370 turns on, the collector voltage increases towards the +35 volt supply and Zener diode D370 conducts to deliver a voltage to the gate of SCR D846. Use the peak response meter on D371's anode for a test point reading.

Troubleshooting Tip

A shorted vertical output transistor, Q301, is the likely cause of excessive current drain from the +35

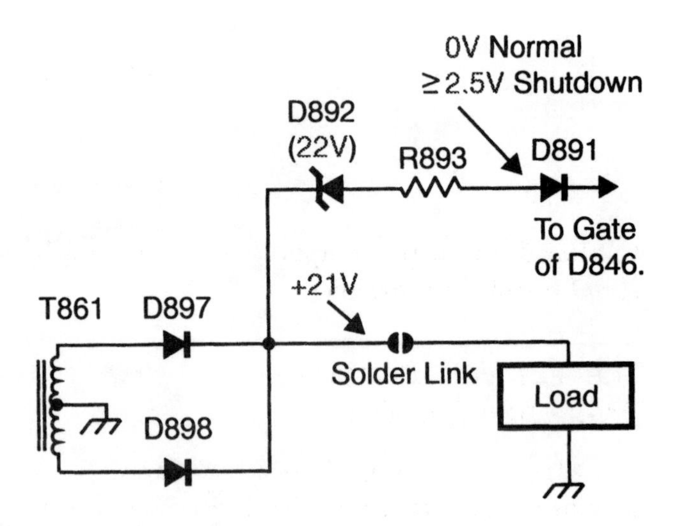

FIGURE 11-42

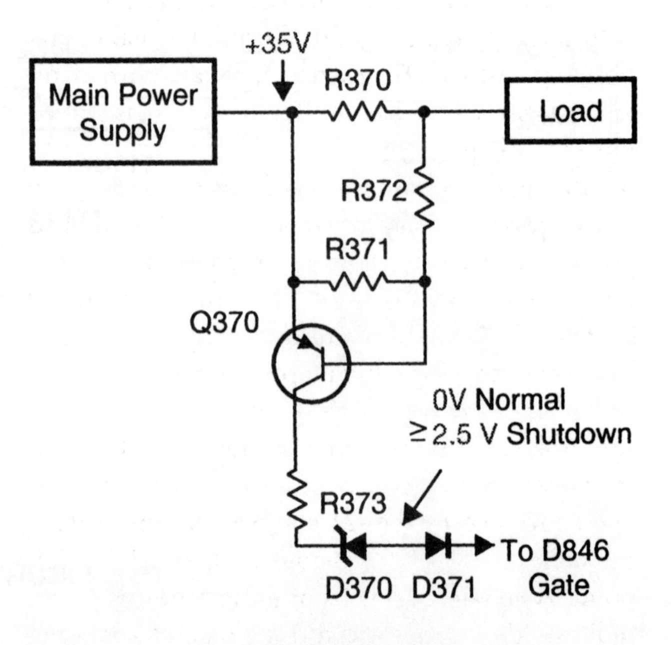

FIGURE 11-43

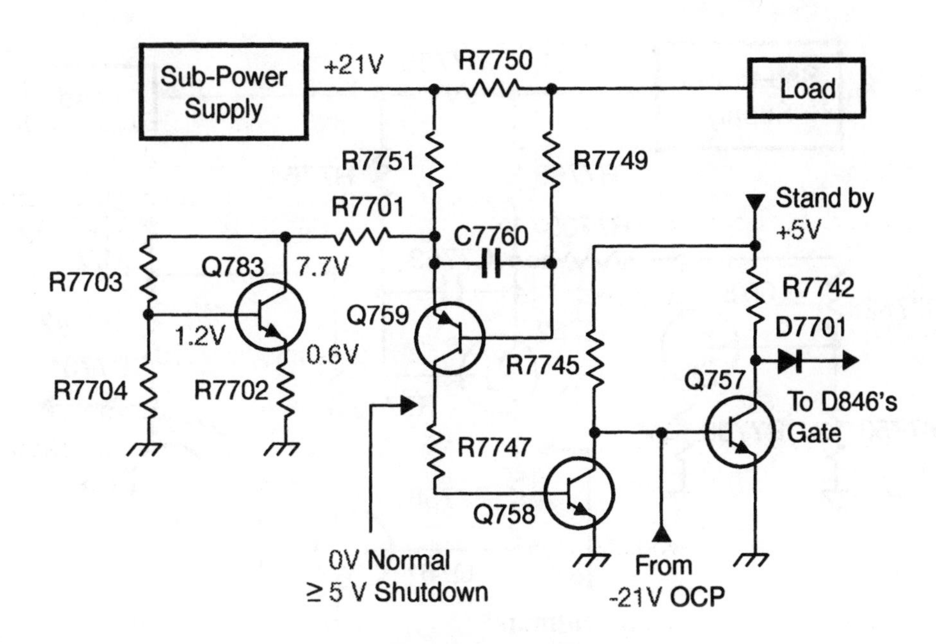

FIGURE 11-44

volt line. If the vertical output fails, usually pins 1, 2 and 6 will short together.

The +21 Volt Overcurrent Protection Circuit

The diagram in **Fig. 11-44** shows the +21 overcurrent protect (OCP) circuit. Resistor R7750 is the overcurrent sensing resistor that monitors the current flow to the convergence outputs Q751 and Q752. An increase in current increases the voltage drop across R7750. During normal operation, transistors Q759 and Q758 are turned off and transistor Q757 is on. Because Q757 is turned on, the voltage at its collector is 0 volts. A slight increase in the voltage across R7750 turns on Q759 and increases its collector voltage. Then Q758 turns on, and its collector voltage drops to ground potential and turns Q757 off. The emitter-collector current of Q757 stops and the voltage on the collector rises to a logic high (approximately 2.1 volts or higher). The logic high is applied to the gate of the SCR D846 through via diode D7701, and shutdown takes place. Because transistor Q757 is also controlled by the -21 volt overcurrent protect, the collector of Q759 should be used as the test point. A voltage of 5 volts or greater, at this point, indicates the transistor is turning on and activating shutdown. Transistor Q783 is always slightly forward biased to reduce the sensitivity of the shutdown circuit and prevent false shutdown.

Troubleshooting Tips

- The overcurrent sensing resistor can increase in value and cause a false or intermittent shutdown. Make certain the current resistor is of the correct value.
- If excessive current is pulled from the power supply, check the convergence output ICs (Q752 and Q751 located on the convergence output PC board) and the surrounding biasing resistors. The digital convergence board can cause Q752 and Q751 to work too hard and pull excessive current. If this is suspected, remove the digital convergence board from the TV receiver, but first make sure the set is unplugged. Plug the TV set back in and turn it on. If the TV receiver comes on, the digital convergence board could be defective. If the TV set still shuts down, Q752, Q751, or any of the associated biasing circuits may be faulty.

NOTE: The TV receiver will power up without the digital convergence board being in place, however the TV set will be out of convergence.

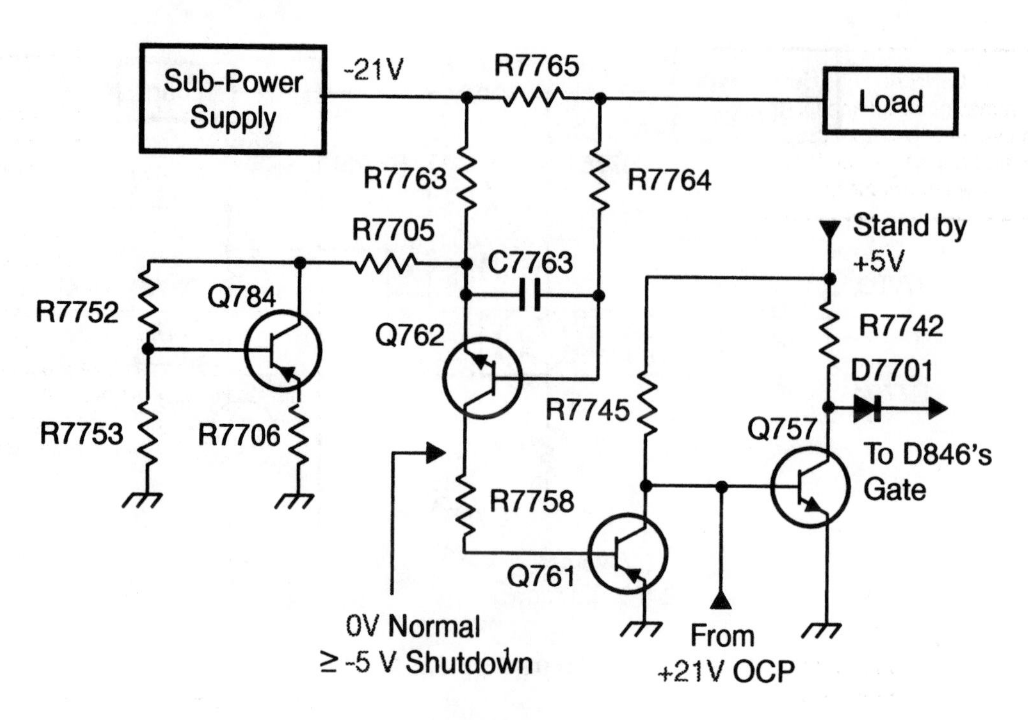

FIGURE 11-45

The raster bows in from all sides because the horizontal and vertical scans are not going all the way to the edge of the CRT's. DO NOT let the TV set run for an extended time in this condition. If left operating long enough, it can burn the screen phosphorous. If additional testing is required in this condition, turn the contrast and brightness all the way down to reduce risk of phosphorous burn.

-21 OVERCURRENT PROTECTION

The -21 volt OCP operates in the same manner as the +21 volt OCP circuit. Refer to the circuit drawing in (**Fig.** 11-45) and the explanation for the +21 volt OCP operation for details.

The shutdown troubleshooting flow chart is illustrated in **Fig.** 11-46.

SWITCH-MODE POWER SUPPLY TROUBLESHOOTING

Now that you have a basic understanding of modern HDTV power supply systems, let's look at ways to troubleshooting them if a problem develops.

The switching transistor output provides the best point to start troubleshooting procedures. Use your scope to analyze the waveform amplitude, frequency, duty cycle, noise, and any ripple. Connect the scope to the terminals on the transistor that connects to the primary of the switching transformer (normally the collector or drain), and connect the ground to the switching power supply's primary or "HOT" ground. Set the scope's volts/div switch to 100 and the timebase frequency to 10 KHz. Now be ready to observe events on the scope screen when the TV receiver is turned on. These results will provide you with valuable clues as to what problems the TV receiver may have.

When the set is powered-up, you should see approximately 150 to 160 volts on the DC voltmeter or scope,. If the voltage is low, or missing, check the unregulated supply and the switching transformer primary, or test for a shorted switching transistor. Also, check out the B+ supply for excessive AC 120-Hz ripple, which may indicate a main filter capacitor problem.

If the unregulated B+ output is good, you are ready to test the switching power supply. With the set turned on, check out the switching pulses, comparing the wave shape thats shown in your Sams Photofact service data. In many cases, some

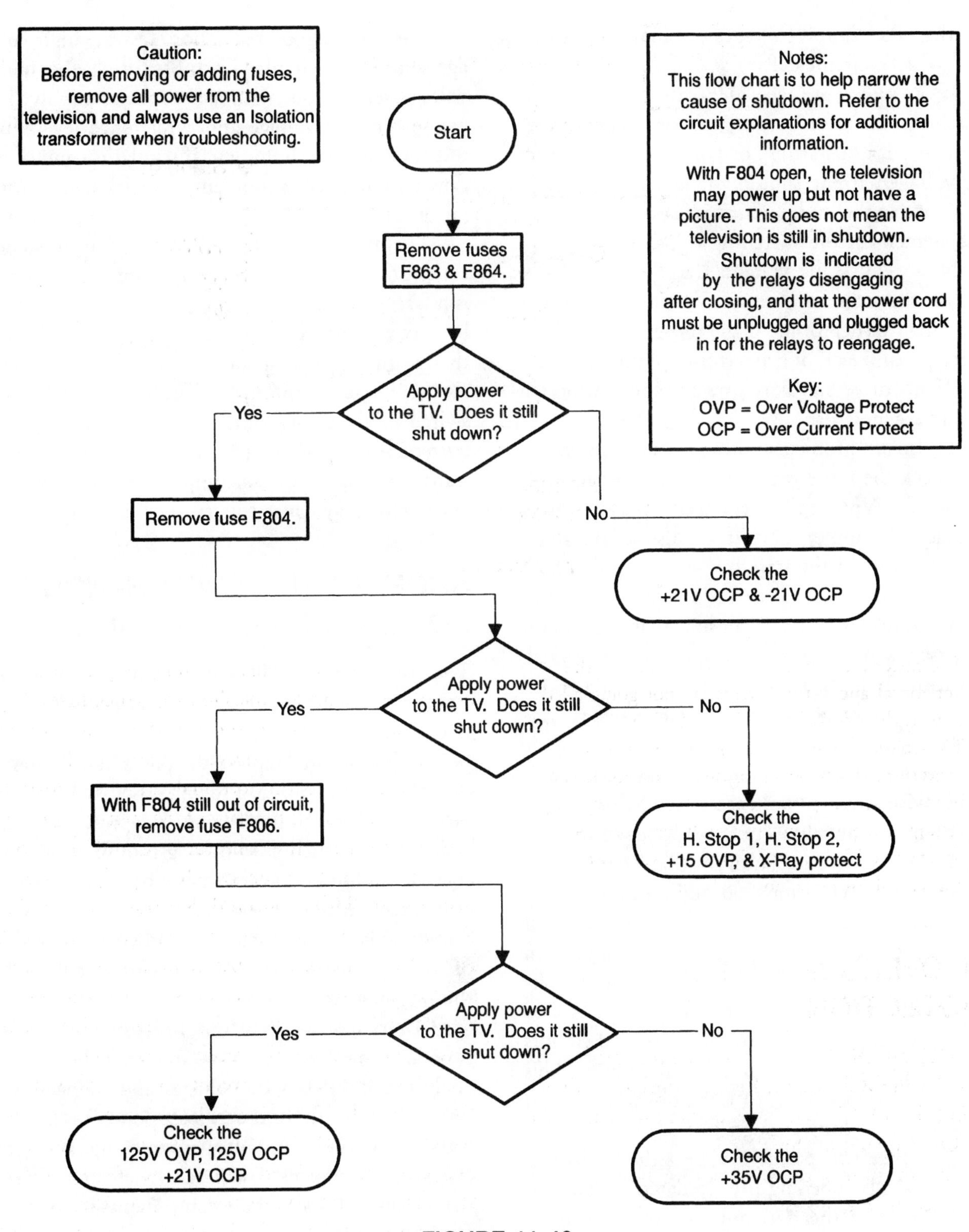
Caution:
Before removing or adding fuses, remove all power from the television and always use an Isolation transformer when troubleshooting.
Start
Remove fuses F863 & F864.
Notes:
This flow chart is to help narrow the cause of shutdown. Refer to the circuit explanations for additional information.
With F804 open, the television may power up but not have a picture. This does not mean the television is still in shutdown.
Shutdown is indicated by the relays disengaging after closing, and that the power cord must be unplugged and plugged back in for the relays to reengage.
Key:
OVP = Over Voltage Protect
OCP = Over Current Protect
Apply power to the TV. Does it still shut down?
Yes
No
Remove fuse F804.
Check the +21V OCP & -21V OCP
Apply power to the TV. Does it still shut down?
Yes
No
With F804 still out of circuit, remove fuse F806.
Check the H. Stop 1, H. Stop 2, +15 OVP, & X-Ray protect
Apply power to the TV. Does it still shut down?
Yes
No
Check the 125V OVP, 125V OCP +21V OCP
Check the +35V OCP

FIGURE 11-46

ringing on the waveform is normal. With your scope, check for the correct frequency of the switching pulses. Confirm that the switching frequency is normal by comparing it with the frequency shown in the Sams Photofact service data.

Next, connect the second channel input for your scope to the switch-models regulated B+ output or to the horizontal output transistor's collector. Also, attach the second scope's probe ground lead to the secondary ground. If the B+ is correct, test the other supply voltages or circuits to isolate the TV set's circuit defect.

If no pulses are viewed on the scope, at the switching transistor output, the switch-mode circuit is dead, not starting, or going into immediate overvoltage/current shutdown. Move the scope's test probe to the input, which is usually the base or gate element of the switching transistor.

Now reduce the voltage/division control setting on the scope, and test for an input drive signal. This will determine if the problem is an open transistor or a lack of drive from the pulse generator. If no drive signal is present, then suspect a defective generator.

To isolate problems with the pulse generator, first check the standby power supply voltages with a scope or DVM. If these voltages are low, or have considerable ripple, they may cause a "no start" problem. Should the standby power supply check out OK, then test power on/off control circuit. Turn the television on to ensure that the microprocessor is providing the proper turn-on signal. The turn-on signal should turn other circuits, including the pulse generator, on. A defective voltage regulator circuit will also prevent the switch-mode from starting. Check for improper voltages in this circuit with a digital voltmeter (DVM) or scope, and then test any suspected components.

Another common symptom that's observed at the switching transistor's output are momentary pulses quickly stopping. This is an indication that the pulse generator is being started up and then shut down. The most likely cause is the high-voltage shutdown control circuits. This can be due to high B+ voltages supplied to the horizontal circuits or problems with in the horizontal circuits. This type of problem is hard to track down because it does not occur long enough to test.

To isolate this problem, defeat the horizontal output stage by opening the regulated B+ path to the collector of the horizontal output transistor. Simulate the horizontal circuit load by connecting a high wattage 500 to 1000 ohm resistor between the B+ point you opened and to ground. Then turn the set ON, and test the regulation of the switchmodels output by varying the ac output line voltage. If the B+ output does not regulate properly, you need to isolate problems in the voltage regulation feedback circuits.

If you have good regulation, test the high-voltage shutdown circuits by substituting the high-voltage sense input with a DC power supply. Increase the DC voltage above reference level, and test for a normal shutdown. If regulation and shutdown are normal, a problem exists in the horizontal output timing or shutdown reference circuits.

The final waveform symptom that is sometimes observed at the switching transistor's output is a fast, short-duration spike. This waveform may be accompanied by a squealing sound or a hotter than normal case temperature of the switching transistor. These symptoms are indications that the switch-mode power supply is quickly going into current limit during each drive pulse. A defective transformer or a shorted component in the secondary circuit may cause this current limiting.

SCOPING THE SWITCH-MODE POWER SUPPLY CIRCUITS

The last step for quickly troubleshooting these circuits is to "shotgun" or replace a suspected part(s) that has not yet been proven defective, but is suspect of being bad. Replace the part(s) and observe the TV set's operation. Proving suspected parts to be good or bad can be difficult. Swapping specially ordered parts, such as switching transformers, and special "high-cost" power transistors, is not always cost effective nor an efficient use of time. Randomly replacing parts can also lead you to mistakes or cause parameteric changes that may compound the original problems.

Components such as optoisolators or IC pulse generators are best tested in-circuit by thoroughly analyzing input/output voltages and scope waveforms. Use a wideband oscilloscope to check these waveforms. Verifying proper working voltages, checking waveforms, and tracing defective signals will isolate defects to these components in the least amount of time.

NOTES ON SPECIAL SOLID-STATE DEVICE REPLACEMENTS

The technician must use care in replacing solid-state devices in the power and sweep stages of the modern HDTV receiver. An exact diode replacement is the safe way to go. Many diodes are special, fast switching devices that cannot be replaced by a "garden-variety" part. If you do not use an exact replacement and have another failure in a short time, you don't know if you have a different failure, or if the generic replacement part is at fault. And, you should not use some generic diodes of a higher voltage and current specifications as "all-purpose" replacements. As an example, you may want to use "ultrafast" diodes to replace diodes that are defective in a switch-mode power supply. Not only are these "fast diodes" more expensive, they will not always perform properly. You will prevent many headaches by finding out the exact specifications of the original component and then replace with the correct generic part.

Some new model electronic devices now have a new generation of fast Schottky components.

Some "fast rectifiers" have switching speeds between 150 and 500 nanoseconds. Breakdown voltage for these diodes is usually a maximum of 1,000 volts. These high-efficiency diodes are a compromise between the Schottky and fast-acting rectifiers.

Soft recovery diodes have a fast forward-switching time, but a comparatively slow reverse-switching time. Soft recovery diodes are used because of the lower cost in making cheaper transformers. Soft recovery diodes let the equipment manufacturer use less expensive magnetics that are more prone to "ringing" problems. These soft reverse recovery diodes are used in the design to help dampen the ringing.

Another new, exotic, but expensive variant are the hybrid diodes, which incorporate the properties of an ultrafast, a high efficiency, and a soft recovery on one-diode package.

If the wrong solid-state device is installed when you make a repair, the replacement part will usually fail prematurely, very quickly, or in a few days or weeks. You lose either in seconds, or days. You end up wasting additional time and labor fixing the set the second time and have an unhappy customer to boot.

Diodes have become very sophisticated and specialized. Thus, stocking just a few diodes with a high current value and fast switching speed will not always meet your needs. Here are some reasons:

1. Most ultrafast diodes have a relatively high forward voltage drops as compared to fast acting diodes. This will cause the ultrafast diodes to operate hotter than the failed fast diode.
2. The forward voltage drop is generally slower for high-voltage rated diodes.
3. The switching speeds are usually slower for high-voltage rated diodes.
4. The forward voltage drop is generally "higher" for high-current diodes.
5. The switching speeds are generally slower for high-current diodes.

It should be noted that the forward voltage drop, V_f, increases for greater breakdown voltages. And, it should be pointed out that the switching speeds become slower for higher amperage diodes. These characteristics are due to a larger semiconductor "die." The large diameter die used for high-current rated diodes will most always result in greater junction capacitance. This results in a slower switching speed. Also, using a high-voltage component will require a longer die. Greater V_f voltage drops results from additional layers in the die needed to make the higher beakdown rating. The forward-bias voltage drop is important because the higher the "Vf," the less efficient and hotter the diode will operate.

In the long run, it will pay to obtain the original, or correct, generic solid-state device.

Portions of this chapter contains information and circuit drawings courtesy of SENCORE,Inc. and TOSHIBA of America

CHAPTER 12

HDTV Satellite Systems and Set-Top Boxes

INTRODUCTION

This chapter begins with some background information of the HDTV DIRECTV satellite system. Industry formats will be explained for satellite receivers and other required HDTV equipment. We will review what HDTV services are now available and some of the upcoming programs that are planned.

Then we'll review various HDTV set-top boxes that are on the market. One unit is a combination digital satellite receiver and HDTV receiver. Some of these boxes will give HDTV viewers the ability to receive standard and HDTV signals from a DIRECTV dish, off-the-air digital broadcasts, and local TV station analog signals.

The chapter concludes with information on Microtune's MM8838 Micromodule RF plug-in chip solution for cable TV set-top boxes.

DIRECTV HIGH-DEFINITION TV SYSTEM

DIRECTV is accelerating the advent of HDTV with the introduction of HDTV sets integrated with the DIRECTV PLUS system receivers and advanced digital set-top receiver boxes.

DIRECTV began delivering coast-to-coast high-definition programming in October 1998. It expanded HDTV programming to two channels on a 24-hour basis in 1999, when HDTV sets with a built-in DIRECTV PLUS System receiver became available. Thus, DIRECTV will be the first and only multichannel TV platform to offer TV viewers two channels of HDTV programming.

HDTV Equipment

There are two ways viewers may receive high-definition programming from DIRECTV. The first is to purchase a DIRECTV/HD digital receiver and connect it to an HD-ready TV set. The second way is to purchase an HDTV set enabled with a built-in DIRECTV PLUS system receiver. In both cases, an 18 x 24-inch oval dish and off-air antenna are installed to receive standard-definition and high-definition DIRECTV programming, as well as local digital terrestrial channels.

HDTV Satellites

DIRECTV delivers high-definition programming from its Tempo 2 satellite at 119 degrees W.L. Both high-definition and standard-definition DIRECTV programming will be received through a single oval-shaped satellite dish.

In October 1999, DIRECTV launched its second HDTV channel featuring pay-per-view movies and selected special events. The HS601 (**Fig. 12-1**) satellite is body-stabilized and is located at 101 degrees west longitude. This satellite is built by the Hughes Space and Communications Company and measures 86 feet from wingtip to wingtip.

The photo in **Figure 12-2** shows engineers at Hughes Space and Communications checking out an HS601 DIRECTV broadcast satellite. The two large antennas are for transmitting signals to Earth, while the smaller, round dish in the center is the uplink receiving antenna. Also visible is the complex wiring for the satellite's 16 transponders that produce 120 watts of signal power per channel.

FIGURE 12-1

Technical Formats

DIRECTV receives television content at its broadcasting facilities from its programmers in a wide variety of technical formats. At the broadcast centers, broadcasting formats that suit the programming material are automatically selected, uplinked to DIRECTV satellites and then downlinked to DIRECTV system receivers, which decode all formats into high-quality audio, video, and high-definition video. The uplink center dishes are shown in **Figure 12-3**.

DIRECTV implements the same policy and accepts high-definition programming in a wide

FIGURE 12-2

FIGURE 12-3

variety of industry standard and ATSC (Advanced Television Systems Committee) broadcasting formats. Programming delivered in high-definition can have up to two million pixels of resolution, resulting in cleaner, crisper, sharper pictures delivered with Dolby Digital Surround Sound audio. These broadcasts, when combined with high-definition displays, will create quality that rivals the movie theater experience.

OVERVIEW OF DIRECTV HDTV SATELLITE SYSTEM

The HDTV satellite system is assigned a unique band of frequencies within the electromagnetic spectrum for its operation. This frequency slot is called the KU-band.

The KU-band satellites are classified into two groups; low- and medium-power KU-band satellites, transmitting signals in the 11.7 to 12.2 GHz FSS band; and the new high-power KU-band satellites transmitting in the 12.2 GHz to 12.7 GHz Direct Broadcast Satellite service (DBS) band. DIRECTV HDTV uses the high-power KU-band satellites for its operating system.

The Satellite System

A satellite system is made up of three basic elements:

- An uplink facility that beams programming signals to satellites orbiting over the equator.
- A satellite that receives the signals and retransmits them back to an earth dish and receiver.
- A receiving satellite dish and HDTV digital receiver. A drawing of this complete HDTV satellite system is shown in **Figure 12-4**.

The picture and sound information originating from a studio or broadcast facility is first sent to an uplink site, where it is processed and combined with other signals for transmission on microwave frequencies. Next, a large uplink dish concentrates these outgoing microwave signals and beams them up to a satellite located 22,247 miles

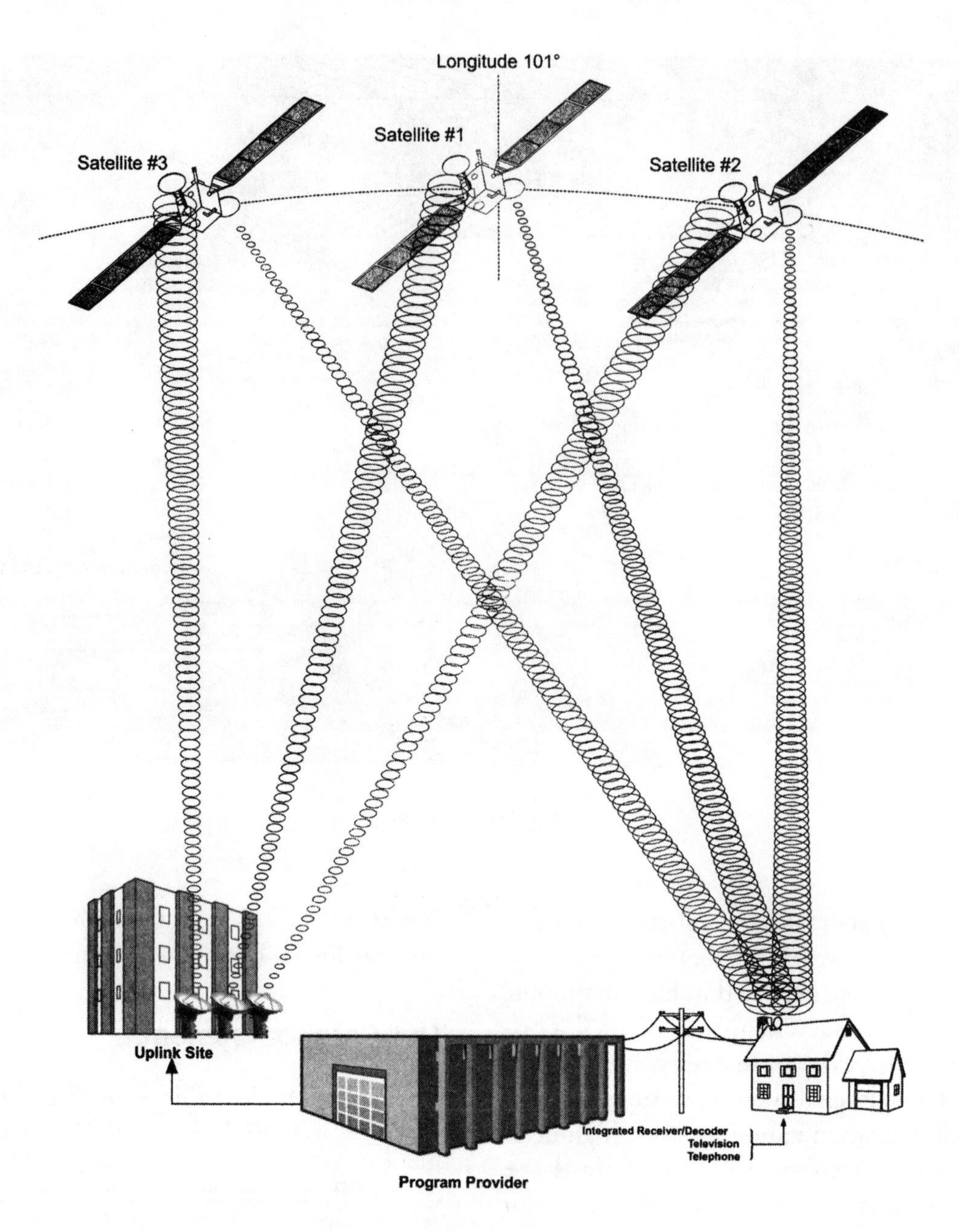

FIGURE 12-4

above the equator. The satellite's receiving antenna captures the incoming signals and sends them to a receiver for more processing. These signals, which contain the original picture and sound information, are converted to another group of microwave frequencies, then sent on to an amplifier for transmission back to earth. This receiver/transmitter package is called a transponder. The outgoing signals from the transponder are then reflected off a transmitting antenna, which focuses the microwaves into a beam of energy that is directed toward the earth. A satellite dish on the ground collects the microwave energy containing the original picture and sound information, and focuses that energy into a low-noise block converter or LBN. The LBN amplifies and converts the microwave signals to yet another lower group of frequencies called the Intermediate Frequency, or IF, that can be sent via conventional coaxial cable to a satellite receiver-decoder within the viewer's home. The receiver tunes the individual transponders and converts the original picture and sound information into video and audio signals that can be received on a conventional TV or HDTV system.

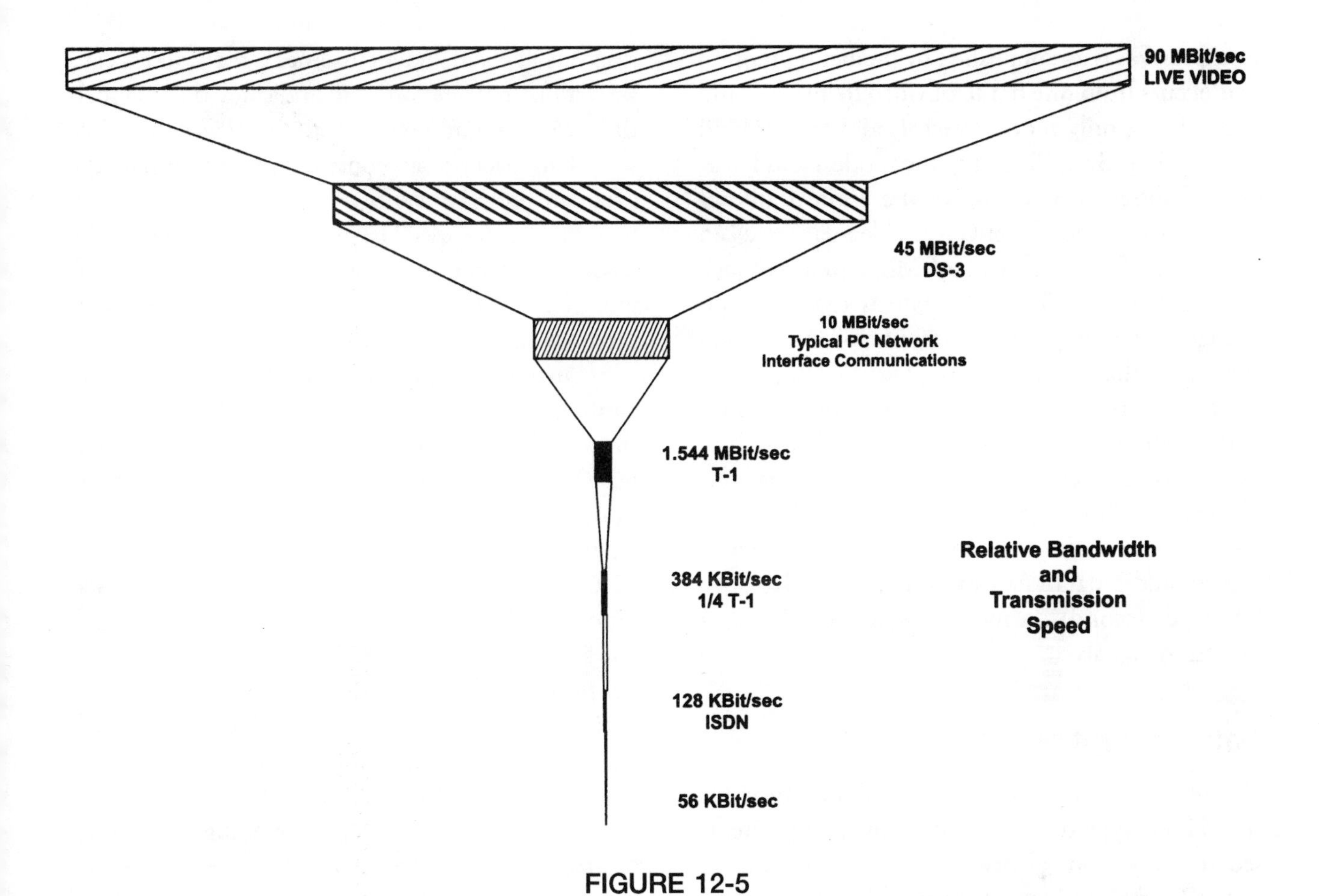

FIGURE 12-5

The uplink signal is transmitted to the satellites orbiting the earth above the equator at 101 degrees west longitude. The signal is then relayed back to earth and decoded on an HDTV satellite receiver. The receiver connects to the phone line for billing information. Refer to Figure 12-4 for complete DIRECTV system setup.

DIRECTV HDTV Uplink Operation

The DIRECTV system transports digital, video, and audio data to a viewer's home via a high-powered KU-band satellite. The program provider sends program material to the uplink site where the signal is digitally encoded. The uplink is the portion of the signal transmitted from the earth to the satellite. The uplink site compresses video and audio, encrypts video, and formats the information into data "packets" that are then transmitted. The signal is transmitted to a satellite, where it is relayed back to earth and decoded by the viewer's TV receiver.

MPEG2 Compression

As shown in **Figure 12-5**, the amount of data required to transmit all the video and audio information contained in a typical program being received by a set-top unit would require a data transfer rate well into the hundreds of Mbps (megabits per second). This is too large and impractical a data rate to be processed in a cost-effective way with current hardware. In order to minimize the data transfer rate, the data is compressed using MPEG2 compression. MPEG (Moving Pictures Experts Group) is a committee that meets under the ISO (International Standards Organization) umbrella to develop a specification for the transportation of moving images (and audio) over communication data networks. Fundamentally, the system is based on the principle that the images contain a lot of redundancy from one frame of video to another. In other words, the background probably stays the same for many frames at a time.

Compression is accomplished by predicting motion that occurs from one frame of video to another and transmitting only motion vectors and background information. By coding only the motion and background differences instead of the entire frame of video information, the effective video data rate can be reduced from hundreds of Mbps to an average of 3 to 6 Mbps. This data rate is dynamic and changes depending on the amount of motion occurring in the video.

In addition to MPEG video compression, MPEG audio compression is also used to reduce the audio data rate. Audio compression is accomplished by eliminating soft sounds that are near loud sounds in the frequency domain. The compressed audio data rate can vary from 56 Kbs (kilobits per second) on mono signals, to 384 Kbps on stereo signals.

Data Encryption

To prevent unauthorized signal reception, the video signal is encrypted (scrambled) at the uplink site. A secure encryption "algorithm," or formula, known as the Digital Encryption Standard (DES) is used to encode the video information. The keys for decoding the data are transmitted in the data packets. The customer's access card decrypts the keys, which allow the receiver to decode the data. When an access card in a receiver is activated for the first time, the serial number of the receiver is encoded on the access card. This prevents the access card from activating any other receiver except the one in which it was initially authorized. The receiver will not function with the access card removed.

Data Packets

The DIRECTV satellite program information is completely digital and is transmitted in data packets. This concept is very similar to data transferred by a computer over a modem. Five different types of data packets are video, audio, CA, PC-compatible serial data, and program guide. Video and audio packets contain the visual and audio information of the program. The CA (conditional access) packet contains information that is addressed to individual receivers. This includes customer e-mail, access card activation information, and indicates the channels the receiver is authorized to decode. PC-compatible serial data packets can contain any form of data the program provider wants to transmit, such as stock reports or software. The program guide maps the channel numbers to transponders and SCID (service channel ID). It also gives the customer TV program listing information.

Illustrated in **Figure 12-6**, is a typical uplink configuration for one transponder. In the past, a single transponder was used for each satellite channel. With digital signals using MPEG compression, more than one satellite channel can be sent on the same transponder. The example in Figure 12-6 shows a live video channel, a satellite-based premium channel, a tape-based program channel, five stereo audio channels (one for each video channel plus two extra for other services such as a second language), and a PC-compatible data channel. Audio and video signals from the program provider are encoded and converted to data packets. The configurations can vary depending on the type of programming. The data packets are then multiplexed into serial data and sent to the transmitter.

Each data packet is 147 bytes long. The first two bytes (a byte is made up of 8 bits) of information contain the SCID and flags. The SCID (service channel ID) is a unique 12-bit number from 0 to 4095 that uniquely identifies the packet's data channel. The flags are made up of 4 bits used primarily to control whether or not the packet is encrypted and which key is used. The third byte of information is made up of a 4-bit packet type indicator and a 4-bit continuity counter. The packet type identifies the packet as one of four data types. When combined with the SCID, the packet type determines how the packet is to be used. The continuity counter increments once for each packet type and SCID. The next 127 bytes of information consist of the "payload" data, which is the actual usable information sent from the program provider. This concept is illustrated in the drawing in **Figure 12-7**.

Although there are only 32 total transponders, the channel capabilities are far greater. Using data compression and multiplexing, the three satellites

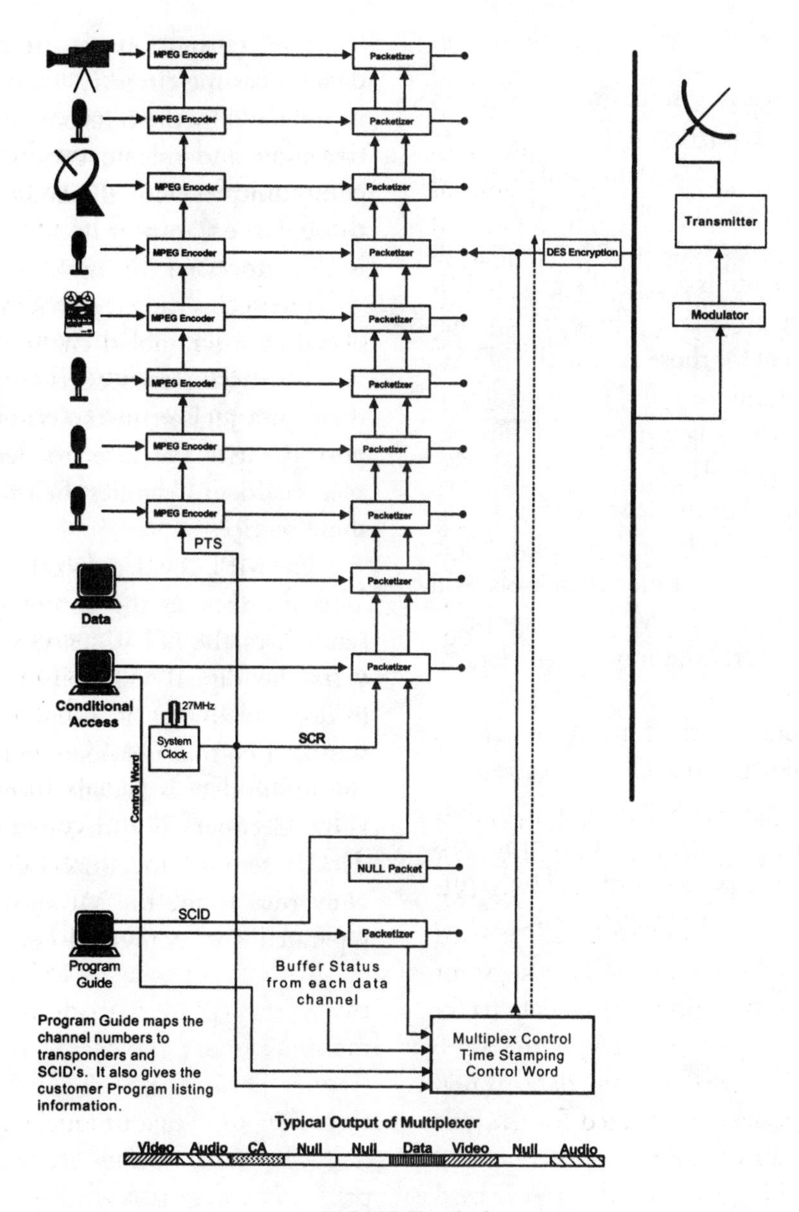

FIGURE 12-6

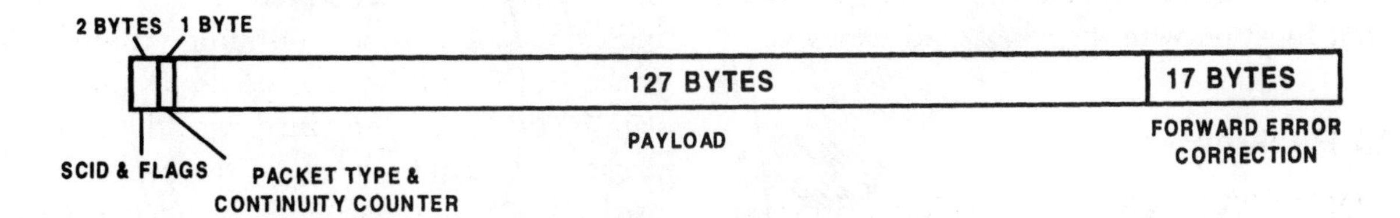

FIGURE 12-7

working together have the possibility of carrying over 150 conventional (non-HDTV) audio and video channels via the 32 transponders.

The "dish" is an 18 inch, slightly oval shaped KU-band antenna. The slight oval shape is due to the 22.5 degree offset feed of the LBN (low-noise block converter), refer to drawing in **Figure 12-8**). The offset feed positions the LBN out of the way so it does not block any surface area of the dish, preventing attenuation of the incoming microwave signal.

Satellite Receiver

The DSS satellite receiver, sometimes called an IRD or integrated receiver/decoder, is a complex digital signal processor. The amount and speed of data the receiver processes rivals even the faster personal computers in use today. The information received from the satellite is a digital signal that is decoded and digitally processed. There are no analog signals found except for those exiting the NTSC video encoder and the audio DAC (digital-to-analog converter). A block diagram of a satellite receiver is shown in **Figure 12-9**.

The downlink signal from the satellite is down-converted from 12.2–12.7 GHz to 950–1450 MHz by the LNB converter. The tuner then isolates a single, digitally modulated 24 MHz transponder. The demodulator converts the modulated data to a digital data stream.

The data is encoded at the transmitter site by a process that enables the decoder to reassemble the data and verify and correct errors that may have occurred during transmission. This process is called forward error correction (FEC). The error-corrected data is output to the transport IC via an 8-bit parallel interface.

The transport IC is the heart of the receiver data processing circuitry. Data from the FEC block is processed by the transport IC and sent to respective audio and video decoders. The microprocessor communicates with the audio and video decoders through the transport IC. The access card interface is also processed through the transport IC.

The access card receives the encrypted keys for decoding a scrambled channel from the transport IC. The access card decrypts the keys and stores them in a register in the transport IC. The transport IC uses the keys to decode the data. The access card also handles the tracking and billing for these services.

The MPEG video decoder processes video data. This IC decodes the compressed video data and sends it to the NTSC encoder. The encoder converts the digital video information into NTSC analog video that is output to the S-video and standard composite video output jacks.

Audio data is likewise decoded by the MPEG audio decoder. The decoded 16-bit stereo audio data is sent to the dual DAC (digital-to-analog converter) where the left and right audio data are separated and converted back into stereo analog

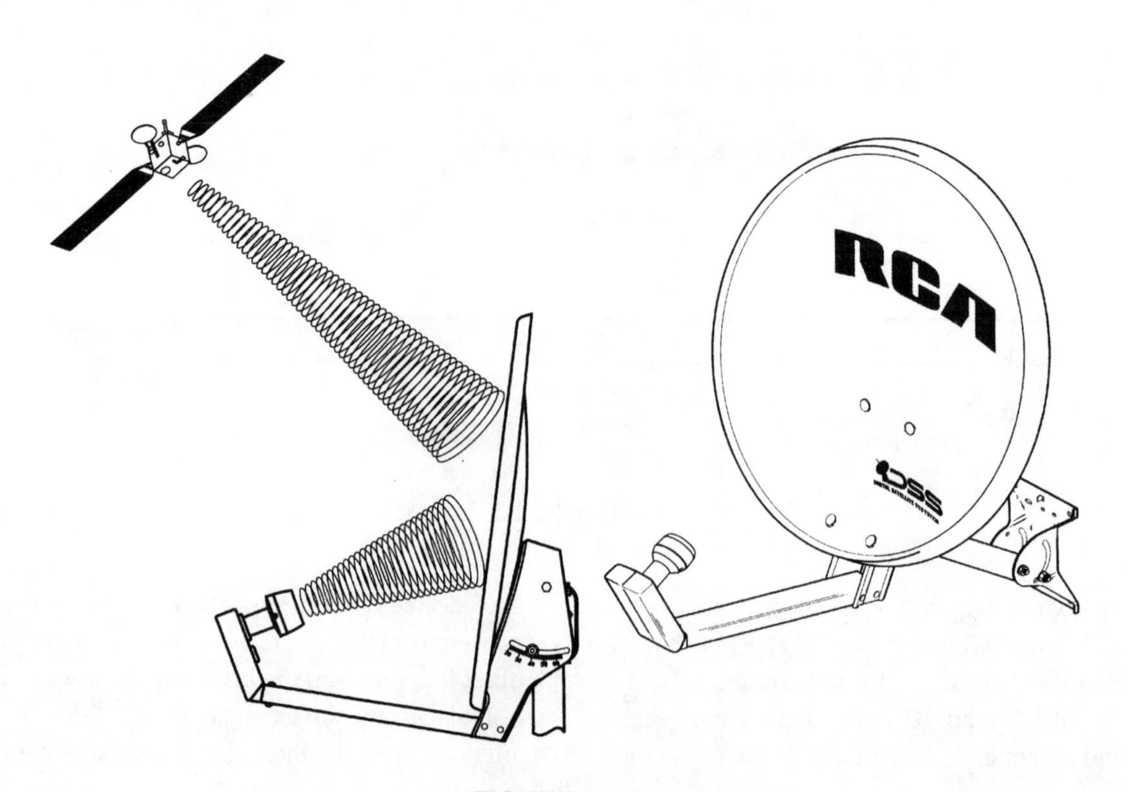

FIGURE 12-8

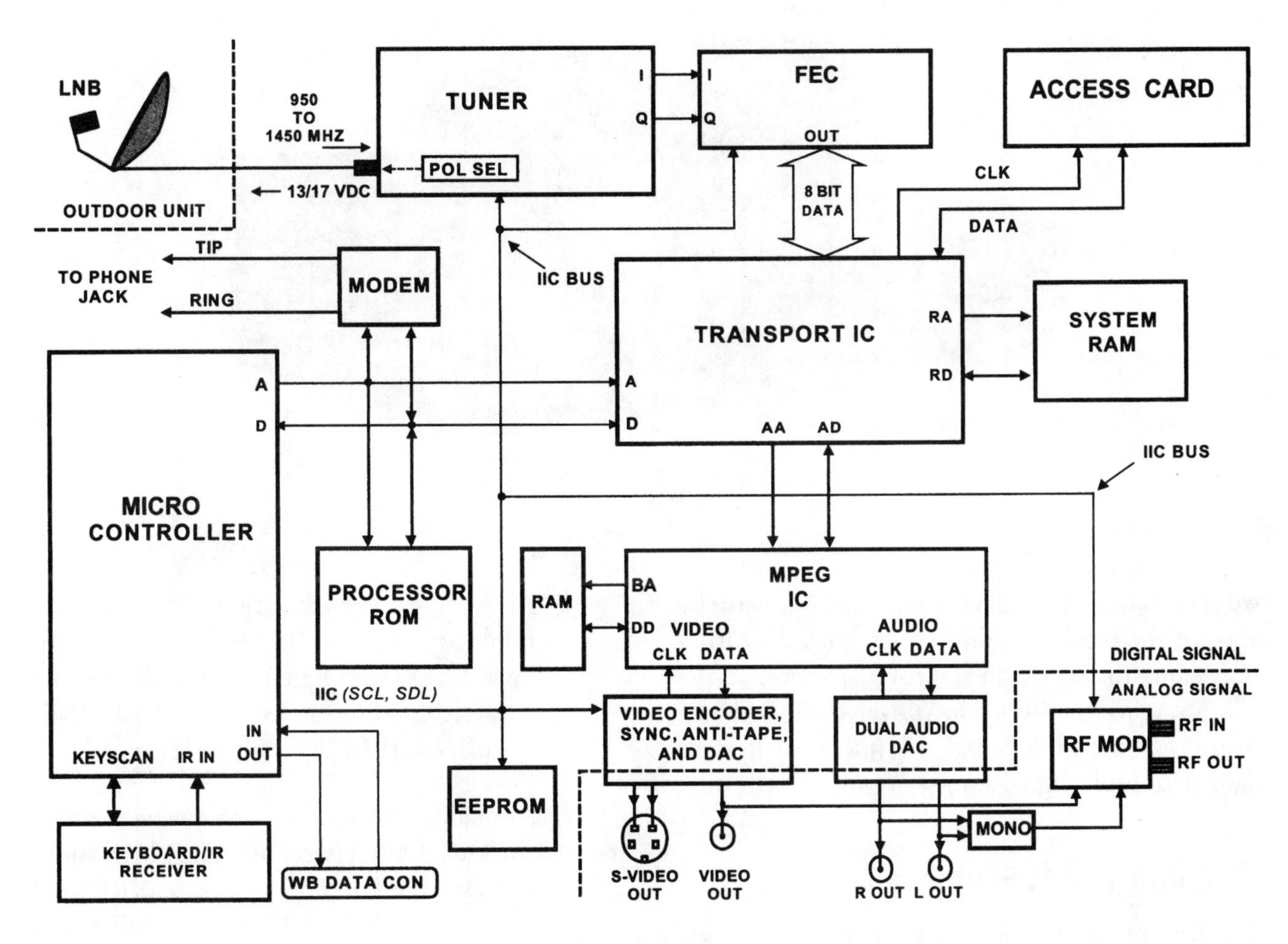

FIGURE 12-9

audio. The audio is output to the left and right audio jacks and is also mixed together to provide a mono audio source for the RF converter.

The microprocessor receives and decodes IR remote commands and front keyboard commands. Its program software is contained in the processor ROM (read only memory). The microprocessor controls the other digital devices of the receiver via address and data lines. It is responsible for turning on the green LED light on the on/off button.

The modem connects to the customer phone line and calls the program provider and transmits the customer program purchases for billing amounts. The modem operates at 1200 bps on early IRDs; 2400 bps on later-model IRDs; and is controlled by the microprocessor. When the modem first attempts to dial, it sends the first number as touch-tone. If the dial tone continues after the first number, the modems switches to pulse dialing and redials the entire number. If the dial tone stops after the first number, the modem continues to dial the rest of the number as a touch-tone number. The modem also automatically releases the phone line if a person in the home picks up another phone on the same line, or an extension phone. The receiver also has two diagnostic test menus for customer and technician troubleshooting.

DIRECTV Plus HDTV Set-Top Receiver

With the PLATINUM HDTV Receiver (**Fig. 12-10**), customers will be able to receive DIRECTV's standard-definition, HDTV programming, local high-definition broadcasts, all available channels, and DIRECTV's Advanced Programming Guide. The HNS (Hughes Network System) receiver includes the most widely adopted video output component for high-definition, YPrPb, enabling it to

FIGURE 12-10

work with most brands on the market. Viewers can also use the product with their standard-definition televisions to receive an improved picture, as the platinum HD set-top box converts off-air high-definition television broadcasts into a 480i standard-definition digital format.

Platinum HDTV Features

The features for the platinum HDTV set-top box are as follows:

- Reception capabilities for DIRECTV, high-definition, and standard-definition digital-quality programming (all 18 off-air high-definition broadcasts — ATSC).
- Reception capabilities for off-air standard-definition broadcasts (NTSC) via an internal tuner
- Ability to convert picture from standard- to high-definition format, and vice versa. (Reception capability for 480i, 480p, 720p, and 1080i formats.
- DIRECTV Advanced Programming Guide, which provides customers with a single guide for both DIRECTV high-definition and standard-definition digital broadcasts.
- Ability to present standard definition (4 x 3 format) on a 16 x 9 TV screen, with either gray sidebars (full picture in less-than-full screen), or with the top and bottom of the picture automatically cropped (less than a full picture in a full screen).
- All features offered with the fifth-generation platinum (standard-definition) receiver, include Dolby Digital audio, RF remote control, one-button record, and all HNS-exclusive DIRECTV system features.

Additionally, the receivers enable viewers to receive the DirecDuo service, providing high-speed internet access from DirecPC as well as DIRECTV programming. An external DirecPC satellite modem and a PC running Windows 98 are required to receive DirecPC service.

SAMSUNG SET-TOP BOXES

Samsung features two digital TV receiver decoders, which are models SIR-T200 and SIR-TS200. Both models utilize Samsung's latest 8VSB Decoder technology to ensure stable reception even under the most difficult conditions of weak signal strength or extreme multipath distortion. The superb performance of this new system will simplify the process of installing and adjusting HDTV antennas for most applications.

The Samsung HDTV technology in these units enables support for advanced data broadcast applications and will be used in field tests already being planned by broadcast affiliates. Both units utilize a flexible I-point cursor control system, making manipulation of data enhancements simple and intuitive.

A unique advantage of this HDTV receiver is the unit's flexibility. With an array of analog

(NTSC) inputs and outputs, the SIR-T200 and SIR-TS200 can be used to upconvert analog, 480-line interlaced signals to either 720P or even 1080i high-definition resolutions for display on a wide variety of HD-ready systems. In effect, each unit contains an advanced virtual line quadruplet in addition to the other features.

The SIR-T200 is a complete HDTV receiver decoder, including an off-air antenna input for both analog and digital broadcasts. Input terminals accept composite, S-video, and NTSC component video signals. Output terminals include wideband component video, RGB-HV on discrete RCA terminals, and a 15-pin RGB connector for use with a computer monitor. Analog composite video and S-VHS terminals are also included. The SIR-T200 allows the user to set the output resolution, regardless of input signal, to 480P, 720P, or 1080i. The SIR-TS200 also adds satellite reception to the powerful features of the SIR-T200. Designed to operate with the DIRECTV PLUS system, the SIR-TS200 will receive hundreds of standard-definition channels from DIRECTV, as well as the latest HDTV satellite broadcasts from DIRECTV.

PANASONIC DIRECTV HDTV SET-TOP BOX

The model TU-HDS20 Panasonic set-top box handles multiple sources of digital and analog TV, including DIRECTV. This is one of many Panasonic products that provide consumers the choice and convenience to enjoy the full extent of digital television entertainment.

The versatile TU-HDS20 affords TV viewers immeasurable viewing choices, and provides a means of enjoying the high-definition digital television experience even where local stations are not yet broadcasting HDTV. In addition to being able to receive all analog and HDTV terrestrial broadcasts in local service areas via a rooftop antenna, the receiver will provide access to a wide range of DIRECTV and digital DIRECTV HD satellite broadcast selections. The receiver will also accept and integrate must-carry NTSC local cable channels into the program guide.

Also available from Panasonic is an optional, 18 x 24-inch oval DIRECTV PLUS antenna designed to enable users to seamlessly view programming from multiple DIRECTV satellites. In addition to using two low-noise blocks, the antenna can be optionally upgraded to receive additional programming by adding a third LNB. The system will also incorporate a multiswitch with four outputs, enabling the viewer to connect up to four independent HD digital receivers in the home.

A key benefit offered by the TU-HDS20 is its Advanced Programming Guide (APG), which seamlessly merges and displays program information from all available sources. For example, a consumer who subscribes to DIRECTV and cable programming, or DIRECTV and terrestrial programming, will be able to view the channel programming in one guide.

Viewers will also appreciate the Roller Guide Menu System, a concept shared with Panasonic's DTV-compatible televisions. With this system, users can select features they want to activate without laboring through a multitude of icons covering the TV screen. Instead, the graphics appear on a "roller" configuration on the left side of the screen. Other useful features include auto-programmable channel scan for NTSC/ATSC signals, rapid tuning between two channels, programmable on/off timers, and V-chip parental guidance system.

Outputs include:

1. Component video (Y, PB, PR)
2. Composite video and audio
3. RGB-HV
4. S-video
5. Dolby Digital audio (optical)
6. Modem PC connection terminal

SAMSUNG DIGITAL SET-TOP BOX

The model SIR-TS200 Samsung set-top box became available in late 2000 and operates with the DIRECTV System to receive standard and HDTV DIRECTV programming, as well as off-air digital signals from local broadcasters. Samsung's HDTV-compatible DIRECTV PLUS set-top box also features the DIRECTV Advanced Program Guide, allowing set viewers to see all available channels

— whether HDTV or standard digital format — in a seamlessly integrated on-screen guide.

The SIR-TS200 allows users to receive programming in a variety of formats, regardless of the original "input" signal. Formats that the HDTV-compatible set-top box receive include 480P, 720P, and 1080i.

In midyear 2000, DIRECTV began broadcasting two channels of high-definition programming, "HBO HDTV" on channel 509 and the DIRECTV-dedicated channel of HDTV pay-per-view movies and events on channel 199.

MICROTUNE'S MM8838 MICROMODULE SET-TOP BOX

The MicroTuner 2030 series is the third generation of MicroTuner's tuner RF-ICS. Like their predecessor, the MicroTuner 2000, the MicroTuner 2030 series devices are single-chip, fully integrated (including low-noise amplifier), dual-conversion tuners that support both analog and digital applications and comply with international standards, including DOCSIS, QAM-256, 8-VSB, NTSC, PAL, DVB and DAVIC. For the first time, the MicroTuner chips are available with performance optimized for specific markets, including interoperable cable set-top boxes, cable modems, PC/TVs, and digital TVs.

Consistent with the OpenCable specifications, the MM8 MicroModule provides the core RF capabilities for analog/digital video, two-way communications, and conditional access. It consists of the MicroTuner 2030-STB, Microtune's new silicon tuner optimized specifically for cable applications, a diplexer, out-of-band tuner, and other associated RF components packaged into a production-ready, credit card-sized module.

The MicroModule functions as the tuning-receiving "heart" of the cable set-top box, providing the broadband "gateway" for delivering video, audio, and data to TVs via a high-speed, two-way cable connection. When integrated into next-generation digital set-top boxes, the MicroModule supports a range of sophisticated applications, including analog broadcast reception, Digital TV, HDTV, Internet browsing, e-mail, cable telephony, on-demand video, closed captioning, and a variety of interactive multimedia services.

The MicroModule 8838 features the MicroTuner 2030-STB single-chip broadband tuner as its technology centerpiece. The dual-conversion device provides the key performance characteristics for IF flatness, phase noise, and linearity. These technical parameters are critical for achieving maximum bandwidth efficiency of the cable system, while delivering high-quality analog and digital signals to TV viewers. The MicroTuner device successfully handles mixed media (video, data, audio) in a single-packed cable system, mitigating background noise and distortion, while successfully managing the interference of analog video with digital data. For the viewer, these technical problems can result in snow and ghosting (analog video), complete loss of signal (digital video), and/ or reduced speed and efficiency of Internet access. The complete MM8838 MicroModule unit is shown in **Figure 12-11**.

The OpenCable spec enables a wealth of new TV applications and services from Web-enhanced video programming to broadband interactive access. The MM8838 places stringent technical requirements on tuner performance, as well as the overall set-top box system design. By integrating the third-generation MicroTuner silicon with other critical functions into a MicroModule, the complicated RF system problems have been solved. In this process, MicroTuner has optimized system performance, quality, and reliability into a turnkey solution.

FIGURE 12-11

Some information in this chapter is courtesy of the following electronics companies: THOMSON MULTIMEDIA; DIRECTV and HUGHES NETWORK SYSTEMS; MICROTUNE ELECTRONICS COMPANY (Kathleen Padula and Stephanie Mayo)

Glossary

The **definitions in this glossary** will be helpful to you as you read this troubleshooting and operation guide for HDTV.

4:2:2: — Term commonly used for component digital video as specified by CCIR-60 1. The numbers refer to the ratio of chrominance sampling rate to luminance sampling rate. There are four luminance samples for every two of each chrominance samples.

4fsc: — Term commonly used for composite digital video. The name is used because the sampling is done at four times the subcarrier frequency, nominally 14.3 MHz for NTSC and 17.7 MHz in PAL.

AES/EBU audio: — Name commonly given to a digital audio standard defined jointly by the Audio Engineering Society and the European Broadcast Union.

ACC (automatic color control) — Used to maintain constant color signal levels.

ACK (automatic color killer) — Turns off the color control if the 3.58 MHz color burst is missing.

ADCAM converter (analog-to-digital converter) — This is the device that converts an analog waveform to a digital representation of the signal through sampling.

Adder, Logical — Switching circuits that generate an output (sum and carry bits) representing the arithmetic sum of two inputs.

Address — A code designation of a particular unit (RAMS, ROMS, and etc. that must respond to an incoming code. When used in HDTV color receivers, it is a digital code that designates the location of information or instructions in memory chips, peripherals, etc.

Aliasing — Undesirable effect caused by sampling at a frequency below the Nyquist limit, which produces spurious waveforms. Also used to describe the jagged stepped lines seen on diagonal lines produced by computer images.

Amplitude Modulation (AM) — A method of conveying information by changing the amplitude of a radio frequency carrier.

Analog — Term describing a system or device that operates on a continuously varying scale rather than increasing or decreasing in fixed steps as a digital system does.

Ancillary data — Term used to describe data, not related to the picture content, that is "imbedded" in the data stream of a digital video signal. Ancillary data is commonly used to transport digital audio, time code, or other picture related auxiliary data.

AND gate — See gate, AND.

Aspect ratio – The ratio of picture width to picture height. For HDTV this is a 16:9 format.

Astable multivibrator – A free-running electronic circuit that generates pulses that can be used as tuning or similar signals.

Asynchronous — Lacking synchronization. In data transfer, the term relates to a signal that is distributed without an accompanying reference clock. Synchronizing information is embedded in the data stream.

Audio — The Latin word for "hear." Used synonymously with the word sound.

Audio carrier — The frequency-modulated RF signal that carries the sound information.

Automatic frequency control (AFC) — A circuit that keeps the HDTV receiver from drifting off of the channel frequency that it is tuned to.

Bandpass filter — A filter that passes a group of continuous frequency and rejects all others.

Bandwidth-1 — The range of frequencies over which a system or circuit can operate with minimal loss. Usually less than 3 dB. The bandwidth of the standard NTSC television is 4.2 MHz. For the PAL television system it is 5.5 MHz.

Bandwidth-2 — The number of continuous frequencies required to convey the information being transmitted, either visual or aural. The bandwidth of a television Channel is 6 MHz.

Beam — The stream of electrons that travel from the electron gun towards the screen in a cathode-ray (CRT) tube.

Binary — A system of numerical representation that uses only two symbols: 0 and 1.

Bit — The smallest unit of information in a binary notation system. An abbreviation for Binary digit. A bit is a single 0 or 1. A group of bits used simultaneously by a digital system (usually 8 bits) is called a byte. In digital video, a group of 8 or 10 bits is often called a word.

Bit parallel — A transmission standard in digital video where whole words are sent simultaneously, each bit on its own pair of wires. This standard is defined by SMPTE 125M.

Bit serial — A transmission standard in digital video where whole words are sent one bit at a time, one after another, down one pair of wires, or more typically, a coaxial cable. This standard is defined by SMPTE 259M.

Bit slip — A condition in serial digital video transmission where word framing is lost. When this occurs, the order of bits is no longer correct and the data is corrupted. This can occur in systems with excessive jitter in the bit stream, which prevents correct decoding of the data.

Bit stream — A series of bits continuously flowing in an unbroken stream.

Black level — The level that represents the darkest part of the image. In a properly adjusted television system, no picture information is at a level lower than this level.

Blanking — The process of turning off the electron beam in a CRT after it reaches the right edge of the screen, while it is being returned to the left edge to begin the next scan line. This is done by setting the level to black, or slightly below black in some systems. If the beam were not blanked, it would leave a trace as it returned to its starting position, which would degrade the image.

Blanking pulse — The pulse used to blank out the electron beam in both the camera and picture tube during the blanking interval.

Burst, color — The portion of a composite video waveform that is placed between the horizontal sync pulse and the beginning of active video information. It is made up of several cycles of sine wave at the frequency of the color subcarrier. It is used by the receiver as a reference to insure correct decoding of the chrominance information in the composite signal.

Burst gate — A circuit that is keyed to conduct during the time of he color burst. It is timed by a pulse from the horizontal sweep transformer.

Byte — Refer to **Bit.**

Cable equalization — The process whereby the frequency response of an amplifier is altered to compensate for high frequency losses in a cable.

CCIR-601 — A recommendation developed by the CCIR for the digitization of component color video signal. It defines the sample rate, horizontal resolution and filters to limit the bandwidth of the signal.

CCIR-656 — A recommendation developed by the CCIR that defines the physical and electrical interface for the exchange of video digitized according to CCIR 601. It describes both a parallel and serial interface, but only the parallel interface has been widely accepted. Instead, an interface described in SMPRE 259M has largely replaced it.

Chroma — A shortened form of chrominance, which is the color part of a video signal. It expresses the hue and saturation of a color but not the luminance or brightness. In a component system, the R-Y, B-Y, or Cr and Cb signals carry the cbroma information. In a composite system, such as PAL, SECAM, or NTSC, the chroma information is encoded.

Chrominance — *See Chroma.*

Clock — A pulse generator that controls the timing of switching circuits and memory states and determines the speed at which the major portion of the computer or timing circuits operate. (The pulse generated is referred to as the *clock pulse.* Sometimes shortened to *clock.)*

Clock jitter — A variation of a digital signal's transition from their ideal position in time. It is the instantaneous difference in phase of a signal to a stable and jitter-free primary clock.

Color bars — A test pattern used to properly calibrate a video system. Brightness, hue, and saturation of a signal can be verified with this pattern and the proper test equipment.

Color burst — An 8-to-10cycle reference of the 3.58 MHz color subcarrier that occurs right after the horizontal sync pulse. It is used as the reference for phase comparison to keep the color oscillator locked to the TV station signal. Also, it references the ACC and ACK circuits.

Color subcarrier — A 3.579545 MHz signal that is modulated to produce the color sidebands that carry the color information. It is suppressed during transmission.

Comb filter — An electronic filter circuit with a pass response that resembles a comb. It separates the video luminance from the color based on their phase relationship.

Component analog video (CAV) — A video format in which separate video signals represents the luminance and chrominance information. The signals are made up of varying analog voltages that represent the picture content. The signals can represent red, green and blue (RGB) output from an image-producing device, or they may be transformed into a luminance signal and chrominance signals as In Y, R-Y, B-Y.

Component digital video — A digital video signal where the signal components are sampled and transmitted in separate channels. It is generally the digitized version of a Y, R-Y, B-Y component analog signals, but can also be a digitized version of RGB.

Composite analog video — An analog video format that contains a mixture of chrominance, sync, and color burst that have been combined by one of the standard encoding processes, NTSC, PAL, or SECAM.

Counter, binary — A series of flip-flops having a single input. Each time a pulse appears at the input, the flip-flop changes state; sometimes called a "T" flip-flop.

CRC — The cyclic redundancy check to verify the correctness of data.

Crossover point — The point, between the grid and pre-accelerator, of a cathode-ray tube where the electrons are emitted by the cathode coverage.

Composite digital video — A digital video signal that results from the digitization of a composite analog video signal. The composite analog signal is usually sampled at four times the frequency of the color subcarrier for the standard being used. Standards have been defined for PAL and NTSC composite digital video.

Dl — A component digital videotape recording standard that confirms to CCIR standards 601 and 656. It uses a cassette containing 19 mm-wide magnetic tape. The term DI is sometimes incorrectly used to indicate the signal or interface format defmed by CCIR 601 and 656.

Demultiplexer — A device or circuit that separates two or more signals that were Previously combined by a multiplexer and sent over a single channel.

Digital word — A group of bits that compose a unit for the purpose of signal Treatment in a system. Each sarnple's value in a digital video system is expressed by eight or ten bits, which make up a digital word for that sample.

DAC, D/A converter — (Digital-to-Analog Converter) It is the device that converts a digital data stream to an analog signal.

Decoded stream — The decoded reconstruction of a compressed bit stream.

Decoder — A device used to convert information from a coded form into a more usable form, ie., binary-to-BCD, BCD-to-decimal, etc.

Decoding (process) — The process defined in the *Digital Television Standard* that reads an input-coded bit stream and outputs decoded pictures or audio samples.

Definition — The ability of a television system to reproduce small details in a picture image.

Delay — Undesirable delay effects are caused by rise and fall times that reduce circuit speed, but intentional delay may be used to prevent inputs from changing while clock pulses are present. The delay time normally is less than the clock-pulse interval.

Demodulation — The process of removing the modulating signal from a modulated radio frequency carrier.

Direct-view receiver — A television receiver in which the image is viewed on the face of the picture tube.

Directional antenna — A antenna designed to receive radio signals better from some directions than from others.

Director — A rod slightly shorter than a dipole that is placed in front of the dipole to provide greater direction.

Discharge tube — A tube used in sawtooth-generating circuits to discharge a capacitor.

Discrete, circuits — Electronic circuits built of separate finished components, such as resistors, capacitors, transistors, etc.

Discrete cosine transformer (DCT) — A mathematical transformer that can be perfectly undone and which is useful in image compression.

Discriminator — A circuit used in FM receivers to convert the frequency-modulated signal into an audio frequency signal.

Dissector tube — A pickup tube containing a continuous photosensitive cathode on which an electron image is formed.

DSM-CC — Digital storage media command and control.

EEPROM — Electrically Erasable Programmable Read-Only Memory.

Enable — A gate is enabled if the input conditions result in a specific output. The specific output varies for different gating functions. For instance, an AND gate is enabled when its output is the same level as its input, whereas a NAND gate is enabled when its output is the complement of its inputs, in some cases, Function-enabling inputs allow operation to be executed on a clock pulse after the inputs are enabled with the correct logic level.

EAV — (End of Analog Video) The timing reference signals that indicates that The active (picture) area of a video line has ended and that the time to begin the Retrace process has begun.

EDH — (Error Detection and Handling) A checksum system for detecting and indicating Bit errors in a digital video system. It is standardized in SMPTE RP-165.

E-E MODE — Electronic to Electronic mode. An equipment where the incoming signal is passed through the equipment for some processing, but returned to the output mostly unmodified. In a VCR the input signal is not recorded, routed directly to the video output of the recorder. Similarly, in some pattern generators, an incoming signal may be routed to the output of the generator in place of a standard pattern.

Encoder — In video, a device that combines the luminance, chrominance, sync And color burst signals into one composite video signal.

Equalizing pulses — Two groups of pulses, one occurring before the serrated vertical sync pulse, the other occurring after. These pulses occur at twice the normal horizontal line rate. These pulses were implemented to insure correct interlace operation.

Error correction — A system for adding additional data to a digital signal to allow transmission errors to be detected and coffected.

Field — One-half of a television picture frame. It is composed of one complete scan of the image. It is made up of 2621/2 lines in the 525-line system. The lines of field one are interlaced with field two to make up a frame or complete picture.

Flicker — Objectionable low-frequency variation in intensity of illumination of a television picture.

Flip-flop — An electronic circuit having two stable states and the capability to change from one state to the other on the application of a signal in a specified manner.

The specific types are as follows:

Flip-Flop, D — A flip-flop with output from the input that appeared one pulse earlier. If a I appears at its input, the output one pulse later will be a 1. Sometimes it is used to produce a one-clock delay. (D stands for data.)

Flip-Flop, JK — A flip-flop having two inputs designated J and K. At the application of a clock pulse, a 1 on the J input will set the flip-flop to the 1 or ON state and 1's simultaneously on both inputs will cause it to change state regardless of what state it has been in. If O's appear simultaneously on both inputs, the flip-flop state remains unchanged.

Flip-Flop RS — A flip-flop having two inputs designated R and S. The application of a 1 on the S input will set the Rip-flop to the I or ON state and a I on the R input will reset it to the

0 or OFF state. It is assumed that l's will never appear simultaneously at both inputs. (In actual practice, the circuit can be designed so that a 0 is required at the S and R inputs.)

Flip-Flop, synchronized RS — A synchronized RS flip-flop having three inputs,RS, and clock (stroke, enable, etc.). The R and S inputs produce states as described for the RS flip-flop. The clock causes the flip-flop to change states.

Flip-Flop, T — A flip-flop having only one input. A pulse appearing on the input will cause the flip-flop to change states. A series of these flip-flop makes up a binary ripple counter.

Fluorescent screen — The chemical coating on the inside face of a cathode-ray tube that emits light when struck by a stream of electrons.

Flyback (sweep transformer) — An abbreviated reference to either flyback transformer or flyback time.

Flyback time — The time when the horizontal output transistor is turned off. The magnetic field in the output stage suddenly collapses to produce a large pulse at the collector of the horizontal output transistor and return the electron beam to the left side of the CRT screen. Also, the period during which the electron beam is returning from the end of a scanning line to begin the next line.

Focus — In a cathode-ray tube, this refers to the size of the spot of light on the fluorescent screen. The tube is said to be focused when the spot is smallest. This term also refers to the optical focusing of camera lenses.

Focusing control — The potentiometer control on the receiver that varies the first anode voltage of an electrostatic tube and focuses the electron beam.

Focusing electrode — A metal cylinder in the electron gun, sometimes called the first anode. The electrostatic field produced by this electrode, in combination with the control electrode and the accelerating electrode, acts to focus the electron beam in a small spot on the screen.

Forbidden — This term, when used for defining coded bit stream, indicates that the value shall never be used. This is usually to avoid emulation of start codes.

FPLL — Frequency and phase-locked loop.

Frame — A complete television picture, consisting of two fields or all 525 lines in the U. S. system. HDTV has more lines depending on mode used. In the NTSC system, 29.97 frames are scanned each second. PAL and SECAM systems scan 25 frames per second.

Frame — A frame contains lines of spatial information of a video signal. For progressive video, these lines contain samples starting from one time instant and continuing through successive lines to the bottom of the frame. For interlaced video, a frame consists of two fields, a top field and a bottom field. One of these fields will commence one field later than the other.

Frame frequency — The number of times per second the picture area is completely scanned. This frequency is 30 times per second in current standard television transmission systems.

Frequency modulation — A system in which the frequency of a radio signal is varied in proportion to the modulating signal in order to transmit intelligence.

Gate — A circuit having two or more inputs and one output, the output depending on the combination of logic signals at the inputs. There are five gates, called AND, exclusive OR, ORNAND, and NOR. The following gate definitions assume that positive logic is being used.

Gate, And — All inputs must have 1-state signals to produce a 1-state output. (In actual practice, some gates require 0-state inputs to produce a 0-state output.)

Gate, exclusive OR — The output is true only when the two inputs are opposite (complementary) and is false if both inputs are the same.

Gate, NAND — All inputs must have 1-state signals to produce a 0-state output. (In actual practice, some gates require 0-state inputs to produce a 1-state output.)

Gate, NOR — Any or more inputs having a 1-state will yield a 0-state output. (In actual practice, some gates require 0-state inputs to produce a 1-state output.)

Gate, OR — Any one or more inputs having a 1-state signal will yield a 1-state output. (In actual practice, some gates require 0-state inputs to produce a 0-state output.)

Genlock — A process whereby a signal is phase locked to a master sync source. Most professional video-generation sources (VTRS, pattern generators, CGs, etc.) can be genlocked to the house sync source.

Ghost — A secondary picture formed on a television receiver by a signal from The transmitter that has reached the antenna by a longer path. Ghosts usually are caused by reflected signals. Some TV sets have circuits that reduce ghosts And the digital HDTV system eliminates most reflected ghost problems.

Horizontal interval — The time allotted for the electron beam in the scanning Device to return to the left edge of the image after a complete video line has been Scanned. It is initiated by the leading edge of the horizontal sync pulse and ends When active video information begins after color burst.

HDTV – High-Definition Television. A system of television having significantly more scan lines per image than current TV systems for increased resolution. In the U.S. such a proposed system would have 1080 active vertical lines with 1920 horizontal samples per line and would transmit a component digital video signal.

Horizontal — The direction of sweep of the electron beam from left to right.

Horizontal blanking — The blanking pulse that occurs at the end of each horizontal line and cuts off the electron beam while it is returning to the left side of the screen.

Horizontal centering control — A control used to move the picture on the TV screen in a horizontal direction.

Horizontal frequency — The rate at which the electron beam makes one scanning cycle across the screen from left to right.

Horizontal interval — The time allotted for the electron beam in the scanning device to return to the left edge of the image after a complete video line has been scanned. It is initiated by the leading edge of the horizontal sync pulse and ends when active video information begins after color burst.

Horizontal output transistor (HOT) — A power transistor that switches current from the B+ supply into the horizontal output stage via the (sweep) flyback transformer primary.

IHVT **(Integrated high-voltage transformer)** — A component that functions as both a flyback tansformer and a high-voltage multiplier. Some IHVTs also provide screen high voltage.

Jaggies — Common name used for a stair-step effect on diagonal lines in an image. This effect is sometimes called aliasing.

Jitter — A variation of a digital signal's transitions from their ideal position in time. It is the instantaneous difference in phase of a signal to a stable and jitter-free primary clock.

Jitter, digital — A variation of digital signal transitions from their ideal position in time. It is the instantaneous difference in phase of a signal from a stable and jitter-free primary clock.

Keystone effect — Distortion of a television image that results in a keystone-shaped pattern.

Line, horizontal scan line — This is one complete unit of video information produced by scanning one horizontal line. It includes the active video information and horizontal sync information. In the U.S. system, there are 15,734 lines each second.

Luminance — The measurable luminous intensity of a video signal. It is one of the components of a video signal. It expresses the brightness and contrast of an image. Its counterpart, chrominance, expresses the hue and saturation of the colors in the picture. A luminance signal is directly viewable on a monochrome TV receiver or monitor.

Multipath reception — The condition in which the RF signal from the transmitter travels by more than one route to a TV receiver antenna, usually because of reflections from obstacles resulting in ghosts in the picture. However, digital TV signals eliminate most of these reflected ghost problems.

Multivibrator — A type of oscillator, using R-C components, commonly used to generate the sawtooth voltages in TV receiver circuits.

Negative ghosts — Ghosts that appear on the screen with intensity variations opposite to those of the picture.

Negative logic — The reverse of positive logic; the more negative voltage represents the 1-state and the less negative voltage represents the 0-state.

Negative transmission — The modulation of the picture carrier by a picture signal, the polarity of which is such that the sync pulses occur in the blacker-than-black level.

Noise — Spurious impulses that modulate the picture or sound signals.

Noninterlace — A scanning method consisting of two fields of horizontal scan lines. The lines from field two are scanned directly on top of the lines from field one.

NRZI (Nonreturn to Zero Inverted) — A digital coding scheme that scrambles the data pseudo-randomly ro render the data stream insensitive to polarity. A change in logic state (from high to low, for example) represents a digital one, while no change in state represents a zero.

NTSC (National Television Systems Committee) — The name of the organization that formulated the standards for the U.S. color television system. NTSC has come to denote the television standard. The system employs a 3.579445 MHz subcarrier whose phase varies with the instantaneous hue of the televised color, and whose amplitude varies with the instantaneous saturation of color. The system operates with 525 lines and 59.94 fields per second.

Nyquist frequency — The lowest frequency that can be used to sample a waveform without significant aliasing when converting an analog waveform to digital. The frequency is usually considered to be twice the highest frequency to be sampled.

Orthogonal sampling — A sampling method where sampling is done in phase with the horizontal line such that the sampling number is taken at the same position in each line. This produces sample positions that are aligned vertically in the image.

PAL (Phase Alternate Line) — The name of the color television system used in most of Europe. It differs from NTSC in several important ways. It uses 625 lines per frame and 50 frames per second. It also uses a subcarrier, but the frequency is 4.43361875 MHz. The color burst is shifted 90 degrees in phase from one line to the next in order to minimize hue errors during transmission.

Parallel transmission — A transmission method that uses a separate physical channel for each bit of the digital word, plus a channel for the synhronizing clock.

Patch panel — A panel containing a quantity of receptacles connected to outputs and inputs of equipment. Signals can be routed by manually connecting short cables, called patch cords, between the receptacles.

Pedestal — The portion of the television video signal used to blank out the beam as it flies back from the right to the left side of the screen.

Phase shift — A change in relative timing between two signals of the same frequency. It is expressed in degrees, one entire period equals 360 degrees.

Pixel — An abbreviation for Picture Element. The smallest visual unit in a digital image system. There is one pixel for each sample taken in the DAC. Thus in component digital video, there are 720 pixels per line.

Preemphasis — The increasing of the relative amplitude of the higher audio frequencies in order to minimize the effects of noise during transmission.

Projection receiver — A television receiver in which the image is optically enlarged and projected onto a screen.

Propagation delay — A measure of the time required for a change in logic level to propagate through a chain of circuit elements.

Quantization — The process of expressing different analog levels with digital levels. Although the analog values vary continuously, the digital values vary in discrete steps, called the quantizing levels.

Random access memory (RAM) — A static or dynamic memory device into to or out of which data can be written from a specific location. The specific RAM location is selected by the address applied via the address bus control lines. Data can be stored in such a manner that each bit of information can be retrived in the same length of time.

Register — A device used to store a certain number of digits within computer circuitry, often one word. Certain registers also may include provisions for shifting, circulating, or other options.

Reserved — This term when used in defining the coded bit stream, indicates that the value may be used in the future for *Digital Television Standard* extensions. Unless otherwise specified within this standard, all reserved bits shall be set to 1.

Resolution — A measure of the ability of a television system to reproduce detail. It is governed by the imaging device, the TV CRT, and the system's bandwidth. Generally an increase in bandwidth results in increased resolution. Resolution in a digital system often refers to the number of levels that the system can express. This is related to the number of bits in the digital word. A digital word can resolve as many levels as the number of bits can express. For example, four bits has a resolution of 2 to the power of 4 (16) different levels, while ten bits can express 2 to the power of 10 (1024) levels. Thus, in expressing levels within the same overall range, longer data words can resolve smaller changes in level, since the overall range is divided into smaller increments.

(Red, Green, Blue) — The three primary colors of light are used in color television systems. Some component video systems treat these three signals in parallel. Sometimes called GBR due to the mechanical arrangement of the connectors in the SNWT interface standard.

Rise time — A measure of the time required for a circuit to change its output from a low to a high level.

Safety capacitor — Name given to the retrace timing capacitor because of the critical role it plays in controlling the pulse amplitude, which in turn, determines the amount of high voltage produced.

Sandcastle — Signal used in many luma/chroma circuits that is a combination of (1) horizontal flyback pulse, (2) horizontal blanking pulse, and (3) vertical blanking pulse. It gets its name from its appearance as a "sandcastle."

Sawtooth — A voltage or current, the variation of which, with time, follows a sawtooth configuration.

Sawtooth voltage — A voltage that varies between two values at regular intervals. Since voltage drops faster than it rises, it gives a waveform pattern resembling the teeth of a saw. Very useful in television receivers to help form the scanning raster for the picture tube.

Scan-derived power supply — A voltage source from a secondary winding on the flyback transformer that is powered by the normal scanning current in the flyback sweep transformer.

Scanning — The process of breaking a picture into elements by means of of a moving electron beam.

Scanning line — A horizontal line, composed of elements varying in intensity, the width of which is equal to the diameter of the scanning electron beam.

Scanning raster — These are lines that sweep across the face of the picture tube that produce varying light and dark images with the electron beams.

Schmidt system — An optical system adapted for television projection receivers in which the light from the image is collected by a concave mirror and directed through a correcting lens on a screen.

Schmitt trigger — A fast-acting pulse generator that produces a constant-amplitude pulse as long as the input exceeds a threshold dc value. Used as a pulse shaper, threshold detector, etc.

Time code — A sequential digital signal that is recorded along with the audio and video information on a videotape to indicate a time position for each video frame on the tape. The information typically gives a readout in hours, minutes, seconds, and frames for each frame of the tape. A coding standard developed by SNWTE has been accepted worldwide. The time-coded information can be recorded on an unused audio track or other longitudinal track of the videotape. In this case, the time code is called LTC (Longitudinal Time Code). Another method records the time code during the vertical blanking interval, which is called VITC (Vertical Interval Time Code). In digital video recorders the time code is recorded as ancillary data during the digital vertical interval.

Time-multiplex — As applied to component digital video, this is the process of interleaving three parallel bit streams so that they can be transmitted on one set of physical channels. Specifically, in component digital video, the eight or ten bits of the CB signal are sent first, followed by the Y sample values for the same pixel, followed by the CR sample. Thus all three sets of values are sent on one channel.

TRS (Timing Reference Signal) — The collective name for the synchronizing words in component digital video. There are two types, the SAV and the EAV.

TRS-ID (Timing Reference Signal — Identification) — A special synchronizing word in component digital video that allows line and field identification for synchronizing purposes, and more importantly, allows the deserializer to establish the word framing in the serial data stream.

Truncate — The process of eliminating the LSBs of a digital word, for example when converting from a ten bit to an eight bit word. This process must be handled with care in equipment designed to avoid introducing distortions into the signal.

Universal asynchronous receiver transmitter (UART) — A device that provides the data formatting and control to interface serial asynchronous data. Input serial data are converted to parallel data for transfer to the microprocessor or microcomputer via the data bus. Conversely, output parallel data, from the data bus, are converted to serial data in the output.

Variable bit rate — Operation where the bit rate varies with time during the decoding of a compressed bit stream.

Vertical blanking — The time after the downward beam movement (field) when the electron beam is turned off.

Vertical blanking pulse — A pulse transmitted at the end of each field to cut off the cathode-ray beam while it is returning to the top of the picture for the start of the next field.

Vertical resolution — A measure of the ability of a system to reproduce fine horizontal lines.

Vertial retrace — The time during which the electron beam returns from the bottom of the CRT to the top of the screen.

Vertical scanning — The motion of the electron beam in the vertical direction.

Vestigal side-band transmission — A type of RF transmission in which one side band is surpressed to limit the bandwidth requirment.

Video — The Latin word meaning "I see."

Video amplifier — A wide-band amplifier for video frequencies. In a television receiver, this term generally refers to the amplifier located after the second detector, the frequency of which extends from approximately 30 Hz to approximately 4 MHz.

Video buffering verifier (VBV) — A hypothetical decoder that is conceptually connected to the output of an encoder. Its purpose is to provide a constraint on the variability of the data rate that an encoder can produce.

Video frequency — The frequency of the digital data signal containing the picture information that is developed from the television scanning process. In the current TV system these video frequencies are to approximately 4 MHz.

Video sequence — A video sequence is represented by a sequence header, one or more groups of pictures, and an end sequence code in the data stream.

Video transmitter — An RF transmitter used to transmit the video picture signal.

Viewing screen — The face of a cathode-ray picture tube or special viewing screen on which a picture image is reproduced.

Waveform — The shape of an electronic signal represented in a graphic form. The representation usually shows increasing voltage or current on the Y axis and time on the X axis. This is most usually shown on an electronic instrument device called an oscilloscope.

Width control — The control in the horizontal sweep circuit that varies the size of the picture in the horizontal direction.

Word — The term *Word* denotes an assemblage of bits considered as an entity. For example, a 16-bit address word contains 16 bits.

Word framing — The process of separating the digital words from an asynchronous serial data stream. In order to correctly extract the digital words, the deserializer must determine correctly where the most significant bit of one word ends and the least significant bit of the next begins. This process is cared *Word framing*.

Y — Abbreviation for luminance signal.

Y/C — An input or output connector that provides the luminance (Y) signal and chroma (C) signal separately.

YCRCB — A name commonly given to component digital video that conforms to SMPTE 125M. The name is derived from the abbreviations for the luminance and each of the color difference channels.

Yoke — A coil of wire specially shaped to deflect the electron beam when a sawtooth current is applied. It consists of a set of vertical windings and a set of horizontal windings. Also, a set of coils placed over the neck of a cathode-ray tube that produces horizontal and vertical deflection of the electron beam when suitable currents are passed through them.

EXPLORING LANS FOR THE SMALL BUSINESS & HOME OFFICE

Author: LOUIS COLUMBUS
ISBN: 0790612291 • **SAMS#:** 61229
Pages: 304 • **Category:** Computer Technology
Case qty: TBD • **Binding:** Paperback
Price: $39.95 US/$63.95CAN

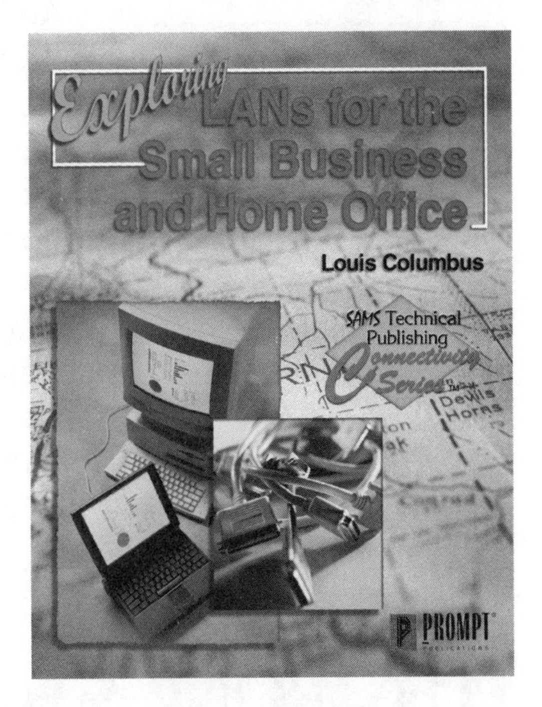

About the book: Part of Sams Connectivity Series, *Exploring LANs for the Small Business and Home Office* covers everything from the fundamentals of small business and home-based LANs to choosing appropriate cabling systems. Columbus puts his knowledge of computer systems to work, helping entrepreneurs set up a system to fit their needs.

PROMPT® Pointers: Includes small business and home-office Local Area Network examples. Covers cabling issues. Discusses options for specific situations. Includes TCP/IP (Transmission Control Protocol/ Internet Protocol) coverage. Coverage of protocols and layering.

Related Titles: *Administrator's Guide to E-Commerce,* by Louis Columbus, ISBN 0790611872. *Administrator's Guide to Servers,* by Louis Columbus, ISBN

Author Information: Louis Columbus has over 15 years of experience working for computer-related companies. He has published 10 books related to computers and has published numerous articles in magazines such as *Desktop Engineering, Selling NT Solutions,* and *Windows NT Solutions.* Louis resides in Orange, Calif.

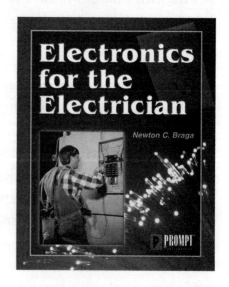

AUTOMOTIVE AUDIO SYSTEMS

Author: HOMER L. DAVIDSON
ISBN: 0790612356 • **SAMS#:** 61235
Pages: 320 • **Category:** Automotive
Case qty: TBD • **Binding:** Paperback
Price: $39.95 US/$63.95CAN

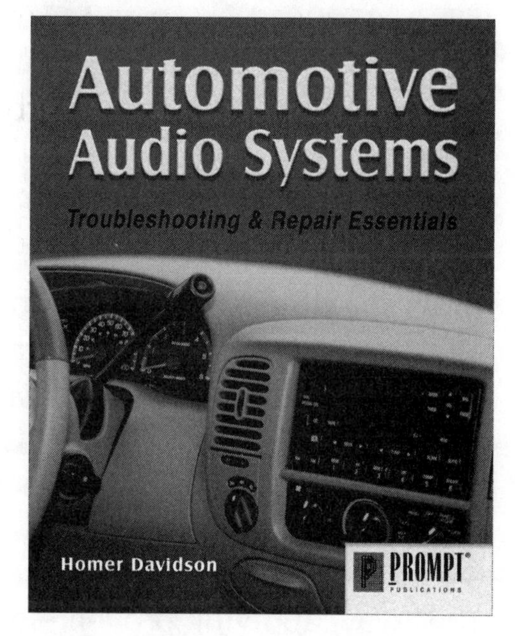

About the book: High-powered car audio systems are very popular with today's under-30 generation. These top-end systems are merely a component within the vehicle's audio system, much as your stereo receiver is a component of your home audio and theater system. Little has been written about the troubleshooting and repair of these very expensive automotive audio systems. Homer Davidson takes his decades of experience as an electronics repair technician and demonstrates the ins-and-outs of these very high-tech components.

Prompt Pointers: Coverage includes repair of CD, Cassette, Antique car radios and more. All of today's high-end components are covered. Designed for anyone with electronics repair experience.

Related Titles: *Automotive Electrical Systems*, ISBN 0790611422. *Digital Audio Dictionary*, ISBN 0790612011. *Modern Electronics Soldering Techniques*, ISBN 0790611996.

Author Information: Homer L. Davidson worked as an electrician and small appliance technician before entering World War II teaching Radar while in the service. After the war, he owned and operated his own radio and TV repair shop for 38 years. He is the author of more than 43 books for TAB/McGraw-Hill and Prompt Publications. His first magazine article was printed in *Radio Craft* in 1940. Since that time, Davidson has had more than 1000 articles printed in 48 different magazines. He currently is TV Servicing Consultant *for Electronic Servicing & Technology* and Contributing Editor for *Electronic Handbook*.

SERVICING RCA/GE TELEVISIONS

Author: BOB ROSE
ISBN: 0790611716 • **SAMS#:** 61171
Pages: 352 • **Category:** Troubleshooting & Repair
Case qty: 26 • **Binding:** Paperback
Price: $34.95 US/$55.95CAN
About the book: Designed to give a detailed overview of the manufacturer and an in-depth analysis of various television chassis. The overview includes a history of RCA/GE/Thomson, discussion of test equipment, technical literature, software available, and a discussion of OEM parts versus generic parts.

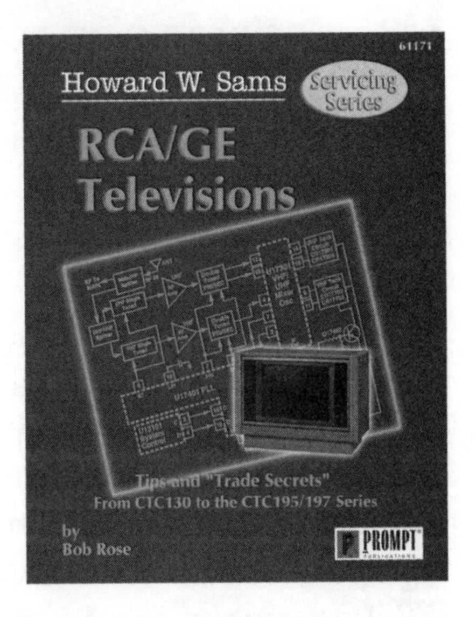

SERVICING TV/VCR COMBO UNITS

Author: HOMER L. DAVIDSON
ISBN: 0790612240 • **SAMS#:** 61224
Pages: 344 • **Category:** Troubleshooting & Repair
Case qty: 26 • **Binding:** Paperback
Price: $34.95 US/$55.95CAN
About the book: Part of Sams Servicing Series, Servicing TV/VCR Combo Units covers the servicing issues surrounding this popular electronic device. TV/VCR combo units have become smaller, more affordable, and functional. They are now used in new ways, including applications in autos, campers, kitchens, and other non-traditional locations. Homer Davidson uses his vast knowledge to cover this subject in a way no one else can.

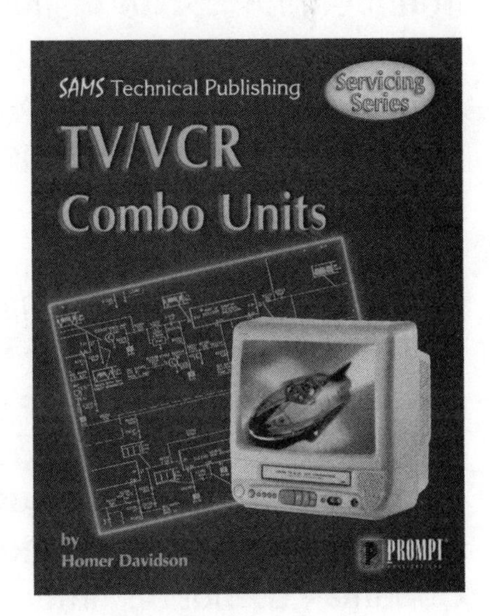

To order today or locate your nearest PROMPT® Publications distributor at 1-800-428-7267 or www.samswebsite.com

Prices subject to change.

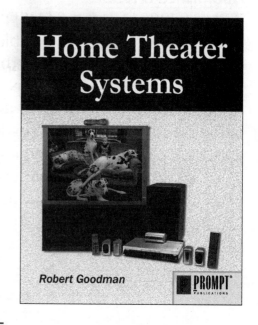